Volume 7

URBANIZATION AND
URBAN PLANNING
IN CAPITALIST SOCIETY

URBANIZATION AND URBAN PLANNING IN CAPITALIST SOCIETY

Edited by
MICHAEL DEAR AND ALLEN J. SCOTT

Routledge
Taylor & Francis Group

LONDON AND NEW YORK

First published in 1981 by Methuen & Co. Ltd

This edition first published in 2018
by Routledge
2 Park Square, Milton Park, Abingdon, Oxon OX14 4RN

and by Routledge
711 Third Avenue, New York, NY 10017

Routledge is an imprint of the Taylor & Francis Group, an informa business

British Library Cataloguing in Publication Data
A catalogue record for this book is available from the British Library

ISBN: 978-1-138-49611-8 (Set)
ISBN: 978-1-351-02214-9 (Set) (ebk)
ISBN: 978-1-138-47863-3 (Volume 7) (hbk)
ISBN: 978-1-138-47864-0 (Volume 7) (pbk)
ISBN: 978-1-351-06800-0 (Volume 7) (ebk)

Publisher's Note
The publisher has gone to great lengths to ensure the quality of this reprint but points out that some imperfections in the original copies may be apparent.

Disclaimer
The publisher has made every effort to trace copyright holders and would welcome correspondence from those they have been unable to trace.

Preface to the 2018 Edition

Towards a New Era of Urbanization and Planning

Since this book was first published, cities have proliferated and grown with astonishing rapidity, and massive city-regions have emerged all over the world. These processes have been underpinned by intensifying economic globalization and the steady integration of major cities into far-flung networks of cooperation and competition. Urban planning and policy concerns have shifted dramatically from their former preoccupation with welfare and redistribution, and are increasingly deployed as instruments for promoting competitiveness, entrepreneurial effort, and economic development in a world of deepening market relationships. Capitalism too, is adjusting in remarkable ways when compared with the earlier fordist regime of accumulation that was a major point of reference for a great many of the book's contributors.

In the twenty-first century, the technological bases of capitalism are increasingly dependent on digital technologies of computation, information storage, and communication, with major consequences for urban development. In this context, human capital needs are increasingly focused on a widening diversity of cognitive and cultural skills. The blue-collar/white-collar division of labor that was such an important feature of fordist metropolitan development is being replaced by a re-stratification of urban society into a privileged upper tier of formally-qualified workers, and a tier of low-wage service workers often composed of migrants from more economically marginal areas. Women account for a dramatically rising percentage of the urban labor force, leading to major shifts in both employment structures and the character of residential space. As a consequence of these and other factors, urban society virtually everywhere is becoming more and more bifurcated in terms of social class and markedly more diverse in terms of the racial, ethnic, and gender relationships inscribed on the urban landscape.

This developmental pathway has further generated new rounds of contradictions and problems that threaten to engulf the social order, including global warming and a deepening environmental crisis. It has also exacerbated and intensified underlying structural problems such as persistent inequality and poverty, and has elicited perverse forms of social control as exemplified by record levels of incarceration that disproportionately impact minority populations.

At the same time, urban theory has continued to mutate in a number of different directions. One notable contemporary line of debate revolves around differences and similarities within the expanding planetary cohort of cities that extends across both the Global North and the Global South. For simplicity's sake, we may characterize the ensuing dilemma in this way: Is it possible to advance a unified concept of the urban that is relevant to all cases in all parts of the world, or must different concepts be developed by reference to different geographical provinces, each with its own peculiar version of an urbanization process? This question goes to the heart of urban analysis and theory in the twenty-first century. Whatever the eventual outcome of this debate, it has enormous ramifications not only for our understanding of capitalist urban dynamics but also for the potential and practice of urban politics and planning.

In spite of these changes in the material and cognitive conditions of urban life, the central themes of this book – commodity production, social reproduction, and planning in capitalist cities – remain as pertinent today as when they first appeared. Their scholarly value resides not only in their detailed focus on how things were in the particular historical episode of fordism, but also in their function as pointers to a present-day analytics of urbanization and urban remediation. In this regard, three overarching arguments in the book must be emphasized. First, cities are embedded in a wider set of social relationships that shape the form and character of intra-urban processes. Second, cities in capitalism are by their nature social constructions that in various ways reflect, support, and resonate with the dynamics of accumulation. Third, if cities derive from a set of wider social relationships, they nevertheless are instrumental in the reproduction of society as a whole. Taking these arguments together, we can say that the pressures of accumulation, as expressed in localized systems of production, labor markets, occupational structures, trade, and all the rest, are fundamental to understanding the fortunes of cities in capitalism. Additionally, these same urban-based processes determine the continuing viability of the accumulation process over space and time.

Of course, urbanization always involves much more than the broad logic of the capitalist economy. Each individual city is also a center of social and political activities that can only be apprehended in terms of their own specific and problematical logics, from neighborhood formation and institution-building to conflicts based on class, race and cultural differences. And once again, such logics reverberate back onto accumulation processes via their effects on material urban form and the constitution of the human subject.

Equally important is the book's emphasis on urban planning operations as reflections of the predicaments and political imperatives that arise as these complex kinetics play out in urban areas. Urban planning is always situated within the framework of existing relations of authority and subordination in capitalist society, and these relations regulate its identity and function. Hence, planning cannot be innocently regarded as the identification and mobilization of "good ideas," or some disembodied "public interest" in abstraction from real urban conditions. Much of the apparatus of urban planning is posited on securing the functional capability of cities as centers of accumulation. This is not to say that creative approaches, innovation, and good design are unrealistic or doomed to failure, but rather to insist on accurate recognition and proper contextualization of the constraints that urban planning must always face.

Urban studies nowadays is characterized by an accelerating and widening flow of research publications, policy briefs, and conference proceedings that together represent an extraordinary diversity of emphasis and perspective. This is as it should be, given the essential complexity of cities and their pivotal role as localized articulations of a globalized capitalism. By providing an essential foundation for understanding the dynamics of the city in capitalism, this book remains a durable and relevant contribution to any understanding of the urban question today.

Michael Dear and Allen J. Scott

Berkeley and Los Angeles, California
November 2017

Urbanization and urban planning in capitalist society

edited by
Michael Dear
and Allen J. Scott

Methuen: London and New York

First published in 1981 by
Methuen & Co. Ltd
11 New Fetter Lane, London EC4P 4EE

Published in the USA by
Methuen, Inc.
733 Third Avenue,
New York, NY 10017

British Library Cataloguing in Publication Data

Urbanization and urban planning in capitalist society.
1. Urbanisation   2. Capitalism
I. Dear, Michael   II. Scott, Allen J
III. University paperbacks
301.36′1   HT151   80–40479

ISBN 0–416–74640–3
ISBN 0–416–74650–0 Pbk (University paperback 692)

Contents

Notes on contributors

JOHN A. AGNEW is Assistant Professor of Geography, Maxwell Graduate School, Syracuse University. Originally from England, he was educated at the University of Exeter and The Ohio State University. His major substantive research interests are in political geography, especially urban conflict and political regionalism, and the geography of development. He has a more general interest in problems of explanation in social science.

MARTIN BODDY graduated from the University òf Cambridge where he also studied for a doctorate. He is now Research Assistant in the School for Advanced Urban Studies, University of Bristol. Research interests in the field of urban political economy include housing finance and policy, financial institutions and urban development, and local state involvement in economic development. Author of *The Building Societies* (London: Macmillan, 1980).

NICHOLAS BUCK is a research associate at the University of Kent at Canterbury, working on a project on an occupational community of dockyard workers and the implications of its collapse. He received his doctorate in Urban Studies, on which the paper in the present volume is based, from the University of Kent in 1980. His other main research interest, arising from this thesis, is in class conflict and local government reform.

HAROLD CHORNEY is a Canadian economist who has published a number of articles on political economy, regional underdevelopment and social theory. His recent works include 'Regional Underdevelopment and Cultural Decay' (in C. Heron (ed.) (1977) *Imperialism, Nationalism and Canada* (Toronto), and 'The Falling Rate of Legitimation: The Problem of the Capitalist State in Canada' (with P. Hansen) in *Studies in Political Economy* (1980). He has taught economics, labour studies, regional economics and planning and social theory at the University of Manitoba and in an adult education 'open university' program. He has also worked as an economist for the Government of Manitoba. He is a member of the editorial advisory board of the *Canadian Journal of Political and Social Theory*.

GORDON L. CLARK is Assistant Professor of City and Regional Planning at Harvard University, Cambridge, Massachusetts. He received his PhD from McMaster University and has research interests in spatial labour markets, regional economic planning and theories of the state.

R. B. COHEN, Senior Research Associate, Conservation of Human Resources, Columbia University and Adjunct Assistant Professor, Division of Urban Planning, Columbia University, has studied the internalization of capital and its impacts on corporate development, financial markets, and the US urban hierarchy. He received his PhD in economics from the New School for Social Research and is the author of *The Corporation and the City* (Cambridge University Press, forthcoming) and *The Impact of Foreign Direct Investment on US Cities and Regions* (New York: Praeger, forthcoming). He is currently involved in studies of the international integration of production in the auto industry, the recent transformation of large US manufacturing firms, structural change in the US manufacturing sector and the impact of changes in international banking in the US economy.

KEVIN R. COX is originally from Warwickshire and was educated at the University of Cambridge and at the University of Illinois. Currently Professor of Geography at The Ohio State University, his primary research interests are urban conflict and urban social thought. His papers have appeared in professional journals in geography, sociology and political science. He is co-editor of *Behavioural Problems in Geography: A Symposium* (Evanston, Ill.: Northwestern University Press, 1969), co-editor of *Locational Approaches to Power and Conflict* (Beverley Hills, Calif.: Sage Publications, 1974) and editor of *Urbanization and Conflict in Market Societies* (London and New York: Methuen, 1978). He is also author of *Conflict, Power and Politics in the City: A Geographic View* (New York: McGraw-Hill, 1973), and *Location and Public Problems* (New York: Methuen, 1979).

MICHAEL DEAR is Associate Professor of Geography at McMaster University, Ontario Canada. Worked as a planner in London before receiving his doctorate in Regional Science from the University of Pennsylvania. Current research interests: social service provision (especially mental health care) and theory of the state in socio-spatial processes.

MATTHEW EDEL is Associate Professor of Urban Studies, Queens College, New York. He received his PhD in economics from Yale University, 1968, and is the author of *Economics and the Environment* and co-editor with Jerome Rothenberg of *Readings in Urban Economics.* He is presently completing a book on the political economy of homeownership and suburbanization in Boston over the past century.

JOACHIM HIRSCH, since 1971, Professor of Political Science in the Department of Social Sciences of the University of Frankfurt/Main. Major research interests: theory of the state, conditions of state intervention, and the state and social movements. Associate editor of *Society: Contributions to a Marxist Theory* (Frankfurt/Main: Suhrkamp Verlag). His major publications include *Scientific and Technological Advance and the Political System* (Frankfurt: Suhrkamp Verlag, 3rd edition, 1973), *State Apparatus and the Production of Capital* (Frankfurt: Suhrkamp Verlag, 1974), and 'Comments on a Theoretical Analysis of the Bourgeois States' (in *Society: Contributions to a Marxist Theory* 8/9 (1976)).

MICHAEL HARLOE is a lecturer in the Department of Sociology, University of Essex. He was formerly a member of the Centre for Environmental Studies, London, and of the London School of Economics. He is the Editor of the *International Journal of Urban and Regional Research* and a board member of the Research Committee on the Sociology of Urban and Regional Development of the International Sociological Association. His publications include *Captive Cities: Studies in the Political Economy of Cities and Regions* (ed.) (New York: Wiley, 1977); *The Organisation of Housing: Public and Private Enterprise in London* (with Issacharoff and Minns) (London: Heinemann, 1974); and *Swindon: A Town in Transition–A Study in Urban Development and Overspill Policy* (London: Heinemann, 1975). At present he is engaged in a long-term comparative study of housing policies and in preparing for publication two further collections of papers on the sociology of urban and regional development.

DAVID HARVEY is Professor of Geography at The Johns Hopkins University in Baltimore. He is the author of *Explanation in Geography* (London: Edward Arnold and New York: St. Martin's, 1970) and *Social Justice and the City* (London: Edward Arnold and Baltimore, Md.: The Johns Hopkins University Press, 1973).

MARC LOS is currently Assistant Professor of Economics at the Université de Montréal and is also affiliated with the Centre de recherche sur les transports of that university. He obtained a PhD in City and Regional Planning from the University of Pennsylvania in 1975. His main research interests are in land use and transportation modelling and also in urban and regional economics.

DOREEN MASSEY is a Social Science Research Council Human Geography Fellow in Industrial Location, formerly based at the Centre for Environmental Studies in London. She is author of *Capital and Land* (with A. Catalano) (London: Edward Arnold, 1978), and *The Geography of Industrial Organization* (Oxford: Pergamon Press, 1980).

JOHN MOLLENKOPF is currently Director of the Urban Studies Program at Stanford University. He has taught previously at the Stanford Graduate School of Business Public Management Program and the MIT Political Science Department. His research interests include the politics of urban development, the management of urban service bureaucracies, and the role of community organizations in urban politics.

CHRIS PICKVANCE was educated at Manchester University and is now Senior Lecturer in Interdisciplinary Studies in the Urban and Regional Studies Unit at the University of Kent. He has published many articles in the field of urban and regional political economy as well as an edited book, *Urban Sociology: Critical Essays* (London: Tavistock, 1976). He has taught at the Universities of Manchester and Toronto, at Concordia University, Montreal and Bogazici University, Istanbul, and is Review Editor of the *International Journal of Urban and Regional Research.*

DAMARIS ROSE, originally from London, attended Sussex University (BA) and the University of Toronto (MA), both in Geography. She is currently completing a doctoral thesis at the University of Sussex on the political significance of homeownership in the context of changing relationships between home and wage-workplace. She is also interested in recent developments in 'inner-city' policy in the context of industrial restructuring, and their bearing on such relationships.

SHOUKRY ROWEIS was born in Egypt and educated at Cairo University, Purdue University and the Massachusetts Institute of Technology. His major current interest is in the dynamics of urbanization in advanced capitalist societies. In 1976 he received the Ontario Confederation of University Faculty Associations award for outstanding contributions to university education. He is presently in the Department of Urban and Regional Planning at the University of Toronto.

ALLEN J. SCOTT was born in England and educated at Oxford University. He has held posts at the University of Pennsylvania, University College London and Toronto University. He is currently Visiting Professor in the Department of Geography, University of California at Los Angeles.

He is the author of numerous publications on urban analysis, regional science, and planning theory.

IVAN SZELENYI, graduate of the University of Economics (MA, 1960) and Hungarian Academy of Sciences (PhD in Sociology, 1973); Ford visiting scholar at Columbia University and University of California, Berkeley (1964–5); Research Fellow, Scientific Secretary, then Head of Department, Institute of Sociology, Hungarian Academy of Sciences (1964–1975); Visiting Research Professor, University of Kent at Canterbury, 1975–6; Foundation Professor of Sociology and Head of Sociology Department, Flinders University of South Australia. Co-author of *The Intellectuals on the Road to Class Power* (with G. Konrad) (New York: Harcourt Brace Jovanovich, 1980); author of *Urban Social Inequalities under State Socialism* (Oxford: Pergamon Press, 1980). Research interests include housing system and housing policy, urban ecological structure, and social problems of urban renewal; urbanization, urban management systems from a comparative and historical perspective; and the sociology of planning, position of planner and intellectuals in the social structure, and social structure of state socialist sociology.

RICHARD WALKER is Assistant Professor of Geography, University of California, Berkeley. He received his BA from Stanford University and his PhD from the Johns Hopkins University, Baltimore, Maryland, in 1977, submitting a dissertation entitled 'The Suburban Solution: Urban Geography and Urban Reform in the Capitalist Urbanization of the United States'. His research and teaching cover environmental issues as well as urban ones; he has written articles on the beginnings of suburbanization in the US, failure of the Clean Air Act, water resources in California, uneven spatial development, industrial siting regulation, energy extravagance, industrial location theory and benefit-cost analysis, and has edited a special issue of *Antipode* on natural resources and environment. He was Visiting Professor in the School of Architecture and Urban Planning, UCLA, in 1979.

The index was compiled by CECIL BEAMISH of McMaster University.

Preface

This book is an attempt to define a general theory of urbanization and planning. It seeks to achieve this goal by bringing together and synthesizing a wide range of critical perspectives on the urban question. In particular, the book attempts to capture and to give form to much of the new urban theory (and its cognate analytical procedures) that is currently emerging in North America and Western Europe. This is a theory that generally insists upon the explicit derivation of contemporary urbanization processes out of the structure of the capitalist mode of production.

The book unfolds in a series of six logical stages. First, the argument opens with a broad *introductory statement* which outlines the main qualities and properties of a critical analysis of urban phenomena in capitalism. Second, the argument proceeds to examine the *conceptual preliminaries* necessary for the establishment of a theory of urbanization and planning as rooted in capitalist social structures. Third, a theoretical exposition is undertaken of the *fundamental logic of urbanization and urban planning*. Fourth, there follows a detailed discussion of *commodity production in cities* and of its effects on urban development. Fifth, a series of analyses is presented of the subtle and controversial problems of *reproduction and social life* in capitalist cities. Sixth, by way of a broad conclusion, a synthesis is made of some of the important political relationships linking *urbanization, social class, and the capitalist State*.

The text as a whole is a joint effort, and each of the chapters that follows was written by a different author or authors. In spite of this, the book was conceived from the outset as a whole, in that a basic outline was defined, and authors were subsequently commissioned to write specific chapters. We have largely kept to this plan, except for the inclusion of two previously published papers which are essential to the development of our argument. In spite of its preconceived structure, the book does not offer any collectively agreed-upon viewpoint. Whilst all the authors share a perspective somewhere to the 'left' of mainstream urban analysis, the arguments found in the various chapters are often at odds with one another. Wherever possible, we have preserved this conceptual tension.

Our objective has been to capture the main currents of a vigorous, on-going discussion rather than to put the finishing touches to a definitive and final statement. Hence, this book provides not only the analytical foundations for a critical and reconstituted urban theory, but also (by provoking new questions) the conditions for the further development of this theory.

As an immediate consequence of these qualities, the book is long; it is often dense in its argument; and it is occasionally repetitious. These qualities have arisen precisely because we have been anxious to provide a *comprehensive* overview of a crucial area of analysis. Furthermore, the universe of discourse from which this book grows addresses arguments which remain in a preliminary and highly unfinished state of development. Unlike, for instance, neoclassical economic theory which has been thoroughly explored and systematically codified, there is little that can be taken as being generally settled in the theoretical framework that is suggested here. Consequently, one of our primary editorial precepts has been to tolerate bulk and repetition of first principles as the inevitable cost of meaningful exposition. One advantage that the reader will immediately enjoy is the unusually high degree of definition and completeness which characterize the discussions of this book.

Urbanization and Urban Planning in Capitalist Society is a closely argued book that extends over an intricate system of ideas. In order to ease the reader's task, we have devoted the greater part of this Preface to a simple exposition of the book's main contents. This exposition is reconstructed at a significantly higher level of generality in Chapter I, where we undertake to outline the basic elements of an urban problematic. Finally, we conclude this preface with some general observations on the use of this book.

Part I   The urban question

Chapter 1   Towards a framework for analysis

Dear and Scott introduce the book as a whole by setting out the broad structure of a theoretical framework. It is shown how the processes of urbanization and planning are embedded in and derived out of the overarching capitalist social formation. It is then demonstrated how a specific urban question (i.e. a nexus of conceptual problems) appears as capitalist social and property relations are projected through the urban land use system. The authors go on to discuss the logics of production and reproduction in cities. Urban life within capitalism is seen as being inherently problematical, and urban planning makes its historical appearance (as a structured element of state apparatus) in order to deal with the in-built predicaments of the capitalist urbanization process.

Part II   Prolegomena to a theory of urbanization and planning

Part II of the book is an attempt to address some basic theoretical issues that are necessary, but also essentially prior, to any investigation of urban phenomena. A description of the structure and internal dynamics of capitalism is established. This is then followed by a critical discussion of the role and scope of the State in capitalism. In the light of these analyses, an epistemological investigation is undertaken of the dual character of urban planning as both a concrete socio-political phenomenon and as a routinized practice that is typified by certain abstract decision rules and procedures.

Chapter 2   Capitalism, accumulation and the explanation of urban phenomena

Edel proceeds to lay down some of the conceptual preliminaries needed to understand the urbanization process as it emerges out of the capitalist mode of production. He explores in considerable detail the relationship of urban phenomena to accumulation. The capitalist mode of production is defined by : (a) domination by a hegemonic class through the control of capital; (b) presence of a proletariat as a subordinate class; (c) extraction of surplus value in the process of production; and (d) monetized exchange of commodities. However, specific capitalist social formations may vary in terms of other dimensions including: (a) landownership arrangements and the presence of non-capitalist social fractions; (b) family and social institutions for the reproduction of labour-power; (c) the role of the State; and (d) the degree of unification and class consciousness within each of the principal classes. Edel goes on to discuss accumulation as a driving force in capitalism and he discusses the genesis of crises within the accumulation process. The final section of the chapter considers the extent to which specific urban phenomena can be explained by reference to the accumulation process and to the general structure of capitalist society. The analysis is then applied critically to an examination of some recent attempts to explain urban policies, social patterns, and built environments.

Chapter 3   The State in capitalism and the capitalist State

Clark and Dear argue that a proper understanding of urban and regional processes requires a comprehensive theory of the State. Two approaches to such a theory are reviewed. The first examines the functions of the State in capitalism focusing on the empirically observable dimensions of state intervention. The second approach analyzes the *capitalist* State as such, and attempts to derive the form and functions of the State from the specificities of the capitalist social order. Four major theories of the

State in capitalism are evaluated and found wanting. These are: (a) the State as simple supplier of public goods and services; (b) the State as facilitator and regulator of the economy; (c) the State as social engineer; and (d) the State as arbiter of social conflicts. In their place, a more complex and historically selfconscious explanation of the capitalist State is sought. A historical materialist theory of the State is derived from analysis of the classical Marxist texts, from the recent literature on the Marxian theory of the State, and from the German debate on the 'derivative State'. The significant emphases and directions posed by a materialist theory of the State are synthesized, and their implications for a theory of urban planning described.

Chapter 4   Some reflexions on epistemology, design and planning theory

In this chapter, Los elaborates upon the implications of the preceding chapters for a general theory of urban planning. To a large extent, Los's chapter is an attempt to address the question of whether, and to what degree, urban intervention constitutes a social process ('planning') or a technical idea ('design'). Los claims that an understanding of this problem can only be achieved if we have a clear understanding of the link between knowledge and action, and of the epistemological concepts that are necessary to determine what that link is. Los discusses the link between knowledge and action in terms of Piaget's genetic epistemology. On this basis, he then examines the problem of the objectivity of theory *in* planning, and the concepts of rationality and optimality. An attempt is made to justify the position that social science and planning cannot be completely objective and value-free, and that both a theory of planning and a theory of design are important. In their proper perspective, abstract concepts of optimality and rational decision-making have an important role to play in planning research and practice, as part of a theory of design. The role of planning theory is then to situate these concepts in a socio-historical context. Planning theory, by its self-reflexive functions and by the importance it attributes to the uncovering of hidden beliefs and assumptions, should make planners more aware of their social responsibilities. This latter theme is picked up with some force at a later stage by Roweis (Chapter 7).

Part III   Urbanization and planning in capitalist society

Having laid the foundations of a focused discourse, the book now moves into an examination of urban phenomena proper. It is shown, in general, how an urbanization process emerges out of the capitalist mode of production and how this process manifests itself geographically as a system of interacting land uses. Four major themes are tackled in this part of the book. First, the complex and crisis-prone relationships linking the built urban environment to the capitalist accumulation process are elucidated. Second, an attempt is made to decipher the social dynamics that govern the peculiar geographical structuring of this environment. Third, the forces underlying the historical emergence of urban planning are examined, and an analysis made of the manner in which the substantive content of urban planning interventions is determined at any given historical moment. Fourth, problems of the comparative analysis of urbanization processes are discussed and the question as to how urbanization varies from social formation to social formation is addressed.

Chapter 5   The urban process under capitalism: a framework for analysis

The objective of this chapter is to sketch a general problematic for the interpretation of the urban process within capitalism. To this end, Harvey develops the two inter-related themes of accumulation and class struggle introduced by Edel (Chapter 2). The Marxist theory of accumulation views the role of investment in the built environment in the light of the internal contradictions of the accumulation process. Hence, investment in the built environment is seen as a response to the different forms of crisis which appear within the capitalist system. The manner in which the built environment itself expresses and contributes to the crisis of capitalism is then explored. It is argued that a permanent struggle exists, in which capitalism attempts to construct an environment in its own image, only to destroy it when a new crisis appears. The analysis then considers the issue of class struggle and how it influences investment in the built environment. Of particular interest is the manner in which class struggle in the workplace is in part transformed, via the urbanization process, into struggles around the reproduction of labour in the living place. Some examples of these struggles are presented in order to illustrate how they derive from the fundamental conflict at the point of production while, at the same time, they influence the direction and form of investment in the built environment.

Chapter 6   The urban land question

Here, Roweis and Scott discuss the logic of urbanization as a system of dense, polarized differential locational advantages embedded in capitalist social and property relations. Urban land is seen as a complex use value produced collectively by the interpenetrating activities of (a) firms and households on the one hand, and (b) urban planners (i.e. the State) on the other hand. The intra-urban locational dynamics of firms and households give rise to persistent breakdowns in the social efficiency and viability of urban space. These breakdowns then call for new rounds of planning intervention. In capitalist society, however, such interventions are necessarily reactive and palliative, and themselves lead to new urban problems and predicaments. In conclusion, Scott and Roweis point toward the relevance of their analysis for a comprehension of urban politics (cf. Part VI).

Chapter 7   Urban planning in early and late capitalist societies: outline of a theoretical perspective

This chapter pursues in greater detail the concept of urban planning developed in Chapter 6. Roweis writes that urban planners tend to lack a systematic knowledge of their discipline-*cum*-profession. Lacking an effective map of social reality, planners tend to vacillate between utopianism and technical pragmatism. The chapter seeks to remedy this situation by deriving urban planning as a concrete social relationship rooted in the overarching structure of capitalist society, and specifically charged with dealing with intra-urban land use predicaments and conflicts. The argument proceeds through a series of inter-woven propositions. First, some methodological questions are addressed to show how planning can be seen specifically as a mechanism of collective (as opposed to private and individual) action in capitalism. Second, the broad structural qualities of early and late capitalist societies are described. Third, the nature of urban planning in these two historical contexts is discussed, and it is shown how planning (and its ideological reflection, planning theory) appears as a response to specific urban problems and social imperatives.

Chapter 8   Notes on comparative urban research

While most of the theoretical chapters in this book are addressed to the question or urbanization as a general process in capitalism, Harloe discusses in detail a number of important issues concerning matters of comparative urban research. The first part of this chapter examines some aspects of the work of the French Marxist urbanists, Castells and Lojkine, in order to illustrate the problems that can appear in the absence of a thorough historical and comparative approach to studies of urban development. The second part of the chapter discusses the meaning and value of comparisons of urban development and policy in Western and Eastern (capitalist and socialist) societies. The last part of the chapter considers some recent arguments advanced by Offe concerning possible limitations to a comparative approach to policy analysis.

Part IV   Commodity production and urban development

Part III of the book provided a general overview of the urbanization process in capitalism. Part IV begins the task of looking at some of the detailed components of this process. Specifically, Part IV deals with the all-important matter of the relations between commodity production and urbanization. It is shown how the structural evolution of the commodity-producing process leads to persistent locational change with powerful repercussions on urban development (including, in particular, a constantly deepening crisis of inner-city areas). This phenomenon of locational change is in part mediated by state policies which in both North America and Western Europe have tended to encourage the movement of industry away from large metropolitan areas and into less developed regions. At the same time, an overview is provided of the capitalist logic of the property sector as that specialized branch of production that is geared to the development and re-development of the built urban environment. Finally, an attempt is made to broach the complex problem of the structure of the urban hierarchy in capitalism via an analysis of the multinational corporation and the international division of labour.

Chapter 9 The UK electrical and electronics industries: the implications of the crisis for the restructuring of capital and locational change

Massey presents some results of an empirical investigation of the locational implications of retrenchment, rationalization and aggregate employment decline. She indicates that the present economic situation in the United Kingdom may be having certain distinctive spatial and regional repercussions. Her essay examines in particular the changing distribution of employment between the depressed and the relatively prosperous regions of the country, and the continuing decline, as centres of manufacturing, of the major conurbations of the country. It is shown that both the intersectoral reorganization of production (the reallocation of capital between branches), and the differential impact of intrasectoral attempts to reduce the cost of variable capital and increase the rate of surplus value, are contributing significantly to these spatial changes. Some of the political implications of these trends are mentioned at the end of the chapter.

Chapter 10 Policies as chameleons: an interpretation of regional policy and office policy in Britain

Pickvance seeks to describe the aims and evolution of regional and office policies in Britain and their effects on urban and regional development. The evolution of these policies is shown to reveal a chameleon-like shift away from their original aims which were to secure more balanced spatial distributions of manufacturing and office employment. It is argued that their actual effects are: (a) to subsidize industrial capital vacating the large conurbations and locating in depressed regions; and (b) to promote central London as an international centre of business. The first effect comes about because incentives to locate in depressed areas rarely influence the basic locational decision, and are allocated irrespective of the number of jobs created. The second effect results from the influx into London of prestigious firms, facilitated by lenient 'control' of new building in central London. This shift from the policies' original aims is due to changing pressures on the State – in particular the growing internationalization of production and the greater leverage of industrial capital in times of economic crisis. The British State is thus increasingly obliged to place the subsidy of industrial capital and the promotion of 'invisible exports' above any concern with the spatial distribution of manufacturing or office employment. The explanation of the two policies as necessary legitimations of capital accumulation is rejected, as is the idea that state policy in general has a necessary 'legitimation function'. In general, this chapter represents a valuable case study of the pressures arising out of the dynamics of modern commodity production, and their effects on urban and spatial development.

Chapter 11 The property sector in late capitalism: the case of Britain

In this chapter, Boddy focuses on the emergence of the autonomous 'property sector' in modern Britain, and on the changing scale and nature of investment of financial institutions in commercial and industrial property. An analytical framework based on the interlocking circuits of industrial, commercial, and interest-bearing capital is developed. This framework is applied at the empirical level to an examination of the particular (and historically determined) articulation of agents involved in the production and finance of property. The growing involvement of financial institutions in the property sector and the increasingly direct nature of this involvement are clarified. It is shown that as a result of various structural changes, the financial system is increasingly exposed to booms and slumps in the property sector. In addition, the form and feasibility of urban and inner-city redevelopment, regional policy, and industrial strategy will be determined by the investment criteria of financial institutions. Finally, it is suggested that strong contradictions remain between public planning and private land ownership (together with the private appropriation of ground rent), and that the widening ownership of land and property by financial institutions has strengthened and altered the structure of opposition to efforts intended to resolve these contradictions and to nationalize land.

Chapter 12 The new international division of labor, multinational corporations and urban hierarchy

In concluding Part IV, Cohen introduces a new and vital theme. His essay analyses how the development of the global system of cities has been linked to changes in the world economy since the Second World War. He focuses on the new international division of labour in the capitalist system and evaluates how changes in multinational corporations and specialized corporate services (banks, law firms, accounting firms, etc.) have contributed to the emergence of a hierarchy of world cities. The dimensions of this hierarchy are explored, with reference to the conflicts and contradictions inherent in it. Probable future trends in the international division of labour and the urban hierarchy are examined.

Part V Reproduction and the dynamics of urban life

Just as the chapters in Part IV of the book dealt with the specialized issue of commodity production and urban development, so Part V now deals with the particular problems of the reproduction of labour in capitalist cities and the dynamics of urban life. The six chapters that make up Part V address three major integrated themes. A first theme concerns the inter-relationships between housing and the accumulation process. A second theme concerns the emergence of individual communities and neighbourhoods within the urban environment and of the conflicts that coalesce around this phenomenon. A third theme concerns the ideological reflections and biases of urban life in terms of the psychology of homeownership, community life, and the containment of dysfunctional urban/social outcomes. Taken collectively, the chapters that make up this part of the book constitute a significant step forward in the theorization of those historical processes which govern the perpetuation of established social relations in capitalism at large and in capitalist cities in particular.

Chapter 13 Community and accumulation

In this chapter, Mollenkopf examines the fundamental concept of accumulation (as developed earlier by Edel and Harvey) within the context of 'community'. According to Mollenkopf, cities concentrate two types of relationships which have an asymmetrical, and sometimes conflicting, interdependency. On the one hand, cities concentrate the mechanisms by which wealth is produced and accumulated; on the other, they concentrate the social relations by which extra-economic communities are formed. While these two tendencies presume each other and can be reinforcing, they operate by distinct logics which can also come into conflict. The characteristically urban conflict is, therefore, not so much within the economic system as between the dictates of the economy and the forms and aspirations of communal groupings. Actors dominated by economic motives have often sought to undermine and dismantle communal patterns which impede their economic ends, and have sought to create new forms more to their liking. Community groupings have resisted these initiatives. Political institutions and alliances, and networks of social ties mediate the differences and conflicts between accumulation and community, and should be studied in this light.

Chapter 14   Accumulation versus reproduction in the inner city: *The Recurrent Crisis of London* **revisited**

Here, Rose attempts to analyze the historical genesis of some of the problems associated with the processes of urban development and redevelopment in London from the mid-nineteenth century onwards. She uses this analysis to clarify the nature of the dilemmas associated with the state housing and planning policies in the metropolitan city. Rose argues that the city and its problems are expressions of a historical process of urbanization shaped in particular ways by the relations of production and reproduction of capitalist society. Specifically it is suggested that the processes of London's development have been structured by the dual imperatives of (a) directly facilitating the circulation and accumulation of capital and (b) reproducing a specific social and technical division of labour. State housing and planning policies have, since the 1890s, attempted to 'manage' the crises periodically produced by the articulation of these imperatives on the terrain of the city, but Rose argues that the very nature of the capitalist mode of production precludes a guaranteed 'solution'.

Chapter 15   A theory of suburbanization: capitalism and the construction of urban space in the United States

Continuing the themes expounded in the previous paper, Walker develops an extended argument about the nature of suburbanization. He suggests that, geographically, suburbanization has three interwoven dimensions: spatial differentiation, decentralization, and waves of urban construction. He affirms that these are best explained by the method of historical materialism. Although the reproduction of capital and capitalism is a structural imperative, the dialectics of past and present, of structure and everyday life, and of spatial and social relations are also important to an understanding of suburbanization. Two other distinctions are also fundamental: that between the use and production of urban space (built environment); and that between work/workplace and leisure/residence. Walker argues that suburbanization has been consistent with the individual, class and systemic interests of American capitalism, i.e. the 'suburban solution' has played a fundamental part in the success of capitalism in overcoming its own inner contradictions. Spatial differentiation is due to such forces as the division of labour, a fragmented mode of living, and the search for monopolistic rents. Decentralization is due to diminishing needs for industrial concentration, repulsion of upper-class residents from the working class at the centre, and profits from property appreciation. Periods of city-building correspond to waves of accumulation, providing a built environment consistent with the changing character of capitalism and outlets for investment capital.

Chapter 16 Capitalism and conflict around the communal living space

In keeping with the focus on conflict and community as developed thus far in Part V, Cox considers in detail the specific conflict between the property development sector and labour with respect to those use-values which are consumed communally, but which are accessible only through residential location. These use-values define the communal living space of labour. However, this living space has a tendency to become increasingly commercialized and to be supplied to ever larger degrees in the commodity form. This leads to urban conflict. On the one hand, the processes of community development and re-development are progressively penetrated (and, in human terms, impoverished) by capital. On the other hand, the human problems sparked off by this penetration create political responses on the part of localized fractions of labour. Hence, the property sector generates barriers to its own reproduction. These barriers must be, and are, suspended as capital exploits the market incentives spontaneously (and perversely) created by neighbourhoods themselves, and as it plays off one fraction of labour against another.

Chapter 17 Homeownership and the capitalist social order

Agnew here concentrates his attention on the importance of homeownership in capitalism. He suggests that prevalent explanations of order and stability in capitalist societies focus either on a societal consensus or coercion by an élite group. Some writers, claiming to follow a Marxist tradition, have proposed an alternative perspective which views the capitalist social order as resting upon the 'practical incorporation' of populations through an ideology derived from their everyday activities. Agnew examines the incorporation thesis in terms of the proposition that homeownership is an important means of practical co-optation in capitalist societies. Basing his argument on empirical studies of homeownership in the United States and Britain, Agnew concludes that: (a) homeownership appears to serve more of an incorporative function in the United States than in Britain although evidence suggests an ambivalent or 'contradictory consciousness' about the rewards and merits of homeownership; and (b) as a consequence, the practical incorporation thesis is more applicable in some contexts than others, according to the degree to which capitalist social relations have/ have not been challenged by new ideas and institutional changes.

Chapter 18 Social and spatial reproduction of the mentally ill

In this essay, the question of reproduction is extended to a rather different situation from those treated above. Dear considers the social and spatial forces involved in the reproduction of two groups, namely, the service-dependent mentally ill on the one hand, and their professional servers on the other hand. The connections of these two groups to the wider

Urbanization and urban planning in capitalist society

social formation are defined, and the subservient/exploitative nature of the client/professional relationship is examined. The persistent tendency towards social and spatial isolation of the mentally disabled in contemporary society is demonstrated. In the modern city, the mental hospital is being replaced by a geographically distinct 'asylum without walls' in the community. Both the community and the State are implicated in this isolationist process.

Part VI   Urbanization and the political sphere

In this final main section of the book, we explore some of the important political relationships linking urbanization, social class and the capitalist State. The discussion is first of all motivated by a very specific historical analysis of municipal labour policy in east London from 1886 to 1914. This is then complemented by a study of Canadian urban political consciousness in the 1910s and 1920s, and (in contrast to some of the ideas developed in the preceding section) of the depoliticizing effects of the urbanization process in capitalism. It is then argued that urbanization creates and calls for many of the conditions leading to the transformation of capitalism into a 'state mode of production'. In a final chapter, our attention is once again drawn to the important issue of the distillation of social conflict at the neighbourhood or community level. Political conflict at this level is rooted in class antagonisms, but also goes beyond traditional concepts of class struggle. It is at this level that one of the most intense conflict mechanisms within modern capitalist society is played out.

Chapter 19   The analysis of state intervention in nineteenth-century cities: the case of municipal labour policy in east London, 1886–1914

Buck's essay assesses the use of historical materialist approaches in the study of class relations and state intervention in nineteenth-century cities. Like Harloe (Chapter 8) Buck uses the work of the recent French Marxist urban sociologists, as well as the British Marxist social historians, as a point of departure. He examines the applicability of the French work which has been primarily developed for the analysis of contemporary cities. Using these concepts, Buck then presents a case study of intervention by local government in West Ham (in east London) during the period 1886 to 1914, focusing on the municipal council's policy on labour issues. He demonstrates the usefulness of the French work, with its conception of the urban as an arena of class conflict, and of urban politics and urban policy as immediate expressions of this conflict. However, the concepts used in analyzing contemporary cities cannot simply be transferred to earlier periods. In particular, it is shown that Castells's work tends towards a restrictive conception of the urban and a static view of the social formation of advanced capitalism.

Chapter 20 Amnesia, integration and repression: the roots of the Canadian urban political culture

Buck's themes are complemented by Chorney's study of Canadian urban political culture. Despite the optimism of Marx that city life would enhance class consciousness, Chorney argues that urbanization has facilitated both the accumulation of capital and the cultural reproduction of capitalist society. Urbanization as a mode of social control and as an engine of capital accumulation has synthesized 'amnesia, integration and repression' as powerful bulwarks against the development of class consciousness. In Canada, this synthesis can be traced: (a) to the repression of radical politics in the first two decades of the twentieth century; (b) to the integration of workers through the medium of trade-union collective bargaining; and (c) to the rise of a city devoted to instrumental social relations, mass consumption and spectacle. The relative underdevelopment of urban class politics is seen as the outcome of factors peculiarly related to the growth of the Canadian urban structure and its situation inside the process of capitalist urbanization.

Chapter 21 The relative autonomy of the State or state mode of production?

In this penultimate chapter, Szelenyi brings the issue of State and politics right into the present. He argues that during the last decade the central concern of the sociological theory of the State was to show the inaccuracy of the 'pluralist' theory of the State and to demonstrate the complex interrelationships between the political State and civil society (cf. Chapter 3). In its first wave of revival, the new Marxist theory of the State (Miliband and Poulantzas) emphasized that under capitalism the State has only a 'relative autonomy' from the class relations of civil society. Later developments in Marxist theory (Habermas and Offe) began to discover that in late capitalism the State begins to transcend capitalist relations of production, and might permanently 'decommodify' certain sectors of production. Neo-Weberian sociology (Pahl) went so far as to speak about the emergence of 'corporatism'. Most recently, Lefebvre even proposed the concept 'state mode of production' to describe the new phase of development in industrial society in which the State powerfully intrudes into civil society, and can therefore no longer be understood as a simple expression of the latter. Szelenyi attempts to clarify the theoretical propositions of Habermas, Offe and Lefebvre and to define the state mode of production as a distinct concept from both the capitalist mode of production and state socialism. He suggests that the state mode of production emerges as the State gains powers to control the flow of surplus into productive investment without necessarily abolishing private ownership. A comparative study of capitalist urbanization and the current urban/regional fiscal crisis shows how the economic and political preconditions of a state mode of production are being created. Szelenyi does not suggest that the state mode of production is an empirical reality, but he identifies those structural conflicts in the sphere of urban and regional development which probably can only be resolved by the emergence of a state mode of production.

Chapter 22   The apparatus of the State, the reproduction of capital and urban conflicts

In the final chapter, Hirsch analyzes the root cause of the crisis of the contemporary State in relation to the politics of urban social movements. His theoretical approach is based upon the Marxist theory of accumulation crises and the materialist theory of the State, as developed by European scholars (cf. Chapters 3 and 21). Hirsch argues that the maintenance of production in advanced capitalism demands increasing state intervention in the processes of production and social change. In this way, economic crises are subsumed as part of the state-controlled reproduction sphere. The consolidation of the apparatus of mass-integration is achieved (a) through state-mediated socio-economic change, and (b) through the need to intercept and limit potential conflict by compensatory and repressive state actions. However, these conflicts tend to become more unstable as a result of such efforts. This instability is intensified because it is concentrated in cities, where autonomous social movements develop as a response to crises in capitalism and in mass integration (including crises in legitimation and administrative rationality). The conflict associated with urban social movements is based in the fundamental class antagonisms inherent in the capitalist social formation, but it has also transcended traditional concepts of class struggle.

We anticipate that this book will be of use to a wide range of students and urban practitioners. In these categories we include geographers, economists, sociologists, political scientists and planners. Some elementary knowledge of cities and urban structure is taken for granted in the chapters that follow. Readers with some background in historical materialist analytical categories and methods will have little difficulty in comprehending the main discussion, and such readers ought to be able to proceed fairly easily through the entire text. However, for those readers without this background we have attempted to incorporate the basic 'vocabulary' necessary for an understanding of this book's major themes. Such readers might prefer to undertake a preliminary overview of the book's argument and basic concepts before returning to the sequential ordering in which the book is deliberately structured. A clear preliminary definition of these concepts is contained in Chapters 1, 2, 7 and 13.

One other point is worth mentioning. We have paid scant attention in the present volume to the critique of mainstream urban theory. This critique has been effectively accomplished elsewhere, and we have little to add to the critical scrutiny to which urban economics and sociology, in particular, have been recently widely submitted. We are convinced that the future development of urban theory must emanate from within the universe of discourse that is outlined below. Accordingly, in this book our attention has been focused exclusively upon describing, securing and expanding *this* line of intellectual enquiry.

The chapters in this book trace out an extraordinarily fertile and exciting discussion which is rich in both scientific and political consequences. We offer this collection to academic and lay audiences in the hope that the discussion will grow and mature, and that its important lessons for urban practice will be increasingly disseminated.

MICHAEL DEAR
ALLEN J. SCOTT

McMaster University/Toronto University
January 1980

I The urban question

1 Towards a framework for analysis

Michael Dear and Allen J. Scott

Introduction

The purpose of this book is to examine the structure, logic and historical manifestations of urban land use and space as outcomes of private and political decision-making within capitalism. Two specific themes are vital to our analysis, and they intersect decisively at various points in the following text. A first theme concerns the organization of urban form and structure as the tangible expression of the private locational decisions of firms and households (the basic behavioural units of civil society). The logic of these decisions is derivative from the logic of the wider capitalist social formation, and it leads persistently to the emergence of social problems and predicaments which in turn call for urban planning. A second theme, therefore, concerns the genesis and character of urban planning (collective urban intervention) as the capitalist State confronts the problems and predicaments of the urban system. As these two themes intersect, there emerges a conceptualization of the urban process as the composite reflection of a system of private and public land-use decisions in the specific context of the capitalist mode of production. This book, in brief, is not just about social processes *in* cities; it is also about the basic historical tendencies and internal order *of* cities.

It is only with the recent emergence of various Marxian, neo-Marxian and critical analyses of the city that a mature, comprehensive treatment of this topic has become possible. Existing mainstream approaches to urbanization and planning have tended to be eclectic and partial, in that they are divorced from any wider theory of capitalist society. At the same time, these approaches have tended to erect artificial barriers between the concepts of urbanization and the concepts of planning that they advocate. The manifest shortcomings of these approaches and their self-imposed limitations have been discussed in several recent critical statements (cf. Castells 1977a; Harloe 1977; Harvey 1973; Tabb and Sawers 1978). Here, we shall proceed directly to an examination of an alternative, historically-rooted view of urbanization and planning as a composite social event. This view is now evolving from the revival of historical materialism that is currently proceeding throughout the social sciences. One of the main points of departure in this revival is the proposition that modern urban phenomena are comprehensible only in the context of some prior analysis of the production and reproduction relations of capitalism. In short, urbanization is decipherable only as a mediated outcome of the social dynamics and imperatives of the capitalist mode of production in specific conjunctural circumstances. This basic viewpoint is beginning to engender a wide and energetic discussion about the urban question and the rational foundations of urban politics.

The intent in this book is to capture and to elaborate upon some of the main axes of this discussion. In what remains of this introductory chapter we provide an elementary guide to the main articulations of the urban question in the light of a historical materialist theory of knowledge and action. The chapters that follow take up the discussion and explore its various ramifications through a wide variety of urban processes and historical situations.

The bases of an analytical orientation

Let us begin with this simple proposition: *neither urbanization in general, nor urban planning in particular, constitute independent, self-determinate occurrences.* On the contrary, they are *social events,* embedded within society, and deriving their logic and historical meaning from the general pattern of society as a whole. These assertions, of course, provide no clues as to the nature of this general pattern. Nor do they (as yet) yield any insights into the ways in which this pattern is mediated and re-ordered by the specific processes of urbanization and planning. What they do affirm, however, is the self-evident (though, in practice, widely overlooked) notion that urbanization and planning can never be effectively treated as objects of theoretical study divorced from some wider theory of society.

Historical materialism and the capitalist mode of production

What, then, are the necessary features of this wider theory of society? Four features in particular would seem to be crucial. These capture essential levels and moments of social and historical reality, and together make conceptually coherent the disparate and fractured nature of social and urban life. First, this theory should establish a definition of society as a total and evolving structure. Second, it must elucidate the mechanisms whereby society is physically reproduced, i.e. it must identify the material foundations of society in terms of a web of forces and relations of production. Third, it must be capable of demonstrating how the life-projects, intentionality and character of individual human beings in society are engendered and maintained. Fourth, it must be capable of illuminating and guiding human action; in other words, it should be policy-relevant in that it is self-concious about matters of social and political change. These four points seem to provide the essential foundations of any reasonably powerful analytical-*cum*-political statement about the function and purpose of human society. There remain, of course, many questions about the manner in which any such statement should proceed, and we do not automatically preclude the possibility that it does not (or cannot) exist. If, however, some viable wider theory of society *cannot* be discovered, then social enquiry must surely fall into eclecticism, disjointedness and an arbitrary empiricism.

Now, of all existing systems of discourse, only one makes any definite claim to attack with coherence and historical self-awareness the four main points outlined above. This is the problematic of historical materialism, with its specific theoretical concept of human history as an interlocking system of modes of production. In particular, the Marxian and neo-Marxian theory of the *capitalist* mode of production constitutes a powerful framework for analysis of contemporary social issues. The various papers in this book attempt, with varying degrees of explicitness, to understand the phenomena of urbanization and planning by situating them in a theoretical context that is either historical materialist in some clearly identifiable way, or is at least consistent with and allied to a historical materialist (as opposed to a mainstream) position. Furthermore, the majority of these papers seek to imbue urbanization and planning with a specific social content by intentionally relating them to the capitalist mode of production.

This effort to derive a theory of urbanization and planning from a more fundamental theory of capitalist society does not presuppose that the debate on this latter issue has been foreclosed. On the contrary, there remains considerable disagreement within and around the problematic of historical materialism as to what constitutes an acceptable theory of capitalist society. (This is exemplified by the debate recently set off by Hindess and Hirst (1977) on the notion of concrete social formations versus abstract modes of production). At the same time, and as a corollary, there is a vital and widely-ranging debate at every level of analysis in the chapters that follow – from questions on the basic theory of knowledge to the political meaning of collective intervention in the urbanization process. What this debate avoids, however, is the rootlessness and capriciousness that characterize so much existing (mainstream) social and urban theory. It is, by contrast, rendered coherent – and analytical and politically productive – by a common agreement that what constitutes an acceptable universe of discourse must always coherently address the four basic points identified above.

The urban question

Despite our earlier assertion that urbanization and planning can only be understood in the context of some global concept of society, this does not mean that effective discussion of these phenomena simply dissolve away into a matrix of more fundamental propositions. To be sure, urbanization and planning are mediated out of a wider system of social processes, but they also retain an authenticity and significance as questions at their own specific level of analysis. The pessimistic view of Castells (1977b: 62), who asserts that 'urbanization is neither a specific real object nor a scientific object' is most certainly wrong in emphasis, if not in substance.

A specifically *urban question* does indeed exist. It is structured around the particular and indissoluble geographical and land-contingent phenomena that come into existence as capitalist social and property relations are mediated through the dimension of urban space. The urban question is composed of a set of integrated facets, each of which poses a further question at *its* own level of resolution. A first facet involves the ways in which the private behavioural entities of modern capitalist society (i.e. firms and households) interact with one another to produce a land use system. A second facet involves consideration of the dynamics of social and institutional breakdown in the urban land-use system, and the concomitant imperative of urban planning. A third facet involves an analysis of the genesis, trajectory and consequences of urban planning. Finally, from these particular questions, there emerges a composite question as to the evolutionary development of the modern city. This question revolves around the interdependent decisions and actions of firms, households and planners in a general urban system consisting of an integrated hierarchy of land-use complexes. It is our contention that, while these various questions are embedded within the wider structure and logic of capitalism, they nevertheless address themselves to analytical problems and human predicaments that *cannot* be automatically read off from the overarching capital-labour relation.

The city, then, is considerably more than a *locale* in which the grand, unmediated events of the class struggle are played out. The city *is* a definite object of theoretical enquiry (though we reaffirm that any thorough urban analysis must be situated within the wider problematic of the historical materialist theory of capitalism). We need, at this stage, to outline the internal order of the capitalist mode of production; and then, from this, to construct detailed statements on the specific properties of capitalist urbanization and planning.

The structure of capitalist civil society and the capitalist State

We here provide a brief, much-simplified description of the form, dynamics and imperatives of capitalist civil society and the capitalist State.

Commodity production in capitalism

The inner core of capitalist society consists of the institution of commodity production. This institution may be characterized as a general social process in which capitalist firms take materials and equipment, combine these with live labour, and then sell the resulting output at prices that secure for producers at least a normal rate of profit. Capitalist society presents itself as a bipartite system of production relations, comprising commodity producers (capitalist firms, together with an associated constellation of directorial, managerial, and stockholder interests) and workers (blue collar and white collar). These production relations coincide, imperfectly but decisively, with the basic allocation of authority and subordination in capitalist society, in that: (a) the meaning and purposes of capitalism (production for accumulation) are ultimately defined by the interests of commodity producers; and (b) the global social structures and processes of capitalism, into which the populace is socialized (e.g. the division of labour, the dynamics of applied technology, the pattern of urbanization, etc.) remain largely outside the domain of deliberate collective decidability. These social structures and processes emerge out of the dynamics of capitalist society, but they are not freely chosen, nor are they usually changeable except after arduous struggle and political conflict. Even so, contemporary capitalism is by no means simply reducible to a rigid model, consisting of a binary social structure of opposing capitalist and proletarian classes. For, around the basic capital-labour relation, there exist many different social groups which enormously complicate the patterns of social and political alliances in capitalism. As the essays in this volume show, one of the significant expressions of this complexity is the modern city, where *territorial* divisions and conflicts consistently breach class divisions and conflicts.

In spite of the constant mutations of capitalist production relations through time, and in spite of the increasing ambiguity of their contingent structural forms, the production of commodities in order to generate profit remains the central motor of capitalist society. It is the key to understanding the dynamics of capitalism. In particular, as a consequence of the competition among commodity producers for markets, the profits earned in commodity production are persistently ploughed back (i.e. accumulated) into expanding the bases of production. Two conditions are essential to the success of this fundamental process, and hence to the continued viability of capitalism. The first is that internally engendered limitations on the processes of production, exchange, and accumulation must be controlled or eliminated. The second is that there must be a constantly available labour force which is effectively socialized into the basic rationality of the production, exchange, and consumption of commodities. In particular, labour must be physically, mentally and morally equipped to perform the tasks of commodity production. *Neither of these vital conditions is spontaneously and automatically guaranteed by purely capitalistic processes of production and exchange.* On the one hand, commodity production itself is latent with self-disorganizing tendencies, such as crises of overproduction, market failure, monopolization, and so on. On the other hand, the reproduction of the labour force depends in part on unpredictable personal and psychological dynamics which constantly threaten to undermine the perpetuation of an effective, compliant and disciplined labour force.

Emergence of the capitalist State

The inability of capitalism spontaneously to regenerate itself is clearly a major dilemma. A further threat to the stability of capitalism derives from the immanent class conflict between commodity producers and the labour force, and this conflict often erupts into overt political struggles. Out of these dissonances, and the concomitant threat of social disorder, there emerges an overarching historical imperative. This is the social necessity for the appearance of some mediating agency, invested with certain powers of social control, and capable of re-establishing vital social institutions when their existence is in some way threatened.

Thus it is that the *State makes its irreversible appearance as the collective guarantor of production and reproduction relations in capitalist society.* On the one hand, the State continually seeks to facilitate accumulation by attempting to ensure that (capitalistically) rational allocation and disposition of resources. On the other hand, it intervenes in the reproduction process in matters of housing, education, medical care, social work and so on. Further, by maintaining a continuous, ideological discourse about its own purposes and functions, and about the positive aspects of social life, the State seeks to legitimate the existing order of things, and to maintain in equilibrium (by physical force, when necessary) the tense internal balance of commodity-producing society. However, although the State is invested with powers of social control, it in no way establishes itself as the ultimate arbiter of all social activity. For the State is embedded in, and takes its meaning from, the general structure of commodity-producing society. Its actions are manifestations of the imperatives of that society, and not of purely self-engendered inclinations. Simply expressed, the State is bound by the very structure of the society that it oversees. Hence, the State in capitalism has no mandate to re-organize the foundations of society. Its mandate is, instead, to maintain those very foundations while engaging in remedial reforms that leave the main structure and purposes of society intact. As a corollary, the capitalist State (existing as it does in a society that is ordered by democratic and market institutions) cannot exist as the private preserve of some privileged or dominant élite. This does not mean, however, that the State in capitalism is somehow perfectly neutral and unbiased. Simply by maintaining the existing social order, the State simultaneously maintains existing relations of authority and subordination in capitalism.

As the capitalist State evolves historically, it interacts with civil society in a process of response and counter-response. Civil society continually encounters internally generated predicaments that require the remedial intervention of the State. Then, as the State intervenes, so society moves forward to a new stage of development, in which new predicaments calling for further state intervention make their appearance. This spiral of events changes society's external form through time, although society's inner logic remains relatively unchanged. As we shall see, such interdependencies between the State and civil society are nowhere more evident than in the domain of the city, which is the ever-changing expression of interactions between private firms and households and urban planners as these interactions are mediated through space.

Urbanization and planning

We are now in a position to make a concise statement about the historical appearance of an urbanization process in capitalism, and about the internal order and dynamics of this process. For the sake of clarity, let us reaffirm that urbanization and planning constitute an integrated social event which is outwardly manifested in the form of a hierarchy of complex, dense and highly polarized land-use systems. Within these systems, civil society (firms and households) and the State (urban planners) interact with each other in highly specific and often analytically puzzling ways.

The urbanization process

Initially, cities in capitalism emerge out of the economic imperatives of commodity production and exchange. Given, in particular, the insistent profit-maximizing drives of commodity producers, the spontaneous development of spatially concentrated clusters of industrial firms is assured in capitalist society. These clusters appear historically at raw material sites and at transport nodes, where the costs of assembling and processing basic inputs is at a minimum. In addition, the concentration of many firms in close proximity to one another helps to reduce the transport costs of shifting secondary inputs and outputs between firms. Workers then assemble in dense residential districts scattered around the emergent industrial nucleus of the city. The market that is thus created attracts yet more firms, and so the city grows partly as a result of its own momentum, up to a point where diminishing returns on further growth begin to set it. The immediate consequence of these growth relationships is a twofold manifestation of a basic urban question.

First, as a system of cities is created, so, via the processes of exchange and migration, there emerges an integrated *hierarchy of centres* of different functions and sizes. We know remarkably little about the basic mechanisms that control the configuration and development of such hierarchies. What is more, there is little political debate about this issue (with the possible exception of occasional asides on the differential spatial distribution of urban amenities, or on the merits of centralized versus decentralized development programmes as a basis for national development). The issue of the urban hierarchy is a theme which permeates the following chapters but, with the exception of Cohen's statement on the international urban hierarchy, it is rarely directly addressed. Cohen has, however, identified a crucial area of enquiry, since the development of a tightly knit nexus of economic relationships, under the aegis of international capital, is steadily forging the cities of the world into a composite system. The economic and political consequences of this development are far-reaching, and we anticipate further research on this matter in the near future.

Second, in each individual city, a complex spatial system materializes, comprising an interdependent assembly of (private and public) functional areas and locations. These areas and locations can be categorized as either *production space* (in which the accumulation process proceeds), or *reproduction space* (in which the regeneration of labour is accomplished). Both of these spaces are mediated by a third, subjacent space, devoted to *circulation needs.* These basic spaces that emerge out of the broad structure of capitalism constitute an intricate land-use pattern expressing the main character of capitalist society. This spatial system is rife with problems, conflicts and predicaments. It is also the source and major target of urban planning. Because of its complexity, and its immense analytical and political interest, this spatial system is the primary object of theoretical enquiry in this book. In short, the arguments in this book crystallize around an urban question that is predicated on the observation that the dense human occupation of land can never, in capitalism, proceed smoothly and unproblematically.

In capitalist society, the urban land-use system is primarily structured by a rent-maximizing land market. To be sure, the dynamics of the land-use system are such that the private appropriation, exchange and utilization of urban land are steadily eroded by the progressive socialization of urban space (via planning). However, contemporary urban land use throughout North America and Western Europe is governed basically by a process of market exchange. From this process emanates the characteristic internal geographical pattern of capitalist cities: a dense commerical core; a tendency to ever-widening peripheral scattering of industry; and socially segregated neighbourhoods. These last are differentiated principally along the lines of cleavage within the prevailing division of labour, i.e. into blue collar and white collar residential areas. At the same time, precisely because urban land *is* privately appropriated, the derivative land-use system is heavily latent with socially deleterious breakdowns and conflicts. These negative outcomes compromise the efficiency of production and the effectiveness of reproduction. Accordingly, they threaten the city as a system, and, beyond this, the continued success of the entire accumulation process. We may, at this stage, attempt to pinpoint the genesis of these breakdowns and problems.

Urban contradictions and the emergence of urban planning

It has already been pointed out that capitalist society has never spontaneously been able to provide all the necessary conditions for its own existence. The central mechanisms within capitalism consist of price signals, exchange of monetary equivalents and profitability criteria, in the context of private and individual decision-making. As long as these mechanisms work effectively, the reproduction of society proceeds smoothly. However, they frequently fail at crucial articulations of society thereby threatening institutional stability, and thus calling for responsive collective action. These mechanisms fail in two significant areas in particular. The first area concerns the provision of major infrastructural artefacts and items of collective consumption. The second concerns the predicament-laden course of urban land-use development and the concomitant need for the social control and management of land-use outcomes.

In the first instance, many outputs necessary for effective production and reproduction in urban space simply cannot be produced spontaneously in pure commodity form. On the one hand, many of these outputs (e.g. streets, subway systems, bridges, and so on) are both extremely capital intensive and highly indivisible. The capital input necessary to produce them, in relation to any practicable scale of production prices, could never secure for producers a normal rate of profit. Their production is therefore abandoned by private producers, and they are produced by the State (if they are socially necessary) out of direct and indirect taxes. On the other hand certain outputs (e.g. cheap housing, cultural and recreational facilities, garbage collection, etc.) *can* be privately produced in the commodity form, but, at capitalistically determined price and wage levels, they would never be consumed in quantities large enough to sustain a socially viable process of reproduction. Once again, therefore, the private production of these outputs is frequently subsidized and/or complemented by the direct entrepreneurial intervention of the State. Since the State is able to produce on a large scale, and hence take advantage of internal scale economies in the production process, it often supplies these outputs in the form of large-scale items of collective consumption.[1]

In the second instance, since the urban land-use system exists in the form of an integrated assembly of interdependent locations, any event at any point in urban space will eventually have some impact on all other locations in that space. In brief, urban land, viewed as a system of differential locational effects, is produced as the joint output of all land users collectively. For the most part, these differential locational effects are highly beneficial, which is the main reason why cities are created at the outset. But, in addition, many of these effects are negative, in that they impose severe penalties on various categories of land users. Such penalties may take a wide variety of forms. For example, they may be the result of incursions of commercial activities into residential neighbourhoods, thus disturbing established patterns of reproduction; or they may be generated by the persistent intensification of business land uses in central city areas – thus causing congestion, parking problems, overloading of public transport facilities, and so on; or they may result from urban sprawl, drawing municipal governments into increasingly costly investments in infrastructure and services, and giving rise to the augmenting need for upgraded expressway systems – a circumstance which in turn triggers off yet further problems; and so on. All these problems are in turn complicated by the contradictory dynamic of accumulation (provoking urban growth and change) versus the inertia and slow convertibility of the built environment (resisting urban growth and change).

Further, because of the slow convertibility of built urban forms, spatial errors and irrationalities in the allocation of urban land are liable to be compounded and re-created through time. It must be noted that these problems are not capricious, in the sense that they emerge out of some purely arbitrary process of urban change. Instead, they emerge organically and necessarily from the land-use dynamics proper to capitalist society; and they call urgently for planning intervention. The fact that this intervention frequently exacerbates the very problems it sets out to resolve is only further evidence of the contradictions and constraints that capitalism imposes upon the interventionist tactics of the State.

The emergence of an urban political sphere

The foundations of a viable conception of an urban political sphere in general, and of urban planning activity in particular, have been essentially established in the preceding discussion. Several of the chapters that follow pursue this important theme in detail. What is needed at the moment is a more explicit development of the principal characteristics of collective urban intervention.

The need for collective action

Recall that the structural core of capitalist civil society consists of the institution of commodity production and exchange, together with a set of derivative social forms: the division of labour, the rate of profit, land markets, the family, residential neighbourhoods and so on. This core functions in conformity with a rationality that is based essentially on price signals, market competition, decentralized production decisions, and such legal arrangements as private property, individual rights, and the contractual equality of persons. These legal arrangements are the formal expression of a social system whose behavioural logic is codified within a system of individual decison-making and action. Civil society, then, can be seen as an ensemble of historically determinate social relationships which are actualized by a behavioural process comprised of a largely *privatized* system of calculations. By contrast, urban planning constitutes a sphere of *collective* political calculations, and it fills a vital decision-making gap within the totality of capitalist society.

As capitalist society finds expression in urban form and process, so it encounters limits to its own further development and viability. In the urban system, these limits are due less to external physical restrictions on the progress of society than they are to internal contradictions in the spatial dynamics of production and reproduction. A successfully functioning capitalism requires a geographical foundation in efficient production and reproduction spaces, but its own immanent logic tends to undermine the essential bases of its success. This logic leads to an urban process that consistently results in multiform breakdowns in production and reproduction space. In short, because of the primacy of individual decision-making in commodity-producing society, a constant stream of pathological outcomes erupts through the urban land-use system. While these pathologies threaten to impair the functional efficiency of society at large, they are nonetheless immune to curative action via the normal (privatized) rationality of civil society. Consequently, when the dislocations, irrationalities and conflicts of the urban system begin to subvert prevailing social relationships, urban planning makes its historical appearance as a means of collectively re-adjusting the spatial and temporal development of urban land use.

Just as the State is a reflection of the political imperatives of civil society, so urban planning acquires and changes its specific goals, emphases and contingent ideologies (planning theory, planning education, professional codes of practice, etc.) in response to specific developments in urban civil society. Hence urban planning is not, and can never be, a simple homeostatic phenomenon (such as an invariant and logical system of decision-making rules and procedures). It is, on the contrary, an ever-changing historical process that is continually being shaped and re-shaped by a broad system of urban tensions. Thus, urban planners in Europe and North America have not turned their attention now to zoning procedures, now to urban renewal, now to expressway construction, simply as a result of the appearance of 'new ideas' within an abstract and self-propelling planning theory. It is only when urban development begins to produce *real* problems and predicaments that planners attempt to counteract them. In summary, planning is a historically-specific and socially-necessary response to the self-disorganizing tendencies of *privatized* capitalist social and property relations as these appear in urban space.

The limits to collective action

The limits of urban planning are set by two inter-related forces. One is the degree of political opposition to, or support for any specific interventionist tactic. The other is the degree of civil disruption engendered by any remedial assault on the functional breakdowns of the urban system. At the same time, planning is an active social force only *within* the prior structures and constraints of commodity-producing society. While planning is undoubtedly necessary to the continued viability of capitalism, it is also constantly resisted by capitalism. Collective control by the State is hence always acquired in a piecemeal and pragmatic fashion, and only with the grudging assent of civil society. Capital, in particular, is perennially unwilling to consent to the extension of state intervention and regulation. Yet, recognizing that its own survival is intimately dependent on some form of collective decision-making, it finally accepts – though always fractiously and only after internal struggle – the curtailment of its own sphere of operations which must occur before planning can function as an effective instrument of public policy.

Thus the reactive and palliative nature of urban planning in capitalism is not simply the result of some technical, analytical or human failure. It is, instead, the inevitable concomitant of a social logic that sets definite barriers around the range and effectiveness of all political action. Urban planning is a response to the imperative of collective action in the urban system, and yet it cannot transgress the very social relationships from which it is derived. It is, in short, a mode of intervention that is only implemented when it serves the specific interests of capitalism.

The conditions under which urban planning is activated or left in abeyance vary as capitalist society evolves through time, and as it encounters new political imperatives. In the early industrial towns, for example, the problems of discordant land uses were largely ignored by the State, with the exception of basic but perfunctory controls geared to matters of public hygiene and the maintenance of social order. At the turn of the century, nothing less than virtually universal zoning could contain urban land-use problems within socially necessary limits. By the 1930s, the State was deeply committed (as a partner of capital in the development of urban land) to massive investments in infrastructure and public housing. At the present time, state control over the process of urban land development is so great that it is virtually everywhere systematized within broad administrative arrangements, such as ministries of planning and/or urban affairs, a body of planning law, the urban general plan, and so on.

At each stage in the unfolding of this historical pattern, planning enters the scene in the form of an indispensable, but always restrained, instrument for overcoming the specific predicaments of the urban system. However, *because it is so limited in its range of operation, planning also emerges as a social phenomenon that compounds the overall problems of capitalist urbanization.* The action of planning itself engenders further rounds of urban predicaments. Thus, housing clearances for urban hygiene purposes in the nineteenth century led directly to the unresolved problem of lodging displaced families; zoning contributed to the overdevelopment of some urban areas at the expense of others, just as it also encouraged urban social segregation; the participation of the State in the physical production of urban land via the provision of complex infrastructure has given rise to the problem of land-use intensification in central business districts, while it has also encouraged uncontrolled outward expansion of cities; and the institutionalization of planning practice within a complex bureaucracy has contributed to the re-politicization of urban planning.

The failures and shortcomings of urban planning in practice are *not*, therefore, the result of a failure of planning research, or the imperfections of planning education, or the professional inadequacies of planners. In the matters of research, education and professional work, established levels of performance are more than equal to the structurally limited tasks that planners are required to perform. The failures of planning in practice are less failures of knowledge than they are inevitable concomitants of collective intervention in a society that at once clamours for and yet restrains such intervention.

Since the political collectivity cannot transcend the structures of civil society (except, of course, by a forced appropriation of the administrative apparatus of the State) it can never secure decisive control over the development of the urban system. Urban planning interventions are, by their very nature, remedial measures generated as reactive responses to urban land use and development pathologies. Planners are frequently able to control the outer symptoms of these pathologies, but they can never abolish the capitalist logic that produces them. Thus, each time that planners intervene to correct a given predicament, so the whole system is carried forward to a new stage of structural complexity in which new predicaments begin to manifest themselves. These, in turn, call for yet further rounds of collective intervention, carrying the urban system forward to a yet more complex state of development, and so on in repetitive sequence.

The urban political sphere

In the capitalist city, the dynamic relationship between civil society and the State assumes the form of an observable private/public partnership in the production and development of urban land. This partnership is the tangible manifestation of the fundamental antithesis between the imperatives of private action (as imposed by the norms and logic of commodity production) and the imperatives of collective action (as imposed by the failures of the institutions of civil society). Furthermore, this partnership between civil society and the State, and, as a corollary, between the market allocation and the political allocation of urban land, is not simply a mechanical relation between two autonomous sets of variables. It is a *dialectical* relation, in the sense that the institutions of civil society give rise, through successive mediations, to the historical need for collective action because they can only reproduce themselves through such collection action; at the same time, however, these same institutions impede and resist the emergence of collective action. In other words, the social and property relations of capitalism create an urban process which repels that on which its continued existence ultimately depends, i.e. collective action in the form of planning. In this way (and notwithstanding the pervasiveness of planning in contemporary cities) the urban system moves forward through time in a pattern of historical development that is ungoverned and, effectively, ungovernable. Beneath the appearance of social control over the evolution of the urban system lies the inexorable dynamic of a complex of land-contingent events that is essentially out of control.

The capitalist State is thus caught up in a constantly escalating spiral of urban interventions. The more it acts, the more it must continue to act. There can be no practical possibility of withdrawing from this process, except at the cost of a dramatic resurgence of those very problems and predicaments that made interventionist tactics necessary in the first place. It therefore seems safe to assume that urban planning, whatever its specific content, will continue to penetrate increasingly into all layers of urban life. This process, however, carries with it severe political penalities. As the State increasingly mediates the process of production and development of urban land and space, so does it visibly modify the distribution of material benefits and costs accruing to various individuals and groups in the city. Concomitantly, discourses on urban planning begin to lose their utopian and apolitical patina (e.g. the conception of planning as that which 'seeks to promote human growth') and the true political nature of planning emerges with ever greater clarity. The more the State intervenes in the urban system, the greater is the likelihood that different social groups and fractions will contest the legitimacy of its decisions. *Urban life as a whole becomes progressively invaded by political controversies and dilemmas.* These controversies and dilemmas are as much related to geographical and territorial divisions of interest (neighbourhoods, suburban versus central city alliances, and so on) as they are to strict class lines of demarcation. In the contemporary city, political conflicts based on class are permeated and frequently submerged by conflicts based on spatial aggregates. In these conflicts, the role of urban planning as an instrument for regulating the institutions of commodity-producing society becomes increasingly apparent; the ideological confusions and distractions (such as mainstream planning theory) that surround the activity of planning start to drop away, as they are confronted with empirical circumstances that are increasingly inexplicable in terms of the received wisdom; urban planning experiences the same incipient crisis of legitimation that haunts the State as a whole in late capitalist society; and planning begins to emerge in its true colours, as one more administrative formation within a state apparatus that is, in its totality, rooted in the logic and predicaments of commodity-producing society.

Notes

[1] Observe, in passing, that these remarks run counter to: (a) Lojkine's (1977) theory of state intervention in urban space, as a mechanism for the devalorization of overaccumulated capital; and (b) Castell's (1977) theory of state intervention, as the socialization of consumption in the interests of accumulation (which is correct in so far as it goes but is unmediated by specific forms of market failure).

References

Castells, M. (1977a) *The Urban Question: A Marxist Approach.* London: Edward Arnold.
—— (1977b) 'Towards a political urban sociology', in M. Harloe (ed.) *Captive Cities.* London: John Wiley, pp. 61–78.
Harloe, M. (ed.) (1977) *Captive Cities.* London: John Wiley.
Harvey, D. (1973) *Social Justice and the City.* London: Edward Arnold.
Hindess, B. and Hirst, P. (1977) *Mode of Production and Social Formation.* London: Macmillan.
Lojkine, J. (1977) *Le Marxisme, l'Etat et la Question Urbaine.* Paris: Presses Universitaires de France.
Tabb, W. K. and Sawers, L. (1978) *Marxism and the Metropolis.* London: Oxford University Press.

II Prolegomena to a theory of urbanization and planning

2 Capitalism, accumulation and the explanation of urban phenomena

Matthew Edel

Introduction

The growth of large cities, the decay of industrial centers and innercity neighborhoods, and the crisis of municipal service provision have suggested to many observers that urban areas are caught up by forces outside of their own control. An increasing number of scholars have turned to a Marxist analysis of broad currents in the capitalist economy, in a search for clues to the origin of urban problems. This paper is meant to serve as background for, and as an examination of, some of the approaches used in this search.

This paper first outlines the general or defining characteristics of capitalism that have been identified by the Marxist tradition of analysis. These include the domination of the economy by a ruling class through the control of capital, the presence of a proletariat, the extraction of surplus value in the process of production and the presence of exchange systems required for production for sale. Additional elements of the capitalist mode of production which allow for variation among capitalist societies, are also discussed. These include land ownership institutions, family and social institutions for the reproduction of labor power, the role of the State, and the degree of unification and class consciousness in each of the two contending classes, the bourgeoisie and the proletariat.

The next section discusses the built-in drive for accumulation and growth inherent in capitalism. This drive gives shape to some of the characteristic features of recent urbanization. Tendencies to crisis originating in the accumulation process are also examined.

The final section discusses the extent to which specific urban phenomena can be explained by reference to the accumulation process and to general features of capitalism. Urban phenomena for which such explanations have been attempted in articles in this volume and elsewhere include the form of the built environment, the institutional structure and policies of local government, and the ideological and organizational patterns which affect the daily life and social reproduction of workers and capitalists in an urban setting.

The discussion is meant as a critical analysis of some of the attempts that have been made at such an explanation. I argue that some of these attempts substitute a general statement that specific policies, institutions or social trends fill certain needs, for a concrete analysis of how these phenomena are generated by the workings of the capitalist accumulation process. I close with some suggestions as to how a more satisfactory explanation can be made, by integrating a historical approach to class struggle with the discussion of the imperatives of capital accumulation.

The capitalist mode of production

Capitalism defined

To link the problems and achievements of modern urbanization to capitalism, one must first decide what one means by capitalism. This section develops the basic features of the Marxist definition of capitalism, which is central to much of the work presented in this book. That definition, which centers on the relationship between owners and workers, is different from other definitions which are sometimes used.

For conservative purists, capitalism is an economy with *no* government interference. A similar view is embodied in the liberal notion that the United States has a 'mixed economy', combining aspects of unregulated private ownership with elements of 'socialist' regulation and even government ownership. In these perspectives, the term 'capitalism' refers to individual ownership and the relation of individual producers through a self-regulating market. On the other hand, when popular magazines claim that Eastern Europe is rediscovering capitalism (because some price incentives are introduced) the implied definition of capitalism is different. It is the existence of any sort of market relations between people, even if these are manipulated by the government. This is a far cry from the definition restricted to purely 'free' enterprise.

Marxism uses yet a third definition, concerned with questions of who produces goods and who controls that production. The issue is not whether individuals or corporations are regulated, or whether they buy and sell in a free market, although these may be important factors in distinguishing kinds or phases of capitalism from each other. The key point is that in capitalism production is controlled by a capitalist class and actually carried out by a class of wage-workers.

In a capitalist system, products are produced for a market, for sale.

More important, they are produced by enterprises that seek a profit. These enterprises, in turn, represent the economic power and goals of their owners, the capitalists. The actual labor in production is primarily performed not by the capitalists, but by workers who do not own the enterprises. These workers form a class, characterized by the fact that they do not own basic assets used in production except their own muscle and brain power. The capitalist class is able to live off profits (or the interest on its capital) precisely because they can use, or exploit, the labor of the working class. What is more, if the efforts of workers, under capitalist direction, are productive enough, more will be produced than is needed for workers to live on and for capitalists' immediate consumption demand. This surplus is invested in new productive capacity by the capitalists. In other words, the capitalists through their investment decisions can control the basic process of economic growth or accumulation. To a considerable extent they can therefore control how society is run.

Whether capitalists are competitive with each other or whether some are monopolists, and whether their decisions are in part restricted by government regulations or not, are secondary considerations. One may have competitive, monopolistic and state-regulated capitalisms. But in all of them, the ownership of capital defines a class, the *bourgeoisie*, which has predominant economic (and political) power. The lack of capital, and the need to find a job working for others, in turn defines the working class, the *proletariat*.

Marxists see capitalism as a 'mode of production', a system with certain fundamental characteristics. 'Mode of production' is an analytic construct, expressing the common elements of the economic driving forces of a related group of actual economies and societies or 'social formations'.[1] These driving forces include both the technical methods by which production is carried out, and the social relationships between classes defined by their role in directing and carrying out production. The meaning of the concept is best seen through a comparison of capitalism with the other basic modes of production that Marx identified in history, including feudalism, primitive communism (or early classless society), ancient modes of production based on slavery, and possible future communism. In each, the distinguishing characteristics concern whether one class works for another, how labor is coordinated, and how the ruling class extracts the surplus produced by the workers. (See Marx 1964, 1967.)

Thus, in feudalism, individual peasants or artisans do the work, deciding individually or in small groups what and how to produce. Their product is, in part, taken from them in various forms of rent, including tithes, feudal dues and the like. What portion of a man's time is spent supporting his ruler is openly apparent in this system. In slavery, one class belongs to another juridically, and does its work at the other's direction in return for food and other goods granted at the sufferance of the owners. In primitive communism no class exists to capture a surplus produced by others. But no large-scale coordination or planning exists to allow much surplus to be created for workers themselves. In an advanced communist mode of production, the workers in a large society would coordinate and plan production above their subsistence needs, and direct the investment of the surplus themselves.

Under capitalism, wage-labor, coordination by capitalist managers and by market relations between enterprises, and the appearance of the surplus in the form of profit are the characteristic features. A division of the workers' time between production for themselves and production for a ruling class exists as in feudalism, but that division is not an openly acknowledged debt of a certain number of days' work per work. Rather it is a division of the product of the workers' labor into wage and profit shares of output, by which the product of part of the workers' working time is left in the owners' hands.[2]

Necessary elements of a capitalist system

The capitalist mode of production may be defined in terms of those features common to capitalism and uncharacteristic of other modes of production. This allows us to distinguish characteristics common to all capitalist social formations and characteristics which may vary among different capitalist economies or societies. It also allows examination of those features of economic performance which must characterize the system, such as the drive for economic growth, and those features which may vary within it. We begin first with the common elements of the capitalist mode of production (see Marx 1967, 1968).

CAPITAL   The first element of capitalism is the domination of the economy by a ruling class through the control of *capital.* The ability to command labor and maintain power over the use of tools, land, and other means of production in a society is fundamental for the ruling class in any mode of production. But in capitalism that control takes the form of ownership and control of a generalized form of wealth referred to by Marx as *Capital.* Capital includes not merely those machines and other goods used in production which conventional economics distinguishes by the terms 'capital' or 'capital goods.' It is not the same as the money net worth of capitalists. It does include all of these things, but fundamentally it is the ability to command labor and to accumulate more wealth through the ownership of wealth.

Capital is distinguished from various forms of wealth which exist under other modes of production, or from 'capital goods' alone, by two aspects. The first is its flexibility. One's wealth under capitalism may be invested in land, in machines, in funds for hiring labor, in money bank accounts, in bonds, in inventories, and so on. Wealth is relatively freely transferable from one form to another. Indeed, profitmaking (to which all capital aspires) requires that capital change forms. A capitalist uses money to hire labor, acquire machines and buy raw materials. He then uses the laborer's work to transform the materials, and sells the goods, thus transfering wealth back into the money form. Such a uniform, flexible type of wealth was not necessary or characteristic in prior modes of production.

The second distinguishing characteristic of capital is its relation to labor. Buying and selling alone, without using labor to transform materials, cannot lead to expanded production and expanded wealth. Capital needs wage-labor to function. Thus part of the definition of capital as a specific form of wealth is that it exists where there is a pool of labor available for hire. Without a proletariat wealth is not capital. The opportunity to use labor by hiring it makes the capitalist's wealth into capital. In a sense then, the labor available to the capitalist, although not owned by him, is as much a part of his capital as are his bank accounts or machines.[3]

THE PROLETARIAT   Since the existence of wealth as capital requires command over labor, the presence of a working class or proletariat is a second fundamental characteristic of the capitalist mode of production. The proletariat is a group of workers who must rely on work for a wage or salary for their livelihood. In the pure case, they have nothing to sell but their ability to labor. But the situation is not changed much by their ownership of some tools or special skills. These may affect the wages a worker gets while employed, but education or tools are usable only if an employer offers work. The worker remains dependent on 'finding' a job. As individuals, workers have to hire out to others, to sell their 'labor-power'.[4]

This sale of labor-power takes place, to be sure, under varying conditions. It may be on terms restricted by law, or affected by custom; its terms may be more favorable to the worker, or to the employer. But one fundamental of capitalist production is that the employee and the employer need not be bound to each other by anything other than the act of hiring. No long-term rights and duties need bind owner and worker, as did feudalism's customs under which being a serf involved long-term rights to land as well as political subjugation to a master. There is no bond of ownership as in slavery, nor permanent deference to a hereditary chief. Within limits, a worker under capitalism can quit to seek another job; an employer can fire the worker. Both are subject only (at times) to certain formal procedures or the granting of severance pay. The labor contract is thus legally a voluntary agreement between a 'free' worker and an equally free employer.

Much of Marx's work was taken up with showing the empty nature of this freedom. Under most circumstances, the need for a job in order to survive left the worker with little bargaining power over the terms of hiring. And once on the job the worker (having sold his labor-power and agreed to work for a specified number of hours) was at the direction of the employer. Marx depicted the worker, having contracted 'voluntarily' with the owner of a factory to do a day's work, following the latter into the plant not as an equal, as they supposedly were in the market, but as a dominated antagonist: 'one who is bringing his own hide to market and has nothing to expect but – a hiding' (1967: 176).

It is interesting to note that in the United States, many industrial contracts explicitly reserve the control of how work is to be done on the job as a 'management prerogative'. The all-important tax code, in giving different pension plan tax breaks to employees and employers, makes a most Marxist sort of distinction at this point:

> 'The basic definition of an employee is one who performs services subject to control by an employer "not only as to the result to be accomplished by the work, but also as to the details and means by which that result is to be accomplished. That is, an employee is subject to the will and control of the employer not only as to what is to be done, but also how and when it shall be done."' (Academic Information Services 1978: 181)

SURPLUS VALUE   The basic form of the exploitation of the proletariat by capitalists is the production and extraction of surplus value. Marx argues that the daily pay received by laborers, and the amount they can produce in a day, are determined by two separate processes.[5] The pay is determined by what living standard is necessary, at a particular time in history, to enable the working class to subsist and reproduce itself. This is not a biologically determined subsistence level; it is determined by 'historical and moral elements', including both the degree to which capital needs better trained and educated workers, and the extent to which labor itself has been able to enforce better minimum standards through class organization and struggle. But once this social minimum, which Marx calls the 'value of labor-power', is established, competition for jobs among individual workers will prevent wages from diverging far above this 'subsistence' floor.

On the other hand, the productivity of a worker in a day is determined in part by the degree to which past development has improved technology or allowed the accumulation of equipment (i.e. by 'the forces of production') and in part by how hard and fast the employer can make the workers work once he has hired them. There is no need for the productivity of a day's work to be any specific multiple of a day's pay, although if revenue falls below labor costs, the employer will soon note the loss and fire the workers. In general, production will exceed pay, and a surplus product will remain as the property of the employer. Thus ownership of capital (including that invested in machines and materials, and that used as cash to hire laborers) allows the owner to end up with products worth more than the sum expended in putting the process into motion. This is what Marx terms *surplus value*.

How much surplus value is produced in society will affect the extent to which capitalists can spend on consumption, on controlling society, and on new forms of production. Within a capitalist economy, it is an important indicator of how strong the system is; the expansion of surplus value is thus a goal of capitalist institutions.

The magnitude of surplus value can be increased in one of two basic ways. 'Absolute surplus value' increases depend on a direct squeeze on labor. The less workers receive, in real terms, for a day of work, or the more work they are directed to do by the capitalists in terms of longer hours or speedups, the more surplus they can produce. 'Relative surplus value' increases, on the other hand, are those brought about by improved technology or by more productive organization. If producing the necessities that workers buy takes fewer hours of work, capitalists can pay their workers less and still leave them able to buy what they need to survive. Relative surplus value increases will be produced also if productivity increases in machine-making or raw material industries, since these producers/good sectors can then reduce the prices to other industries (if they are not controlled by monopoly). These decreases in turn add to the margin of surplus in the industries that use the producers' goods. (Productivity increases in the industries that produce luxury goods for capitalist consumption will not increase the rate of surplus value, although capitalists, as consumers, may benefit.) Both absolute and relative surplus value changes can be affected by capital-labor conflict, since that may influence both the pace and the organization of work as well as workers' living standards.

The capitalist enterprise takes its profits out of surplus value. It may also make payments to other nonworkers: interest payments for the use of borrowed money, rent for the use of land, and perhaps taxes. All of these revenues are considered shares in the total surplus value produced, because they are not payments for actual labor done.[6]

COMMODITY EXCHANGE The existence of capital as a flexible, transformable type of wealth requires that capitalists be able to sell the products their workers produce. If this were not so, each worker would have to receive as wages the goods he himself produced; surplus value would take the form of unsold inventories. The system would collapse. This indeed is a possibility which emerges in economic crises, when large inventories do remain unsold, and production is decreased for want of buyers. Capitalists must 'realize' their profits by selling their goods in order to keep on hiring workers and producing. Thus capitalism requires a 'sphere of commodity exchange' in which goods are bought and sold.[7]

Exchange does more than facilitating the transfer of inventories of goods into generalized wealth. It also affects how people perceive their economic system. People acquire goods by buying them. Employers hire workers. And so people tend to focus on prices and wages as the important economic facts in their lives. Prices and wages *are* important, but in focusing primarily on them, people may lose sight of the class relationships that underlie prices and wages. Because most purchases or sales are 'voluntary' acts by buyers and sellers, it appears that everything sells for what it is worth and that the worker gets a 'fair day's pay for a fair day's work'. This at least seems to hold if there is some competition in the economy to ensure against arbitrary price setting or gouging by monopolists. Attention is thus diverted from the fact that 'a fair day's pay' (or the value of labor power) is determined by class conflict and not really by productivity. The exploitative character of capitalist production, according to Marx, is masked by the apparent fairness of market exchange.

Capitalist exchange has one other important feature: the use of money.

A system using money as an intermediary commodity in exchanges is more efficient than barter. Thus the exchange sphere takes the form of a market in which capitalists sell their products for money, pay workers with money, and use money to buy new equipment. Use of money, however, creates some problems for the system. Fluctuations in the supply of money may interrupt change, causing inflation or recessions. Furthermore, money creates possibilities of making loans for future repayment. This in turn creates pressures on capitalists and workers, and risks of default, because debts must be repaid. From the capitalist perspective, these may be necessary evils, since without money and generalized exchange the system itself could not exist. But it does mean that vulnerability to monetary crisis is a necessary aspect of capitalism.

Variable characteristics within the capitalist mode

The existence of generalized capital, a proletariat, surplus value, and commodity exchange are all necessary elements in capitalist systems. But capitalist systems may also have characteristics which vary as among different countries or over time. Some of this variation occurs because what we see as capitalist systems mix elements of capitalism with elements of other modes of production (Harvey 1973: 203). But even within the definition of capitalism, functions may be specified for which some institution must exist, but for which a variety of very different specific institutions are possible. Four of these areas are important enough to merit notice alongside of the defining characteristics already discussed.

LAND OWNERSHIP Any economic system requires procedures for managing the use of land and other natural resources. Capitalism requires rules of land ownership for two reasons. First, to control and maintain a proletariat in its subservient position, it is necessary to limit its free access to land. Although in some cases capitalism has existed and even thrived in the presence of free frontier land, movement to the frontier has never been an option for all workers. What is more, capitalist societies have soon developed mechanisms to select some individuals, buyers or homesteaders, as the owners of these lands. Second, some rules are required as to which capitalist owns what resources. However, the forms that property rights take may vary, as may the class composition of the landowning group.

Property is held privately in most but not all cases. Land may be held primarily by 'small farmer' or 'peasant' proprietors who farm it themselves, by capitalist corporations ('agribusiness'), by individual capitalists who hire workers, or by members of an aristocracy whose existence predates capitalism. Furthermore, land nationalization can be compatible with capitalism as long as the government does not use its land to allow workers to escape the necessity of wage-labor. State ownership of mineral deposits, forests, transportation routes, housing projects and certain plantations is not uncommon, although it is rare to have (most) farmland owned by government (Massey and Catalano 1978).

Who owns the land and how they use it, makes an important difference in the day-to-day functioning of a capitalist system. Ownership patterns clearly influence the distribution of income. They may determine how much surplus value is taken in rents. They affect the extent to which agricultural, mineral, energy, and housing production are responsive to new demands or to new technology. They will affect the prices of these vital commodities. And they will have an important impact on the political structure of society. Indeed, the conflict between capitalists and aristocratic landlords was a predominant feature of the early stages of capitalism in most countries.

SOCIAL REPRODUCTION   For capitalism to survive both the bourgeoisie and the proletariat must themselves survive and perpetuate themselves from generation to generation. They must not only reproduce themselves physically, but also pass on class positions, skill and attitudes, either to their own children or to new recruits. But the institutions which take part in or foster this social reproduction can vary greatly among different capitalist societies.

Something called 'the family' is important in all societies, but what a family is can vary greatly. Extended families, the parent-child nuclear family, large tribes and even communes may fit its definition. Social relationships may vary greatly even within one of these family systems. Capitalism seems to have a tendency to take functions away from the family and to turn them over to businesses or government agencies. (Thus, for example, less food production and education now occur in the home than occurred there a century ago, while care of the elderly and very young children, and some aspects of housekeeping, are also being commercialized or bureaucratized.) However, the extent to which this has been done, and the very form the family takes, differ among capitalist countries. Both pre-capitalist institutions and national experiences since the rise of capitalism may affect what the family does within a functioning capitalist system, but it is generally the case that much of the training of new workers is still done by worker's families.

Apart from the family, other institutions such as schools, clubs, churches, the mass media, social work bureaucracies, and the police may play a role in preparing the new generations for their 'station in society'. These agencies may be operated by the government or by private groups. They will perform different specific tasks in different societies. Informal social relationships at the neighborhood level may also play a role in the reproduction of different classes, a point important to planners.

The variety possible in these institutions, and in particular the variation in the degree of direct control capitalists have over them, is an important source of differences among capitalist societies. The extent to which the work force is seen as flexible or mobile, the skill level of labor, the extent of patriarchal relations between men and women, and the degree of paternalism or *noblesse oblige* in ruling class attitudes – all will be related to how social reproduction is organized.

THE STATE   The role of the State or government may also vary considerably among capitalist countries. But certain criteria must be met for capitalism to exist. The State must provide a framework for exchange, a guarantee of capitalists' property rights, and some degree of control to keep workers in their subordinate class position. In a crisis situation, the government must be controllable by capitalists if their vital interests are at stake. (If this is not so, the government will fall apart if there is any challenge to capitalist institutions, and either capitalism will be overthrown or the capitalists will construct a new, and probably repressive, state apparatus compatible with their interests). Thus the State has been called an organ of the ruling class, a specifically capitalist government.

However, the ways in which the government is constituted and performs its roles may vary. The government may be democratic or dictatorial, centralized or decentralized in form. In its normal operations (outside of periods of crisis) it may express not only some general common interests of capitalists, but also the particular interests of subgroups of capitalists, large or small. It may be more or less sensitive to pressures and demands by labor, or by 'pre-capitalist' classes such as peasants or a landed aristocracy. Further the state functionaries – administration, bureaucracy and army – may have the power to impose some policies in their own specific interests. There is also variation in the State's role in society, in terms of how much property it owns, what share of production it takes in taxes, or which areas of production, exchange or reproduction it administers or regulates. As long as it does not interfere with the most vital interests of capital, defends these interests in crises, and fulfills its other essential roles for capital, government can take on many forms.

CLASS SOLIDARITY   A final feature which may vary among capitalist systems is the degree of unification within each of the two major contending classes, the bourgeoisie and the proletariat. The internal unity of each class, in turn, involves economic dimensions like the degree to which profit rates are unequal in different industries or wages are equal for different workers. It involves institutional dimensions like the existence of monopoly or labor unions. And it involves an important dimension of class consciousness by either class.

Capitalists always have common interests in the preservation of capitalism. But competition always sets individual capitalists against each other. 'Fractions of capital' – groups with interests in different industries or kinds of investment – may favor different policies. In some countries, differences between domestic and foreign capitalists may be important. The degree to which capitalists can act in their common interests, rather than just competing with each other, may depend on a number of historic factors. The size, experiences and specific interests of the capitalist class will all be important in affecting capitalist class consciousness. Over time, a number of developments may increase this consciousness: concentration of capital into fewer hands, the growth of firms with interests in a variety of industries or activities, and an accretion of past experiences with common action may all make it easier for capitalists to act together for common goals. But the extent of class unity by capitalists is still widely variable among different countries.

Similarly, the degree of unity within the working class also varies among capitalist societies. Individual laborers may be unprotected by labor organizations, then some of them may be organized at a craft or industry level, or all may be united nationally by unions or political parties. Labor action may achieve higher wages for some workers with others left unprotected, or it may affect and unify the wages of all workers. Labor unity may be directed only to immediate economic issues like wages, or it may be directed to broader issues of political power and revolutionary change.

Like capitalist class activity, labor solidarity may increase over time, although there is no automatic mechanism to enforce this (Edel 1979). Divisions of the working class based on racial, religious, regional or other lines may impede this solidarity, as may differences in wage levels in different industries. Past class experiences, and the tendency of capital accumulation to break down localized or protected enclaves of higher wages, all tend to unify labor. But the extent to which the unification is actually accomplished, and the extent to which it results in either economistic or revolutionary class consciousness are likely to vary between countries.

The existence of all these variable characteristics – class consciousness forms of state organization, institutions of social reproduction, and systems of property ownership – means that capitalist systems may be very diverse in many aspects. Frontier America, Nazi Germany, laissez-faire England of the nineteenth century, modern European welfare states, and the dependent capitalist nations of the Third World are clearly unlike each other in many important aspects. However, there are fundamental ways in which they are comparable to each other. All operate on the basis of the fundamental capitalist institutions – existence of capital and of a proletariat, production of surplus value, and monetized exchange. And their very variation occurs along dimensions set by the organization of capitalist production. It may, that is, be measured in terms of the way in which the variable characteristics of capitalism discussed above, are combined. The concept of a common 'mode of production' is useful, therefore, both in determining common elements of capitalist systems, and in locating precisely their factors of difference.

The accumulation process

If there is a common characteristic to capitalism, apart from the defining qualities discussed above, it is the need to expand. Accumulation is a persistent drive in such a system. Capital accumulation in the orthodox sense means an increase in the available physical means of production; in the Marxist sense it means an increase in the generalized wealth controlled by capital.

Strong pressures in any capitalist system encourage accumulation in both senses. Nonetheless, the very process of accumulation also generates contradictory forces, which lead to crisis, depression and the destruction of capital. At times such counterforces have dominated the capitalist system as an international whole. At other times they have led to more localized stagnation. Capitalism thus has been marked by regional and general economic cycles, and by geographic unevenness of development.

The drive to accumulate

'Accumulate! Accumulate! That is their Moses and their Prophets.' Marx's well known aphorism about the capitalist class expresses the general point that capitalists both want and need growth. But varying interpretations can be given to the remark. According to some interpreters of development process, a desire to accumulate is the important factor. Some 'Protestant ethic', 'need for achievement', 'entrepreneurial spirit' or 'greed' drives part of the population to become capitalists (cf. Edel 1970). Once established in that vocation they pursue the grail of growth. This interpretation, however, misses Marx's main point. In his view the bourgeois ethos of accumulation, although real enough, is not the major cause of growth. Strong economic necessities – competition and class struggle – impel the capitalist enterprise, and at times the capitalist State, to seek growth. The need for growth is taught to the capitalist (and in some cases the workers, too) by experience. Several factors that compel growth may be identified.

REINVESTABLE SURPLUS   In the first place, capitalists have the means to generate economic growth. They receive most of the surplus value created in production, and thus have control over what is to be done with it. Since the surplus value absorbs all the excess produced over workers' consumption needs, it is the only significant source of investable funds in a capitalist economy. Whether there will be growth depends, to a large extent, on whether capitalists (and the rentiers and landlords who also receive shares of surplus value) spend it all on their own consumption, or spend part on new facilities or more workers. This power does not guarantee that investments will be made, but it cannot but contribute to the capitalists' view that they can influence the future by their actions. Even in the absence of other specific pressures, some capitalists will want to expand their activities.

COMPETITIVE PRESSURES   Second, and going beyond the simple possibility of expansion, is the pressure to invest that stems from competition. As Marx argued, competition forces businesses to invest because those that do not grow and invest in advanced equipment will lose markets to their competitors. Their owners may lose their property and be forced into the proletariat, a fate they seek to avoid through new investment. In an economic downturn, declining profit rates will be fought with new investments designed to cut costs by saving labor. In prosperity, firms seek to consolidate their position by maintaining or expanding their share of growing markets. Thus the competitive pressure for growth exists whether the economy is growing, or not.

Even when the economy is controlled by large corporations, seemingly immune to fears of bankruptcy and to price competition, a competitive pressure for growth is still strong. For one thing, even firms which have monopolistic control of particular industries face foreign competition, competition from other products, and threats of future competition if other large firms enter their industries. For another, when corporations are large enough, competition exists *within* them as different divisions compete for precedence, and their managers compete for promotions. If an official's promotion is determined by his division's growth, strong pressure for growth in the organization as a whole is created. Finally, growth allows a compromise between the interests of shareholders and managers over the division of earnings. Shareholders will veto excessive managerial salaries; managers will limit the disbursement of dividends. Reinvestment for growth is an obvious compromise, allowing managers to expand their power internally, while increasing the worth of the shareholder's assets.

DEBTS   A third factor which may impel capitalist growth is the structure of debts that generally develops in a capitalist economy. Capitalists often borrow to expand their businesses and to take advantage of investment opportunities not financeable out of current incomes. Once this borrowing becomes generalized, as it frequently does, a strong pressure to seek means for further growth is created. For each firm, growth becomes a means to pay off debts without reducing income available for other purposes or selling off shares in his business. Thus as debts increase, capitalists must push harder to expand their revenues.

CAPITALIST CLASS DEFENSE   These forces operate at the level of individual firms. But if capitalists can work together as a class through private associations or through influence on government, they promote growth as desirable as a protection for their common interests.

Investment booms stave off periods of depression, which would affect all capitalists' incomes adversely. Thus government stimulation of the economy is sometimes favored by the capitalists.

What is more, growth may protect the power of a capitalist class when it is challenged abroad or at home. For example, growth provides new surplus resources for national defense weapons for defense of capitalists' interests abroad. But growth also allows workers' discontent to be met with increases in wages or social benefits without cutting into profits. It is thus useful for capitalists as a response to class struggle.

WORKING-CLASS INTERESTS   These last three factors, competition, debt, and class defense also exert strong pressures on workers in a capitalist economy to favor growth. Although growth may increase the wealth and power of capitalists in the long run, workers have a stake in supporting it on many occasions.

Competitions for jobs give workers a stake in the competitive success of their employers, and a stake generally expanding the number of jobs available. Historically, periods of growth have been associated with rising living standards. What is more, growth removes some of the pressures workers may feel from debt burdens or from employers' attempts to cut costs at the expense of their employees.

Of course, the working class cannot act directly in favor of growth under capitalism, because they do not control the investable surplus value. However, to the extent that they can urge general growth policies on the government, or to the extent that workers can demand or oppose specific investments through union, political party or pressure group actions, they can exert an indirect influence. This worker pressure has at times been an important factor in generating growth, by forcing government action when capitalists were too disunited, or too afraid of state involvement, to push for policies necessary for growth.

The foregoing pressures suggest why the drive to accumulate is inherent in capitalism. It is fostered by the action of individual capitalists, by the organized pressure on policy of the capitalist class, and (sometimes) by working-class activities. Reinvestment of profits by capitalists and their businesses, and public policy in favor of continued growth, are the normal tendencies of the system.

The existence of these tendencies does not mean, however, that growth will either be continuous or tranquil under capitalism. Growth itself creates such problems as congestion and pollution. The same competition that breeds growth can also lead to bankruptcy of artisans, farmers and some other small capitalists.

Growing proletarianization and sometimes greater unemployment are among its results. Both competition and government growth policy create larger and larger concentrations of capital and power on the one hand, and of poverty on the other, leaving the society even more rent by inequality. Inequalities between neighborhoods, between regions, and between nations, may also be increased. And periods of growth give rise, too, to the conditions for economic crises that periodically disrupt accumulation, bringing suffering and sometimes war in their wake. Capitalism's power as an engine of growth, in short, does not obliterate its existence as a system of class domination, in which there are losers as well as winners.

Contradictions in the accumulation process

Capitalism may need growth, it may generate forces for growth, but it does not always achieve growth. Indeed the accumulation process itself undermines its own bases, generating conflict and crisis. Theories of economical crisis are among the most controversial topics in present-day Marxist analysis. They certainly cannot be treated exhaustively in a short introduction to the capitalist mode of production. But some idea of the sorts of contradictions generated by capitalist accumulation can be presented.[8]

In Marx's analysis of capitalism the most fundamental contradiction created by accumulation is that it changes the division of labor in society, thus undermining the social formation that is doing the accumulating. Accumulation may be the 'job' (although not the labor) of capitalist investors, but it also reduces the number of capitalists. Technological change, competition and the need to control labor all combine to favor larger and larger capitalist businesses. Concentration and centralization of capital, into larger and fewer units, is a strong tendency in capitalist development. Because this centralization bankrupts small capitalists, and at the same time because the growth of capitalism spreads capitalist relations into more and more sectors of the economy (displacing people from their old forms of livelihood), the proletariat is continually expanding. In the United States, for example, the number of small farmers and self-employed artisans has been greatly reduced over the past century. Marx states as a 'general law of capitalist accumulation' that the accumulation of capital *is* the expansion of the proletariat (Marx 1967, Vol. I., Chapter 25).

In the long run, accumulation creates the possibility of a fundamental change in power relations. The proletariat can eventually aspire to seize control over society from the capitalists. Even in the short run, changes in the division of labor and the quantitative class composition of society can undermine the basis of specific economic expansions, leading to a boom-bust pattern.

Historically, sharp downturns have always occurred after periods of prosperity and growth in capitalist economies. Whether these downturns appear at regular intervals (as 'cycles') or whether they are unpredictable as to time and duration, is debated. But analysts agree that growth does not continue uninterrupted forever. And its interruption is painful. A slowing down of expansion, in which improvements might be consolidated and absorbed, and in which people could relax and enjoy the fruits of recent progress, seems to be excluded by capitalism.

What causes crises or downturns? In a period of prosperity, a large amount of surplus value is generated and reinvested. Expansion of employment, both in investment industries (construction, machinery, mineral exploration, etc.) and in production of workers' and capitalists' consumption goods, is generated by that investment. The new employees themselves form a market for much of the new production. But this situation does not last. Several imbalances begin to occur, rooted in shifts in the division of labor, which can upset the pattern. The order of importance of these different factors is debated, but what the problems are can be identified.

PROFIT 'SQUEEZE' The expansion of the proletariat, which accompanies successful growth, creates new pressures on the system. The bankrupting of small businesses, the destruction of traditional sectors of the economy by corporate competition force more people into the working class. The disruption of old ways creates discontents. Then, as labor becomes more used to new patterns of work, it also becomes better organized, demanding the fruits of growth. Thus profits, the basis of accumulation, may be reduced by higher wages, or spent on social control in an attempt to cope with discontent.

'UNDERCONSUMPTION'   In an attempt to resist wage increases, the capitalists may attempt to automate, reorganize work processes, and otherwise cut labor costs. The result of this, and of the destruction of older economic sectors, is to create more unemployed workers. The presence of this 'reserve army' of labor can ease shortages of workers, and can generally depress wages, which is initially favorable to profit and accumulation. However, it may also undercut the market for significant sectors of capitalist production. As unemployment rises, or wages fall, many workers cannot afford new appliances, cars, homes or other goods, so these industries are forced to lay off workers too, and to leave capacity unutilized. A crisis of 'underconsumption' may occur.

DIMINISHING REVENUE, RISING RENTS   As surplus value is reinvested, and as sectors of the economy expand accordingly, some resources needed in production may become 'scarce' relative to the new demand for them. The prices of these resources, or more precisely the rents their owners can demand, will rise. In addition, increasing production creates more competition for the markets provided by existing consumers. The result may be falling profitability, if new products and new technologies do not come along rapidly enough.

'FISCAL' AND REPRODUCTION CRISES   Growth may raise requirements for social investments or public spending to cope with some of the by-products of accumulation: urban congestion, pollution, social tensions, etc. Maintaining services in regions which are losing jobs or population to new competitors may increase the average cost of these services. Maintaining the growth process may bring forth demands for complementary state investments in education, transport systems, and the like. The result is the diversion of more resources to expenses which, while they provide employment, are 'unproductive' in terms of any direct increase in capital investments and the production of surplus value.

'DISPROPORTIONALITY'   Accelerated growth requires 'capital goods' industries, like machine tools and heavy equipment, along with construction industries, to grow more rapidly than other industries. Because their demand is dependent on continued growth, the prominence of these industries makes the economy more vulnerable to slowdowns in the growth rate. Even a small decline in growth and investment can be disastrous for *new* plant and equipment production. Layoffs in these industries, in turn, can trigger a general recession.

'FALLING RATE' OF PROFIT   The attempt to stave off rising costs using new technologies may increase the ratio of equipment to labor used in production. A larger portion of the revenue from sales must thus go to amortizing equipment and materials costs, which can depress the profit margin. Meanwhile, the book value of companies will have risen, reflecting their increased stock of machinery. Unless the new technologies increase productivity sufficiently, or unless labor income can be squeezed to reduce the wage share of revenues, profit rates (profits divided by book value) will decline.

Not all Marxists agree whether this will affect investment and reduce economic activity. Keynesian and neoclassical economics suggest a falling *expected* profit rate will increase either cash hoarding or consumption at the expense of investment. Some Marxists seem to accept one or the other of these views, suggesting that the falling rate of profit will put some damper on reinvestment plans. They rarely spell out the mechanism by which this occurs, however. Other Marxists suggest that if debts have been contracted based on past profit rates, the profit-rate decline will trigger a financial crisis. At least in this special case, the falling rate of profit does seem to be a problem, although financial collapse may be staved off by inflation which raises nominal, although not real, profits.

The complex interactions of these six factors, which partially counteract and partially exacerbate each other, leave the accumulation process vulnerable to interruption. In some sense, the process itself creates the conditions that frustrate it. Both the continuing pressures for growth, and the contradictory forces which are generated, can be important as influences on the urban phenomena that concern the authors in this collection.

The explanation of urban phenomena

In the urban arena, a wide variety of phenomena are important subjects for Marxist analysis, either because they affect the living conditions, employment or power of the working class, or because they have become 'policy problems' over which conflict has emerged. The form of the built urban environment, the geographical concentration of employment and incomes, and the distribution of powers to local units of government are traditional objects of urban analysis. But problems of the social reproduction of the working class – i.e. the sustenance, training and ideological formation of workers and their families and communities – are now also considered part of, and perhaps the essence of, the 'urban question' (Castells 1978).

Relating the general characteristics of capitalism discussed above to urban struggles and outcomes is not an easy task, but it is a necessary step in developing any Marxist analysis of urban phenomena. Marxists have typically criticized orthodox economics and sociology for amassing empirical evidence of correlation between different phenomena, without considering that all of the factors observed may be shaped by some underlying process. A proper explanation, it is held, must take the characteristics of the reigning mode of production into account. The problem, though, is *how* this can be done.

The modes-of-production controversy

An important current controversy has challenged whether specific events and phenomena can be explained by the characteristic of modes of production (cf. Hindess and Hirst 1975, 1977; Asad and Wolpe 1976; Murray

1977; Wolff 1977). It is increasingly recognized that Marxism must go beyond a simple economic determinism in which the economic base, narrowly defined as goods production, is seen as rigidly determining all aspects of society. However, in seeking better explanations, Marxists have often slipped into one of two alternative traps.

The first trap is a mere listing of causes, involving both economic and 'superstructural' (e.g. political, ideological, cultural) elements, without an adequate theory of how their interaction is determined. This perspective can lead to the definition of increasing numbers of supposed 'modes of production' through the mere listing of empirical characteristics that are observed together. This leaves unclear what is or is not a mode of production, and how the mode of production (defined as a set of characteristics) can explain any particular characteristic.

The alternative trap is a 'structuralist' notion that the whole determines all of its parts, or more precisely that 'all dimensions or aspects of (a mode of production) are determined effects of its structure' (Wolff 1977: 90). In this view, all aspects of the mode of production are thought to be determined by the mode itself. However, there is no way of knowing whether particular aspects are necessary or variable, nor how the structure itself came into existence or may generate conditions for change. As Hindess and Hirst put it, this approach fails because it assumes 'the persistence of the structure through its own action' (1975: 319).

In criticizing these two approaches, Hindess and Hirst have seemed to deny the usefulness of the mode-of-production concept for any explanation of specific phenomena or of change. In their initial work (1975) the relationship of the mode of production to political and other aspects of a society is either unspecified, or falls back into the structuralist trap the authors criticize (cf. Murray 1977). Hindess and Hirst indeed claim that one cannot derive an explanation of how a mode of production is overthrown from the defining characteristics of that mode. If that crucial aspect of Marxist concern is left unexplainable, so too are more minor urban contradictions.

In a later work, however, the authors suggest a way out of the impasse. A dynamic approach is introduced by specifying a mode of production as having *conditions of existence* – not necessarily created by the mode itself – which allow its relations of production to be reproduced. In some cases, it is suggested, these conditions may not be reproduced in stable form, leading to 'a struggle between those seeking to reestablish the necessary conditions of existence and those seeking to exploit their absence by establishing new class relations' (Wolff 1977: 92, summarizing Hindess and Hirst 1977). There is no need, in this view, for a general proof that all aspects of a system are determined by the system – only that they are not incongruent with its reproduction. Thus there can be variable elements as well as necessary elements within a mode of production, as was posited in the description of capitalism on p. 24 above.

Economic determinism is thus redefined. The claim that there is 'some privileged, necessarily determinant position to the economic level within a social formation' (Wolff 1977: 92) is replaced by a methodological stance for seeking explanations. Marxist theory is seen as 'that conceptual framework in which class relation is the concept with which analysis begins, around which it constructs an understanding (knowledge) of a social formation' (Wolff 1977: 92). The concept of a mode of production is thus a reference point which specifies possible conditions under which a set of relations of production, involving appropriation of surplus labor, will survive.

Reproduction, accumulation and urban phenomena

This view suggests that, for any discussion of urban phenomena, the starting point should be the place of those phenomena in the reproduction or interruption of the social relations of production. In the case of capitalism, this reproduction normally occurs on an expanded scale, rather than simply replicating a static system. That is, reproduction is accumulation. This is so because growth itself becomes one of the conditions of existence of capitalism.

The Marxist explanation of urban phenomena thus sets itself a more defined task than simply (or somehow) looking at the relationships between different characteristics of cities. It must relate these characteristics to such aspects of the capitalist accumulation process as:

1. the way in which labor is employed in production to create values and surplus value;
2. the way in which labor power is reproduced; and
3. the way in which surplus value is 'realized' through sales of goods, and is circulated to allow new investment.[9]

The analysis must recognize that at any of these points, the reproduction of capitalist relations and the accumulation of capital may be interrupted or may be affected by the ongoing struggle between capitalists and the working class. (Indeed, the historic aim of Marxist thought is the identification of ways to intervene in these interruptions and struggles, to create revolutionary change).

The linking of phenomena to the accumulation process can take one of two forms. In some cases, an attempt is made to explain phenomena as effects of a preexisting process of accumulation – as by-products of growth and/or depression. In other cases, however, explanation takes the form of considering the phenomena as themselves part of the accumulation process. The first approach is compatible with a conceptualization of the system as involving a narrowly defined economic 'base' which affects urban and other 'superstructures'. The second, however, requires base and superstructure to be considered as inseparable in practice, or at least linked by close two-way relationships. Analyses may be found which are based in either of these two approaches.

SIMPLE CAUSALITY   The one-way link, leading from the basic accumulation process to specific phenomena, is sufficient to explain some attributes of modern cities. Investment of surplus, competition, centralization of capital, and the growth of the proletariat can account for some urban social patterns. The instability of urban communities is thus related to the underlying instability of capitalist employment, and to the frequent taking of land for new profit-making uses. The fiscal weakness of local tax bases is explicable as a result of the mobility of capital in a system in which reinvestment is constantly occurring – and can occur away from areas which raise taxes. The 'heaping up' of workers into large towns, and the segregation by class of housing markets, discussed by Marx (1967) and Engels (1973), and by many others since then, can be explained, at least in general, on this basis.

In these cases, the Marxist explanation of immediate causes may not be too distinct from other simple economic determinisms like neoclassical economics or social ecology. Marxists will be more attuned to problems raised by growth and crises, or to differences between the actions of different classes, than are orthodox observers who assume static and harmonious conditions. Generally, what the orthodox perspectives take as givens (ownership and income distribution among classes, and the acceptance of the market as part of the 'moral order') are shown by Marxists to be features of capitalism which may be criticized rather than assumed to be natural. The implication that market outcomes are optimal is discarded in the Marxist view. But the terms of analysis of specific events are, at least, translatable from one system of analysis into the other.

COMPLEX INTERRELATIONSHIPS TO ACCUMULATION   On the other hand, some phenomena are less easily explainable as direct products of economic decisions or self-interest. In general, these are cases involving complex forms of class interest involved in the patterning of social programs or government policy. These are the cases which orthodox analysis usually assigns as the result of 'autonomous preferences' or 'tastes' – for discrimination, for public policies, for lifestyles, etc. It is here that Marxist analysis can make a unique contribution, that goes beyond claiming short-term psychological approval of policies or actions as their cause. Basing itself on the relationship of phenomena to accumulation, Marxism can learn much about their causes and their consequences.

The analysis is frequently not a simple one, because the relationship of phenomena to accumulation can be conceptualized either in global, structural terms, or in terms of the relation of class struggle to policy outcomes. Different studies, involving different conceptual approaches to the same basic relationship between accumulation and policy, are reviewed in the following sections.

APPROACHES EMPHASIZING CAPITAL'S REQUIREMENTS   One approach which is currently in vogue is to attempt to demonstrate that the phenomena being discussed are favorable or necessary for accumulation. A suggested implication is that they are in some sense brought into existence because they aid accumulation. Thus Samuel Bowles and

Herbert Gintis (1976) analyze educational systems in terms of the preparation they give to students for docile acceptance of a role in capitalist production. Francis Piven and Richard Cloward (1971) analyze welfare systems as programs to cool out unrest or to channel workers into lowpaying jobs. David Gordon (1972, 1977) suggests job structures and residential patterns are designed to disrupt labor unity.

These explanations, at times, seem to accept the 'structuralist' view, cited before, that superstructural phenomena are determined by the totality of the accumulation process to which they contribute. The authors cited are not necessarily committed to a general structuralist position as articulated by Poulantzas (1975) or Althusser (1970), and they do attempt to go beyond its limitations. But their argument is easily read as a claim that policies or social trends exist because they are needed for accumulation.

Some objections to this approach have already been suggested. If phenomena exist because they are needed by an accumulation process, and they are themselves part of the accumulation process, what is being said is that they exist because they need themselves. How the accumulation process can exist as a thing apart to determine the characteristics of its component parts is unclear. Apart from a broad philosophic claim that the whole may determine its specific aspects, or the organic analogy that parts of the body may be explainable by their necessity to the entire organism, some more common-sense explanations are required.

One way out of the structuralist quandary is to posit that the capitalist class is conscious of its joint interest in sustaining the accumulation process, and is able to use the State, and perhaps other institutions of pressure or propaganda, to impose the fundamental feature of its program. Another possible approach is more evolutionary: it posits that any phenomena that detract from the accumulation process will cause crises in the system. As a result, whether capitalists are a cohesive ruling group, or indeed rule directly at all, the offending arrangements will be eliminated. In this view, even if 'the ruling class does not rule', and a worker-dominated government comes to power within a capitalist sytem, it will have to do what is needed for accumulation to proceed in the capitalists' interest – unless it can transform the basic economic system completely (Block 1977).

Even these more tenable approaches, however, assume that phenomena must always conform to the needs of the capitalist class, with any variations from this being transitory. Either the capitalist class rules directly and knows its needs exactly, or an evolutionary mechanism ensures all aspects of a capitalist sytem fit these needs. Gone is any question of whether forces contradictory to accumulation may be generated in specific areas of society. Gone is any question of struggle over outcomes, between or within classes, short of the final struggle over whether capitalism should exist. From the viewpoint of working-class organizing and struggle, these theories may be useful for debunking views that particular institutions are designed in the workers' interests. But they are of little use as explanations, or in designing strategy or tactics.

APPROACHES EMPHASIZING CLASS STRUGGLE   A less fatalistic approach, still grounded in the logic of accumulation, is possible. This approach would begin with the hypothesis that more than one combination of institutions or policies may be compatible with accumulation, at least in some historical circumstances. That the definition of the capitalist mode of production (as it is explained on p. 24 above) includes specifications of variable characteristics, would indicate that some variability is possible, and exists. Mentioning only examples from the macropolitical sphere: fascism, individualist democracy, military dictatorship, and social democracy have all been forms of rule used within the capitalist State. The system which reigns at any time does make a difference for workers' living conditions, for possibilities of future working class gains, and for the details of accumulation itself.

This is not to say any random set of attributes may be taken on by a capitalist system. Rather, the different institutions in existence at any time may have to fit together to form a coherent 'social structure of accumulation' – a full set of integrated institutions necessary for individual capital accumulation to continue (Gordon, 1978). But alternative institutional sets may be posited as possible at times. Thus, for example, fascism and social democracy may both be viable alternatives for continued accumulation in situations where concentration of capital is great, although individualist democracy may not fit any viable pattern in such a situation of monopoly. Presumably, different forms of institutions for social reproduction, including housing patterns and forms of local government, would have to fit into or be part of such an institutional set. Within any one social structure of accumulation, different options may be open for the arrangement of some but not all urban institutions. In studying any set of institutions in relation to the accumulation process, it would be necessary to argue which details were invariant within a structure, and what alternative structures or details of structures were possible but not used.

The existence of different possible structures in turn requires consideration of class struggle in the explanation of urban phenomena. For if several 'social structures of accumulation' are possible, working-class pressure may determine which one will emerge. Thus class struggle must be taken as a variable to consider in explaining observed urban phenomena. More important, from a political perspective, the effect of struggle on urban institutions may be considered in developing working-class strategy. Working-class political action may be evaluated in terms of its role in enforcing alternatives more or less favorable to labor in the short or long run (cf. Gorz 1964).

An emphasis on class struggle is, I believe, the proper focus in studying urban and other institutions. I have elsewhere argued (Edel 1977) that American suburbanization was instituted as a result of class conflicts and rising costs of social reproduction which had created an impasse for accumulation in the 1880s. Suburbanization was not the only possible solution to the urban crisis of the time. In effect, suburbanization was selected through class conflict: working-class struggles eliminated other proposed solutions. (A *status quo* solution of declining housing standards might have resolved the problem of costs had the working class not demanded some action. Another proposed solution, the company town, was restricted it its use because its introduction at Pullman became the occasion of a major strike.) Working-class movements in fact demanded suburban housing. The Workingmen's political campaigns and urban union agitation of 1886, and the rise of 'municipal socialism' as an electoral movement in the 1890s and after, were important in opening inexpensive housing options in the suburbs to at least an upper stratum of workers. However, the working class failed to press for other solutions, such as high-quality cooperative, union-sponsored or municipal housing, which could have solidified a working-class political base instead of dispersing it among new suburbs. In this sense, the suburban outcome or compromise was at best a partial victory for labor, improving living standards but in the long run fostering class fragmentation and political weakness (Edel 1977). The key point is that suburbanization cannot be considered as the only possible outcome of the crisis of the late 1880s that would have been compatible with some form of capitalist accumulation.

Conclusion

To restate the point more theoretically: urban phenomena, and other details of a capitalist system, are created historically, with working-class organization, consciousness and pressures affecting which alternatives are chosen. Specific outcomes must be consistent with accumulation, unless the working class is mature enough to overthrow capitalism. However, the specific outcomes are not logically deduceable from the existence of the capitalist mode of production or from the tendency to accumulation. Concrete analysis of the links involved, and identification of opportunities for working-class intervention, become the tasks of radical urban analysis.

Notes

[1] Hindess and Hirst (1975) in a formal analysis of the mode-of-production concept, define it as a structure that articulates forces and relations of production. They treat the concept in a logical manner, rather than as a crystallization of aspects of actual societies, so that in their view a mode of production can be defined to which no actual society corresponds.

[2] Some problems of definition occur in the limiting cases of 'petty commodity production' and 'state capitalism'. In the former case, everyone is the owner of a farm or artisan business, and there is no propertyless working class. State capitalism would be a situation in which government is the owner of the productive resources, but acts as employer of wage-labor. These may be thought of as unstable situations. Petty commodity production may evolve toward an unambiguous form of capitalism as larger and smaller business differentiate themselves, and some people go bankrupt and begin working for others. State capital-

ism may also evolve toward more normal capitalist forms, as the government begins seeking profit and bureaucrats become a new capitalist class.

Alternatively, these systems may evolve away from capitalism, replacing market relations and wage-labor with other relations of production. But it is unlikely they will remain the same for long, whereas capitalism or other pure modes of production before it, such a feudalism, lasted for centuries despite internal contradictions that always exerted some pressures for change.

3 It is in this sense that Marx considered capital a 'social relation' rather than a set of things.

4 Nor is the situation fundamentally affected if some unemployed workers receive transfer payments (welfare, social security, unemployment compensation, etc.) from the State. They are still essentially without a source of income of their own. In effect, the State hires them *to not work*, under terms similar to those under which others are hired to work. Limitations on earned income from other sources mean that the worker's time is effectively sold to the State.

5 In considering wage and productivity determination as separate processes, Marxist economics diverges from the orthodox view that labor will be paid an amount related to productivity. In the orthodox view, the amount paid is the worker's marginal product, i.e. the amount contributed to production by the worker under the assumption that he was the last hired or first fired. Profit in this view is determined by the productivity of the owner's equipment, similarly calculated as a marginal contribution.

6 The production of value, in Marx's definition, is measured net of the replacement cost of materials or equipment used up in production. Thus, even if no surplus value is produced, wages in an industry will not equal its total gross production, because part of the product is used to acquire replacement materials and equipment from other industries. This expense will, however, be matched by the costs of labor (plus surplus value and replacement of materials and equipment) in the supplying industry. Marx referred to the expense of replacing equipment and materials as expenditures on *constant capital* (C). He called the expense of wages to cover workers' necessary living costs *variable capital* (V). Gross value produced in the economy is thus $C+V+S$ where S represents surplus value.

Using these terms, several different measures of profit can be defined. $S/C+V+S$ is the proportion of total production that is surplus value (the 'profit share'). $S/C+V$ is a measure of the 'profit rate', the ratio of profits to capital advanced by the capitalist, although this Marxian measure is not the same as an accountant's measure since C and V are measured in annual expenditure rather than total book-value terms. S/V, the ratio of surplus value to wages, is used as a measure of the rate of exploitation.

These ratios are all for the economy as a whole. In dealing with individual industrial sectors, the analysis is more complex. Marx describes some industries, which produce higher than average amounts of surplus value (in relation to total product or to variable capital), as transferring some of their surplus value to other industries by selling at prices below value. This divergence of prices from values in particular industries is enforced by competition if all industries must operate at the same profit rates ($S/C+V$), but different industries have different ratios of labor cost to materials and equipment costs (C/V). The appropriateness of Marx's analysis here is the subject of the complex debate over the 'transformation problem', the relationship between values and prices. (Marx 1967, Vol. 3).

7 Marx, indeed, refers to exchange and consumption as 'moments' of production, that is, as necessary aspects of and phases in the production process.

8 For discussion of Marxian crisis theory see Alcaly (1978) and Weisskopf (1978).

9 These different necessities of capitalist production are used, with varying emphasis, in the existing literature on urban development and problems written from a Marxist perspective. Thus Harvey (1973) places a relatively greater emphasis on realization problems than do other writers; Castells (1978) focuses discussion of urban problems on social reproduction; and Hymer (1972) and Gordon (1977) place more emphasis on the production of surplus value, and particularly, the spatial and occupational divisions of labor introduced to allow capitalist control over production. These concerns should not be considered as mutually exclusive, but integrating them is difficult.

References

Academic Information Services, Inc. (1978) *Tax Guide for College Teachers.* Washington: AIS.

Alcaly, R. (1979) 'An introduction to Marxian crisis theory', in *URPE, US Capitalism in Crisis.* New York: Union for Radical Political Economics, pp. 15–21.

Althusser, L. (1970) *For Marx.* New York: Vintage.

Asad, T. and Wolpe, H. (1976) 'Concepts of modes of production', *Economy and Society* V, (November): 470–505.

Block, F. (1977) 'The ruling class does not rule: notes on the Marxist theory of the state', *Socialist Revolution* 33 (May–June): 6–28.

Bowles, S. and Gintis, H. (1976) *Schooling in Capitalist America.* New York: Basic Books.

Castells, M. (1978) *The Urban Question.* Cambridge, Mass.: MIT Press.

Edel, M. (1970) 'Innovative supply: a weak point in economic development theory', *Social Science Information* 9,3 (June).

—— (1977) 'Rent theory and labor strategy: Marx, George and the urban crisis', *Review of Radical Political Econ.* 9,4 (Winter): 1–15.

—— (1979) 'Collective action, Marxism and the prisoner's dilemma: comment', *Journal of Economic Issue* 9 (September), 751–61.

Engels, F. (1973) *The Condition of the Working Class in England.* Moscow: Progress Publishers.

Gordon, D. (1972) *Theories of Poverty and Underemployment.* Lexington, Mass.: D.C. Heath.

—— (1977) 'Capitalism and the roots of urban crisis', in R. Alcaly and D. Mermelstein, *The Fiscal Crisis of American Cities.* New York: Random House.

—— (1978) 'Up and Down the Long Roller Coaster', in *URPE, US Capitalism in Crisis.* New York: U.R.P.E., pp. 22–35.

Gorz, A. (1964) *Strategy for Labor.* Boston: Beacon Press.

Harvey, D. (1973) *Social Justice and the City.* Baltimore: Johns Hopkins Press.

Hindess, B. and Hirst, P. (1975) *Pre-capitalist Modes of Production.* London: Routledge and Kegan Paul.

—— (1978) *Mode of Production and Social Formation.* Atlantic Highlands, N.J.: Humanities Press.

Hymer, S. (1972) 'The multinational corporation and the law of uneven development', in J. N. Bhagwati (ed.) *Economics and World Order.* New York: The Free Press, pp. 113–40.

Marx, K. (1964) *Pre-capitalist Economic Formations.* New York: International Publishers.

—— (1967) *Capital.* Three vols. New York: International Publishers.

—— (1968) 'Wages, Price and Profit', in K. Marx and F. Engels, *Selected Works.* New York: International Publishers, pp. 186–229.

Massey, D. and Catalano, A. (1978) *Capital and Land.* London: Edward Arnold.

Murray, M. J. (1977) 'The Marxist theory of modes of production', *Review of Radical Political Econ.* 9,4 (Winter): 61–86.

Piven, F. and Cloward, R. (1971) *Regulating the Poor.* New York: Pantheon.

Poulantzas, N. (1975) *Political Power and Social Classes.* London: New Left Books.

Weisskopf, T. E. (1978) 'Marxist perspectives on cyclical crisis', in *URPE, US Capitalism in Crisis.* New York: U.R.P.E., pp. 241–60.

Wolff, R. D. (1977). 'Marxist theory and history: a Review Essay', *Review of Radical Political Econ.* 9,4 (Winter): 87–92.

3 The State in capitalism and the capitalist State*

Gordon Clark and Michael Dear

The role of the State in urban and regional spatial processes has been curiously neglected. It is usually completely ignored, or (at best) interpreted in isolation from market forces in analyzing spatial outcomes. This is true of analysts of all persuasions (including monetarists, Keynesians and Marxists) in spite of their professed interest in the increasing level of state intervention in market societies. It has been suggested that the growth of state power has been directly parallelled by a decline in explicit theorizing about the role of the state (Wolfe 1977). Yet even the most cursory glance at the historical evidence suggests a wealth of unanswered research questions. For instance, is the State itself neutral with respect to spatial outcomes, or is it biased in favour of some particular group(s)? How can we account for extraordinary resilience of the State to weather crises and to mutate under changed historical circumstance? What actually *is* the State?

The purpose of this paper is to provide a critical review of the role of the State in the capitalist space economy. An important distinction is made between studies of the 'State in capitalism' and of the 'capitalist State'. The former studies are concerned largely to explain the observed *functions* of the State in a given society; the latter studies focus in addition on the *form* of the capitalist State – i.e. how a given social organization gives rise to a particular structure or apparatus. The significance of this distinction is elaborated in the first section of this paper, which examines

* This article is based on a paper produced for the Annual Meeting of the Regional Science Association, Chicago, November 1978.

analytical alternatives for research into the State. This first section concludes with a list of desirable criteria for a viable theory of the State. The paper then reviews alternative theories of the State in capitalism; these are found to be wanting in the light of our criteria. Hence, in the paper's third section, the potential contributions of the various theories of the capitalist State are explored. Finally, the paper concludes with some remarks on the significant theories in a materialist theory of the capitalist State.

Alternative analytic approaches to the theory of the State

The character of the State

In the absence of any consensus on the appropriate point of departure for analyzing the theory of the State, let us begin with Mandel's conclusion that the State acts for 'the protection and reproduction of the social structure (the fundamental relations of production), in so far as this is not achieved by the automatic process of the economy' (Mandel 1975: 474).

Most viewpoints could ultimately be reduced to this simple assertion, although as expressed, it deliberately leaves unresolved the questions of the purpose, method and degree of state intervention to maintain the social structure. These questions can only be addressed if an analytical distinction is made between state *power* and state *apparatus* (Althusser 1971: 140). State power is the authority-relation mediating between the State itself and other social-class forces. It is a power which is expressed in the content of state policy. However, this translation of power into policy requires a state apparatus, which is the material organization for the exercise of state authority. Therborn (1978, Chapter 1) recognizes four types of apparatus, which reflect various state functions:

> 'the governmental apparatus (i.e., the rule-making legislative and executive bodies, both central and local), the administration, the judiciary, and the repressive apparatus (police, military, etc.)' (Therborn 1978: 41).

The multidimensional character of the state apparatus is widely conceded (cf. Miliband 1973), although many scholars have preferred to focus less on apparatus and more on state power. For example, Poulantzas (1969) places more emphasis on the State as a relation for the exercise of power.

The recent resurgence of interest in the State in the Marxist literature has extended our appreciation of (and sometimes, our confusion over) the character of the State. The powerful *ideological* purposes of the State have been emphasized by Althusser (1971: 143 ff.) *inter alia*. These include the institutional apparatuses of religious, educational and political organizations. The *evolutionary* character of the State has also been recognized, including the inherent tendency for growth in elements of the state apparatus. As Gramsci (1971: 13) notes:

'The democratic-bureaucratic system has given rise to a great mass of functions which are not at all necessitated by the social necessities of production, though they are justified by the political necessities of the fundamental group.'

Analytical alternatives

Previous studies of the State (or, more accurately, of political power) have tended to adopt one of three approaches (Therborn 1978: 130–1). First, there is a 'subjectivist' approach, which seeks to locate who has power in society. Studies of the power élite and pluralism fall under this rubric. Secondly, the 'economic' approach has attempted to describe a representative theory of democracy based primarily upon exchange relationships in society. Thirdly, and in profound contrast, is the 'historical-materialist' approach, which seeks to define the nature of power, rather than its subject or quantity. It is an approach which views the State as one component of an ongoing social process of production and reproduction.

Both the subjectivist and economic approaches are modes of analysis which address the *functional* aspects of the State. The former describes the structure and characteristics of the distribution of power in society, and in whose interest power is wielded. The latter describes the exchange relationships between groups with qualitatively different kinds of power. It is only the historical-materialist approach, derived from Marxist theory and practice, which plunges more deeply into the structure of social relations in order to explain why power is distributed in the observed manner, and why it is wielded in the interests of certain groups. In short the historical-materialist approach is concerned with the derivation of the *form* of the State, as well as its functions.

We propose to *characterize those modes of investigation which focus on state function as theories of the State in capitalism. Conversely, those modes which focus on form and function will be designated theories of the capitalist State.* This is a distinction which derives from a central epistemological distinction in Marxist theory – that between the 'level of appearances' and the 'social reality' underlying and causing those appearances. The preferable scientific method is one which allows the underlying structural processes to be linked systematically to empirically observable phenomena (Wright 1978, Chapter 1). However, this requirement, so simply stated, is one of the most difficult methodological hurdles in social science. It is also a core issue in social science theory and philosophy. Some brief digression into these wider issues is vital to the progress of this paper.

It has recently been suggested that geography and other spatial sciences have been committed to a positivist epistemology which has profoundly isolated them from society and its culture. This isolation is cause for concern, since it makes ' . . . social science an activity performed *on* rather than *in* society' (Gregory 1978: 51). Progress toward a critical social science, and away from the 'empirical-analytic' traditions of positivism, depend upon the application of what Habermas has called the 'historical-hermeneutic' approach in social science (cf. Gregory 1978, Chapters 2,5).

This approach is founded in rejection of the possibility of an autonomous science, insisting instead that concepts of science depend upon determinate social context(s) and practice(s). Scientific progress thus depends upon an emancipatory dialogue between internal and external theoretical frameworks: the reciprocal dialectic between these two frames of reference is the source of knowledge (Habermas 1974).

In searching for an appropriate mode of analysis for the State, the choice between the empirical-analytic and historical-hermeneutic approaches is crucial. On the one hand, the former offers a well-defined analytical methodology which succeeds in describing only the functions of the State. On the other hand, the latter offers a powerful heuristic for interrogating the relationship between the state apparatus and the underlying social formation. However, the historical-hermeneutic approach proceeds in essence by contradiction and extensive debate is not conducive to the establishment of scientific 'findings' or 'results'. This is best understood by recognizing that the criteria of validity in empirical-analytic knowledge is a successful *prediction*; in historical-hermeneutic knowledge, the criteria is successful *interpretation* (Gregory 1978, Table 1).

In spite of the difficulties presented by the historical-hermeneutic methodology, we are persuaded that it is the only option for analysis of the theory of the State. This is because it is a methodology which requires the derivation of state functions and form from the wider social formation. Furthermore, because spatial structure is to be interpreted as an expression of social process (cf. Castells 1976), analysis of the spatial outcome of state intervention *requires* a theory of society as well as a theory of space.

Large methodological difficulties remain. To get below the level of appearances clearly requires much more than developing a series of 'testable hypotheses' with respect to state intervention. Indeed, such a methodology is anathema to Marxist analysts, since it is firmly grounded in positivist epistemology and ideology. On the other hand, those analysts who have attempted to generate empirical studies using Marxist categories appear to have lost much of the dialectical character of Marxist theory in the process. As Wright (1978: 10) suggests, however, there is another alternative, involving

> 'the attempt to develop empirical research agendas firmly rooted within . . . Marxist theory. Such an approach would reject the positivist premise that theory construction is simply a process of empirical generalization of law-like regularities but would also insist that Marxist theory should generate propositions about the real world which can be empirically studied.'

This sentiment is echoed by Hirsch, in his study of the capitalist State: 'it will be vital for the theory of the State not to derive the State . . . as an abstract form, but to come to grips with it as the concrete social organizational nexus which it represents in practice' (Hirsh 1978: 107).

We are now in a position to clarify the desirable qualities of a theory of the capitalist State. In the remainder of this paper, various historical and contemporary views of the State in capitalism and the capitalist State will be evaluated according to these criteria. A viable theory of the capitalist state is one which:

1. derives the form of the State from its relationship to the wider structural relations of the capitalist society;
2. identifies the range of state production and reproduction functions necessary for the continuation of the capitalist social structure, as well as the purpose and source of non-necessary functions;
3. specifies the relationship between the economic and the political spheres, particularly identifying class and non-class determinants of state action;
4. accounts for the different functional organizations of the State, in both sectoral and spatial terms, and explains the actions of these different organizations;
5. allows for the historical evolution of different state apparatuses; and
6. permits the generation of tractable analytical propositions about the form and function of the State in the real world.

While many of these criteria can be tracked back to the preceding discussion, the significance of others will become fully apparent as the paper progresses.

The State in capitalism

Theories of the State in capitalism focus on the functions of the state apparatus. Four particular characterizations of the State have been previously noted, based upon different functional interpretations of the State's role (Dear and Clark 1978). These identify the State as: (i) 'supplier' of public or social goods and services; (ii) 'regulator and facilitator' of the operation of the market place; (iii) 'social engineer', in the sense of intervening in the economy to achieve its own policy objectives; and (iv) 'arbiter' between competing social groups or classes. Although it is analytically convenient to distinguish among these four viewpoints, the categories are by no means mutually exclusive. There is a considerable overlap and certain categories may subsume others.

The simplest view of the State is as a *supplier of public goods and services.* Three particular reasons for public good provision are normally noted: the existence of external effects associated with a particular good; other kinds of market failure which are unrelated to the good's characteristics (e.g. monopolistic tendencies); and a preference for certain standards in community affairs. Public good provision is predominantly regarded as an allocative function of government. The analytical task in the allocation problem focuses upon the proper criteria for government intervention, and the optimal allocation rules for public good provision. However, although it may be analytically convenient to isolate the allocative function, the distributive consequences of allocative decisions cannot be ignored. The separation of the two issues implicitly favours a given income distribution within society, focusing instead on ameliorating the impacts of private market failure.

The State may also be viewed as a means of *regulating and facilitating* the operation of the private market. The State's intervention to maintain the market is based on two assumptions: first, that a well-regulated and efficient market will create the best possible allocation of resources; and second, that the market may not inevitably achieve optimal conditions and systemic equilibrium. The Keynesian revolution encouraged the State to assume responsibility for socio-economic stabilization, and to maintain the market's efficiency through its use of fiscal, budgetary, monetary and competitive policies. However, stabilization may not be the only measure involved in regulating the economy. The use of state policy to enforce market regulation has also been important; e.g. through anti-monopoly and anti-trust legislation. The possible scope of this role is limited to maintaining the 'rules' of the market game. This may involve state intervention to improve information flows in both time and space, as well as intervention designed to facilitate growth and competitive development.

A significant element in the State's behaviour in advanced capitalist economies is in adjusting market outcomes to fit its own normative policy goals; i.e. in *social engineering*. Such intervention involves a judgement about what society ought to be rather than what it is. This is an important difference, since distribution itself becomes an element of legitimate concern for the State. Thus the State may operate to ensure distributive justice, although this is qualified by the acceptance of the market as the means of distribution. In this regard, the State as social engineer seeks to redress socio-economic imbalances and maintain fairness for disadvantaged groups in a market society.

The notion that the State holds a mandate to adjust outcomes in favour of particular social groups introduces a view of the State as *arbiter* of inter-group conflicts in a society. Considerable ambiguity surrounds the possible approaches the State may adopt in its mediation efforts. Dye (1972) has summarized five simple models of public decision-making. State arbitration can be viewed as rational, that is, based on some logical criterion of choice; incrementalist, being founded mainly in slight shifts in position from existing practices; élitist, reflecting the interests of the ruling power groups; group-biased, implying some genuine efforts at compensation amongst all interested parties; and institution-based, suggesting that the State may act in its own interests, and may possess hidden objectives in entering group conflict. Whichever view (or combinations of views) are eventually accepted as descriptive of the State's arbiter role, the change of emphasis from decisions in the market place to decisions in the political forum is an important distinguishing characteristic of this concept. It reflects, more than the previous alternatives so far considered, the reality of power and its use by the State. Alternatively, a number of studies have attempted to construct representative theories of democracy (Buchanan and Tullock 1962). In this series of models the voter is assumed to voice his/her preference for a group of social goods on the basis of democratic voting. Such voting is assumed to represent 'true' preference rankings in society. Votes then become *market signals,* and much emphasis is placed on the ability of the State to recognize and react to changes in consumer preferences; the State derives its role from a consumer mandate, and limits to collective action reside in the discretionary voting of consumers. In a conflict-oriented view of society Dahrendorf (1959, Chapter VII) has argued that the distribution of political authority in society defines a clearly visible *authority structure.* At the bottom rung are the mere citizenry who occupy no political position other than that common to all members of the polity. In contrast, the three branches of authority in society (the legislative, judiciary and executive branches of government) represent the peak of the authority structure. The responsibility of the State as arbiter involves the resolution of many kinds of social conflicts which may involve all branches of authority.

Summary

The analyst who views the State as a supplier of public goods and services will direct attention at market failures in the economic system. Relevant policy questions include the proper criteria for state intervention and for allocative efficiency. The research problem in this category is concerned with issues of *allocation.* In contrast, the view of the State as regulator and facilitator focuses more on the *stabilization* function of government. It is a view which allows for rather greater, but still selective, intervention in the market. The analyst of the State as social engineer recognizes that the State can intervene on its own behalf in the market. Policy questions are predominantly of a *distributive* nature. The State as arbiter allows that the State is in a powerful position to *arbitrate* amongst conflicting group interests in society. More importantly, the approach changes the focus from decision-making in the market system to decision-making in the political arena. It is possible for issues of allocation, stabilization, and distribution to be conveniently subsumed within the arbiter framework.

The capitalist State

The only truly self-aware studies of the form of the capitalist State are to be found in the Marxist literature. However, Marx himself did not develop any consistent theory of the capitalist State. Significant progress in a Marxist theory of the State has been made more especially during the past two decades. During the same period, there have been a number of studies which, although using the historical-materialist approach, have gone beyond the more traditional Marxist categories. This resurgence of interest has resulted in a literature which is at once both stimulating and confusing. This section attempts to systematize the contemporary debate on the capitalist State by pursuing three themes: the classical Marxist theory of the State; the resurgence of contemporary Marxist debate; and the current theory of the 'derivative' State.

The classical Marxist theory of the State

The point of departure for all studies of the form of the capitalist State is to analyze the genesis and development of the State with respect to the wider set of social relations from which it derives. The historical-materialist methodology thus proceeds by interrogating the relationship between the capitalist State and the form of production in capitalist society.

In his review of Marxist theories of the State, Jessop concludes that '. . . nowhere in the Marxist classics do we find a well formulated, coherent and sustained theoretical analysis of the state' (Jessop 1977: 357). This conclusion should not be surprising, given the famous dictum of the *Communist Manifesto,* that 'The executive of the modern state is but a committee for managing the common affairs of the whole bourgeoisie' (Marx and Engels 1967 edition: 44).

Subsequently the view tended to be taken that the capitalist State was fundamentally the coercive instrument of the ruling classes, and was a product of the irreconcilability of class antagonisms (Lenin 1949 edition: 9). However, Jessop points out that these early texts contain much historical insight and form the basis for later, more rigorous analysis. Specifically, Jessop (1977: 354–7) recognizes six different approaches to the theory of the State in the classical Marxist literature. These are:

1. the State as a *parasitic* institution, with no essential role in economic production;
2. the State and state power as *epiphenomena,* i.e., superficial reflections of an independent economic base;
3. the State as a *factor of cohesion* in society, regulating class conflict predominantly in the interests of the dominant class;
4. the State as an *instrument of class rule,* as a consequence of its 'capture' by a dominant class;
5. the State as a *set of institutions,* which tends to avoid assumptions about the class character of the State, focusing more on the empirical manifestations of the state apparatus; and
6. the State as a *system of political domination,* with special attention to the characteristics of political representation and state intervention (e.g., democracy as the best political setting for capitalism).

Contemporary Marxist theory and its extensions

Recent Marxist research has attempted to extend our understanding of the function of the State in capitalism (Gold, Lo and Wright 1975; Harvey 1976). Three approaches have evolved: the 'instrumentalist', in which the links between the ruling class and the state élite are systematically described and analysed; the 'structuralist', which examines why the State functions particularly with respect to class conflict and contradictions inherent in the system; and finally, a more explicitly 'ideological' perspective which places emphasis on the 'consciousness' and ideology through which the State pursues class exploitation and control. This last approach necessarily involves the State in propagandizing the dominant ideology, and in an attempt to 'mystify' reality. The three approaches are not mutually exclusive, although their distinctions have been much debated. As Gold, Lo and Wright (1975) argue, the different approaches tend to highlight different aspects of the State's role as agent within the capitalist economy.

Miliband's (1973) work is probably the best example of the *instrumentalist* approach. He explores the 'conspiracy' between the ruling class and the State's bureaucratic élite. This conspiracy has, as its objectives, the maintenance of the system, and the development of the primary institutions to serve the capitalist interest. The focus of this approach has been to document the extent of the State's capital links. Little attempt has been made to clarify whether the direct participation of the ruling class in the State is a cause or effect (Figure 3.1a).

Diagrammatic representation of theories of the capitalist state

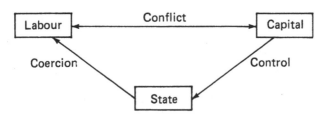

Figure 3.1a Instrumentalist

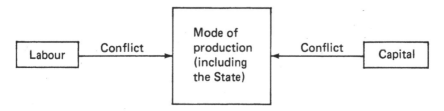

Figure 3.1b Structuralist

In contrast, the *structuralist* view is that the functions of the State are broadly determined by the structure of society itself, rather than the people who occupy positions of power (Gold, Lo and Wright 1975, 1: 36). The State attempts to alleviate persistent class contradictions, generate accumulation and accomodate contradictions within society as the balance of power between classes shifts (Figure 3.1b). Thus, the State is not an autonomous entity, but reflects the balance of power among classes at any given time. As Poulantzas (1969: 73) comments:

'The direct participation of members of the capitalist class in the state apparatus . . . is not the important side of the matter. The relation between the
bourgeois class and the state is an objective relation. This means that if the
function of the state in a determined social formation and the interests of
the dominant class in this formation coincide, it is by means of the system
itself: the direct participation of members of the ruling class in the state
apparatus is not the cause but effect.'

A third Marxist approach reflects a wider concern with the *ideology*
and mystification of capitalist reality. The ideologists ask the question:
'What is the State?' The conclusion, derived from a Hegelian-Marxist
perspective, is that:

' . . . the state is a mystification, a concrete institution which serves the
interests of the dominant class, but which seeks to portray itself as serving
the nation as a whole, by obscuring the basic lines of class antagonism.'
(Gold, Lo and Wright 1975 1: 40)

Thus, the State seeks to misrepresent itself in order to maintain class
divisions and inequality, and to defuse different class aspirations by relating
the 'national interest' to one mythical broad homogeneous community
(Bedeski 1977; Jessop 1977: 357).

The key to resolving the respective merits of these three approaches is
definitional: What exactly is the State? Poulantzas has usefully polarized
the debate by clarifying the distinction between the state as 'subject' and
as 'object'. The subject viewpoint is structuralist, in that the State is regarded as ' . . . the condensate of a *relation of power* between struggling
classes' (Poulantzas 1976: 74, italics added). This viewpoint argues for a
materialist theory of class relations, with the State signifying one dimension
of those relations. In contrast, the 'object' approach implies an instrumentalist view of the State (cf. Miliband 1973). The State is an autonomous
institution which is effectively 'captured' by the ruling class and, hence,
becomes the object and tool of class control and repression.

Serious criticisms have been offered of both the instrumentalist and
structuralist viewpoints. It has been argued that Miliband's instrumentalist
approach has done little more than take account of changes in the nature
of the ruling class. An empiricist focus on how the State acts, and the
absence of a theoretical framework for the empirical investigation of state
organization ' . . . limits his work to a redirection of the "plurality of
élites" into a theory of the *influence* of the ruling class' (Offe and Ronge
1975: 137). Moreover, while power élites undoubtedly exist, the instrumentalist approach fails to identify the logic whereby the élites themselves
are constituent elements of a wider social order which is independent of
specific institutions and personalities. This wider logic is the specific concern of the structuralists, who conceive of the State as a permanent, necessary constituent of a class-based society (Poulantzas 1969, 1976). However,
such an approach fails to take into consideration either the historical evidence for change in the State entity or its concrete empirical reality and
function within society (Offe and Ronge 1975). Moreover, the structuralist

view detracts from the historical evolution of the contemporary State. It seems unconcerned with distinguishing between those state characteristics which are a necessary result of class conflict and those which are a product of contemporary conditions.

Apart from the limitations inherent in the instrumentalist and structuralist approaches, it has recently been argued that the apparent dichotomy presented by the two approaches is false and misleading. For instance, Jessop (1977: 357–8) argues that Miliband merely treats the state apparatus in isolation from its articulation with economic forces – in our terms, Miliband is concerned with the State in capitalism. In contrast, Poulantzas accepts the necessity of analysing the capitalist State within the structure of capitalist society. However this early focus on the relative autonomy of the political with respect to the economic tends to detract from the central concerns of the form and structural limits of the capitalist State (Holloway and Picciotto 1978: 5–7).

From a concern with the economic dimensions of the State has sprung a more recent debate between the 'New-Ricardians' and the 'Fundamentalists' (Holloway and Picciotto 1978: 10–14; Jessop 1977). The former tend to subordinate the political to the economic, and emphasize how the State acts in the interest of capital. The latter, Fundamentalist group firmly reject this emphasis in favour of an approach which interprets the economic and political spheres in the light of the need for capital accumulation. Hence, state interventions arises directly from the needs of capital. Yet such an approach tends to by-pass the problems of ' . . . the specificity of the political and the role of the political system' (Holloway and Picciotto 1978: 11).

The attempt to develop a specific theory of the political is a characteristic of the work of Habermas and Offe. An important effort has been made by Claus Offe to relate state structure and function to *historical changes* in advanced capitalist societies, as well as analysing the concrete *empirical realities* of state activity (Offe 1976; Offe and Ronge 1975). Important insights follow from this effort. For example, while (arguably) aiming for crisis-free stabilization and integration in capitalist economies, the expanded functions of the State may themselves be a source of dysfunction and crisis.

The important point of departure in Offe's work has been emphasized by Habermas (1976: 375). The State can be regarded as an input-output mechanism (Figure 3.1c). Its output consists of 'autocratically-executed administrative decisions', for which it requires an input of 'mass loyalty, as little attached to specific objects as possible'. This approach has the important consequence of removing the necessity for the subject-object distinction. The State may be regarded as a 'thing' and an 'invisible hand' *simultaneously*. The State exists as an independent, identifiable entity, with its own functions and objectives; and, at the same time, it is clearly situated as a constituent element of a wider set of power relations within society. According to this reasoning, the State will act in the interests of all members of a capitalist class society, and many policies will not directly serve the

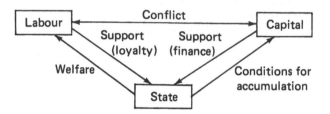

Figure 3.1c Input – output

interest of the capitalist class (Offe and Ronge 1975: 144). Hence, most state intervention may be viewed as long-term strategies of crisis avoidance, necessitated by the inevitable incidence of class antagonisms (Habermas 1976: 392; Offe 1976: 415). The political input-output system itself is unique in its situation and competence to deal with such antagonisms, which are usually characterized as crises of output or of input. Output crises, relating to the State's administrative decisions, take the form of crises in *rationality*. Input crises take the form of crises in *legitimation* – the system simply does not succeed in maintaining the necessary level of mass loyalty (Habermas 1976).

The current *fiscal* crisis of the State is intimately linked to crises in rationality and legitimation. As O'Connor (1973: 6) has argued, much state effort is devoted toward maintaining societal conditions under which profitable capital accumulation is possible. At the same time, the maintenance of social harmony is necessary. However, the capitalist State which openly uses its coercive power to enable óne class to profit at the expense of others loses its legitimacy and risks undermining the basis of its mass support. This structural contradiction is compounded during inflationary periods when rising costs and public expenditure cutbacks cause state output to fall below expectations. A crisis in legitimacy is thus initiated which imposes increased pressure upon the rationality crisis, and so on. As the legitimation, rationality and fiscal crises compound one another, crucial variations are evidenced in the relative autonomy of the State, i.e. the degree of separability between the economic and political systems. For instance, in times of fiscal crisis, the State may impose a prices-and-incomes control policy, thus altering the autonomy gap by strengthening the political dimension. Only later may reflation policies, in the light of rising unemployment, reconcede autonomy toward the economic dimension.

The state derivation debate

Perhaps the single most important new trend in the State debate has been the emergence, in the German literature, of the 'state derivation' debate. The major essays in this debate have been collected by Holloway and Picciotto (1978), who note that:

'Instead of simply reiterating the connection between capital and the state, however, the contributions to the debate have accepted the separation of the economic and the political and have tried to establish, logically and histori- cally, the *foundation of that separation in the nature of capitalist production.*' (Holloway and Picciotto 1978: 14–15; italics added)

This need to situate the separation of the political from the economic in the analysis of capital has produced two major themes in the derivation debate, (i) a 'capital logic' school; and (ii) a 'materialist' theory of the State. Analysts in both themes have proceeded from a critique of those theorists (e.g. Habermas and Offe) who have divorced politics from capital accommodation. The derivationists ask: Why do social relations in capital- ist societies take *separable* forms of political relations and social relations? In Pashukanis' words:

'Why does the dominance of a class not continue. . . . [as] the subordination in fact of one part of the population to another part? Why does it take on the form of official state domination? Or, which is the same thing, why is not the mechanism of state constraint created as the private mechanism of the dominant class? Why is it disassociated from the dominant class – taking the form of an impersonal mechanism of private authority isolated from soci- ety?' (Pashukanis, quoted in Holloway and Picciotto 1978: 18–19)

In order to answer this core question of the form of the capitalist State, the *capital logic* analysts insist first upon the separation of the State and civil society. This separation is vital in the provision of the general condi- tions of capital accumulation and reproduction, because no individual or competing capital is able to ensure the reproduction of the whole (i.e. the total social capital). An autonomous State is thus necessary to the reproduction of the capitalist social formation. Accordingly, the form of the capitalist State and its concomitant functions are concerned with cor- recting the deficiencies of private capital, and with organizing individual capitals into a viable aggregate (Figure 3.1d). The concerns translate into four specific functions: the provision of general material conditions of production; establishing general legal relations; regulation and suppression of conflict between capital and wage labour; and safeguarding total national capital on the world market (Altvater 1978: 42).

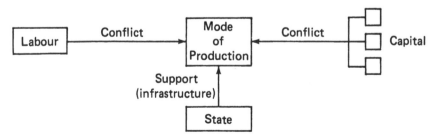

Figure 3.1d Capital logic

The capital logic approach does not conceive of the State merely as an instrument established and controlled by capital. The State will intervene against wage-labour and individual capitals when either threaten the interests of capital as a whole (Jessop 1977: 363). However, in spite of this significant theoretical distinction, the school has been subject to intense criticism. More especially, it has been noted that the capital logic approach again tends to reduce the political to an epiphenomenon of the economic base, without clarifying the conditions for reciprocal influence. Equally importantly, the capital logic approach is fundamentally ahistorical. It is only able to indicate probable forms of the State, and to specify broad limits of variation within which the process of capital accumulation will not be threatened (Holloway and Picciotto 1978: 21–2; Jessop 1977: 364).

In response to these difficulties with the capital logic approach, several scholars have attempted to introduce a greater historical specificity, and an analytical focus on class struggle, in the study of the capitalist State. An emphasis upon the logical implications of competing capitals is replaced, in the *materialist theory* of the State, by a focus on the antagonistic relation between capital and labour in the process of accumulation (Holloway and Picciotto 1978: 26).

As formulated by Hirsch (1978), the specific form of the State derives from the social relations of domination in capitalist society. The coercive, exploitative nature of the capitalist social relations is obfuscated by the constituting of discrete 'political' and 'economic' spheres. Hence, official state domination abstracts the relations of force from the immediate process of production (Hirsch 1978: 61–4). State intervention is thus primarily conditioned by the emergent crises of capitalism, which find their source in the process of accumulation, especially in the tendency of the rate of profit to fall (Figure 3.1e). As Holloway and Picciotto (1978: 26) observe:

> 'This approach, which takes as its starting point the antagonistic relation between capital and labour in the process of accumulation, thus provides us with a framework for an historical and materialist analysis of the state.'

State intervention, in short, is regarded as a response to the *political* repercussions of accumulation. Again, there is no necessity for such intervention

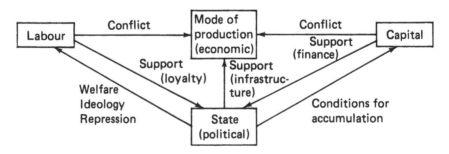

Figure 3.1e Materialist

always to be in the interests of capital. Instead, the form of state intervention could be more consistently interpreted as *crisis-management,* as determined by changes in the balance of class forces (cf. Habermas 1976; and Jessop 1977: 306).

Conclusions

The purpose of this paper has been to provide a review of the State in capitalism and the capitalist State. Theories of the State in capitalism tend to focus upon superficial analyses of state functions as they are empirically observable. In contrast theories of the capitalist State question both the functions and the forms of the State as they are based in the specificities of the capitalist social formation. There seems little doubt that this latter group provides a more powerful basis for a fundamental interrogation of the role of the State, although the former group of theories tend to be more tractable empirically. However, not all theories of the capitalist State possess the same potential. It now seems that the furore surrounding the instrumentalist vs. structuralist debate in the Marxist literature has been somewhat ill-founded and confusing. Much the clearest direction has been outlined by the recent German debate on a materialist theory of the State. In concluding this paper, we offer our assessment of the significant emphases and directions proposed by a materialist theory of the State.

1. In spite of protestations to the contrary (e.g. Holloway and Picciotto 1978: Introduction; Therborn 1978: 30), the materialist approach does provide for a specific theory of the political, which focuses on the derivation of the form and functions of the State from the capitalist social formation. The relationship between the economic and the political spheres is properly regarded as one of mutual interaction.
2. In the materialist theory, emphasis is placed upon an historical interpretation of the State's development. In the longer term, the form of the State can be expected to alter as conditions of capital accumulation change. Hence, strategies of intervention (i.e. state functions), including appropriate forms of political representation and legislation, will also change. This notion that the institutional structure and its mode of intervention will be transformed as the social formation changes is the core of Wolfe's recent characterization of the six 'temporary' forms of the American state apparatus (the accumulative, harmonious, expansionist, franchise, dual, and transnational state; cf. Wolfe 1977). The precise form of the State may be derived, in theory, from the stage of development of the relevant mode of production, and the relationship of that mode to others in the international order (Therborn 1978, Chapter 3).
3. The materialist theory isolates crises in capital accumulation as the catalyst for state intervention, which is intended to cope with the political repercussions of such crises in order to ensure the reproduction of social relations (Habermas 1976; Offe and Ronge 1975). Such

an emphasis lays stress on the role of the State as a system of political domination, although the separation of the economic and the political obscures the form and purpose of state domination (Hirsch 1978).

4. The materialist theory places new emphasis on the ideological and repressive functions of the State. As Therborn (1978: 173) suggests, the process of social reproduction is a synthesis of ' . . . economic, political and ideological processes'. Ideology tells individuals what exists, what is possible, and what is right and wrong. The State plays a vital role in convincing the ruled of the legitimacy of the rulers. The aspirations and interests of the 'people' must therefore be incorporated into the dominant ideology if the social order is to be maintained (cf. Jessop 1977: 367–8). Where opposition is encountered, varying forms of state repression might be anticipated, including ' . . . *prohibition* of opposition, *restriction* of . . . opposition, *harassment and terror,* and *surveillance*' (Therborn 1978: 222).

5. A materialist theory of the State allows for a clear focus on the functions of the State. Following Offe, emphasis is placed on the processes of transformation whereby demands upon the state apparatus (inputs) are translated into administrative decisions (outputs), and on how crises of legitimacy and rationality are managed in these processes (cf. Habermas 1976). Henceforth, observed problems in specific categories of state intervention, such as urban and regional planning, should no longer be regarded as failures of particular mechanisms. Instead, they should be traced to systematic, structural contradictions in the form and function of the State, and their relationship to the capitalist mode of production.

6. As well as the broader questions of the form and function of the capitalist State, future research ought to direct particular attention toward two relatively unexplored issues: first, the politics of the State, especially the language with which the political debate is conducted, and the limits which the vocabulary imposes (cf. Habermas 1976); secondly, the role of the local State in support of, and in contradiction to, the federal state apparatus and the rest of civil society.

References

Althusser, L. (1971) *Lenin and Philosophy and Other Essays.* London: New Left Books.
Altvater, E. (1978) 'Some problems of state interventionism' in J. Holloway and S. Picciotto (eds.) (1978).
Bedeski, R. E. (1977) 'The Concept of the State: Sun Yat-sen and Mao Tse-tung.' *China Quarterly* 70:338–54.
Castells, M. (1976) 'Theory and ideology in urban sociology' in C. G. Pickvance (ed.) *Urban Sociology: Critical Essays.* London: Tavistock.
Dahrendorf, R. (1959) *Class and Class Conflict in Industrial Society.* Stanford: Stanford University Press.
Dear, M. and Clark, G. (1978) 'The State and geographic process: a critical review.' *Environment & Planning A* 10:173–83.
Dye, T. (1972) *Understanding Public Policy.* New Jersey: Prentice-Hall.

Gold, D. A., Lo, C. Y. H., Wright, E. O. (1975) 'Recent developments in Marxist theories of the capitalist State.' *Monthly Review* 27, 5:29–43; 6:36–51.

Gramsci, A. (1971) *Selections from the Prison Notebooks.* New York: International Publishers.

Gregory, D. (1978) *Science, Ideology and Human Geography.* London: Hutchinson U.P.

Habermas, J. (1974) *Theory and Practice.* Boston: Beacon Press.

—— (1976) 'Problems of legitimation in late capitalism' in P. Connerton (ed.) *Critical Sociology.* Harmondsworth: Penguin Books.

Harvey, D. (1976) 'The Marxian theory of the State.' *Antipode* 8:80–98.

Hirsch, J. (1978) 'The state apparatus and social reproduction: elements of a theory of the bourgeois State' in J. Holloway and S. Picciotto (1978).

Holloway, J. and Picciotto, S. (eds) (1978) *State and Capital: A Marxist Debate.* London: Edward Arnold.

Jessop, R. (1977) 'Recent theories of the capitalist State.' *Cambridge Journal of Economics* 1:353–74.

Lenin, V. I. (1949) *The State and Revolution.* Moscow: Progress Publishers.

Mandel, E. (1975) *Late Capitalism.* Translated and revised Ed. London: New Left Books.

Marx, K. and Engels, F. (1967) *The Communist Manifesto.* Moscow: Progress Publishers.

Miliband, R. (1973): *The State in Capitalist Society.* New York: Basic Books.

O'Connor, J. (1973) *The Fiscal Crisis of the State.* New York: St. Martin's Press.

Offe, C. (1976) 'Political authority and class structures' in P. Connerton (ed.) *Critical Sociology.* Harmondsworth: Penguin Books.

Offe, C. and Ronge, V. (1975) 'Theses on the theory of the State.' *New German Critique* 6:137–47.

Poulantzas, N. (1969) 'The Problem of the capitalist State.' *New Left Review* 58:67–78.

—— (1976): 'The capitalist State: a reply to Miliband and Laclau.' *New Left Review* 95:63–83.

Therborn, G. (1978) *What Does the Ruling Class Do When It Rules?* London: New Left Books.

Wolfe, A. (1977) *The Limits of Legitimacy: Political Contradictions of Contemporary Capitalism.* New York: The Free Press.

Wright, E. O. (1978) *Class, Crisis and the State.* London: New Left Books.

4 Some reflexions on epistemology, design and planning theory*

Marc Los

Theories of planning and epistemology

The current debate in planning theory

The current renewed interest for planning theory has several origins: the desire of practicing planners to give a meaning and some usefulness to their role in society; the need felt by academic planners to understand the link between methodology and theories on one hand and actual planning practice on the other hand, possibly with planning education in mind (see Faludi 1973a, 1973b, 1978b; Harris 1975, 1980); the need felt by some planners to improve by means of planning the distribution of power in society (see Davidoff 1965; Friedmann 1973); and finally the need of the adepts of a critical or radical view of planning to give more theoretical foundation to their dissatisfaction with current practice and current methodology (see Scott and Roweis 1977; Goldstein 1975; Goldstein and Noyelle 1977). Following Harris (1967a) and Faludi (1973a), we will make the distinction between 'theory *in* planning', i.e. the social sciences that can be used in planning (urban economics is an example) and 'theory *of* planning', i.e. a theory concerned with the action of planning as such and not with what is being planned. The theory *of* planning itself can be subdivided into two theories: a procedural (or normative) theory of planning

---

* This article is based on a paper delivered at the 25th Anniversary Conference of the Department of City and Regional Planning, University of Pennsylvania, April 1977, entitled 'Epistemology and Planning Theory: A Piagetian Paradigm'.

concerned with how planning *ought* to be done and a positive (or explanatory, or empirical) theory of planning concerned with how planning *is* being done or with the social nature of planning itself. There seems to be a consensus about these logical distinctions among different authors although the precise meaning attached to each concept, the connections between theories *in* planning, procedural and positive theories *of* planning, as well as the relative importance given to one theory or the other depend on each author's particular views.

There is by now a voluminous literature on planning theory and, at the risk of simplifying and of not doing complete justice to the richness of each author's views, we can make the following typology of these works: firstly, the theories of planning conceived as general theories of decision-making (see Davidoff and Reiner 1962; Faludi 1973b; Harris 1975, 1980); secondly, the 'advocacy planning' theory (see Davidoff 1965); thirdly, the 'new humanists' (see Friedmann 1973); fourthly, the 'critical theorists' (see Goldstein 1975); and fifthly, the 'Marxists' (see Scott and Roweis 1977; Goldstein and Noyelle 1977). With the exception of the fifth group, more concerned with explaining 'the facts of planning' than with procedural theory, all other theorists are concerned with the latter. Harris's recent works (1975, 1980) represent an excellent expression to date of the theory of planning conceived as a general theory of decision-making in a social context, but not necessarily as a specifically urban theory. It is the continuation of a long and respectable tradition of which Simon is a well-known representative (see March and Simon 1958; Simon 1969). In his book, Harris presents an original way of looking at social decision-making by using mathematical programming as a paradigm, rather than as a tool. This allows him to point at some shortcomings of different styles of planning, to characterize 'intrinsic' difficulties of planning and to indicate simultaneously the possibilities and limitations of such well-known techniques as mathematical programming, simulation models, econometric techniques and cost-benefit analysis. For the 'advocacy planning' movement (see Davidoff 1965) urban planning is conditioned by the distribution of power in society, with the consequence that powerless groups can find themselves without a technical expertise. It is proposed that planners put their technical expertise at the service of disadvantaged groups so as to counterbalance the unequal distribution of power. The emphasis of this approach to normative planning is on public participation. For the 'new humanists' (see Friedmann 1973) planning should be done through mutual learning. These authors deemphasize any scientific approach and emphasize personal knowledge, communication and dialogue as a way of solving society's problems. Goldstein (1975) shares Davidoff's concern for the inequalities in the distribution of power in society and Friedmann's distrust of scientific or technological approaches to planning. His ideas are inspired by the Frankfurt School's pessimistic view of science and technology in modern society (see Jay 1972; Habermas 1970, 1971, 1973; Horkheimer 1972) and lead him to recommend political action. For Goldstein and Noyelle (1977) or Scott and Roweis (1977), a theory of planning should be concerned with explaining the origin and evolution of planning by replacing it in the larger context of the historical evolution and transformation of the capitalist mode of production. These authors base their views on a Marxist social theory: urban planning is one particular aspect of the intervention of the State in the functioning of the capitalist mode of production. As such, its role is either to help maintain the 'social and property relations' of the capitalist system (see Scott and Roweis 1977) or to help avoid crises in the accumulation of capital by maintaining effective demand (see Goldstein and Noyelle 1977). Goldstein and Noyelle's views are based on Baran and Sweezy's Keynesian interpretation of Marx (see Baran and Sweezy 1966). This group of planning theorists considers the decision-making tradition either as vacuous (see Scott and Roweis 1977) or apologetic (see Goldstein and Noyelle 1977).

Necessity of an epistemological viewpoint to assess alternative theories of planning

As expressed by the title of a guide to the literature on planning theory (see Friedmann and Hudson 1974) the focus of attention of the renewed interest in planning theory is on the relationship between knowledge and action, and more particularly on the relationship between social knowledge and social action. And indeed, a major concern of a theory of planning should be the link between knowledge and action. At this level of generality this view can encompass and indeed has encompassed widely divergent conceptions of planning theory: as we have seen, the general view is that planning theory is concerned with what *ought* to ·be the link between knowledge and action. However one cannot solve that problem without first finding what *is* in general the link between knowledge and action, i.e. what knowledge is and what the role of action in its constitution is. This is the proper concern of epistemology. Therefore a theory of planning is dependent on a conception of epistemology, and indeed this paper will attempt to answer questions about epistemology in general before answering questions about planning theory.

Some authors are quite explicit about their epistemological assumptions. For instance, in their classical article, Davidoff and Reiner (1962) make an explicit reference to logical-positivist philosophers like Ayer and Carnap and to their belief that facts and values are distinct, and Goldstein (1975) to the views of the Frankfurt School on the role played by the sciences in society. Faludi (1978b) makes an explicit reference to, and borrows from, Karl Popper (1972) and his ideas on the methodology of science. For some other authors only implicit assumptions of an epistemological nature are made. For instance the 'new humanists' make use of concepts like *mutual learning* and *personal knowledge* and deemphasize the scientific methods of deduction and controlled experimentation. This raises the problem of the existence of a mode of knowledge distinct from scientific knowledge: this is indeed postulated by parascientific epistemologies such as Bergson's intuitionism or Husserl's phenomenology but there is no explicit reference to such philosophers in Friedmann's book (1973).

This author agrees with Faludi's statement (see Faludi 1978b) that planning theorists should make known their philosophy of science. Different epistemologies can have an impact on planning in different ways. Firstly, there exist today different theories of urban form or of urban growth. These theories can have a *direct* impact on policy (for instance, if they state that the private market allocates land efficiently among alternative uses, this suggests *no* planning as the optimal policy) or an *indirect* one by being embedded in predictive models and used for predicting the consequences of different measures (rent control, housing allowances, housing subsidies, etc.). Policies based on theories of urban form will be assessed differently if one believes in the neutrality of social science or if one believes that ideologies and values influence theories directly or indirectly. Secondly, different epistemologies may lead to different 'procedural' theories of planning. For instance, if, following the phenomenological tradition, one believes that the human sciences should not use the same analytical methods as the natural sciences, one is led in particular to the complete suppression of the use of mathematical techniques in planning, and to the rejection of the so-called 'rational paradigm' of the planning process.

Why Piaget?

The epistemological point of view chosen here will be based on Jean Piaget's genetic epistemology. Several reasons justify an interest in Jean Piaget's epistemological ideas. Firstly, his merit has been to show that epistemology could be an activity subject to the same kinds of mutual deductive and experimental controls that exist in the sciences in general. Traditionally epistemology was considered part of philosophy: it was a matter of philosophical taste if, for instance, one gave mathematics a Platonician, empiricist or Kantian interpretation. If we follow Piaget, it is no longer a matter of taste but an object of scientific investigation. Not surprisingly, as we will see in the next section, the answer obtained by Piaget's investigation to the question of the origin of mathematics is none of the traditional interpretations. One can understand Piaget's new approach to epistemology as a consequence of his first training both as a biologist and as a philosopher. Secondly, an interest in Piaget is justified by the fact that his epistemological views are largely supported by some works in the history of the sciences (see Kuhn 1962; Dobb 1973). Thirdly, his findings are not proved but at least made plausible by some works done in the field of artificial intelligence, in particular in the Artificial Intelligence Laboratory at the MIT (see Winston 1970; Winograd 1971; Hewitt 1971; Minsky and Papert 1972): his ideas on the development of knowledge are plausible in the sense that they can be 'implemented' in computer programs. To anticipate, for Piaget, knowledge is embodied in *schemes,* i.e. the patterns of action an individual has in his interaction with the external world, and the concept of *scheme* is analogous to the concept of *procedure* in computer science. Therefore, there is a natural transition between a structuralist approach to the psychology of cognitive processes such as Piaget's and artificial intelligence as conceived by the researchers of the MIT team. This is acknowledged by Minsky and Papert themselves (1972) in one of their progress reports. Fourthly, Piaget's views are consistent with the dialectical tradition in the social sciences in general (see Goldmann 1966, 1967, 1970). And, last but not least, his genetic epistemology appears to reconcile and synthesize in many ways several of the major traditional philosophies of science, such as Kantism, Empiricism and Hegel's dialectics, although it is not to be confused with any of them.

Nevertheless, one has to admit that no epistemology can be shown to be absolutely 'true', Piaget's epistemology included, if only for the simple reason that criteria for truth in epistemology remain to be defined and it is doubtful whether a consensus on such criteria could ever be reached. It would be consistent with the point of view adopted in this paper to say that one can only approximate 'truth' rather than obtain it, and that many such approximations are *a priori* plausible.

Organization of the paper

Some questions to be dealt with in this paper will be concerned with epistemology and theory *in* planning, others with the theory *of* planning, normative and explanatory.

The first set of questions is concerned with the nature of knowledge in general: what are the respective roles of the subject and of the object? Is there a clearcut distinction between knowledge and action? What role, if any, does language play in the constitution of knowledge? What is the role of experimental psychology in answering questions related to the nature of knowledge? A second set of questions has to do with the role of mathematics in causal explanations: what is the status of experimental knowledge and its relationship with logico-mathematical knowledge? This question itself raises the problem of the respective origins of logico-mathematical knowledge and of experimental knowledge. A third set of questions is concerned with the nature of the distinction that should be made between the social sciences and the natural sciences: can theories in the social sciences have the same 'objectivity' as in the physical sciences? Do the social sciences differ from the natural sciences by their methods? The phenomenologists for instance make a distinction between *comprehension* as a mode of knowledge valid for the *Geisteswissenschaften* and *explanation* as a mode of knowledge valid for the *Naturwissenschaften*. How valid is this distinction? Many modern theories of urban form are based on microeconomic theory. Is economic theory objective? Does ideology play an important role in economic theorizing as Marxist economists generally think? A fourth set of questions is concerned with the role that the sciences play in society: are they simply a means of societal control in the hands of the dominant classes of society, as suggested by the Frankfurt School or do they have some autonomy? A fifth set of questions is concerned with the status of the concept of rationality in planning theory. What kind of rationality is planning aiming at? Is it a search for optima? What is the status of the so-called 'rational paradigm' of the planning process? What is the role of techniques such as systems analysis, operation research, cost-benefit analysis? Can one delimitate a frontier between a theory of planning and a theory of design? A sixth set of questions is concerned with the nature of a theory of planning. What functions would a theory of planning serve? What approaches would it use? What should be the role of historical methods and of formalization in the development of a theory of planning?

This paper will make a few attempts at answering these questions and is organized as follows: the second section (pp. 68–74) presents Piaget's views on epistemology; after a definition of the nature and methods of a scientific epistemology, it presents traditional viewpoints and then that of Jean Piaget's genetic epistemology. The consequences of Piaget's findings for causality, logic, mathematics and experimental knowledge are presented in detail. The third section (pp. 75–8) is concerned with the implications of genetic epistemology for the social sciences in general and economics in particular. It will focus on the problem of the 'objectivity' of the social sciences. The fourth section (pp. 78–84) is concerned with planning theory. After stating the consequences of Piaget's viewpoint for planning theory, it discusses the concept of rationality, the respective roles of a theory of design and of a theory of planning, outlines the functions, nature and methods of a theory of design and of a theory of planning. The point of view taken in this paper is then compared with the views of other authors. The last section concludes the paper.

Piaget's views on epistemology[1]

Nature and methods of a scientific epistemology

The nature of epistemology is defined as follows (see Piaget 1967b): epistemology is the study of the constitution of valid knowledge. This definition refers to the conditions of accession to knowledge as well as to the conditions which are constitutive of valid knowledge and has several consequences:

1 It refers to valid knowledge, which is the area of logic, but also to the conditions of accession, which is a question of fact.
2 The conditions of validity or accession are not necessarily the same for different kinds of knowledge: e.g. mathematics and biology.
3 The word accession shows that knowledge is considered as a process and not as a state.
4 The constitutive conditions of knowledge refer to conditions of formal validity as well as to experimental validity.
5 In order to determine the respective roles of the object and of the subject in knowledge, it is necessary to know the conditions of accession to knowledge because the role of the subject is not apparent in a given 'achieved' state of knowledge.

Hence the importance given by Piaget to the historical and genetic methods in epistemology. These remarks lead to a second definition of epistemology, a genetic definition: epistemology is the study of the passage from states of lesser knowledge to states of greater knowledge.

From the preceding definitions of the object of epistemology, three conditions on the methods of epistemology can be derived.

1 Nothing valid can be said about the nature of the principles, methods or concepts at stake without knowing their actual use in the field concerned and without criticizing them directly within that field itself.
2 Any question of formal validity belongs to the realm of logic.
3 In addition to questions of formal validity, a large number of questions of fact arise concerning the role and the activities of the subject in the process of knowledge.

Three kinds of methods are associated to these three conditions:

1 Methods of direct analysis: these methods attempt to state by reflexive analysis, the conditions of knowledge at stake, when a new body of scientific theories arises, or when a crisis arises.
2 Methods of formal analysis: these methods add to the direct analysis of the processes of knowledge, a critical examination of the conditions of their formalization, and also of the coordination between this formalization and experience.
3 Genetic methods: they attempt to understand the process of scientific knowledge, as a function of its development (or genesis). There are two kinds of genetic methods: the methods concerned with the sociogenesis of knowledge (or historical development among human societies) and its cultural transmission; and the methods concerned with the psychogenesis of the concepts and operative structures formed during the development of individuals. The methods of the first kind are historical methods: they attempt to study the origin and development of a present body of knowledge. The methods of the second kind are the methods of genetic epistemology: by a combination of psycho-genetic analysis and formalization of structures they attempt to uncover the psychological conditions of the formation of knowledge and to coordinate these results with the study of the conditions of formalization.

Epistemological traditions – the problems that they raise

Following Piaget, we will distinguish three categories of theories of knowledge:

1 The *metascientific* theories which start from a reflexion on the sciences of their time and extend this reflexion into a general theory of knowledge.
2 The *parascientific* theories which are based on a criticism of the sciences and attempt to reach a mode of knowledge distinct from scientific knowledge (or in opposition to scientific knowledge).
3 The *scientific* theories which remain within a reflexion on the sciences.

Metascientific theories of knowledge were elaborated by Plato, Aristoteles, Descartes, Leibniz, Kant, Hume and Hegel, among others. We will discuss the epistemologies of the last three philosophers because of their importance for modern thought. Kant's theory was based on a reflexion on Newtonian physics. His main problem was the origin of the harmony between the deductive instruments of mathematics and the data of experience. His answer was that the subject possesses *a priori* synthetic structures (space, time, causality . . .) in addition to analytic structures forming the logico-mathematical apparatus. These *a priori* structures impose to perceptions and experiments in general a structure compatible with mathematical deductions. They are pregiven in some sense which raises the psychological question of the nature of these 'pregiven' structures. An example of an *a priori* structure in modern science is Chomsky's concept of linguistic competence. For the empiricist theory of knowledge, all knowledge is derived exclusively from experience, sensations (Condillac) or repeated associations (Hume). It is the implicit epistemology of common sense and therefore the epistemology of most planners or planning theorists. This view of knowledge, which assumes the existence of pure facts, presupposed results of an experimental psychology that was yet to come. Hegel's dialectics was based on a reflexion on history. For him, knowledge proceeds by thesis, antithesis and synthesis. His philosophy is at the origin of Karl Marx's own philosophy of science and in fact anticipates Piaget's own findings on the development of cognitive processes. *Parascientific* theories such as Bergson's intuitionism and Husserl's phenomenology were elaborated as a reaction against materialist and positivist metaphysics and attempted to delimit clearly a domain for scientific knowledge and a domain for another form of knowledge (intuition). The central idea of phenomenology is the following: there is an intuition of *'essences'*. These *'essences'* are inseparable from the *'phenomena'* or *'facts'* and we go from the fact to the essence by a process of *reduction,* consisting in freeing ourselves from our natural limitations. Phenomenology attempts also to duplicate experimental psychology by a phenomenological psychology that would be founded on 'intentionality'. There is a radical separation between *Geisteswissenschaften* and *Naturwissenschaften: comprehension* is adapted to mental phenomena while *explanation* is relevant to the natural sciences. *Scientific* epistemologies attempt only to explain scientific knowledge and not knowledge in general, either because they consider scientific knowledge as the only possible form of knowledge or because they consider it their duty to interpret it for itself. The fundamental fact which characterizes these epistemologies is that they were born because of crises and not because science was considered achieved. Two sources (among others) can be assigned to the scientific epistemologies: firstly, the scientists themselves have more and more been concerned by epistemological problems, not for the pleasure of philosophizing, but to overtake the crises occuring in science, when these crises are not simply the result of divergences on experimental results, but when the very concepts or principles used to obtain the results are challenged. Secondly, given the continuous transformation of the sciences, some epistemologies have attempted to fix once and for all the principles of science in a system of norms: this is the origin of positivism.

The epistemology internal to the sciences ceases to be simply a reflexion on science; it becomes an instrument of the progress of the sciences as an internal organization of the foundations of the sciences and it is elaborated by the very same people who will base their work on this reflexion. For example, Hilbert, Bernays and Ackermann wanted to demonstrate the non-contradiction of arithmetic and lay the foundations of geometry on a purely axiomatic basis, forbidding any recourse to intuition. The dream of Hilbert was to make this demonstration using only instruments belonging to arithmetic. This dream turned out to be unachievable, because of Gödel's theorem, proving, loosely speaking, 'the impossibility of proving the noncontradiction of a system by means of only the instruments belonging to that system or weaker means.'[2]

Because of its modern importance much attention will be devoted to positivism. The purpose of this epistemological approach is to stabilize the sciences, by delimiting frontiers between science and metaphysics and by fixing once and for all the principles and methods of the sciences. This was already the program of Auguste Comte. The neo-positivism of the Vienna Circle and of its direct and indirect disciples have made some progress since Comte. In particular, logic plays a dominant role in logical positivism. Since it is also the case with Piaget's genetic epistemology, and since logical positivism is still predominant in many countries, it is worth comparing it with Piaget's approach and examining in more detail its basic assumptions.

Logical positivism was born in Vienna from the convergence of two streams of thinking: firstly, Mach's empiricism according to which any experiment can be reduced to sensations; secondly, the logicist school, founded by Whitehead and Russell. Shlick, who founded the Vienna Circle in 1924, wanted to synthesize empiricism and logic: logico-mathematical structures are considered as a language, which is tautological and whose only function is to describe adequately the truths of experience. The program of this school of thought was to reach the unification of scientific knowledge and to find a method common to all sciences, which can provide guarantees against the accumulation of meaningless concepts or pseudo-problems. One of the most influential modern adherents to this school of thought is Popper. Two conditions must be satisfied for this program to be possible: firstly, the respect for the facts; secondly, the elaboration of an exact and common language. Logic is supposed to be that language. It has no other power than to combine signs (logical syntax) and meanings (logical semantics) according to certain rules (pragmatics). This reduction of thought to language tends to reduce to a minimum the role of the subject in the process of knowledge: his role is only to combine signs. Given these programs and methods the logical positivists solve the problem of the adequation of our logico-mathematical apparatus to the external world by two related answers: all logico-mathematical truths can be reduced to tautologies.[3] There is a clearcut distinction between *analytic* (or tautological) relations and *synthetic* relations, the latter concerning exclusively experimental knowledge. This means that any formalization of logical or mathematical truths will eventually reach 'tautological expressions', which by a suitable choice of axioms or inference rules, can be transformed into identities. Consequently, there is a radical separation or dualism, between experimental or empirical knowledge and logico-mathematical knowledge: on one hand is a language, which is purely tautological; on the other hand, are the facts, which cannot be reduced to the language and produce a synthetic knowledge depending on experience alone, and not at all on the logico-mathematical language. The language is used to describe the facts but does not play any role in their constitution. Without being concerned by a genetic analysis, logical positivism nevertheless assumes two psychogenetic facts without verification. (i) The origin of analytic (or logico-mathematical) relations is to be found in language. (ii) The origin of synthetic (or experimental) relations is to be found in perception. These assumptions should not rely on a formal analysis but on questions of fact: they call for a genetic analysis. And in fact an essential result of the genetic analysis of logico-mathematical structures (see Beth and Piaget 1961) is that these structures have roots which appeared before language; they have their origin in the coordination of actions. These actions are the starting point of what will become operations of thought. On the other hand, genetic analysis confirms the distinction between logico-mathematical knowledge and experimental knowledge, but it attributes to this opposition a very different sense, as will be discussed further below. Concerning perception, a result of genetic analysis shows that at all levels of development, perception is linked with action: if it is true that there exists a 'pure' logico-mathematical knowledge, there is no 'pure' physical or experimental knowledge, completely independent of logico-mathematical knowledge, however elementary. As a result, the duality analytic knowledge/synthetic knowledge is untenable.

The logical positivists have had the merit of insisting on formal analysis as an important tool to epistemological research, but they have transformed the method into a doctrine, or even a dogma, the dogma that science can be 'closed'. This is a very important drawback of the logical positivist school of thought and it comes from the fact that it has taken a synchronic point of view on scientific knowledge: science is a state instead of a process, and therefore the role of the subject disappears. The perspective changes as soon as the question of the origin and development of the sciences is raised. To the essentially static methods of logical positivism, a *dialectical point of view* must be substituted, in order to reconcile the genesis of structures and the possibility of their formalization at each equilibrium stage and in order to understand, at each stage of cognitive development, the respective roles of the object and of the subject.

Jean Piaget's genetic epistemology and its implications for logico-mathematical knowledge, experimental knowledge and causality

As mentioned in the preceding paragraph, all theories of knowledge whatever their origin, make assumptions of a psychological nature which cannot be accepted without further experimental verification, i.e. without recourse to psychology. This idea was at the origin of Piaget's genetic epistemology.

The main hypothesis of Piaget's biological theory of knowledge is that the development of knowledge prolongs without discontinuity the functioning of living organization (see Piaget 1967e): 'Cognitive mechanisms are an extension of the organic regulations from which they are derived. These mechanisms constitute specialized and differentiated organs of such regulations in their interactions with the external world.' Knowledge 'adapts' to the external world through the twin processes of *assimilation* and *accommodation.*

In order to understand these two processes one has to understand first the concept-of-action *scheme.* By definition it is what can be transferred, generalized or differentiated from one situation to the next in an action, i.e. what is common to several repetitions or applications of the same action. Analogous concepts are the concept of *procedure* in computer science and the concept of *algorithm* in mathematical programming.[4] Any knowledge of an object implies an action on this object which incorporates this object-to-action schemes of the subject (assimilation) and which can transform the action schemes themselves (accommodation).[5]

Knowledge therefore is viewed as a dialectial process of interaction between the subject and the object, in which the structures of the subject are constructed progressively. There are successive equilibria corresponding to the integration of the structures at level n into the structures at level (n + 1) so that part of a structure at level (n + 1) is isomorphic to an entire previous structure at level n. These successive equilibria are the result of a process of self-regulation. This view of the development of knowledge in children carries over to the evolution of scientific knowledge in the natural sciences, also characterized by a progressive differentiation and reconstruction of theories with integration of the previous theories in the subsequent ones: one well-known example of this process is the integration of Newtonian physics into Einsteinian physics. Piaget's ideas on the evolution of the natural sciences are different from Kuhn's (1962, 1970a, 1970b, 1974) for whom the succession of scientific paradigms does not appear to represent a clear sense of progress. However Piaget's conception of individual learning and of the development of knowledge and Kuhn's conception of the development of the sciences have in common their insistence on discontinuities (Kuhn's scientific revolutions and Piaget's stages of intelligence).

These hypotheses about human intelligence have been confirmed experimentally by Piaget and his collaborators (see Furth 1969; Piaget 1972, 1978) and we will present below the results of their research concerning the nature of logic, mathematics and causality. However, several answers can be made at this point concerning the distinction between the subject and the object in the development of knowledge as well as the role of action in the constitution of knowledge: no clear-cut distinction exists between the subject and the object of knowledge because no knowledge occurs without actions of the subject on the object and neither empiricism nor a priorism represent valid theories of knowledge.

The origin and nature of logic and mathematics is fundamental for an understanding of the nature of scientific knowledge because logic is a formalization of the rules of deduction and because the external world seems to obey mathematical laws, with the consequence that this harmony has to be explained. The main result of the studies done by Piaget is that there is a real continuity between the way logico-mathematical structures are constructed by children and the way mathematicians and logicians construct new structures. He calls this common process *reflective abstraction*. The latter is characterized by a progressive dissociation of form from content, for instance the construction of operations on operations. An example of such a process in mathematics is the passage from the integer numbers to the rational numbers. The first structures constructed by the child find their origin in the laws of the general coordination of actions. They are discovered by means of logico-mathematical experiments, utilizing concrete objects in the early childhood; but soon these concrete objects are replaced by abstract objects. The difference between form and content in logico-mathematical experiments is relative to the stage of development. A form for the structure at level n, becomes a content for the structure at level (n + 1). A logico-mathematical experiment proceeds by abstraction from the actions on the objects and not by abstraction from the objects themselves as in physical experiments.[6]

Successive logico-mathematical structures are, strictly speaking, neither invented nor discovered but constructed: given a structure, many possible new structures can be constructed by reflective abstraction from this structure but it takes a mathematician's work to actualize some of them. This constructivist interpretation of logico-mathematical knowledge leads to the following questions: where does it start and where does it end? The answer to the first question is that the origin of logico-mathematical structures is to be found in the very functioning of living organization: since there is no living organization without a fundamental interaction with the physical environment, this in turn explains the harmony between mathematics and the physical environment (see Piaget 1967e). The answer to the second question is that there is no end: the theorems on the limits of formalization (Gödel's theorem in particular) show that even logic cannot be fixed once and for all (see Beth and Piaget 1961) but is in a process of construction and therefore that there is nothing absolute in the norms of logic.

Concerning causality, Piaget's research shows that it consists in attributing mathematical structures to the objects. Since these mathematical structures are themselves in continuous construction there is no absolute objective truth about the external world, even in the natural sciences, except in a limit sense.

Several conclusions can be drawn from these results. First, logico-mathematical knowledge is not a pure language but is genetically anterior to language and is in a process of continuous construction. Second, any experimental knowledge has to be interpreted within logico-mathematical structures. Thus, although there may exist a pure logic and pure mathematics, there cannot be a pure experimental science.

Implications for the social sciences, for theory *in* planning

Implications for the social sciences in general

Firstly, Piaget's view of the process of knowledge is that of a dialectical process of interaction between the subject and the object. This epistemological view, as noticed by Goldmann (1966, 1967, 1970) is similar to Marx's philosophy of science by its emphasis on human action: in Marxist terminology, social knowledge depends on social practice. Secondly, Piaget's structuralist epistemology emphasizes the importance of formalization in causal explanations (see Piaget 1967a,f). The social sciences do not differ from the natural sciences by their methods as was thought by the phenomenologists. In fact the physical sciences have borrowed methods from the social sciences as much as the social sciences have borrowed methods from the physical sciences (see Piaget 1967f): information theory provides an example of a formal theory which found applications in the two domains. Thirdly, genetic epistemology demonstrates the relative autonomy of the development of the sciences, contrary to the Frankfurt School's gloomy view of science. This autonomy is not contradictory to the dependence on social practice: social practice has an influence on the development of the social sciences but is not its unique determinant.

The last two implications of Piaget's views, i.e. the role of formalization and the autonomy of the sciences, are shared with the logical positivist school. However Piaget's dialectical conception of social knowledge implies that there need not be *one* social science as there is *one* physical science. Human actions at the origin of the construction of causal structures in physics have their roots in the biological constitution of the human species and therefore an advanced degree of objectivity can be reached in physics: it can be the same in different societies or different cultures. On the other hand, social practice varies with social groups, societies, cultures, and, since the subject of the social sciences is simultaneously their object, the same degree of objectivity cannot be reached. The 'decentering' process, i.e. the effort towards finding laws or structures by mutually controlled deduction and experimentation, is more difficult for the social scientist than for the natural scientist: values and ideologies play a greater role and they vary in particular with the social class or the culture of the social scientist. This is far from implying that *some* degree of objectivity cannot be reached through mutually controlled deduction and experimentation. This means however that the kind of objectivity that one can reach in the social sciences is *conditional* on a certain world-view, itself in dialectical interaction with a social practice. We may conclude from there that one may have to accept the possibility of several competing paradigms in the social sciences.

The case of economics and of urban economics in particular

The history of economic thinking provides an excellent example of competing paradigms dependent on social practice. This in itself would justify devoting some attention to economics, but another reason is that economics as a social science has more impact on policy decisions in the most diverse areas of society than any other social science. In particular urban economics, as a subfield within the general field of economics, may have had some impact on policies or the absence of policies in urban planning, directly, or indirectly by the conception of the city which it provides.

The possibility of competing paradigms in economics is well documented by the history of theories of value and distribution (see Dobb 1973; Benetti 1975; Benetti, Berthomieu and Cartelier 1975). For the 'neo-classical' school with its culmination in the Walrasian concept of a general equilibrium, the remuneration of the factors of production, which include labour and capital, is determined in simultaneity with prices in general. The general equilibrium is an optimal and harmonious social state resulting from the attempt by each individual in society to reach his best interest: the Walrasian equilibrium (especially in its modern formalizations) is an aesthetically appealing way of expressing Adam Smith's 'invisible hand' idea. Some modern criticisms of the Walrasian theory, and in particular critical assessments of the logical necessity of the existence of an 'auctioneer', come either from the neo-Keynesian disequilibrium school (see Benassy 1976) or from the Marxist school (see Fradin 1976, 1977). For the 'classical' school represented by David Ricardo (1970) and especially its modern disciples, such as Piero Sraffa (1960), the distribution of income between wages and profits is logically prior to the determination of prices. For the neo-classical school as well as for the classical one only *relative* prices have meaning. Marx, however, defended a concept of *absolute* value of commodities and explained that profits originate in surplus value, the difference between the value of commodities and the value of the labor necessary to produce them. His theory differs from Ricardo's theory and from the neo-classical theory but it is fair to say that there is still an intense debate among Marxist economists about which interpretation of Marx's theory of value is the correct one.[7]

The importance of social practice in the shaping of these different theories is evident. Maurice Dobb shows very clearly the importance that the debate on free trade and the conflict between aristocrats and capitalists had on Ricardo's theory of rent, and more generally on his theory of income distribution.[8] Maurice Dobb (1973:37) aptly summarizes the link between theory and practice in economics in a way that is very close to Piaget's assimilation and accommodation concepts:

> 'New concepts and theorems have to be envisaged simultaneously as being fashioned in response to (and hence patterned upon) older ones as critical assessment of their adequacy to fulfill the role for which they have been cast – and as a reflection of changing human experience and of the problems and conflicts involved in human social activity that is itself motivated by the use of abstract notions applied to human beings in general, to their artifacts and to things.'

The origin of modern urban economics (see Richardson 1977) is Von Thünen's (1826) agricultural land use theory. His concept of land rent, expressing savings in transport costs, as an allocator of land to the most efficient usage is the direct ancestor of the modern concept of 'bid rent' developed by Alonso (1964) in an urban context, on the basis of modern microeconomic theory. Most works following Alonso attempted to apply Walrasian general equilibrium theory to explain urban form: in all these theories land rent is conceived as an index of scarcity allocating urban land between alternative uses. The question is of course raised of the social optimality of the urban form resulting from the market land allocation. Since the existence of 'externalities' in cities is widely recognized, many of the modern theories attempt in an ad hoc and piecemeal fashion to provide guidance towards the correction of 'market imperfections': for instance by determining optimal congestion taxes one would try to transform a market equilibrium which is not socially optimal into a socially optimal equilibrium; one would thereby 'solve' the social problems resulting from traffic congestion. Theoretical works in urban economics trying to explain urban form or some aspects of it, such as the functioning of the housing market, within a different paradigm than the general equilibrium paradigm, are extremely rare.

The dialectical interaction between theory and practice is quite clear in urban economics: one will not have the same land use policy if one believes that the price of land basically fulfills its role of allocating land optimally among alternative uses (granted that some taxes or subsidies have to be created to take into account externalities such as traffic congestion, air pollution, etc.) or if one does not believe so (see Roweis and Scott 1977).

Similarly, the interaction between theory and practice appears obvious if we consider again the theory of income distribution. A national incomes policy is not going to be the same if one believes that the market pays labor at its 'just' price, as the neo-classical school thinks, or if one believes that income distribution (the distribution between profits and wages as a first approximation) is exogenous to the functioning of the economic system and is the result of a conflict between 'capitalists' and 'unions' possibly arbitrated by the State, as is believed by the neo-Ricardian school (the so-called 'Cambridge School').

A value-free social science is impossible

In contrast to the positivist ideal, we have to accept the possibility of different competing paradigms in economics, sociology and all the other social sciences. Max Weber's ideal of a value-free social science is impossible. This state of affairs may be unpalatable for somebody who, following the logical positivist ideal, would like to believe in the Unity of Science, but is unavoidable in a dialectical epistemology.

Some consequences on urban planning are worth mentioning at this point. A theory can guide policy *directly* and the practical impact will differ with the theory chosen. In addition to inspiring policy directly, different economic theories can have an *indirect* effect on policy through the use during the planning process of predictive models in which they are embedded: the predictions and therefore the policies based on those predictions may differ with the underlying theories. One will not hold the same opinion of the proposed policies if one believes in the 'objectivity' of social science or if one believes that the policies are *conditional* on a certain paradigm.

Implications for the theory *of* planning

Implications of Piaget's epistemology for theories **of** *planning*

Given the dialectical interaction between social knowledge and social action, the distinction between theory *in* planning and theory *of* planning has to be qualified. As indicated in the previous section, social practice, values and ideologies play an important role in the development of theories *in* planning. A fortiori, values and ideologies play a decisive role in theories *of* planning concerned with social action as such.

Differences in epistemological viewpoints have an impact either on the relative emphasis on procedural and on positive theories of planning, or on the very conception of a methodology of planning and of an explanatory theory of planning.

A positivist viewpoint will tend to emphasize a normative theory of planning because it tends to accept the objectivity of facts, problems, models and theories: problems tend to be viewed as technical. An 'empirical' theory of planning is important only insofar as it can improve day-to-day planning practice and make it closer to the ideal of rational planning. For Faludi, for instance, 'normative theory is concerned with how planners ought to proceed rationally' while 'behavioral approaches focus more on the limitations which they are up against in trying to fulfill their programme of rational action' (see Faludi 1973a). Thus positive theory is, at worst anecdotal, at best remedial or even put forward to show how limited in its effects urban planning really is (see Meyerson and Banfield 1955). On the other hand, for a dialectical point of view, a positive theory is fundamental because it explains the origin, context and therefore the *possibilities* of planning and does not abstract from social theory in general, as if urban planning could be studied in isolation from the study of the functioning of society at large.[9]

Different epistemological views will lead to different conceptions of a 'normative theory'. For Friedmann (1973) the belief in a mode of knowledge distinct from the deductive and experimental methods of the sciences in general, entails a distrust for formal methods or theories and a reliance on the virtues of communication. For the 'Critical Theory' of Goldstein (1975) influenced by the Frankfurt School and its belief that science is not autonomous but a tool in the hands of the ruling class, all the methods of operations research, systems analysis, econometrics, etc. are harmful, and planners should rely on political action. On the other hand, for a logical positivist viewpoint as well as for a dialectical viewpoint based on Piaget's genetic epistemology, scientific approaches are an indispensable part of any planning methodology.

Wrong epistemological views can have adverse effects on planning. The positivist view leads to the artificial separation between values and facts and therefore between Politics and Science, in the tradition of Max Weber. It leads to the exclusive emphasis on methodology and to the belief in the possibility of 'absolute objectivity' if one could know 'all the facts'. Planning theory tends to be considered as reducing to a theory of rational decision-making. The intuitionism of the 'new humanism' (see Friedmann 1973) can lead to subjectivism and irrationality and to almost any kind of metaphysics, religious or other.[10] The emphasis of 'Critical Theory' on the use of the scientific method for increased societal control of the people and the maintenance of the existing social order is quite healthy after a long period of too much hope (especially in the USA) in the tools of systems analysis and operation research for curing all social ills. However, the case is pushed too far: political action without some theory and some methodology may produce counterproductive effects.

This author's views on positive planning theory are close to those of Goldstein and Noyelle (1977) or Scott and Roweis (1977). However, for reasons that will be developed in the next paragraph, these authors tend to neglect the importance of a 'normative theory' of planning. This could lead practicing planners to the belief that they have no degrees of freedom in their profession.

Rationality, planning theory and theory of design

The preceding discussion leads to a more thorough examination of the concept of rationality, in particular in the so-called 'rational paradigm' of the planning process, which can be depicted as follows:

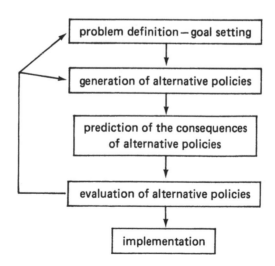

Several types of mathematical models are proposed in this paradigm: design models for the 'generation of alternatives' phase, predictive models for the 'prediction of consequences' phase, evaluation techniques for the 'evaluation' phase. Optimization techniques such as linear programming can be used as design models, simulation models or econometric techniques can be used for prediction and cost-benefit analysis or multicriteria analysis can be used as evaluation techniques. The problem definition and goal-setting phase is usually viewed as non-technical, and sometimes outside the professional role of the planner.

This paradigm conceives of the planning process as a general decision-making procedure independent of the particular subject matter, whether it is urban planning, corporate planning, transportation planning, educational planning, etc. It has a long history (see Simon 1969; Davidoff and Reiner 1962; Harris 1967a,b, 1968). Britton Harris's recent book (1978) constitutes a brilliant exposition of this paradigm.

This paradigm has been widely criticized. Most of the criticisms are directed implicitly at the notion that, once the problem and the goals have been stated, some form of absolute objectivity can be reached in the form of an optimal solution to the planning problem. This would imply a positivist epistemology and it will be argued in the rest of this section that the major limits of the 'rational paradigm' are not due to the paradigm as such but to the 'common sense' epistemology of most people supporting it, using it or even criticizing it.

The criticisms to the 'rational paradigm' seem to fall into four categories. The first category of critics assumes that a mode of knowledge such as intuition, distinct from scientific knowledge, can allow us to solve planning problems by some form of dialogue between individual intuitions leading through some magic mutual learning to a solution to the problem. For them we can dispense with the paradigm altogether. We will not dwell on this point of view any further. The second category of critics attacks the universality (or generality) of the paradigm although for diverse reasons. A third category of critics attacks the assumption of perfect information about the goals and the consequences and the assumption of exhaustive search among alternatives. This category of criticisms leads to the idea that we can reach only a 'bounded rationality': instead of finding the optimal solution to a problem, we can only *satisfice* and obtain a good, if not optimal solution (see March and Simon 1958). A fourth category of critics attacks the 'rational paradigm' on the ground that goals, supposed to be somehow the outcome of the political process, are taken as given in this view of the planning process: some think that the planners should be concerned more with the definition of goals than with the other parts of the process; some propose a cyclical planning process whereby the goals would be revised after each iteration by some institutional arrangement involving the experts and some form of public participation. Others reject the paradigm altogether in favor of pure public participation. We will examine in turn each of the last three categories of criticisms.

The universality of the paradigm is criticized by Goldstein and Noyelle (1977) who are concerned with Harris's views and Scott and Roweis (1977) who are addressing their criticisms to the paradigm in general. For Goldstein and Noyelle, 'having abstracted his model from social and historical practice, Harris implicitly presents the kind of economic rationality which governs capitalism and capitalist planning as a universal, structural basis of the organization of every society'. For Scott and Roweis,

'mainstream planning theory engages in a form of generalization that we might designate as indeterminate *abstraction* and given this proclivity, mainstream theory appears not so much to be incorrect as it is simply trivially true. . . . As a result of its indeterminate abstraction, immunizing it against refutability, mainstream theory tends to impede scrutiny of the real world of urban planning, and consequently, to license false consciousness and cognitive myopia.'

These criticisms are not valid: they fail to distinguish between normative and positive theories of planning. This is clear in the two quotations above. Goldstein and Noyelle's argument does not resist close examination: it is hard to see how a normative theory in *itself* can be apologetic of anything; it is only by *omission,* i.e. if no other planning theory was taught for instance, that it could become *indirectly* apologetic by its exclusivity. Similarly, the allusion to *refutability,* borrowed from Popper, makes clear that Scott and Roweis criticize the 'rational paradigm' as if it were a positive theory, which it is not. It seems to me that Goldstein and Noyelle's or Scott and Roweis's arguments are misplaced. What these authors are criticizing implicitly is the positivist epistemology of most people proposing or using this paradigm, and the related belief in the 'objectivity' or 'rationality' of the solutions obtained. They are also criticizing implicitly the *exclusive* emphasis on methodology at the expense of a positive theory that would try to explain the origin of planning institutions and of planning problems as they are perceived by planners. The origin of the emphasis on methodology is the belief that the major failures of urban planning today are failures in thought processes and in methodology rather than the natural results of economic and social forces in society at large. The first belief appears to be that of Britton Harris (1975, 1980) while the second is shared by Roweis and Scott (1977). To summarize, these critics fail to differentiate the rational paradigm as such from a 'package' of which it seems to be a part: this package would contain the rational paradigm, the idea that failures of planning are essentially failures of methodology, and a positivist epistemology. In fact one can abandon the last two items of this 'package' without abandoning the first.

The last two categories of criticisms on the other hand seem valid, as well as the accompanying ideas of 'bounded rationality' and 'cyclical planning process', but again, they can be viewed in different lights depending on one's epistemology. In order to discuss this more thoroughly, we will follow Britton Harris (1975, 1980) and use mathematical programming as a paradigm for the rational planning process. Within this terminology the fourth class of critics says that the objective function and the constraints come from outside the planning process and the third class of critics says in effect that the true dimensionality of the problem-space is not known and that in general there is no exhaustive search procedure. However, most of the latter critics tend to think that if, somehow, one could have all the relevant dimensions of the problem-space and could search the latter exhaustively, one could reach the optimal solution, given the objective function and the constraints: the planning problem is viewed as purely technical. This assumes implicitly an empiricist epistemology, in the sense that the dimensions are seen as non-problematical, the only problem being to be aware of all of them. For a dialectical epistemology however, not only the objective function and the constraints are problematical, but the problem-space as well, in the sense that it has no objectivity: different theories of urban form would lead to different ways of structuring the problem or to not stating it at all. In this view, design models and predictive models are not only mutually dependent, but paradigm-dependent. The problem-space and the underlying theories depend on the world-view and the values of the expert or of the social group to which he belongs.

To sum up, we can only reach a 'bounded rationality' within a 'conditional rationality': within a certain world-view (or paradigm), and the resulting planning problem, i.e. once the problem-space, the objective function and the constraints are stated, the 'rational paradigm' keeps its validity: the problem must still be solved. For instance, once it has been decided that a new town should be built, with such and such objectives in mind and assuming such and such theories of human behavior, it may still not be a trivial problem to solve (combinatorial explosion, non-convexities, existence of many local optima, etc.). This question, however, belongs to the theory of design rather than to the theory of planning.

Indeed the theory of design, also called the theory of problem-solving, has attracted the attention of researchers in many fields, for instance researchers in artificial intelligence such as Simon and Newell (see Simon 1969; Simon and Newell 1972). Recently Harris (1980) has thoroughly explored the implications of using mathematical programming as a paradigm for design and his book constitutes the basis for a valid theory of design in a social context. In addition to exploring some of the 'intrinsic' difficulties of planning such as the existence of local optima, he shows how certain styles of problem-solving are really 'truncated' problem-solving methods. This suggests ways in which planning styles could be improved. For instance, different professions adopt different styles of solving problems, according to their training and to their usual way of thinking: the focus on constraints constitutes the architects' approach which consists in setting 'design standards' towards which to strive; the focus on prediction and evaluation leads to the 'cost-benefit' tradition of planning, in which professionals with an economic training are prominent. In the first case tradeoffs between objectives are not permitted. In the second case the problem-space is not sufficiently explored.

Role and nature of a theory of planning

The preceding discussion makes explicit the difference of focus between a theory of planning and a theory of design. The former would be essentially an explanatory theory concerned with substantive issues while the latter would be dealing with methodological issues such as search procedures, predictive techniques or evaluation techniques.

A theory of planning would be a reflexion on the practice of planning. It would therefore be an *epistemology* of planning which would attempt to explain the manner in which planning is practiced, how 'problems' and institutions appear. Since knowledge and action are in dialectical interaction with each other, it would be concerned with explaining models and theories as well. It would be aimed, for instance, at uncovering the assumptions implicit in the way in which planning problems are stated and solved. It would replace planning within its general social context, i.e. it would relate it to the political, economic, philosophical or ideological spheres as well as to the state of the natural and social sciences. Like epistemology in general it would be concerned with questions of fact and questions of validity and would therefore have to use two kinds of methods: genetic methods and formalization.

For the genetic part, such a theory of planning would have to rely on a theory of ideologies and metaphysical systems and on the sociology of knowledge, as well as on political science and economic history. It would explain the evolution of practice and of different theories (or metaphors) of the city. In addition to using a genetic or historical approach,[11] planning theory would attempt to formalize the manner in which planning problems are formulated and solved in the present as well as in the past. The advantage of formalization would be to make explicit belief systems as well as the underlying assumptions of planning practice such as different theories (or metaphors) of the city.

Comparison with other points of view

This author believes in the joint necessity of a theory of design and of a theory of planning. Since as far as 'design' is concerned, this author's ideas are closest to those of Harris and since as far as 'planning theory' is concerned, they are closest to those of Scott and Roweis, Goldstein and Noyelle, it may be worthwhile at this point to state clearly the areas of agreement or of disagreement with these different authors. In addition, given the importance of Faludi's books in the planning field, this author's ideas will be contrasted with his.

Unlike Faludi, we do not believe in the 'generality' of a positive theory of planning. For Faludi (1973a),

'a positive theory of planning treats its subject-matter as a researchable phenomenon: it assumes that planning shows enough constancy to warrant the formulation, in the long run, of "general laws", which, ideally at least, would explain its recurring aspects and allow making predictions about planning. Such general laws of planning might also be transformed into "prescriptions" concerning improvements to the institutions and the style of planning based on planning knowledge.'

There are no such general laws. We should take a historical approach and consider that different societies generate planning institutions and theories differently. The correct concept would be that of a 'planning community' analogous to Kuhn's concept of a 'scientific community'.

On the other hand we agree with Scott and Roweis's (1977) following statement on the nature of a theory of planning:

> 'analytical derivation of the nature, content and trajectory of urban planning out of those relatively durable social phenomena in which urban planning is embedded as opposed to a purely inductive derivation of urban planning, out of the myriad short run circumstances that encompass observable planning events.'

We would essentially add to this definition the epistemology of theories, models and metaphors, as well as the study of ideas about the planning process itself.

As discussed in a previous paragraph, this author does not agree with the rejection of a 'normative theory' of planning. It is neither vacuous nor apologetic and Harris's book (1980) provides such a normative theory. However Harris claims too much scope for his theory of design. For instance he states (see Harris 1975) 'that the optimal design of processes and organizations is perhaps even more complex than the optimal design of policies, but the use of the paradigm can be fruitful in defining more accurately the nature of these difficulties.' It is only in a utopian and idealistic sense that institutions, organizations and processes are amenable to the same kind of 'optimizing' as, say, a transportation network: there may not be any 'subject' for deciding on the appearance of fundamental planning institutions and to whom Harris's paradigm could be useful. Goldstein and Noyelle (1977) distinguish between planning at level I, concerned with 'the reorganization and design of institutions with a more or less well defined function' and planning at level II which is the 'design, implementation and administration of programs within these basic institutional structures'. This distinction is quite valid: Harris's paradigm is essentially useful for planning at level II, not for planning at level I. Planning at level I cannot be 'planned' but only 'explained'. There are discontinuities in the 'facts of planning' which have to be recognized: only for specific problems, already well situated in time and space, within the context of existing institutions, is there a 'subject' that 'plans'. This fact does not prevent the theory of design from being as important as the theory of planning for the progress of the urban planning field.

Conclusion

In this paper we have presented the implications of Piaget's genetic epistemology for the relationship between knowledge and action. This laid the foundation of our belief that social science and planning cannot be completely objective and value-free and that a theory of design and a theory of planning are both extremely important. Put in their proper perspective, i.e. their *conditional* nature, concepts of optimality and rational decision-making have an important role to play in planning practice and research, as part of a theory of design. It is the role of planning theory to situate these concepts. By its self-reflexive function and the importance it attributes to uncovering hidden beliefs and assumptions as well as by the importance it attributes to social theory, planning theory should make the planner more aware of his social responsibilities. A proper *balance* between the development and teaching of planning theory and design theory is necessary for the progress of urban planning.

Notes

[1] This section was largely inspired by Piaget (1967a).

[2] In its second part, Gödel's theorem (which applies to formal systems which can be interpreted as arithmetic), states that, in these systems, *there exist undecidable propositions,* i.e. propositions which are neither derivable, nor refutable. (A proposition is derivable, when it can be derived from the axioms by the transformation rules; a proposition is refutable when its negation is derivable in the system.) (See Delong 1970.)

[3] The logical positivists use the word tautology in two meanings: identities (dogs are dogs); judgements considering identically all the possibilities (either x is a dog or x is not a dog).

[4] Indeed many recent works in artificial intelligence are a proof of the fruitfulness of the analogy.

[5] To pursue the analogy with mathematical programming, the process of accommodation means that the algorithm itself can be modified whenever it is applied to a different problem.

[6] This distinction is explained by Piaget himself (see Beth and Piaget 1961): 'When a child discovers that a big rock is heavier than a small one, it is a physical experiment, because . . . the subject discovers a property which was in the rocks before he weighed them: it is an abstraction from the objects themselves. On the other hand, when he puts five rocks in a row and finds out that the number five remains the same when counting from left to right or from right to left, he performs a logico-mathematical experiment because the experiment is not concerned with the objects as such but with the relationship between the action of ordering and the action of adding. The linear order was not in the rocks before action ordered them in a row. . . .'

[7] Gilles Dostaler (1978a,b) gives an excellent historical account of the debates among Marxist economists and others that took place at the end of the nineteenth century and at the beginning of the twentieth century: he shows that some of the theoretical discussions that took place among economists of diverse schools at the time anticipated many of the present debates, though in a less formalized way.

[8] David Ricardo was very much involved in the conflicts of his time: 'Although the doctrines of the classical school were very abstract, especially in the form

given to them by Ricardo, they were related very closely to practical issues of their day, indeed surprisingly closely'. 'A germ of his [Ricardo's] theories of value and distribution emerged in a topical pamphlet of February 1815 directed towards the debate of that same month in the House of Commons on the new Corn Law and towards establishing the theoretical case for free import of corn' (see Dobb 1973, Chapter I).

[9] For instance, Goldstein and Noyelle (1977) explain the origin of planning by means of Baran and Sweezy's 'surplus absorption' theory (see Baran and Sweezy 1966), while Scott and Roweis (1977) see urban planning as arising from the need to maintain the capitalist 'social and property relations'.

[10] It may not be completely by accident that this enthusiasm for the virtues of communication and dialogue occurred in California.

[11] Good examples of this approach are to be found in Goldstein and Noyelle (1977) and Wilson (1975).

References

Alonso, W. (1964) *Location and Land Use.* Cambridge, Mass.: Harvard University Press.

Apostel, L. (1967) 'Logique et dialectique', in Jean Piaget (ed.) (1967a).

Baran, P. and Sweezy, P. (1966) *Monopoly Capital.* New York: Monthly Review Press.

Benassy, J. P. (1976) 'Théorie du déséquilibre et fondements microéconomiques de la macro-économie.' *Revue Economique* 755–804.

Benetti, C. (1975) *Valeur et répartition.* Grenoble: Presses Universitaires de Grenoble/François Maspero.

Benetti, C., Berthomieu, C. and Cartelier, J. (1975) *Economie classique, économie vulgaire.* Grenoble: Presses Universitaires de Grenoble/François Maspero.

Beth, E. W. and Piaget, J. (1961) *Epistémologie mathématique et psychologie.* Paris: Presses Universitaires de France.

Boutaud, D. (1978) 'Quelques réflexions critiques sur les théories de la rente foncière urbaine.' Université de Montréal: Publication # 89, Centre de recherche sur les transports.

Davidoff, P. (1965) 'Advocacy and pluralism in planning.' *Journal of the American Institute of Planners* 31, 4:331–8.

Davidoff, P. and Reiner, T. A. (1962) 'A choice theory of planning.' *Journal of the American Institute of Planners* 28:102–15.

De Gaudemar, J. P. (1976) *Mobilité du travail et accumulation du capital.* Paris: François Maspero.

Delong, H. (1970) *A Profile of Mathematical Logic.* Reading, Mass.: Addison-Wesley.

Dobb, M. (1973) *Theories of Value and Distribution Since Adam Smith. Ideology and Economic Theory.* Cambridge: Cambridge University Press.

Dostaler, G. (1978a) *Valeur et prix – Histoire d'un débat.* Paris: François Maspero/Presses Universitaires de Grenoble/Montréal: Les Presses de l'Université du Québec.

—— (1978b) *Marx, la valeur et l'économie politique.* Paris: Editions M8-Anthropos.

Dumont, L. (1977) *Homo aequalis – Genèse et épanouissement de l'idéologie économique.* Paris: Editions Gallimard.

Faludi, A. (ed.) (1973a) *A Reader in Planning Theory.* Oxford: Pergamon.

—— (1973b) *Planning Theory.* Oxford: Pergamon.

—— (1978a) *Essays in Planning Theory and Education.* Oxford: Pergamon.

—— (1978b) 'Beyond planning theory.' 18th European Congress of the Regional Science Association, Fribourg, Switzerland, 29 August – 1 September 1978.

Fradin, J. (1976) *Les fondements logiques de la théorie néoclassique de l'échange.* Grenoble: Presses Universitaires de Grenoble/François Maspero.

—— (1977) 'Recherches sur l'analyse éconimique de l'ordre social.' Document de travail no. 8, Département de sciences économiques, Faculté de droit et des sciences économiques et politiques de Besançon, Besançon, France.

Friedmann, J. (1973) *Retracking America: A Theory of Transactive Planning.* Garden City, N.Y.: Doubleday.

Friedmann, J. and Hudson, B. (1974) 'Knowledge and action: a guide to planning theory.' *Journal of the American Institute of Planners* 40,1:2–16.

Furth, H. G. (1969) *Piaget and Knowledge – Theoretical Foundations.* New Jersey: Prentice-Hall.

Gale, S. (1975) 'On a metatheory of planning theory' in R. H. Wilson and T. Noyelle (eds) (1975).

—— (1977) 'Paradigms and counter-paradigms or "Does neatness count in planning theory?" ' in T. Noyelle (ed.) (1977).

Goldmann, L. (1966) *Sciences humaines et philosophie.* Paris: Gonthier.

—— (1967) 'Epistémologie de la sociologie' in J. Piaget (ed.) (1967a).

—— (1970) *Marxisme et sciences humaines.* Paris: Gallimard.

Goldstein, H. (1975) 'Towards a critical theory of planning' in R. H. Wilson and T. Noyelle (ed.) (1975).

Goldstein, H. and Noyelle, T. (1977) 'Planning and social practice: a theory of capitalist planning in the US' in T. Noyelle (ed.) (1977).

Grize, J. B. (1967) 'Historique – Logique des classes et des propositions – Logique des prédicats – Logiques modales' in J. Piaget (ed.) (1967a).

Habermas, J. (1970) *Towards a Rational Society.* Boston: Beacon Press.

—— (1971) *Knowledge and Human Interests.* Boston: Beacon Press.

—— (1973) *Theory and Practice.* Boston: Beacon Press.

—— (1975) *Legitimation Crisis.* Boston: Beacon Press.

Harris, B. (1966) 'The uses of theory in the simulation of urban phenomena.' *Journal of the American Institute of Planners* 32,5:258–73.

—— (1967a) 'The limits of science and humanism in planning.' *Journal of the American Institute of Planners* 33,5:324–35.

—— (1967b) 'The city of the future: the problem of optimal design.' *Papers and Proceedings of the Regional Science Association* 19:185–95

—— (1968) 'People, problems and plans: the purpose and nature of design.' *Transaction of the Bartlett Society* 7:9–53.

—— (1975) 'A fundamental paradigm for planning' in R. H. Wilson and T. Noyelle (ed.) (1975).

—— (1980) *A Paradigm for Planning* (in manuscript).

Harvey, D. (1973) *Social Justice and the City.* London: Edward Arnold.

Herbert, D. J. and Stevens, B. H. (1960) 'A model for the distribution of residential activity in urban areas.' *Journal of Regional Science* 2:21–36.

Hewitt, C. (1971) 'Procedural embedding of knowledge in PLANNER' in Proceedings of the 2nd Conference on Artificial Intelligence, British Computer Society, London.

Horkheimer, M. (1972) *Critical Theory.* New York: Herder and Herder.

Jay, M. (1972) *The Dialectical Imagination.* Boston, Mass.: Little, Brown, and Co.

Kuhn, T. S. (1962) *The Structure of Scientific Revolutions.* Chicago: University of Chicago Press.

—— (1970a) 'Logic of discovery or psychology of research' in I. Lakatos and A. Musgrave (ed). *Criticism and the Growth of Knowledge.* Cambridge: Cambridge University Press.

—— (1970b) 'Reflections on my critics' in I. Lakatos and A. Musgrave (ed.) *Criticism and the Growth of Knowledge.* Cambridge: Cambridge University Press.

—— (1974) 'Second thoughts on paradigms in F. Suppe (ed.) *The Structure of Scientific Theories.* Urbana, Ill.: University of Illinois Press.

Ladrière, J. (1967) 'Limites de la formalisation' in J. Piaget (ed.) (1967a).

Lipietz, A. (1974) *Le tribut foncier urbain.* Paris: François Maspero.

—— (1977) *Le capital et son espace.* Paris: François Maspero.

March, J. G. and Simon, H. A. (1958) *Organizations.* New York: John Wiley.

Meyerson, M. and Banfield, E. (1955) *Politics, Planning and the Public Interest.* Glencoe: The Free Press.

Minsky, M. and Papert, S. (1972) 'Artificial intelligence progress report.' Cambridge, Mass.: A. I. Memo no. 252, M.I.T.

Noyelle, T. (ed.) (1977) *1976 Symposium on Planning Theory: Papers in Planning #2.* Philadelphia: Department of City and Regional Planning, University of Pennsylvania.

Piaget, J. (ed.) (1967a) *Logique et connaissance scientifique.* Paris: Gallimard.

—— (1967b) 'Introduction et variétés de l'épistémologie' in J. Piaget (ed.) (1967a). Paris: Gallimard.

—— (1967c) 'Les méthodes de l'épistémologie' in J. Piaget (ed.) (1967a).

—— (1967d) 'Epistémologie de la logique' in J. Piaget (ed.) (1967a).

—— (1967e) *Biologie et connaissance.* Paris: Gallimard.

—— (1967f) 'Les deux problèmes principaux de l'épistémologie des sciences de l'homme' in J. Piaget (ed.) (1967a).

—— (1972) *L'épistémologie génétique.* Paris: Presses Universitaires de France.

—— (1978) *L'équilibration des structures cognitives – Problème central du développement.* Paris: Presses Universitaires de France.

Popper, K. R. (1972) *The Logic of Scientific Discovery.* London: Hutchinson.

Ricardo, D. (1970) *Principes de l'économie politique et de l'impôt.* Paris: Calmann-Lévy.

Richardson, H. W. (1977) *The New Urban Economics: and Alternatives.* London: Pion.

Roweis, S. T. and Scott, A. J. (1977) 'The urban land question' in K. Cox (ed.) *Urbanization Processes and Conflict in Market Societies.* Chicago: Maaroufa and London: Methuen.

Scott, A. J. (1976) 'Land use and commodity production.' *Regional Science and Urban Economics* 6, 2:147–60.

—— (1978) 'Urban transport and the economic surplus: notes toward a distributional theory' in A Karlqvist, L. Lundqvist, F. Snickars, and J. Weibull (eds) *Spatial Interaction Theory and Planning Models.* Amsterdam: North-Holland.

Scott, A. J. and Roweis, S. T. (1977) 'Urban planning in theory and practice: a reappraisal.' *Environment and Planning* 9:1097–119.

Simon, H. A. (1969) *The Sciences of the Artificial.* Cambridge, Mass.: M.I.T. Press.

Simon, H. A. and Newell, A. (1972) *Human Problem-Solving.* New Jersey: Prentice-Hall.

Sraffa, P. (1960) *Production of Commodities by Means of Commodities.* Cambridge: Cambridge University Press.

Suppe, F. (1974) 'The Search for philosophic understanding of scientific theories' in F. Suppe (ed.) *The Structure of Scientific Theories.* Urbana, Ill.: University of Illinois Press.

Thünen, J. H. von (1966) *Der Isolierte Staat in Beziehung auf Nationalökonomie und Landwirtschaft.* Stuttgart: Gustav Fischer. (Reprint of 1826 ed.)

Wilson, R. H. (1975) 'Has planning theory forgotten its history?' in R. H. Wilson and J. Noyelle (eds) (1975).

Wilson, R. H. and Noyelle, J. (eds) (1975) *1975 Symposium on Planning Theory: Papers in Planning 1.* Philadelphia: Department of City and Regional Planning, University of Pennsylvania.

Winograd, T. (1971) *Procedures as a Representation for Data in a Computer Program for Understanding Natural Language.* Cambridge, Mass.: MAC TR-84, M.I.T.

Winston, P. (1970) *Learning Structural Descriptions from Examples.* Cambridge, Mass.: MAC TR-76, M.I.T.

III   Urbanization and planning in capitalist society

5 The urban process under capitalism: a framework for analysis*

David Harvey

My objective is to understand the urban process under capitalism. I confine myself to the capitalist forms of urbanization because I accept the idea that the 'urban' has a specific meaning under the capitalist mode of production which cannot be carried over without a radical transformation of meaning (and of reality) into other social contexts.

Within the framework of capitalism, I hang my interpretation of the urban process on the twin themes of *accumulation* and *class struggle*. The two themes are integral to each other and have to be regarded as different sides of the same coin – different windows from which to view the totality of capitalist activity. The class character of capitalist society means the domination of labour by capital. Put more concretely, a class of capitalists is in command of the work process and organizes that process for the purposes of producing profit. The labourer, on the other hand, has command only over his or her labour power which must be sold as a commodity on the market. The domination arises because the labourer must yield the capitalist a profit (surplus value) in return for a living wage. All of this is extremely simplistic, of course, and actual class relations (and relations between factions of classes) within an actual system of production (comprising production, services, necessary costs of circulation, distribution, exchange, etc.) are highly complex. The essential Marxian insight, however, is that profit arises out of the domination of labour by capital and that the capitalists as a class must, if they are to reproduce

* This article is reproduced from *International Journal of Urban and Regional Research* 2,1 by kind permission of Edward Arnold (Publishers) Ltd.

themselves, continuously expand the basis for profit. We thus arrive at a conception of a society founded on the principle of 'accumulation for accumulation's sake, production for production's sake'. The theory of accumulation which Marx constructs in *Capital* amounts to a careful enquiry into the dynamics of accumulation and an exploration of its contradictory character. This may sound rather 'economistic' as a framework for analysis, but we have to recall that accumulation is the means whereby the capitalist class reproduces both itself and its domination over labour. Accumulation cannot, therefore, be isolated from class struggle.

The contradictions of capitalism

We can spin a whole web of arguments concerning the urban process out of an analysis of the contradictions of capitalism. Let me set out the principal forms these contradictions take.

Consider, first, the contradiction which lies within the capitalist class itself. In the realm of exchange each capitalist operates in a world of individualism, freedom and equality and can and must act spontaneously and creatively. Through competition, however, the inherent laws of capitalist production are asserted as 'external coercive laws having power over every individual capitalist'. A world of individuality and freedom on the surface conceals a world of conformity and coercion underneath. But the translation from individual action to behaviour according to class norms is neither complete nor perfect – it never can be because the *process* of exchange under capitalist rules always presumes individuality while the law of value always asserts itself in social terms. As a consequence, individual capitalists, each acting in their own immediate self-interest, can produce an aggregate result which is wholly antagonistic to their collective class interest. To take a rather dramatic example, competition may force each capitalist to so lengthen and intensify the work process that the capacity of the labour force to produce surplus value is seriously impaired. The collective effects of individual entrepreneurial activity can seriously endanger the social basis for future accumulation.

Consider, secondly, the implications of accumulation for the labourers. We know from the theory of surplus value that the exploitation of labour power is the source of capitalist profit. The capitalist form of accumulation therefore rests upon a certain violence which the capitalist class inflicts upon labour. Marx showed, however, that this appropriation could be worked out in such a way that it did not offend the rules of equality, individuality and freedom as they must prevail in the realms of exchange. Labourers, like the capitalists, 'freely' trade the commodity they have for sale in the market place. But labourers are also in competition with each other for employment while the work process is under the command of the capitalist. Under conditions of unbridled competition, the capitalists are forced willy-nilly into inflicting greater and greater violence upon those whom they employ. The individual labourer is powerless to resist this onslaught. The only solution is for the labourers to constitute themselves as a class and find collective means to resist the depredations of capital. The capitalist form of accumulation consequently calls into being overt and explicit class struggle between labour and capital. This contradiction between the classes explains much of the dynamic of capitalist history and is in many respects quite fundamental to understanding the accumulation process.

The two forms of contradiction are integral to each other. They express an underlying unity and are to be construed as different aspects of the same reality. Yet we can usefully separate them in certain respects. The internal contradiction within the capitalist class is rather different from the class confrontation between capital and labour, no matter how closely the two may be linked. In what follows I will focus on the accumulation process in the absence of any overt response on the part of the working class to the violence which the capitalist class must necessarily inflict upon it. I will then broaden the perspective and consider how the organization of the working class and its capacity to mount an overt class response affects the urban process under capitalism.

Various other forms of contradiction could enter in to supplement the analysis. For example, the capitalist production system often exists in an antagonistic relationship to non- or pre-capitalist sectors which may exist within (the domestic economy, peasant and artisan production sectors, etc.) or without it (pre-capitalist societies, socialist countries, etc.). We should also note the contradiction with 'nature' which inevitably arises out of the relation between the dynamics of accumulation and the 'natural' resource base as it is defined in capitalist terms. Lack of space precludes any examination of these matters here. But they would obviously have to be taken into account in any analysis of the history of urbanization under capitalism.

The laws of accumulation

We will begin by sketching the structure of flows of capital within a system of production and realization of value. This I will do with the aid of a series of diagrams which will appear highly 'functionalist' and perhaps unduly simple in structure, but which nevertheless help us to understand the basic logic of the accumulation process. We will also see how problems arise because individual capitalists produce a result inconsistent with their class interest and consider some of the means whereby solutions to these problems might be found. We will, in short, attempt a summary of Marx's argument in *Capital* in the ridiculously short space of three or four pages.

The primary circuit of capital

In volume one of *Capital,* Marx presents an analysis of the capitalist production process. The drive to create surplus value rests either on an increase in the length of the working day (absolute surplus value) or on the gains to be made from continuous revolutions in the 'productive forces' through reorganizations of the work process which raise the productivity of labour power (relative surplus value). The capitalist captures relative surplus value from the organization of cooperation and division of labour within the work process or by the application of fixed capital (machinery). The motor for these continuous revolutions in the work process, for the rising productivity of labour, lies in capitalist competition as each capitalist seeks an excess profit by adopting a superior production technique to the social average.

The implications of all of this for labour are explored in a chapter entitled 'the general law of capitalist accumulation'. Marx here examines alterations in the rate of exploitation and in the temporal rhythm of changes in the work process in relation to the supply conditions of labour power (in particular, the formation of an industrial reserve army), assuming all the while, that a positive rate of accumulation must be sustained if the capitalist class is to reproduce itself. The analysis proceeds around a strictly circumscribed set of interactions with all other problems assumed away or held constant. Figure 5.1 portrays the relations examined.

The second volume of *Capital* closes with a 'model' of accumulation on an expanded scale. The problems of proportionality involved in the aggregative production of means of production and means of consumption are examined with all other problems held constant (including technological change, investment in fixed capital, etc.). The objective here is to show the potential for crises of disproportionality within the production process. But Marx has now broadened the structure of relationships put under the microscope (see Figure 5.2). We note, however, that in both cases Marx assumes, tacitly, that all commodities are produced and consumed within one time period. The structure of relations examined in Figure 5.2 can be characterized as the *primary circuit of capital.*

Much of the analysis of the falling rate of profit and its countervailing tendencies in volume three similarly presupposes production and consumption within one time period although there is some evidence that Marx intended to broaden the scope of this if he had lived to complete the work. But it is useful to consider the volume three analysis as a synthesis of the arguments presented in the first two volumes and as at the very least a cogent statement of the internal contradictions which exist within the primary circuit. Here we can clearly see the contradictions which arise out of the tendency for individual capitalists to act in a way which, when aggregated, runs counter to their own class interest. This contradiction produces a tendency towards *overaccumulation* – too much capital is produced in aggregate relative to the opportunities to employ that capital. This tendency is manifest in a variety of guises. We have:

1 Overproduction of commodities – a glut on the market.
2 Falling rates of profit (in pricing terms, to be distinguished from the falling rate of profit in value terms which is a theoretical construct).
3 Surplus capital which can be manifest either as idle productive capacity or as money-capital lacking opportunities for profitable employment.
4 Surplus labour and/or rising rate of exploitation of labour power.

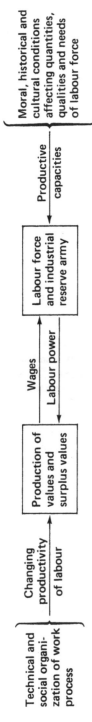

Figure 5.1 The relations considered in 'the general law of accumulation'. Source: volume 1 of *Capital*.

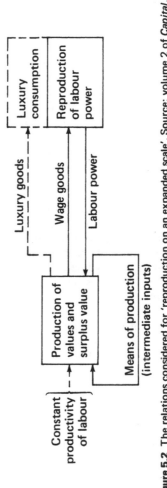

Figure 5.2 The relations considered for 'reproduction on an expanded scale'. Source: volume 2 of *Capital*.

One or a combination of these manifestations may be present at the same time. We have here a preliminary framework for the analysis of capitalist crises.

The secondary circuit of capital

We now drop the tacit assumption of production and consumption within one time period and consider the problems posed by production and use of commodities requiring different working periods, circulation periods, and the like. This is an extraordinarily complex problem which Marx addresses to some degree in volume two of *Capital* and in the *Grundrisse*. I cannot do justice to it here so I will confine myself to some remarks regarding the formation of *fixed capital* and the *consumption fund*. Fixed capital, Marx argues, requires special analysis because of certain peculiarities which attach to its mode of production and realization. These peculiarities arise because fixed capital items can be produced in the normal course of capitalist commodity production but they are used as aids to the production process rather than as direct raw material inputs. They are also used over a relatively long time period. We can also usefully distinguish between fixed capital enclosed within the production process and fixed capital which functions as a physical framework for production. The latter I will call the *built environment for production.*

On the consumption side, we have a parallel structure. A *consumption fund* is formed out of commodities which function as aids rather than as direct inputs to consumption. Some items are directly enclosed within the consumption process (consumer durables, such as cookers, washing machines, etc.) while others act as a physical framework for consumption (houses, sidewalks, etc.) – the latter I will call the *built environment for consumption.*

We should note that some items in the built environment function jointly for both production and consumption – the transport network, for example – and that items can be transferred from one category to another by changes in use. Also, fixed capital in the built environment is immobile in space in the sense that the value incorporated in it cannot be moved without being destroyed. Investment in the built environment therefore entails the creation of a whole physical landscape for purposes of production, circulation, exchange and consumption.

We will call the capital flows into fixed asset and consumption fund formation the *secondary circuit of capital.* Consider, now, the manner in which such flows can occur. There must obviously be a 'surplus' of both capital and labour in relation to current production and consumption needs in order to facilitate the movement of capital into the formation of long-term assets, particularly those comprising the built environment. The tendency towards overaccumulation produces such conditions within the primary circuit on a periodic basis. One feasible if *temporary* solution to this overaccumulation problem would therefore be to switch capital flows into the secondary circuit.

Individual capitalists will often find it difficult to bring about such a switch in flows for a variety of reasons. The barriers to individual switching of capital are particularly acute with respect to the built environment where investments tend to be large-scale and long-lasting, often difficult to price in the ordinary way and in many cases open to collective use by all individual capitalists. Indeed, individual capitalists left to themselves will tend to under-supply their own collective needs for production precisely because of such barriers. Individual capitalists tend to overaccumulate in the primary circuit and to under-invest in the secondary circuit; they have considerable difficulty in organizing a balanced flow of capital between the primary and secondary circuits.

A general condition for the flow of capital into the secondary circuit is, therefore, the existence of a functioning capital market and, perhaps, a State willing to finance and guarantee long-term, large-scale projects with respect to the creation of the built environment. At times of overaccumulation, a switch of flows from the primary to the secondary circuit can be accomplished only if the various manifestations of overaccumulation can be transformed into money-capital which can move freely and unhindered into these forms of investment. This switch of resources cannot be accomplished without a money supply and credit system which creates 'fictional capital' *in advance* of actual production and consumption. This applies as much to the consumption fund (hence the importance of consumer credit, housing mortgages, municipal debt) as it does to fixed capital. Since the production of money and credit are relatively autonomous processes, we have to conceive of the financial and state institutions controlling them as a kind of collective nerve centre governing and *mediating* the relations between the primary and secondary circuits of capital. The nature and form of these financial and state institutions and the policies they adopt can play an important role in checking or enhancing flows of capital into the secondary circuit of capital or into certain specific aspects of it (such as transportation, housing, public facilities, and so on). An alteration in these mediating structures can therefore affect both the volume and direction of the capital flows by constricting movement down some channels and opening up new conduits elsewhere.

The tertiary circuit of capital

In order to complete the picture of the circulation of capital in general, we have to conceive of a *tertiary circuit of capital* which comprises, first, investment in science and technology (the purpose of which is to harness science to production and thereby to contribute to the processes which continuously revolutionize the productive forces in society) and second, a wide range of social expenditures which relate primarily to the processes of reproduction of labour power. The latter can usefully be divided into investments directed towards the qualitative improvement of labour power from the standpoint of capital (investment in education and health by means of which the capacity of the labourers to engage in the work process will be enhanced) and investment in cooptation, integration and repression of the labour force by ideological, military and other means.

Individual capitalists find it hard to make such investments as individuals, no matter how desirable they may regard them. Once again, the capitalists are forced to some degree to constitute themselves as a class – usually through the agency of the State – and thereby to find ways to channel investment into research and development and into the quantitative and qualitative improvement of labour power. We should recognize that capitalists often *need* to make such investments in order to fashion an adequate social basis for further accumulation. But with regard to social expenditures, the investment flows are very strongly affected by the state of class struggle. The amount of investment in repression and in ideological control is directly related to the threat of organized working-class resistance to the depredations of capital. And the need to coopt labour arises only when the working class has accumulated sufficient power to require cooptation. Since the State can become a field of active class struggle, the mediations which are accomplished by no means fit exactly with the requirements of the capitalist class. The role of the State requires careful theoretical and historical elaboration in relation to the organization of capital flows into the tertiary circuit.

The circulation of capital as a whole and its contradictions

Figure 5.3 portrays the overall structure of relations comprising the circulation of capital amongst the three circuits. The diagram looks very 'structuralist-functionalist' because of the method of presentation. I can conceive of no other way to communicate clearly the various dimensions of capital flow. We now have to consider the contradictions embodied within these relations. I shall do so initially as if there were no overt class struggle between capital and labour. In this way we will be able to see that the contradiction between the individual capitalist and capital in general is itself a source of major instability within the accumulation process.

We have already seen how the contradictions internal to the capitalist class generate a tendency towards overaccumulation within the primary circuit of capital. And we have argued that this tendency can be overcome temporarily at least by switching capital into the secondary or tertiary circuits. Capital has, therefore, a variety of investment options open to it – fixed capital or consumption fund formation, investment in science and technology, investment in 'human capital' or outright repression. At particular historical conjunctures capitalists may not be capable of taking up all of these options with equal vigour, depending upon the degree of their own organization, the institutions which they have created and the objective possibilities dictated by the state of production and the state of class struggle. I shall assume away such problems for the moment in order to concentrate on how the tendency towards overaccumulation, which we have identified so far only with respect to the primary circuit, manifests itself within the overall structure of circulation of capital. To do this we first need to specify a concept of productivity of investment.

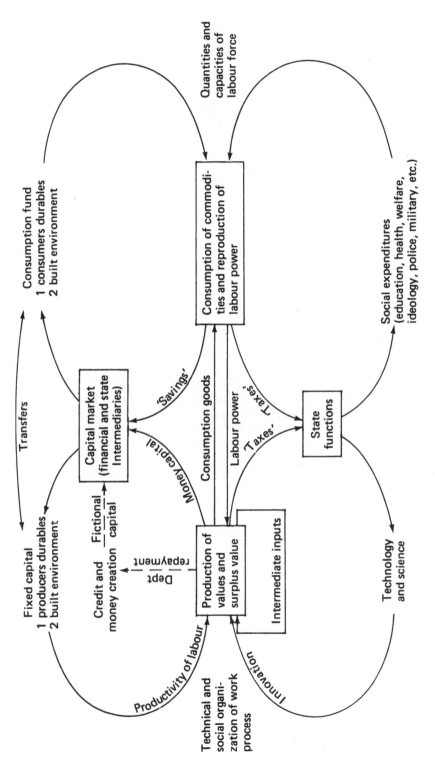

Figure 5.3 The structure of relations between the primary, secondary and tertiary circuits of capital

On the productivity of investments in the secondary and tertiary circuits

I choose the concept of 'productivity' rather than 'profitability' for a variety of reasons. First of all, the rate of profit as Marx treats of it in volume three of *Capital* is measured in value rather than pricing terms and takes no account of the distribution of the surplus value into its component parts of interest on money capital, profit on productive capital, rent on land, profit on merchant's capital, etc. The rate of profit is regarded as a social average earned by individual capitalists in all sectors and it is assumed that competition effectively ensures its equalization. This is hardly a suitable conception for examining the flows between the three circuits of capital. To begin with, the formation of fixed capital in the built environment – particularly the collective means of production – cannot be understood without understanding the formation of a capital market and the distribution of part of the surplus in the form of interest. Second, many of the commodities produced in relation to the secondary and tertiary circuits cannot be priced in the ordinary way, while collective action by way of the State cannot be examined in terms of the normal criteria of profitability. Third, the rate of profit which holds is perfectly appropriate for understanding the behaviours of individual capitalists in competition, but cannot be translated into a concept suitable for examining the behaviour of capitalists as a class without some major assumptions (treating the total profit as equal to the total surplus value, for example).

The concept of productivity helps to bypass some of these problems if we specify it carefully enough. For the fact is that capitalists as a class – often through the agency of the State – do invest in the production of conditions which they hope will be favourable to accumulation, to their own reproduction as a class and to their continuing domination over labour. This leads us immediately to a definition of a productive investment as one which directly or indirectly expands the basis for the production of surplus value. Plainly, investments in the secondary and tertiary circuits have the *potential* under certain conditions to do this. The problem – which besets the capitalists as much as it confuses us – is to identify the conditions and means which will allow this potential to be realized.

Investment in new machinery is the easiest case to consider. The new machinery is directly productive if it expands the basis for producing surplus value and unproductive if these benefits fail to materialize. Similarly, investment in science and technology may or may not produce new forms of scientific knowledge which can be applied to expand accumulation. But what of investment in roads, housing, health care and education, police forces and the military, and so on? If workers are being recalcitrant in the work place, then judicious investment by the capitalist class in a police force to intimidate the workers and to break their collective power, may indeed be productive indirectly of surplus value for the capitalists. If, on the other hand, the police are employed to protect the bourgeoisie in the conspicuous consumption of their revenues in callous disregard of the poverty and misery which surrounds them, then the police are not acting to facilitate accumulation. The distinction may be fine but it demonstrates the dilemma. How can the capitalist class identify, with reasonable precision, the opportunities for indirectly and directly productive investment in the secondary and tertiary circuits of capital?

The main thrust of the modern commitment to planning (whether at the state or corporate level) rests on the idea that certain forms of investment in the secondary and tertiary circuits are potentially productive. The whole apparatus of cost-benefit analysis and of programming and budgeting, of analysis of social benefits, as well as notions regarding investment in human capital, express this commitment and testify to the complexity of the problem. And at the back of all of this is the difficulty of determining an appropriate basis for decision-making in the absence of clear and unequivocal profit signals. Yet the cost of bad investment decisions – investments which do not contribute directly or indirectly to accumulation of capital – must emerge somewhere. They must, as Marx would put it, come to the surface and thereby indicate the errors which lie beneath. We can begin to grapple with this question by considering the origins of crises within the capitalist mode of production.

On the forms of crisis under capitalism

Crises are the real manifestation of the underlying contradictions within the capitalist process of accumulation. The argument which Marx puts forward throughout much of *Capital* is that there is always the potential within capitalism to achieve 'balanced growth' but that this potentiality can never be realized because of the structure of the social relations prevailing in a capitalist society. This structure leads individual capitalists to produce results collectively which are antagonistic to their own class interest and leads them also to inflict an insupportable violence upon the working class which is bound to elicit its own response in the field of overt class struggle.

We have already seen how the capitalists tend to generate states of overaccumulation within the primary circuit of capital and considered the various manifestations which result. As the pressure builds, either the accumulation process grinds to a halt or new investment opportunities are found as capital flows down various channels into the secondary and tertiary circuits. This movement may start as a trickle and become a flood as the potential for expanding the production of surplus value by such means becomes apparent. But the tendency towards overaccumulation is not eliminated. It is transformed, rather, into a pervasive tendency towards over-investment in the secondary and tertiary circuits. This over-investment, we should stress, is in relation solely to the needs of capital and has nothing to do with the real needs of people which inevitably remain unfulfilled. Manifestations of crisis thus appear in both the secondary and tertiary circuits of capital.

As regards fixed capital and the consumption fund, the crisis takes the form of a crisis in the valuation of assets. Chronic overproduction results in the devaluation of fixed capital and consumption fund items –

a process which affects the built environment as well as producer and consumer durables. We can likewise point to crisis formation at other points within our diagram of capital flows – crises in social expenditures (health, education, military repression, and the like), in consumption-fund formation (housing) and in technology and science. In each case the crisis occurs because the potentiality for productive investment within each of these spheres is exhausted. Further flows of capital do not expand the basis for the production of surplus value. We should also note that a crisis of any magnitude in any of these spheres is automatically registered as a crisis within the financial and state structures while the latter, because of the relative autonomy which attaches to them, can be an independent source of crisis (we can thus speak of financial, credit and monetary crises, the fiscal crises of the State, and so on).

Crises are the 'irrational rationalizers' within the capitalist mode of production. They are indicators of imbalance and force a rationalization (which may be painful for certain sectors of the capitalist class as well as for labour) of the processes of production, exchange, distribution and consumption. They may also force a rationalization of institutional structures (financial and state institutions in particular). From the standpoint of the total structure of relationships we have portrayed, we can distinguish different kinds of crises:

a *Partial crises* which affect a particular sector, geographical region or set of mediating institutions. These can arise for any number of reasons but are potentially capable of being resolved within that sector, region or set of institutions. We can witness autonomously-forming monetary crises, for example, which can be resolved by institutional reforms, crises in the formation of the built environment which can be resolved by reorganization of production for that sector, etc.

b *Switching crises* which involve a major reorganization and restructuring of capital flows and/or a major restructuring of mediating institutions in order to open up new channels for productive investments. It is useful to distinguish between two kinds of switching crises:

1 *Sectoral switching crises* which entail switching the allocation of capital from one sphere (e.g. fixed capital formation) to another (e.g. education);

2 *Geographical switching crises* which involve switching the flows of capital from one place to another. We note here that this form of crisis is particularly important in relation to investment in the built environment because the latter is immobile in space and requires interregional or international flows of money-capital to facilitate its production.

c *Global crises* which affect, to greater or lesser degree all sectors, spheres and regions within the capitalist production system. We will thus see devaluations of fixed capital and the consumption fund, a crisis in science and technology, a fiscal crisis in state expenditures, a crisis in the productivity of labour, all manifest at the same time across all or most regions within the capitalist system. I note, in passing, that there have been only two global crises within the totality of the capitalist system – the first during the 1930s and its Second World War aftermath; the second, that which became most evident after 1973 but which had been steadily building through the 1960s.

A complete theory of capitalist crises should show how these various forms and manifestations relate in both space and time. Such a task is beyond the scope of a short article, but we can shed some light by returning to our fundamental theme – that of understanding the urban process under capitalism.

Accumulation and the urban process

The understanding I have to offer of the urban process under capitalism comes from seeing it in relation to the theory of accumulation. We must first establish the general points of contact between what seem, at first sight, two rather different ways of looking at the world.

Whatever else it may entail, the urban process implies the creation of a material physical infrastructure for production, circulation, exchange and consumption. The first point of contact, then, is to consider the manner in which this built environment is produced and the way it serves as a resource system – a complex of use values – for the production of value and surplus value. We have, secondly, to consider the consumption aspect. Here we can usefully distinguish between the consumption of revenues by the bourgeoisie and the need to reproduce labour power. The former has a considerable impact upon the urban process, but I shall exclude it from the analysis because consideration of it would lead us into a lengthy discourse on the question of bourgeois culture and its complex significations without revealing very much directly about the specifically capitalist form of the urban process. Bourgeois consumption is, as it were, the icing on top of a cake which has as its prime ingredients capital and labour in dynamic relation to each other. The reproduction of labour power is essential and requires certain kinds of social expenditures and the creation of a consumption fund. The flows we have sketched, in so far as they portray capital movements into the built environment (for both production and consumption) and the laying out of social expenditures for the reproduction of labour power, provide us, then, with the structural links we need to understand the urban process under capitalism.

It may be objected, quite correctly, that these points of integration ignore the 'rural-urban dialectic' and that the reduction of the 'urban process' as we usually conceive of it to questions of built environment formation and reproduction of labour power is misleading if not downright erroneous. I would defend the reduction on a number of counts. First, as a practical matter, the mass of the capital flowing into the built environment and a large proportion of certain kinds of social expenditures are absorbed in areas which we usually classify as 'urban'. From this standpoint the reduction is a useful approximation. Second, I can discuss most of the questions which normally arise in urban research in terms of the categories of the built environment and social expenditures related to the reproduction of labour power with the added advantage that the links with the theory of accumulation can be clearly seen. Third, there are serious grounds for challenging the adequacy of the urban-rural dichotomy even when expressed as a dialectical unity, as a primary form of contradiction within the capitalist mode of production. In other words, and put quite bluntly, if the usual conception of the urban process appears to be violated by the reduction I am here proposing then it is the usual conception of the urban process which is at fault.

The urban-rural dichotomy, for example, is regarded by Marx as an expression of the division of labour in society. In this, the division of labour is the fundamental concept and not the rural-urban dichotomy which is just a particular form of its expression. Focusing on this dichotomy may be useful in seeking to understand social formations which arise in the transition to capitalism – such as those in which we find an urban industrial sector opposed to a rural peasant sector which is only formally subsumed within a system of commodity production and exchange. But in a purely capitalist mode of production – in which industrial and agricultural workers are all under the real domination of capital – this form of expression of the division of labour loses much of its particular significance. It disappears within a general concern for geographical specialization in the division of labour. And the other aspect to the urban process – the geographical concentration of labour power and use values for production and reproduction – also disappears quite naturally within an analysis of the 'rational spatial organization' of physical and social infrastructures. In the context of advanced capitalist countries as well as in the analysis of the capitalist mode of production, the urban-rural distinction has lost its real economic basis although it lingers, of course, within the realms of ideology with some important results. But to regard it as a fundamental conceptual tool for analysis is in fact to dwell upon a lost distinction which was in any case but a surface manifestation of the division of labour.

Overaccumulation and long waves in investment in the built environment

The acid test of any set of theoretical propositions comes when we seek to relate them to the experience of history and to the practices of politics. In a short paper of this kind I cannot hope to demonstrate the relations between the theory of accumulation and its contradictions on the one hand, and the urban process on the other in the kind of detail which would be convincing. I shall therefore confine myself to illustrating some of the more important themes which can be identified. I will focus, first, exclusively on the processes governing investment in the built environment.

The system of production which capital established was founded on a physical separation between a place of work and a place of residence. The growth of the factory system, which created this separation, rested on the organization of cooperation, division of labour and economies-of-scale in the work process as well as upon the application of machinery. The system also promoted an increasing division of labour between enterprises, and collective economies-of-scale through the agglomeration of activities in large urban centres. All of this meant the creation of a built environment to serve as a physical infrastructure for production, including an appropriate system for the transport of commodities. There are abundant opportunities for the productive employment of capital through the creation of a built environment for production. The same conclusion applies to investment in the built environment for consumption. The problem is, then, to discover how capital flows into the construction of this built environment and to establish the contradictions inherent in this process.

We should first say something about the concept of the built environment and consider some of its salient attributes. It is a complex composite commodity comprising innumerable different elements – roads, canals, docks and harbours, factories, warehouses, sewers, public offices, schools and hospitals, houses, offices, shops, etc. – each of which is produced under different conditions and according to quite different rules. The 'built environment' is, then, a gross simplification, a concept which requires disaggregation as soon as we probe deeply into the processes of its production and use. Yet we also know that these components have to function together as an ensemble in relation to the aggregative processes of production, exchange and consumption. For purposes of exposition we can afford to remain at this level of generality. We also know that the built environment is long lived, difficult to alter, spatially immobile and often absorbent of large lumpy investments. A proportion of it will be used in common by capitalists and consumers alike and even those elements which can be privately appropriated (houses, factories, shops, etc.) are used in a context in which the externality effects of private uses are pervasive and often quite strong. All of these characteristics have implications for the investment process.

The analysis of fixed capital formation and the consumption fund in the context of accumulation suggests that investment in the built environment is likely to proceed according to a certain logic. We presume, for the moment, that the State does not take a leading role in promoting vast public works programmes ahead of the demand for them. Individual capitalists, when left to their own devices, tend to under-invest in the built environment relative to their own individual and collective needs at the same time as they tend to overaccumulate. The theory then suggests that the overaccumulation can be syphoned off – via financial and state institutions and the creation of fictional capital within the credit system – and put to work to take up the slack in investment in the built environment. This switch from the primary to the secondary circuit may occur in the course of a crisis or be accomplished relatively smoothly depending upon the efficiency of the mediating institutions. But the theory indicates that there is a limit to such a process and that at some point investments will become unproductive. At such a time the exchange value being put into the built environment has to be written down, diminished, or even totally lost. The fictional capital contained within the credit system is seen to be just that and financial and state institutions may find themselves in serious financial difficulty. The devaluation of capital in the built environment does not necessarily destroy the use value – the physical resource – which the built environment comprises. This physical resource can now be used as 'devalued capital' and as such it functions as a free good which can help to reestablish the basis for renewed accumulation. From this we can see the logic of Marx's statement that periodical devaluations of fixed capital provide 'one of the means immanent in capitalist production to check the fall of the rate of profit and hasten accumulation of capital-value through formation of new capital'.

Since the impulses deriving from the tendency to overaccumulate and to under-invest are rhythmic rather than constant, we can construct a cyclical 'model' of investment in the built environment. The rhythm is dictated in part by the rhythms of capital accumulation and in part by the physical and economic lifetime of the elements within the built environment – the latter means that change is bound to be relatively slow. The most useful thing we can do at this juncture is to point to the historical evidence for 'long waves' in investment in the built environment. Somewhere in between the short-run movements of the business cycle – the 'Juglar cycles' of approximately ten-year length – and the very long 'Kondratieff's', we can identify movements of an intermediate length (sometimes called Kuznets cycles) which are strongly associated with waves of investment in the built environment. Gottlieb's recent investigation[1] of building cycles in 30 urban areas located in eight countries showed a periodicity clustering between 15 and 25 years. While his methods and framework for analysis leave much to be desired, there is enough evidence accumulated by a variety of researchers to indicate that this is a reasonable sort of first-shot generalization. Figures 5.4, 5.5 and 5.6 illustrate the phenomenon. The historical evidence is at least consistent with our argument, taking into account, of course, the material characteristics of the built environment itself and in particular its long life which means that 'instant throw-away cities' are hardly feasible no matter how hard the folk in Los Angeles try.

The immobility in space also poses its own problematic with, again, its own appropriate mode of response. The historical evidence is, once more, illuminating. In the 'Atlantic economy' of the nineteenth century, for example, the long waves of investment in the built environment moved inversely to each other in Britain and the United States (see Figures 5.7 and 5.8). The two movements were not independent of each other but tied via migrations of capital and labour within the framework of the international economy at that time. The commercial crises of the nineteenth century switched British capital from home investment to overseas investment or vice versa. The capitalist 'whole' managed, thereby, to achieve a roughly balanced growth through counterbalancing oscillations of the parts, all encompassed within a global process of geographical expansion.[2]

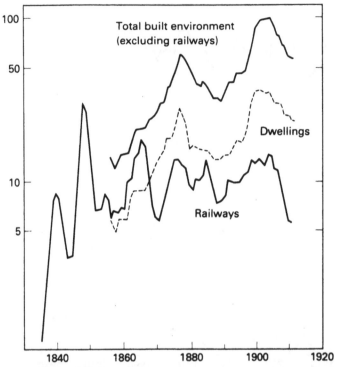

Figure 5.4 Investment in selected components of the built-environment in Britain (million £ at current prices)

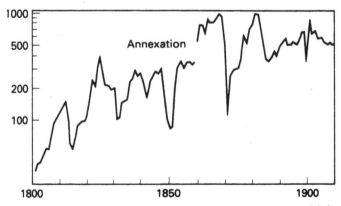

Figure 5.5 Construction activity in Paris – entries of construction materials into the city (millions of cubic metres). Source: after Rougerie

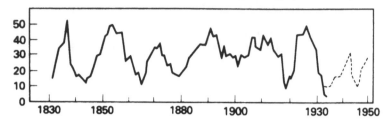

Building activity per capita in the United States (1913
dollars per capita). Source: after Brinley Thomas

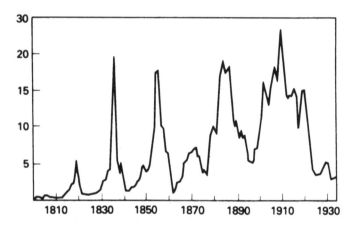

Sales of the public lands in the United States (millions of acres
of original land entries: US Dept of Agriculture figures)

Figure 5.6 'Long-waves' in investment in the built-environment of the United States

Uneven spatial development of the built environment was a crucial element
to the achievement of relative global stability under the aegis of the *Pax
Britannica* of the nineteenth century. The crises of this period were either
of the partial or switching variety and we can spot both forms of the
latter – geographical and sectoral – if we look carefully enough.

The global crises of the 1930s and the 1970s can in part be explained
by the breakdown of the mechanisms for exploiting uneven development
in this way. Investment in the built environment takes on a different meaning at such conjunctures. Each of the global crises of capitalism was in
fact preceded by the massive movement of capital into long-term investment
in the built environment as a kind of last-ditch hope for finding productive
uses for rapidly overaccumulating capital. The extraordinary property
boom in many advanced capitalist countries from 1969–73, the collapse
of which at the end of 1973 triggered (but did not *cause*) the onset of
the current crisis, is a splendid example. I append some illustrative materials in Figure 5.9.

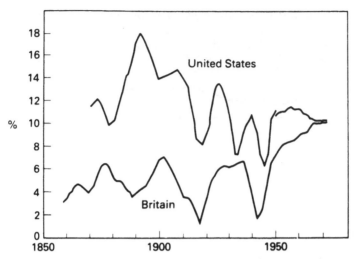

Figure 5.7 Different rhythms of investment in the built-environment – Britain and the United States (per cent of GNP (USA) and GDP (Britain) going to investment in the built-environment – five-year moving averages)

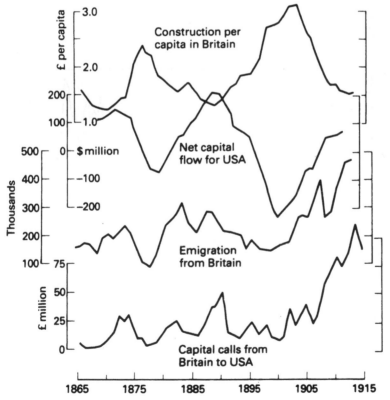

Figure 5.8 Uneven development in the Atlantic economy – Britain and the United States. Source: after Brinley Thomas

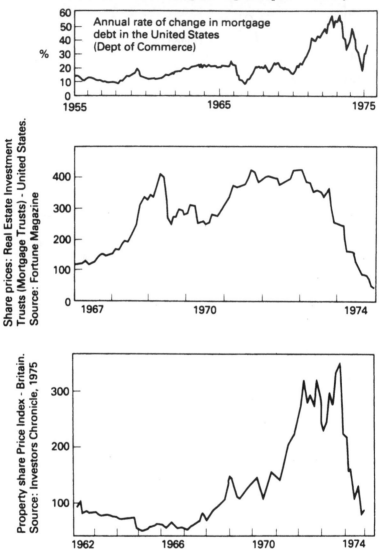

Figure 5.9 Some indices of the property boom – Britain and United States

While I am not attempting in any strict sense to 'verify' the theory by appeal to the historical record, the latter most certainly is not incompatible with the broad outlines of the theory we have sketched in. Bringing the theory to bear on the history is in fact an extraordinarily difficult task far beyond the scope of a short paper. But rather than make no argument at all I will seek to illustrate how the connections can be made. I will therefore look a little more closely at the two aspects of the theory which are crucial – overaccumulation and devaluation.

The flow of investment into the built environment depends upon the existence of surpluses of capital and labour and upon mechanisms for pooling the former and putting it to use. The history of this process is extremely interesting. The eighteenth century in Britain was characterized, for example, by a capital surplus much of which went into the built environment because it had nowhere else to go. Investment in the built environment took place primarily for financial rather than use-value reasons – investors were looking for a steady and secure rate of return on their capital. Investment in property (much of it for conspicuous consumption by the bourgeoisie), in turnpikes, canals and rents (agricultural improvement) as well as in state obligations were about the only options open to rentiers. The various speculative crises which beset investment in the turnpikes and canals as well as urban property markets, indicated very early on that returns were by no means certain and that investments had to be productive if they were to succeed.[3]

It would be difficult to argue that during this period the surplus of capital arose out of the tendency to overaccumulate as we have specified it. The latter is, strictly speaking, a phenomenon which arises only in the context of the capitalist mode of production or in capitalist social formations which are relatively well-developed. The 'long waves' of investment in the built environment pre-date the emergence of industrial capitalism and can be clearly identified throughout the transition from feudalism.[4] We can see, however, a strong relationship between these 'long waves' and fluctuations in the money supply and in the structure of capital markets. Perhaps the most spectacular example is that of the United States (Figure 5.6) – when Andrew Jackson curbed land deals in paper currency and insisted on *specie* payment in 1836, the whole land development process came to a halt and the financial reverberations were felt everywhere, particularly by those investing in the built environment. The role of 'fictional capital' and the credit and money supply system has always been fundamental in relationship to the various waves of speculative investment in the built environment.

When, precisely, the tendency towards overaccumulation became the main agent producing surplus capital and when the 'long waves' became explicitly tied to overaccumulation is a moot point. The evidence suggests that by the 1840s the connections had been strongly forged in Britain at least. By then, the functioning of the capital market was strongly bound to the rhythms imposed by the development of industrial capitalism. The 'nerve centre' which controls and mediates the relations between the primary and secondary circuits of capital increasingly functioned after 1830 or so according to a pure capitalist logic which affected both government and private activity alike. It is perhaps symptomatic that the fall of the July monarchy in France in 1847 was directly related to the indebtedness of that regime incurred in the course of promoting a vast programme of public works (many of which were not very productive). When the financial crisis, which had its origins in England and the extraordinary speculation in railroad construction, struck home in late 1846 and 1847, even the state debt of France could not withstand the shock.[5] For good reason, this crisis can perhaps be regarded as the first really solid and all-pervasive crisis in the capitalist world.

And what of the devaluation which inevitably results? If the devaluation is to function effectively, according to our theory, then it must leave behind a use value which can be used as the basis for further development. When many of the American states defaulted on their debts in the early 1840s, they failed to meet their obligations on the British capital market but kept the canals and other improvements which they had built. This was, in effect, expropriation without compensation – a prospect which the United States government treats with great moral indignation when some third-world country threatens it today. The great railroad booms of the nineteenth century typically devalued capital while littering the landscape with physical assets which could usually be put to some use. When the urban mass transit systems went bankrupt at the turn of the century because of chronic overcapitalization, the mass transit systems were left behind as physical assets. Somebody had to pay for the devaluation of course. There were the inevitable attempts to foist the costs onto the working class (often through municipal expenditures) or onto small investors. But big capital was not immune either, and the problems of the property companies in Britain or the real-estate investment trusts in the United States at the present time are exactly of this sort (although the involvement of pension funds and insurance companies affects individuals). The office space is still there, however, even though the building that houses it has been devalued and is now judged a non-earning asset. The history of devaluations in the built environment is spectacular enough and fits, in general, with the theoretical argument.

The contradictory character of investments in the built environment

We have so far treated the process of investment in the built environment as a mere reflection of the forces emanating from the primary circuit of capital. There are, however, a whole series of problems which arise because of the specific characteristics of the built environment itself. We will consider these briefly.

Marx's extensive analysis of fixed capital in relation to accumulation reveals a central contradiction. On the one hand, fixed capital enhances the productivity of labour and thereby contributes to the accumulation of capital. But, on the other hand, it functions as a use value and requires the conversion of exchange values into a physical asset which has certain attributes. The exchange value locked up in this physical use value can be re-couped only by keeping the use value fully employed over its lifetime, which for simplicity's sake we will call its 'amortization time'. As a use value the fixed capital cannot easily be altered and so it tends to freeze productivity at a certain level until the end of the amortization time. If new and more productive fixed capital comes into being before the old is amortized, then the exchange value still tied up in the old is devalued. Resistance to this devaluation checks the rise in productivity and, thus, restricts accumulation. On the other hand the pursuit of new and more productive forms of fixed capital – dictated by the quest for relative surplus value – accelerates devaluations of the old.

We can identify exactly these same contradictory tendencies in relation to investment in the built environment, although they are even more exaggerated here because of the generally long amortization time involved, the fixity in space of the asset, and the composite nature of the commodity involved. We can demonstrate the argument most easily using the case of investment in transportation.

The cost, speed and capacity of the transport system relate directly to accumulation because of the impacts these have on the turnover time of capital. Investment and innovation in transport are therefore potentially productive for capital in general. Under capitalism, consequently, we see a tendency to 'drive beyond all spatial barriers' and to 'annihilate space with time' (to use Marx's own expressions).[6] This process is, of course, characterized typically by 'long waves' of the sort which we have already identified, uneven development in space and periodic massive devaluations of capital.[7]

We are here concerned, however, with the contradictions implicit in the process of transport development itself. Exchange values are committed to create 'efficient' and 'rational' configurations for spatial movement at a particular historical moment. There is, as it were, a certain striving towards spatial equilibrium, spatial harmony. On the other hand, accumulation for accumulation's sake spawns continuous revolutions in transportation technology as well as a perpetual striving to overcome spatial barriers – all of which is disruptive of any existing spatial configuration.

We thus arrive at a paradox. In order to overcome spatial barriers and to annihilate space with time, spatial structures are created which themselves act as barriers to further accumulation. These spatial structures are expressed in the form of immobile transport facilities and ancillary facilities implanted in the landscape. We can in fact extend this conception to encompass the formation of the built environment as a whole. Capital represents itself in the form of a physical landscape created in its own image, created as use values to enhance the progressive accumulation of capital. The geographical landscape which results is the crowning glory of past capitalist development. But at the same time it expresses the power of dead labour over living labour and as such it imprisons and inhibits the accumulation process within a set of specific physical constraints. And these can be removed only slowly unless there is a substantial devaluation of the exchange value locked up in the creation of these physical assets.

Capitalist development has therefore to negotiate a knife-edge path between preserving the exchange values of past capital investments in the built environment and destroying the value of these investments in order to open up fresh room for accumulation. Under capitalism there is, then, a perpetual struggle in which capital builds a physical landscape appropriate to its own condition at a particular moment in time, only to have to destroy it, usually in the course of a crisis, at a subsequent point in time.

The temporal and geographical ebb and flow of investment in the built environment can be understood only in terms of such a process. The effects of the internal contradictions of capitalism, when projected into the specific context of fixed and immobile investment in the built environment, are thus writ large in the historical geography of the landscape which results.

Class struggle, accumulation and the urban process under capitalism

What, then, of overt class struggle – the resistance which the working class collectively offers to the violence which the capitalist form of accumulation inevitably inflicts upon it? This resistance, once it becomes more than merely nominal, must surely affect the urban process under capitalism in definite ways. We must, therefore, seek to incorporate some understanding of it into any analysis of the urban process under capitalism. By switching our window on the world – from the contradictory laws of accumulation to the overt class struggle of the working class against the effects of those laws – we can see rather different aspects of the same process with greater clarity. In the space that follows I will try to illustrate the complementarity of the two viewpoints.

In one sense, class struggle is very easy to write about because there is no theory of it, only concrete social practices in specific social settings. But this immediately places upon us the obligation to understand history if we are to understand how class struggle has entered into the urban process. Plainly I cannot write this history in a few pages. So I will confine myself to a consideration of the contextual conditions of class struggle and the nature of the bourgeois responses. The latter are governed by the laws of accumulation because accumulation always remains the means whereby the capitalist class reproduces itself as well as its domination over labour.

The central point of tension between capital and labour lies in the workplace and is expressed in struggles over the work process and the wage rate. These struggles take place in a context. The nature of the demands, the capacity of workers to organize and the resolution with which the struggles are waged, depend a great deal upon the contextual conditions. The law (property rights, contract, combination and association, etc.) together with the power of the capitalist class to enforce their will through the use of state power are obviously fundamental as any casual reading of labour history will abundantly illustrate. What specifically interests me here, however, is the process of reproduction of labour-power in relation to class struggle in the workplace.

Consider, first, the quantitative aspects of labour power in relation to the needs of capitalist accumulation. The greater the labour surplus and the more rapid its rate of expansion, the easier it is for capital to control the struggle in the workplace. The principle of the industrial reserve army under capitalism is one of Marx's most telling insights. Migrations of labour and capital as well as the various mobilization processes by means of which 'unused' elements in the population are drawn into the workforce are manifestations of this basic need for a relative surplus population. But we also have to consider the costs of reproduction of labour power at a standard of living which reflects a whole host of cultural, historical, moral and environmental considerations. A change in these costs or in the definition of the standard of living has obvious implications for real-wage demands and for the total wage bill of the capitalist class. The size of the internal market formed by the purchasing power of the working class is not irrelevant to accumulation either. Consequently, the consumption habits of the workers are of considerable direct and indirect interest to the capitalist class.

But we should also consider a whole host of qualitative aspects to labour power encompassing not only skills and training, but attitudes of mind, levels of compliance, the pervasiveness of the 'work ethnic' and of 'possessive individualism', the variety of fragmentations within the labour force which derive from the division of labour and occupational roles, as well as from older fragmentations along racial, religious and ethnic lines. The ability and urge of workers to organize along class lines depends upon the creation and maintenance of a sense of class consciousness and class solidarity in spite of these fragmentations. The struggle to overcome these fragmentations in the face of divide-and-conquer tactics often adopted by the capitalists is fundamental to understanding the dynamics of class struggle in the workplace.

This leads us to the notion of *displaced* class struggle, by which I mean class struggle which has its origin in the work process but which ramifies and reverberates throughout all aspects of the system of relations which capitalism establishes. We can trace these reverberations to every corner of the social totality and certainly see them at work in the flows of capital between the different circuits. For example, if productivity fails to rise in the workplace, then perhaps judicious investment in 'human capital' (education), in cooptation (homeownership for the working class), in integration (industrial democracy), in persuasion (ideological indoctrination) or repression might yield better results in the long run. Consider, as an example, the struggles around public education. In *Hard Times*, Dickens constructs a brilliant satirical counterpoint between the factory system and the educational, philanthropic and religious institutions designed to cultivate habits of mind amongst the working class conducive to the workings of the factory system, while elsewhere he has that archetypal bourgeois Mr. Dombey, remark that public education is a most excellent thing provided it teaches the common people their proper place in the world. Public education as a right has long been a basic working-class demand. The bourgeoisie at some point grasped that public education could be mobilized against the interests of the working class. The struggle over social services in general is not merely over their provision, but over the very nature of what is provided. A national health-care system which defines ill health as inability to go to work (to produce surplus value) is very different indeed from one dedicated to the total mental and physical well-being of the individual in a given physical and social context.

The socialization and training of labour – the management of 'human capital' as it is usually called in the bourgeois literature – cannot be left to chance. Capital therefore reaches out to dominate the living process – the reproduction of labour power – and it does so because it must. The links and relations here are intricate and difficult to unravel. I will consider various facets of activity within the dwelling place as examples of displaced class struggle.[8]

Some remarks on the housing question

The demand for adequate shelter is clearly high on the list of priorities from the standpoint of the working class. Capital is also interested in commodity production for the consumption fund provided this presents sufficient opportunities for accumulation. The broad lines of class struggle around the 'housing question' have had a major impact upon the urban process. We can trace some of the links back to the workplace directly. The agglomeration and concentration of production posed an immediate quantitative problem for housing workers in the right locations – a problem which the capitalist initially sought to resolve by the production of company housing but which thereafter was left to the market system. The cost of shelter is an important item in the cost of labour-power. The more workers have the capacity to press home wage demands, the more capital becomes concerned about the cost of shelter. But housing is more than just shelter. To begin with, the whole structure of consumption in general relates to the form which housing provision takes. The dilemmas of potential overaccumulation which faced the United States in 1945 were in part resolved by the creation of a whole new life style through the rapid proliferation of the suburbanization process. Furthermore, the social unrest of the 1930s in that country pushed the bourgeoisie to adopt a policy of individual homeownership for the more affluent workers as a means to ensure social stability. This solution had the added advantage of opening up the housing sector as a means for rapid accumulation through commodity production. So successful was this solution that the housing sector became a Keynesian 'contra-cyclical' regulator for the accumulation process as a whole, at least until the *débâcle* of 1973. The lines of class struggle in France were markedly different (see Houdeville 1969). With a peasant sector to ensure social stability in the form of small-scale private property-ownership, the housing problem was seen politically mainly in terms of costs. The rent control of the inter-war years reduced housing costs but curtailed housing as a field for commodity production with all kinds of subsequent effects on the scarcity and quality of housing provision. Only after 1958 did the housing sector open up as a field for investment and accumulation and this under government stimulus. Much of what has happened in the housing field and the shape of the 'urban' that has resulted can be explained only in terms of these various forms of class struggle.

The 'moral influence' of suburbanization as an antidote to class struggle

The second example I shall take is even more complex. Consider in its broad outlines, the history of the bourgeois response to acute threats of civil strife which are often associated with marked concentrations of the working class and the unemployed in space. The revolutions of 1848 across Europe, the Paris Commune of 1871, the urban violence which accompanied the great railroad strikes of 1877 in the United States and the Haymarket incident in Chicago, clearly demonstrated the revolutionary dangers associated with the high concentration of the 'dangerous classes' in certain areas. The bourgeois response was in part characterized by a policy of dispersal so that the poor and the working class could be subjected to what nineteenth-century urban reformers on both sides of the Atlantic called the 'moral influence' of the suburbs. Cheap suburban land, housing and cheap transportation were all a part of this solution entailing, as a consequence, a certain form and volume of investment in the built environment on the part of the bourgeoisie. To the degree that this policy was necessary, it had an important impact upon the shape of both British and American cities. And what was the bourgeois response to the urban riots of the 1960s in the ghettos of the United States? Open up the suburbs, promote low-income and black homeownership, improve access via the transport system . . . the parallels are remarkable.

The doctrine of 'community improvement' and its contradictions

The alternative to dispersal is what we now call 'gilding the ghetto' – but this, too, is a well-tried and persistent bourgeois response to a structural problem which just will not disappear. As early as 1812, the Reverend Thomas Chalmers wrote with horror of the spectre of revolutionary violence engulfing Britain as working-class populations steadily concentrated in large urban areas. Chalmers saw the 'principle of community' as the main bulwark of defense against this revolutionary tide – a principle which, he argued, should be deliberately cultivated to persuade all that harmony could be established around the basic institutions of community, a harmony which could function as an antidote to class war. The principle entailed a commitment to community improvement and a commitment to those institutions, such as the Church and civil government, capable of forging community spirit. From Chalmers through Octavia Hill and Jane Addams, through the urban reformers such as Joseph Chamberlain in Britain, the 'moral reformers' in France and the 'progressives' in the United States at the end of the nineteenth century, through to model cities programmes and citizen participation, we have a continuous thread of bourgeois response to the problems of civil strife and social unrest.

But the 'principle of community' is not a bourgeois invention. It has also its authentic working-class counterpart as a defensive and even offensive weapon in class struggle. The conditions of life in the community are of great import to the working class and they can therefore become a focus of struggle which can assume a certain relative autonomy from

that waged in the factory. The institutions of community can be captured
and put to work for working-class ends. The Church in the early years
of the industrial revolution was on occasion mobilized at the local level
in the interests of the working class, much as it also became a focus for
the black liberation movement in the United States in the 1960s and is
a mobilization point for class struggle in the Basque country of Spain.
The principle of community can then become a springboard for class action
rather than an antidote to class struggle. Indeed, we can argue that the
definition of community as well as the command of its institutions is
one of the stakes in class struggle in capitalist society. This struggle can
break open into innumerable dimensions of conflict, pitting one element
within the bourgeoisie against another and various fragments of the working class against others as the principles of 'turf' and 'community autonomy' become an essential part of life in capitalist society. The bourgeoisie
has frequently sought to divide and rule but just as frequently has found
itself caught in the harvest of contradictions it has helped to sow. We
find 'bourgeois' suburbanites resisting the further accumulation of capital
in the built environment, individual communities in competition for development producing a grossly inefficient and irrational spatial order even
from the standpoint of capital, at the same time as they incur levels of
indebtedness which threaten financial stability (the well-publicized current
problems of New York are, for example, typical for the historical experience
of the United States). We find also civil disorder within the urban process
escalating out of control as ethnic, religious and racial tensions take on
their own dynamic in partial response to bourgeois promptings (the use
of ethnic and racial differences by the bourgeoisie to split the organization
in the workplace has a long and ignoble history in the United States in
particular).

Working-class resistance and the circulation of capital

The strategies of dispersal, community improvement and community competition, arising as they do out of the bourgeois response to class antagonisms, are fundamental to understanding the material history of the urban
process under capitalism. And they are not without their implications
for the circulation of capital either. The direct victories and concessions
won by the working class have their impacts. But at this point we come
back to the principles of accumulation, because if the capitalist class is
to reproduce itself and its domination over labour it must effectively render
whatever concessions labour wins from it consistent with the rules governing the productivity of investments under capitalist accumulation. Investments may switch from one sphere to another in response to class struggle
to the degree that the rules for the accumulation of capital are observed.
Investment in working-class housing or in a national health service can
thus be transformed into a vehicle for accumulation via commodity production for these sectors. Class struggle can, then, provoke 'switching
crises', the outcome of which can change the structure of investment flows
to the advantage of the working class. But those demands which lie within

the economic possibilities of accumulation as a whole can in the end be conceded by the capitalist class without loss. Only when class struggle pushes the system beyond its own internal potentialities, is the accumulation of capital and the reproduction of the capitalist class called into question. How the bourgeoisie responds to such a situation depends on the possibilities open to it. For example, if capital can switch geographically to pastures where the working class is more compliant, then it may seek to escape the consequences of heightened class struggle in this way. Otherwise it must invest in economic, political and physical repression or simply fall before the working-class onslaught.

Class struggle thus plays its part in shaping the flows of capital between spheres and regions. The timing of the 'long waves' of investment in the built environment of Paris, for example, is characterized by deep troughs in the years of revolutionary violence – 1830, 1848, 1871 (see Figure 5.5). At first sight the rhythm appears to be dictated by purely political events yet the typical 15–25-year rhythm works just as well here as it does in other countries where political agitation was much less remarkable. The dynamics of class struggle are not immune to influences stemming from the rhythms of capitalist accumulation, of course, but it would be too simplistic to interpret the political events in Paris simply in these terms. What seems so extraordinary is that the overall rhythms of accumulation remain broadly intact in spite of the variations in the intensity of working-class struggle.

But if we think it through, this is not, after all, so extraordinary. We still live in a capitalist society. And if that society has survived then it must have done so by imposing those laws of accumulation whereby it reproduces itself. To put it this way is not to diminish working-class resistance, but to show that a struggle to abolish the wages system and the domination of capital over labour must necessarily look to the day when the capitalist laws of accumulation are themselves relegated to the history books. And until that day, the capitalist laws of accumulation, replete with all of their internal contradictions, must necessarily remain the guiding force in our history.

A concluding comment

I shall end by venturing an apology which should properly have been set forth at the beginning. To broach the whole question of the urban process under capitalism in a short article appears a foolish endeavour. I have been forced to blur distinctions, make enormous assumptions, cut corners, jump from the theoretical to the historical in seemingly arbitrary fashion, and commit all manner of sins which will doubtless arouse ire and reproach as well as a good deal of opportunity for misunderstanding. This is, however, a distillation of a framework for thinking about the urban process under capitalism and it is a distillation out of a longer and much vaster work (which may see the light of day shortly). It is a framework which has emerged as the end-product of study and not one which has been arbitrarily imposed at the beginning. It is, therefore, a framework in which I have great confidence. My only major source of doubt, is whether I have been able to present it in a manner which is both accurate enough and simple enough to give the correct flavour of the potential feast of insights which lie within.

Notes

[1] Gottlieb (1976) provides an extensive bibliography on the subject as well as his own statistical analysis. The question of 'long waves' of various kinds has recently been brought back into the Marxist literature by Mandel (1975) and Day (1976).

[2] The main source of information is Brinley Thomas (1972 edition) which has an extensive bibliography and massive compilations of data.

[3] The whole question of the capital surplus in the eighteenth century was first raised by Postan (1935) and subsequently elaborated on by Deane and Cole (1967). Recent studies on the financing of turnpikes and canals in Britain by Albert (1972) and Ward (1974) provide some more detailed information.

[4] The best study is that by Parry Lewis (1965).

[5] The study by Girard (1952) is truly excellent.

[6] I have attempted a much more extensive treatment of the transport problem in Harvey (1975).

[7] See Isard (1942) for some interesting material.

[8] The account which follows is a summary of Harvey (1977).

References

Albert W. (1972) *The Turnpike Road System in England.* Cambridge: Cambridge University Press.
Day, R. (1976) 'The theory of long waves: Kondratieff, Trotsky, Mandel.' *New Left Review* 99:67–82.
Deane, P. and Cole, W. A. (1967) *British Economic Growth, 1688–1959: Trends and Structure.* Cambridge: Cambridge University Press.
Girard, L. (1952) *Les politiques des travaux publics sous le Second Empire.* Paris: Armand Colin.
Gottlieb, M. (1976) *Long Swings in Urban Development.* New York: NBER.
Harvey, D. (1975) 'The geography of capitalist accumulation: a reconstruction of the marxian theory.' *Antipode* 7, 2:9–21.
Harvey, D. (1977) 'Labour, capital and class struggle around the built environment in advanced capitalist societies.' *Politics and Society* 6:265–95.
Houdeville, L. (1969) *Pour une civilisation de l'habitat.* Paris: Editions Ouvrières.
Isard, W. (1942) 'A neglected cycle: the transport building cycle.' *Review of Economics and Statistics* 24:149–58.
Lewis, J. Parry, (1965) *Building Cycles and Britain's Growth.* London: Macmillan.
Mandel, E. (1975) *Late Capitalism.* London: New Left Books.
Marx, K. edn. (1967) *Capital* (three volumes). New York: International Publishers.
—— edn. (1967, 1968 and 1971) *Theories of Surplus Value.* Moscow: Progress Publishers.
—— edn. (1973) *Grundrisse.* Harmondsworth: Penguin Books.
Postan, M. (1935) 'Recent trends in the accumulation of capital.' *Economic History Review* 6:1–12.

Rougerie, J.(1968) Remarques sur l'histoire des salaires à Paris au dix-neuvième siècle. *Le Mouvement Sociale* 63:71–108.

Thomas, B. edn. (1973) *Migration and Economic Growth: A Study of Great Britain and the Atlantic Economy.* Cambridge: Cambridge University Press.

Ward, J. R. (1974) *The Finance of Canal Building in the Eighteenth Century.* London: Oxford University Press.

6 The urban land question*

Shoukry T. Roweis and Allen J. Scott

Introduction

Our objective in this paper is to discuss the phenomenon of contemporary urban land development and the characteristic problems with which it is associated. We situate this phenomenon between two polarities: the process of spontaneous urbanization and the process of deliberate urban intervention.

Our discussion opens with a purely descriptive account of contemporary urban land problems and policies, and we then summarize, and subsequently criticize, the theoretical/analytical underpinnings which seem to sustain those policies. Later, we attempt to lay the foundations of a formal and critical alternative theory of urban land. On the basis of this theory, we present some assertions and hypotheses about the urban land development process as it exists today. Lastly, we identify a number of policy questions that emerge, we hope, not from a gratuitous attitudinizing on our part, but from the fundamental logic of our analysis.

Current urban land problems and policies

In order to initiate our main discussion we now attempt, in a very general way, to survey the main problems and policies that have seemed to domi-

* This paper is a considerably revised version of a paper originally commissioned by the Canadian Council on Urban and Regional Research and presented to the Ontario Forum on Land Management (Toronto, October 1976).

nate the North American urban scene in recent decades. We undertake, first, to construct a general catalog of problems; and, second, to characterize generally the kinds of policies that have developed in response to these problems. In the conclusion to our paper, we shall build upon and extend the basic notions developed in this initial elementary survey. Observe, at the outset, that our survey is restricted to those problems that have their specific origins in the process of urban land development. This is ultimately the same as saying that, out of the range of problems that occur *in* urban areas, we restrict our attention to problems *of* urban areas.

Urban land problems

In the twentieth century, the urban problem in North America has been, par excellence, the problem of urban expansion consequent upon general economic growth. This growth confronts society with grave difficulties concerning the collective production and utilization of urban land. Already, by the end of the nineteenth century, these difficulties had made themselves known. They were, thus, not new difficulties, although they were taking on new forms, just as the economic system itself was taking on new forms. The old methods of treating these difficulties were, hence, becoming increasingly inoperative, while effective new methods had not yet been developed.

By the 1920s a prototypical response to the overall problem of urban growth and expansion had evolved in most North American cities. It was a doubly faceted policy involving both aggressive, peripheral urban expansion and the rationalization of land uses by zoning. An early wave of peripheral expansion at the end of the nineteenth century had taken the form of streetcar suburbs; that is, compact, dense, and geographically distinct clusters of suburban settlements. By the 1920s, however, the automobile had become a pervasive and dominating mode of personal transport. Thus, the early pattern of suburban expansion gave way to a new pattern of interminably sprawling, low-density housing. This was associated with the general widening and paving of county roads and the outward extension of utilities, particularly sewers, frequently well in advance of actual housing construction. Annexation as a typical nineteenth-century administrative procedure was gradually abandoned in favor of a system of independent municipal governments in the growing suburbs. At the same time, there were the beginnings of the suburbanization of industrial activities. All of this occurred within a regime of land-use zoning (or restrictive covenants) which encourage the formation of geographically homogeneous land-use districts, at least in the new suburban areas.

Contrary to much received opinion, this double policy of peripheral expansion and land-use zoning was, by and large, effective in dealing with the urban problem. It allowed for the provision of ample space for industrial expansion; it partly succeeded in rationalizing the geographical configuration of urban infrastructure; it encouraged expansion of the supply of housing, whether directly to high and middle-income families or indirectly (by filtering) to low-income families; and it expedited the takeover of highly accessible downtown locations by the burgeoning tertiary – and especially quaternary – sectors of economic activity. Furthermore, throughout the period stretching from the beginning of the 1920s to the end of the 1950s, the phenomenal expansion in the supply of serviced land staved off excessive land hoarding, speculation, and the formation of oligopolies in the land development industry. Land-use zoning gave rise to significant improvements and economies in the provision of basic urban infrastructure, and it eliminated many of the uncertainties that property owners and developers had hitherto faced. Ironically, this very success became the prime source of most manifest urban land problems in the 1960s and the 1970s.

By the 1950s, the aggressive policy of facilitating urban expansion was taking the form of massive intraurban highway construction, thereby encouraging further suburbanization and further utilization of the private automobile. This same process gave rise to formidable traffic and parking problems, particularly in the central city. It also provoked significant economic difficulties for intraurban public transport companies. A syndrome was thus created whereby expressways gave rise to the need for yet more urban expressways. Suburban expansion itself began to appear increasingly inefficient. Low densities of development meant relatively wasteful expenditures on basic infrastructure. Thus, the social costs of each incremental development on the urban periphery were significantly higher than those of previous increments. As commuting distances increased, so the need for faster (and yet safer) highway travel made itself felt. With these demands for improved levels of service, highway construction costs began to mount. Low population densities in the suburbs further meant that the provision of social and community services involved a considerable degree of waste. Under the pressures of ever-mounting fiscal obligations, municipalities resorted to discriminatory land-use controls designed to maximize property tax revenues and to minimize fiscal obligations. These practices resulted in intermunicipal rivalries and confrontations, and they tended to impose disproportionate fiscal burdens on the central cities. These burdens were all the more onerous in the circumstances of a progressive flight of middle-class families, as well as industrial activities, from the central cities to the suburbs. A further syndrome was, therefore, created. The erosion of the tax base of central cities inevitably produced a deterioration in their levels of social and community services, and this produced yet further rounds of out-migration. In the process, central business districts, nevertheless, typically retained, and even added to, their quaternary office functions (administration and finance). In this way, the central business districts continued to exist as important white-collar employment centers. Hence a typical, and economically wasteful, pattern of daily intraurban travel began to make its appearance. One the one hand, white-collar workers commuted from the suburbs to the central city. On the other hand, low-income, blue-collar workers commuted from the central city outward to the peripheral industrial districts. Yet more highway links and parking facilities had to be added in the central cities in order to accommodate this wasteful transport pattern. As a consequence, portions of the already scarce stock of central low-income housing were destroyed, a phenomenon exacerbated by the expansion of office activities. Overcrowding and an acute shortage of low-income housing were the result.

Although fairly typical, the scenario described above was neither invariable nor universal among North American cities. In some cities, urban land development tended, in certain respects, to follow rather different lines. In particular, in Canada and in some parts of the United States, intraurban expressway construction did not occur on anything like the massive scale that it did elsewhere. In these cases, there tended to be a relative lag in the supply of peripheral, developable land, and this situation led in turn to an overall inflation of land and housing prices. In these circumstances, urban land hoarding became attractive, if not inevitable, and a significant group of professional hoarders made its appearance in the form of large, vertically integrated land development firms. The unusually high land prices, combined with a general land shortage, made highrise residential developments feasible, if not economically necessary, especially in the central city. However, in the process of assembling land units for the purpose of constructing highrise apartment blocks, land developers typically triggered a series of escalating confrontations with neighborhood organizations. Members of these organizations correctly recognized the difficulty of finding viable substitutes for their inner-city homes, and thus tended to hold on tenaciously to them. By contrast, under conditions of increasing demand pressures on the housing market combined with rising land prices, it was economically logical for the development companies (which, in addition, had the necessary financial and organizational capability) to redevelop urban land at maximum feasible densities. The confrontations of developers and neighborhood groups have frequently implicated a variety of government agencies in difficult and costly political and administrative dilemmas.

Under these conditions of a general land shortage due to restricted urban expansion, central business districts retain their hegemony over retail and office activities. The price of land at central locations escalates continually. This escalation of land prices results in two apparently contradictory outcomes: on the one hand, an insistent intensification of land uses; and, on the other hand, land hoarding (typified by the pervasive parking lot in the case of Toronto). By diminishing effective supply, land hoarding drives land prices still higher, producing yet more intensification of land uses. Given that central business district firms tend to be highly labor intensive, land-use intensification soon results in some serious transport problems, such as overloaded transit systems, congested central streets, a scarcity of parking facilities, and the like. Political pressures thus mount over the issue of inadequate downtown transport facilities, and this produces planning intervention in the form of improvements in service and capacity. These improvements augment the locational advantages of the central business district, and this leads to further increases in land prices and in land-use intensification.

In the old residential neighborhoods surrounding the business core, two sorts of pressures characteristically come in to play. First, as land prices escalate in the core, pressure mounts to redevelop surrounding low-income residential properties into office and commercial space. Second, as the ratio of white-collar to blue-collar employment in the core rises, so more and more middle-income families are prompted to purchase and renovate properties in old, low-income neighborhoods in close proximity to the core. Both types of pressure threaten to dispossess low-income families of their housing, a situation that is made all the more urgent, given the low rate of expansion of substitute low-income housing on the urban periphery. These pressures then result in political confrontations, the disruption of neighborhoods with a high degree of social cohesion, and further rises of land and housing prices.

In the long run, the complex unfolding of these various forces and outcomes diminishes both the political and economic prospects of more rational urban development programs in the future. This assertion is based on two general observations: first, piecemeal increments to existing infrastructure and social overhead capital to cope with bottlenecks as they arise (reactive planning) preempts more general and far-reaching policy options in the future; second, massive, but premature, private investment in localized renovation and redevelopment in conformity with a purely private economic calculus further blocks future possibilities of more socially progressive, collective redevelopment schemes.

The descriptions presented above are meant to outline only the most general tendencies in the evolution of urban land problems in North American cities in this century. Of course, particular cities will exhibit particular and idiosyncratic deviations from these tendencies. Nevertheless, the tendencies are, in one form or another, virtually everywhere observable in North America. In the face of the pervasive and serious problems raised in this discussion, the question immediately arises: what policies, programs, and modes of intervention have been developed to counter them?

Urban land policies

In general, urban governments in North America have seemed to intervene in urban affairs in three rather different, yet interrelated, ways: (a) by the application of a variety of fiscal devices; (b) by legally restricting private rights to use urban land in certain ways; (c) by direct physical undertaking of urban development (and/or redevelopment) programs.

The first type of urban intervention, the application of various sorts of fiscal devices, includes policy instruments such as the property tax, controls on the pricing of urban goods and services, and direct governmental subsidies and grants. Let us deal with each of these instruments in turn.

The property tax seems to be ubiquitous in North America. Further, in recent years various mutations of the simple property tax as such have appeared: these include a variety of development levies, speculation taxes, and site value taxes. But, for all their differences, these taxes all have one fundamental characteristic in common: namely, the fact that they have left largely intact the operation of those macroforces (whether economic or social) that determine the structure of the land market as a whole. For this reason, property-tax policies have had remarkably little palliative effect on the general urban land problems outlined earlier.

Controls on the pricing of urban goods and services include matters such as rent control, road pricing, legislated limits on mortgage rates, and so on. These policies have definite effects on urban land prices and land uses. However, they tend to be eclectic 'plugging' devices, applied in a purely piecemeal way and independently of any global, urban planning strategy. At the same time, they tend to overlook the fundamental structural roots of the problems they seek to counter and to treat only their superficial symptoms; rent control is a perfect example of this tendency. In this sense, depending on their conjunctural relationships, they often, in the end, only complicate the problems that they seek to resolve.

Direct governmental subsidies and grants involve a whole gamut of policy instruments, including, for instance, the subsidization of mass transit, the allocation of funds to low-income housing programs, the provision of grants to community services, and so on. Many kinds of urban activities would have been impossible in the absence of this public financial support. Moreover, so long as public revenues were forthcoming in a quantity sufficient to cope with reformist demands, very few individuals, with the exception of a small minority of conservatives, could question this type of intervention.[1] However, urban intervention via subsidies and grants tends to have a two-fold drawback. In the first place, as in the case of controls on the pricing of urban goods and services, it has tended to deal with the symptoms, rather than with the fundamental causes, of urban land problems. For example, a fiscal policy (such as the subsidization of nonprofit housing schemes) that seeks to counter the endemic shortage of low-income housing in modern capitalist cities may, indeed, help to alleviate the problem in the short run; but it can never attack the persistent root causes of that shortage: a permanent imbalance in the distribution of the economic surplus. In the second place, urban reform via subsidies and grants tends to be ultimately self-negating. The continued feasibility of these measures requires that, in the long run, aggregate expenditures keep pace with aggregate public revenues. This in turn requires that the demands of revenue-absorbing sectors keep pace with revenue-producing sectors. However, recent empirical experience in North American cities has amply demonstrated that this equilibrium is far from being satisfied. On the contrary, public expenditures tend persistently to outrun public revenues, as the State is called upon, to an ever-increasing degree, to fulfill its role as guarantor of last resort in an unending escalation of urban land problems. Not only do fiscal reforms fail to deal with these problems in any fundamental sense; but, again, in some instances, they actually exacerbate them. Thus, continued subsidization of certain inefficient urban services may very well contribute only to these same services being offered at yet higher levels of inefficiency. Fiscal reforms tend to be both merely palliative and, finally, self-paralyzing.

The second type of urban intervention, legal restrictions on land uses, includes planning devices such as official plan provisions, zoning ordinances, subdivision controls, building codes, etc. It also includes certain recent experiments with legal transfers of development rights. These various types of restrictions all possess the *technical potential* of significantly modifying the operation of the urban land market; that is, the determination of the spatial configuration of urban land prices and uses. However, in practice, and for *political* reasons, they have tended to be formulated in a way such that their impacts on the land market have been deliberately restrained. It is clear that if tight and durable controls were imposed, for example, on office building heights at downtown locations, then this would eventually depress land prices at these locations. But the controls must, indeed, be tight and durable. If landowners and developers sense that the controls are liable, sooner or later, to be modified and diluted, then they will continue, more or less, to exchange and develop land as if the controls did not exist. Admittedly, there are innumerable cases in which such legal controls have been quite definite, and they have accordingly had an influence on the urban land market. In other cases, they have not been able to withstand the political pressures to change them. We may ask: what accounts for the observed tightness and durability of controls in certain cases and their ephemeral existence in others?

Cases where legal restrictions on land uses have been significantly durable tend to share a common characteristic: they are typically cases where land prices are *raised* (or at least stabilized), rather than lowered, by the restrictions. The example of land zoned for suburban residential activity is outstanding in this respect. On the other hand, cases where legal restrictions on land uses have been continually relaxed are typically cases where these restrictions would otherwise threaten to depress land prices. Either they threaten to depress land prices, generally, or at least they threaten the interests of a politically significant group of owners. The mutations of planning regulations in the business cores of large metropolitan centers represent an object lesson in this process.

The third, and last, type of public urban intervention, direct physical land development (or redevelopment), has probably been of more widespread significance to the urban land question than have either of the other two. It includes activities such as the provision of various types of urban infrastructure and social overhead facilities, public housing construction, urban renewal, the laying out of industrial estates, land banking, and so on. Through this type of intervention, government plays a key role in producing privately developable urban land and in shaping the spatial configuration of urban land prices and uses. Yet, ironically, and despite its crucial role, this type of intervention remains virtually incapable of dealing with *real* urban land problems. Publicly serviced land is left to be exchanged and utilized by innumerable private owners and users, all of them following specific private interests and oblivious to the collective consequences of their actions. This anarchical process leads to uncontrolled, unexpected, and unintended spatial configurations of differential locational advantages; and, hence, of urban land prices, uses, and problems.

Despite the definite potential effectiveness government might have through this type of intervention, it, in fact, finds itself participating in the creation of the very problems it seeks to combat.

From this very sketchy review of contemporary urban land policies, four fundamental conclusions emerge. First, whereas a wide variety of policies is employed to guide and regulate private rights to the use of urban land, none has seemed to have had very great and unambiguous, long-run effectiveness. Second, whereas governments have assumed a major (and rapidly expanding) responsibility for, and control over, the production of developable land, the spontaneous utilization of this land in the private realm is largely left to follow its own momentum. Third, there are no effective policies seeking to control the *exchange* of urban land. Fourth, and as a consequence of the preceding three points, there remains a large and intractable body of concrete social, political, and economic problems around the issue of the urban land nexus.

In the succeeding section of this paper we consider some of the current theoretical formulations underpinning this lamentable state of affairs.

Current theoretical underpinnings of urban analysis and their policy implications

Current urban land theory in North America seems to fall largely into two opposing approaches. On the one hand, there is a clearly dominant approach that is rooted in conventional North American social science, and especially in marginalist neoclassical economics. On the other hand, there is a minor approach, highly critical of the former, and growing out of it by a sort of mechanical negation. This approach is, self-professedly, radical and reformist, but we prefer (for reasons that will become apparent) to identify it by its central sophism: the manipulated city hypothesis. Much of our purpose in the present account is to describe these two conflicting approaches, to demonstrate their policy consequences, to criticize them, and then to propose what we feel is a more adequate theoretical foundation for thinking about the urban land question.

The land market as a social harmonizer

As we have already indicated, urban economics is currently dominated by a marginalist, neoclassical epistemology.[2] The point of departure of this epistemology is the assertion that individuals construct, for themselves, internalized sets of ordered consumption preferences. Each individual has a fixed and given budget. Then, each individual, in conformity with his specific set of preferences, will select an actual consumption program that maximizes his level of total satisfaction, subject to the constraining influence of his budget.

This general point of departure has the following implications for an analysis of the urban land development process. The level of satisfaction of any individual residing in the urban system is a function of three different components: (a) the quantity of residential space consumed by that individual; (b) the total distance traveled from the individual's residence to various destinations in the urban area; and (c) the aggregate of all other goods and services consumed. In terms of expenses, the individual's budget is allocated to three different items: (a) residential land rent; (b) transport costs; and (c) all other goods and services. In conformity with this global characterization of the nature of economic being in urban space, each individual then seeks a residential location that maximizes his total satisfaction. This involves finding an equilibrium location that fully exhausts the individual's budget, so that any further reallocation of the budget among the three expenditure items listed above results only in a lowering of the individual's total level of satisfaction. Under appropriate assumptions about the structure of tastes and preferences, the residential mobility of urban population, and the fluidity of built urban structures, these processes can be shown to stimulate a pattern of urban settlement to the extent that both population density and land prices decline steadily from the city center outwards. An elementary generalization permits extension of this schema from the case of individuals to the case of firms. In brief, firms will seek a location such that profits are maximized, where profits are identified as total revenues minus total costs (including land rent).

Of particular interest, for our purposes, is the role that land rent plays in this general neoclassical model. Land in this model is a 'good,' much like any other good. Any plot of land (or, alternatively, any location in urban space) commands a rent or price that is essentially a function of two forces. In the first place, the price is a function of the relative scarcity of similar plots or locations. In the second place, the price is a function of the buyer's willingness to bid for land and the seller's willingness to sell. What this amounts to, then, is a conventional market or supply and demand model, despite the curious geographical heterogeneity of the urban land base. As a corollary, the model covertly insinuates itself as representative of the best of all possible worlds. Thus, urban land in this model is a scarce resource, and the market perfectly harmonizes and coordinates its allocation among competing users. Bidding for land *secures an agreement* between buyers and sellers about mutually satisfactory prices which also guarantee that the quantity of land offered for sale will be identical to the quantity actually bought. In addition, this market equilibrium must be socially optimal (in the context of the model) in the sense that realized land uses (i.e. final resource allocation) will be Pareto efficient.

These properties of the neoclassical model do, indeed, seem to be convincing (overwhelmingly so), and they emerge directly from a logic that is perfectly consistent within its own terms of reference. Unfortunately, as we hope to demonstrate in due course, those same terms of reference are hopelessly inadequate and misleading.

The manipulated city hypothesis

In recent years a certain body of critical writings has accumulated in opposition to the dominant neoclassical, or liberal, ideology. We might classify these writings as variously contributing to the *manipulated city hypothesis.*[3] In this view, urban land development is not the outcome of the myriad decisions of atomized individuals, each seeking to secure only his own satisfaction. Rather, urban society is seen as an amalgam of special interests, various social formations, neighborhood communities, and the like. Members of these interest groups act in concert. Moreover, social relations are generally dichotomized into relations between *exploiters* and *exploited.* These relations are underpinned by the degree of *social power* that each group possesses.

In the urban context, the main power group is seen as emerging from a coalition between finance capital (banks, trust companies, etc.) and the real-estate interest (developers, construction companies, landlords, etc.).[4] This coalition acts as a 'class monopoly.' By reason of its control over the whole urban land development process, and especially the housing market, it is able to manipulate both the spatial form of the city and (concomitantly) to extract enormous super-profits (or, in Harvey's terms, 'class-monopoly rents'[5]) from the mass of powerless users of urban land. The instruments whereby the coalition effects this state of affairs are diverse means such as red-lining; blockbusting; fixed mortgage rates; excessive profit markups; political influence; and, from time to time, simple corruption and swindling. In this process, the coalition, with equal indifference, destroys old neighborhoods, ghettoizes ethnic groups, and herds an unwilling populace into sterile highrise apartments. In short, it rapes and then dehumanizes the city, all in a merely pecuniary interest.

Policy implications

Theory is neither neutral nor passive. It embodies *human interest* and, hence, possesses *policy consequences.* Thus, both of the theoretical paradigms outlined above have their characteristic political flavors and, accordingly, their characteristic impacts upon government interventions and policy decisions.

The neoclassical paradigm is associated with a single, dominating norm that emanates from its doctrinaire belief in the naturalness and effectiveness of the market mechanism. That norm is: *Let the market work as long as possible; assist the market, or stimulate it, through public intervention, if, and only if, it fails.* Over a century ago, John Stuart Mill advocated this imperative (with the explicit recommendation that it represents the 'normal' state of human affairs) in the following terms: '. . . throw, in every instance, the burden of making out a strong case, not on those who resist, but on those who recommend government interference. Laissez-faire, in short, should be the general practice: every departure from it, unless required by some great good, is a certain evil. . . .'[6]

This market norm is, in fact, still widely promulgated, despite the persistent and ever-increasing evidence of 'market failure' in the urban system. Consider, for example, the pervasiveness of state-controlled activities such as subsidized housing, municipal transport systems, public utilities, and the rest. Given the phenomenon of market failure, the neoclassical ideology suggests two further operating principles in the matter of urban public policy. In the first place, whenever possible, administratively induce some form of competitive pricing process (e.g. road pricing).[7] In the second place, if direct administrative control is inevitable, then attempt to arrange the outcome as a simulacrum of a free market operation (cf. the public housing model suggested by Herbert and Stevens[8]). This latter principle is, of course, the basis of all modern welfare economics. In any case, the market norm is sustained by an abstract standard of performance that is in one sense totally convincing: the Pareto, or Pareto-Kaldor, criterion. At least it is convincing as a criterion of judgment in a world of perfectly atomized, independent, and ahistorical human beings. The only problem remains: does it have any relevance to practical policy decisions in the real universe of concrete social and economic processes? We shall argue that it does not because it abstracts away from the historically specific property relations of capitalist society. In the end, the neoclassical paradigm seems to lead only to a general, ameliorative philosophy ultimately eclectic, shortsighted, and self-restricting.[9]

By contrast, the theorists of the manipulated city see (no doubt correctly) the neoclassical doctrine of universal harmonies as only an ideological smoke screen masking the exploitative social relations of modern technological society. Of all these relations, those of particular concern to these theorists are the ones that emanate from the nexus of real-estate interests. Such interests are seen as being guided in their actions uniquely by an implicitly unscrupulous and mysteriously subjective profit motive. Thus, having rejected market rationality as a norm, these theorists (and their activist colleagues) seem to deny its empirical existence as well. To them, there is no immanent structural logic (to be revealed by analysis) that will account for urban land problems as a general social phenomenon. Rather, it is all simply a matter of a greedy, powerful, and unprincipled clique of *individuals* who have chosen, out of a purely idiosyncratic ethical *lapsus,* to pursue a set of private interests which are, by faith and definition, opposed to those of 'the people'.

The policy consequences of this position are, of course, evident. Stop urban growth; stop large-scale developments; preserve what exists; control rents and impose taxes on speculation; substitute the moral principle of 'equity' for the economic principle of efficiency (as if the two were mutually exclusive). The positions taken in recent years by many rate-payers' associations and by the so-called reform aldermen of the city of Toronto and their theoretical spokesmen represent a perfect example of this philosophy in action.

A critique and evaluation

Of course, there is something of value and interest in current theories of the land development process, both in the neoclassical mainstream and in the reformist critique. However, both approaches seem to us inadequate, mystified, and, in various respects, misleading. The following evaluation, therefore, deals with current theoretical underpinnings of urban analysis from the point of view of their inadequacies as statements of social theory and their failures as guidelines for policy making.

In recent years much has been made of the rather artificial, and often unrealistic, assumptions that are so pervasive in neoclassical analysis: perfect competition and perfect knowledge; the absence of externalities and of legal and social restraints; ubiquitous transport facilities and monocentric cities; instantaneous and costless residential relocation; and so on. In fact, to criticize the theory from this formalistic perspective is to miss the whole point: namely, *the fundamental epistemological inadequacy of the neoclassical problematique*. In this paper we consider and criticize three especially crucial aspects of the general neoclassical approach as it is applied to the analysis of urban land. These are: (a) the question of consumers' tastes and preferences and the formation of residential space; (b) the social and property relations that govern the production, exchange, and utilization of urban land; and (c) the nature of the State and the logic of public policy formulation.

First, then, 'tastes and preferences,' as such, are purely epiphenomenal. They are part of a larger cultural momentum that is itself embedded in, and grows out of, a global, social/historical process. This is the evolutionary development of a *mode of production,* by which phrase we mean not only a set of production techniques, but, more importantly, a web of social, political, and legal relations governing human interactions in the processes of production and exchange. It is in the context of a specific mode of production, and a derivatively specific social formation, that humans acquire and develop their tastes and preferences. The latter emerge, not from a capricious and abstract subjectivity, but from a series of collective human intentions, aspirations, and life projects made realistically realizable within a specific social formation. In a word, the system of production, in the sense given above, both historically and epistemologically precedes the system of consumption. Concomitantly, any attempt to establish a general theory of demand as an *exogenous* determinant of the processes of production and exchange is to engage in a *petitio principii* in the classical sense of taking what is to be proved as part of the would-be proof.[10] Neoclassical urban theory *á la* Alonso, Muth, and Solow, engages in the same kind of logical error.[11] It takes the spatial pattern of land prices and uses as a direct outcome of exogenously given consumers' tastes and preferences, whereas both are certainly part of a single explicandum. This scientific nullity is made finally evident by the fact that neoclassical urban models *can never in practice refute the alleged relationship* between land prices and uses, on the one hand, and tastes and preferences, on the other, for in these models 'tastes and preferences' are always themselves taken to be revealed in the realized pattern of land prices and uses. This is a process of explaining the thing by itself with a vengeance. But, more importantly, all such analyses are profoundly quietistic. For, if current urban patterns are nothing more than the expression of the consumer's preferences, then on what grounds (in a democratic society) can public intervention ever be justified?

Second, it is certainly true that a private land market is *one* way of allocating land to competing users, and in certain senses an administratively efficient way. However, to leave matters there, with the self-serving addendum that competitive market allocation is Pareto efficient (if not also eminently 'fair', according to certain enthusiasts) is to lose sight of a whole underlying level of analysis. The neoclassical paradigm abstracts away from the social and property relations of production and exchange. Thus, the so-called 'factors of production' – capital, labor, and land – lose their socio-historical specificity as articulations of human relations and interests. They are simply reduced to neutral and mutually substitutable *technical inputs* to an abstract economic process that rewards them according to their marginal productivity. In particular, this mystified conception conceals the peculiar nature of urban land and, hence, fails to grasp its twofold contradictory status, namely: (a) as a human product that is collectively produced, yet whose specific use values depend upon the uncontrollable ways in which it is privately utilized; and (b) as a human product that is collectively produced, yet whose benefits are privately appropriated in the form of land rent.

We shall return to this important issue in due course. For now, it is necessary to point out that this failure to grasp the peculiar nature of urban land has grave policy implications. Neoclassical analysis suggests that there are *no* fundamental irrationalities in the ways in which urban land is produced, exchanged, and utilized, and that those problems that do exist are the consequence of simple market imperfections that can be corrected in a series of simple, *ad hoc,* 'fine tuning' operations.

Third, we see no reason whatever to accept, either as a statement of fact or as a *desideratum,* that liberal notion of the State (hence, public policy and intervention) that is implicit in the neoclassical paradigm: namely, the State as a disinterested referee above and outside of society, intervening in the affairs of society only when some abstract, formalistic criterion of social optimality requires it. On the contrary, *the rationality of the State derives from the rationality of the civil society within which the State is historically embedded.* By ignoring this entirely self-evident premise, and by seeking to construct a series of abstract and disembodied criteria of social optimality and state intervention, the neoclassical paradigm (particularly as it is applied to urban land analysis) has thrown itself into something approaching a crisis of credibility; palpably, its policy prescriptions become ever more irrelevant.

The reformist critique has the merit of suggesting an analysis whose purely descriptive accuracy is frequently all too painfully evident. But what it gains at the level of surface appearances, it loses at the level of essence. From an epistemological point of view, the singling out of the property development (and finance capital) interests as the villain of the piece seems to us strained and artificial. It is the role and *raison d'être* of capital as a whole that are at issue here. In particular, the property development industry generally is a capitalist branch of production like any other capitalist branch and is *subject to the same general, overarching capitalist structure and logic.* From the observation that the activities of the property development industry generate uniquely problematical outcomes, the manipulated city theorists have been lured into the belief that the fundamental logic leading to these outcomes must itself be unique. This is precisely where the manipulated city theorists are in error. For the uniqueness – real as it is – of the outcomes is not due to some unique logic governing the activities of the property development industry, but rather to the application of a perfectly general capitalistic logic to a unique object of production and exchange: urban land. For this reason, the repeated attempts on the part of the manipulated city theorists to attack the property development industry *as such* have always failed fundamentally. This proposition is reinforced by consideration of the evident fact that landlords (and their cohorts) no longer in North America form a distinctive and identifiable social *class* apart. Along with land-holding and land-development activities themselves, landlords have been generally assimilated into the life and momentum of capital at large. Thus, any analysis of urban land and property processes must begin with an analysis of late capitalism as a general structure. And assertions of theft and conspiracy (to call things by their names) scarcely offer a very promising foundation from which to undertake such an analysis. Long ago Engels, in his attack on Proudhonist philosophy, showed, in *The Housing Question,* that the rhetorical banner of that philosophy – 'Property is theft' – is scientifically and politically null: scientifically because owners of production units extract surplus value in the form of unpaid labor *not* by theft, but in the perfectly normal course of events; politically because, to the degree that property owners are indeed swindlers, then to the same degree can they be dealt with by *existing* juridical arrangements.[12] The fundamental core and the central question of capitalism is not the pervasiveness of theft and conspiracy, but the process of the extraction and appropriation of surplus value. This process is hidden in the deep structure that lies below the commonly accepted social relations of capitalist society, and it is not peculiar – as the reformists seem to suggest – to the real-estate interests but is spread out over all spheres of productive activity.[13] In spite of these castigations, the spirit of Proudhon lives on, and, for example, in the form of the recent antideveloper movement in Toronto, it has succeeded only in exacerbating an already severe housing shortage. More recently, overtaken by a fit of moral indignation at the alleged cupidity of landlords, it has given us a policy of rent control that is surely destined ultimately to produce many more problems than it solves.[14]

The theory that we propose below is, we suggest, more far reaching than current theoretical formulations of whatever tendency. It paints a view of the world that is more real (hence progressive) than the view of the neoclassical theorists but, at the same time, infinitely more intractable (hence historically meaningful) than the world of the New Proudhonists.

The urban land nexus

We now attempt to set up a general theoretical framework for the elucidation of the urban land question as we see it. The inextricability of human knowledge and human interests makes it imperative, indeed inescapable, that we adopt a specific 'point of view' on this question. In any case, limits on human time, knowledge, and the ability to conceptualize problems impose this imperative.

Our dominant interest in the urban land question is in policy and its action consequences. The issue to us is not only what *can* be done about urban land problems, but also what *must* be done. At the same time, this point of view is not merely capricious or idiosyncratic; it is given to us historically. We do not live in a peasant society or in a science fiction utopia. We live within a capitalist socio-economic system that is markedly and specifically *urban* in character, and, as we have shown above, that is beset by a number of difficult congenital problems. Our consciousness of these problems is less a matter of personal choice than it is a matter of a controlled response to a given historical circumstance. Further, consciousness of these urban problems is not contingent upon the prior formulation of some abstract, speculative, ideal image of urban life. On the contrary, it is contingent upon the concrete structure and development of capitalist society itself. North American society has not engaged in frequently bitter urban controversies and disputes over property rights, institutionalized zoning, urban renewal, highway construction, and the rest simply because existing urban realities did not measure up to certain abstract urban visions. It was only when urban growth and development began to produce real problems that society started to act consciously to counteract them. It is precisely in this way that the sense and purpose of urban analysis are historically given. By the same token, a historically concrete urban analysis that is self-conscious about the course of policy and urban intervention will automatically withhold itself from posing intellectual puzzles that are purely abstract, formal, and pretendedly universal (such as, for example, the notion of a general urban equilibrium under perfect market assumptions). It will, rather, actively seek out those specific and concrete questions that appear in the context of socially and historically given modes of the organization of production (i.e. the necessities of social being) and of the conditions of reproduction (i.e. of human attitudes, job skills, and simple physical existence). To abstract away from these historically determinate parameters, relations, and problems is to rob analysis of its real social power and meaning. For this reason we initiate our discussion of the urban land nexus by mapping out the fundamental structure of the global socio-economic order within which urbanization processes and urban land problems are embedded.

Commodity production: a brief statement

Capitalist society is organized around the general social relations of commodity production and exchange. This presupposes – historically and analytically – the presence of three conditions: first, that humans have achieved a stage of technological productivity in a way such that they can produce an economic surplus over and above what is needed for immediate consumption; second, that production has become sufficiently articulated and advanced to allow for the emergence of specialized commodity producers;

and third, that wage labor, involving the exchange of labor power for wages determined on a labor market, has become the customary mode of organizing work relations. Given these conditions, commodity production emerges in the manifest form of entrepreneurial organization, where capitalist firms hire labor to work on materials and means of production (capital). In this way *commodities* are produced and sold for money prices. The revenues thus received are then spent partly to pay for labor, materials, and depreciated capital. Whatever is left after these payments are made is profit. Payments made to labor in the form of wages are spent on consumption commodities, and this consumption process ensures that commodity production as a whole is sustained. Profits are reinvested to enlarge the sphere of production. Thus proceeds commodity production and with it, the accumulation of capital.[15] These various, interlocking relations are demonstrated graphically and enlarged upon in Figure 6.1.

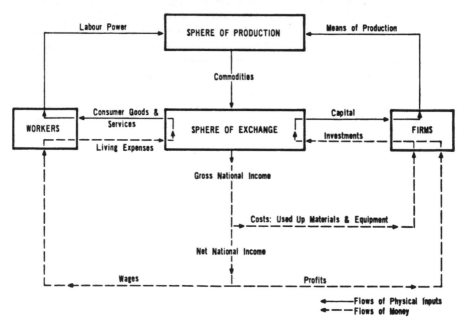

Figure 6.1 Commodity production: a simple schema of the interrelations of capital and labor

Commodity production, however, does not occur in a 'wonderland of no dimension', to borrow a phrase from Isard.[16] It is articulated in geographical space. It takes place *on land*. Concomitantly, the overall production process is mediated by a transport system. Nucleated clusters of firms and households come into being in order to minimize the friction of distance. More specifically, within each urban nucleation a configuration of relative spaces and locations emerges, and with it, a definite pattern of land uses. In particular, we can distinguish three principal kinds of urban land uses: namely, land used for commodity production and exchange (production space); land used for residential activity (reproduction space); and land used for transport (circulation space). The first two kinds of uses generally appear on privately owned land, while the last generally appears on publicly owned land. Depending on its relative location, any plot of land will be more or less valuable in exchange. A pattern of urban land rents (hence, prices) then emerges as a reflection of these differential locational advantages. Privately owned urban land is exchanged accordingly. On commodity-producing land, rent is paid out of profits, and it thus represents to firms an explicit reduction of profits. On residential

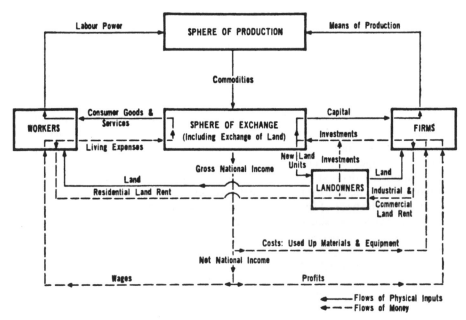

Figure 6.2 Commodity production: a simple schema of the interrelations of capital, labor and land

land, rent is paid out of wages, and it represents to workers a reduction in their standard of living. In both cases, however, rent is ultimately, in aggregate, a simple proportion of net national income, specifically appropriated by owners' land. These various propositions are clarified by Figure 6.2, which also shows, by comparison to Figure 6.1, how land inserts itself into the process of commodity production, generally.

Underlying these evident market relations there exists a deep structure of social and property relations. This structure underpins market exchanges, though it is not itself always clearly visible through them. At its core is an ongoing socio-political conflict (hence an ongoing, theoretical-*cum*-ideological controversy) over the distribution of the total social surplus (the net national income in Figures 6.1 and 6.2) into profits, wages, and land rents. For a given surplus, the higher the share of any one of these claims upon that surplus, the lower the shares of the other two. The socio-political conflict over income shares is thus structural, intrinsic, and ineluctable within the capitalist system of production. Furthermore, *the market does not eliminate (i.e. dissolve) this conflict;* quite the contrary, for continued market exchange is possible only in the circumstance where there exists a political power capable of maintaining the tense social balance between the three antagonistic claimants to the social surplus: capital, labor, and land. Observe that for the present we do not suggest, a priori, that these claimants inevitably constitute distinct, concrete social classes, though they do represent definite and distinguishable economic interests. We will return to this question in the final section of this paper.

Now, the neoclassical doctrine that free market exchange leads to an efficient (if not just) distribution of the social surplus by rewarding each production factor according to its marginal productivity is essentially a theoretical and ideological diversion. In fact, market prices (the *sine qua non* of exchange) are themselves nothing but the observable *results* of that central socio-political conflict over the distribution of the social surplus. For, as the work of the so-called Cambridge School has shown, commodities cannot be priced *before* the distribution of the social surplus is known.[17] In short, prices are empirically and analytically subsequent to profits, wages, and rents, with these economic categories in turn dependent upon the deep structure of capitalist property relations. Clearly, a scientific and, hence, useful political economy (and urban analysis) must start out on the basis of those historically given social and property relations which underlie and shape observable exchange relationships. In relation to urbanization processes and urban land problems, then, *the essential point of departure for us is not the phenomenon of competitive bidding for land, but the deep structure of urban property relations in relation to which the competitive bidding for land is only the faintest and most superficial pulsation.* However, in .order to pursue this matter of urban property relations, we must first elucidate certain essential characteristics of urban land. We must distinguish urban land as an object of exchange from commodities in the strict sense and, thereby, clarify those peculiar social relations that govern the production, exchange, and utilization of urban land.

The nature of urban land

Unlike raw land, urban land is only partly a primeval natural endowment. It is, in addition, *serviced* land. It is land on which humans have expended labor materialized in the form of infrastructural and structural artifacts. In this sense, urban land is a product of human labor. But is it similar to other human products such as wheat, typewriters, or haircuts? To answer this question we need to consider a crucial, analytical distinction between goods and services which *are* producible in the commodity form, and goods and services which are not.

Producibility in the commodity form is taken here to mean *the possibility of continued (unshackled) production of use values by individual capitalist firms as products capable of realizing their exchange values in the market, and hence capable of yielding at least the ongoing rate of profit to their producers.* The stability and viability of capitalist commodity production, as such, require the absence of external, that is, nonmarket barriers to private production. For the system to work, supply must be responsive to price signals. There are two complementary dimensions to that condition. First, it must be institutionally feasible for private producers to produce commodities on a continuing basis by the application of labor and materials. Second, private producers must have *no option* but to produce in the commodity form, and this limitation is enforced by competition among firms. In brief, commodity production requires that supply be responsive to price signals; that same responsiveness requires that goods and services be producible *in the commodity form.* It is important to point out that the notion of producibility in the commodity form does not signify producibility in the simple, physical, technical sense, but rather the possibility of production within the system of historically specific social relations of capitalism.

In capitalist society, however, certain goods and services may well be socially and economically useful, or even essential, and may be readily producible in the simple technical sense, yet are not producible in the commodity form. This may occur for one or more of the following reasons. First, whenever production in the strict commodity form entails the necessity of output levels at scales so massive as to lead to the emergence of monopolies or oligopolies (e.g. mass transport and communications), the State will frequently intervene and banish the commodity form, either by regulating or taking over production. This, of course, will occur only if the monopolistic or oligopolistic practices begin to have serious and deleterious effects on other key sectors of the economy. Second, goods and services which are essential to the survival of capitalist society, and yet which cannot be sold in such quantities or at such prices as will yield their producers at least the ongoing rate of profit, will not be produced in the commodity form, but in some other form (e.g. low-income housing). Third, the State will take over, or severely regulate, the production of any goods and services whenever the social stability of capitalist society requires that those goods and services be supplied in conformity with delicate political or ideological criteria (e.g. education, police, alcoholic beverages). This, we may point out, will occur even if such goods and services *can* be produced in the commodity form. Fourth, some economically useful by-products of human activity (i.e. externalities, such as concentrations of shoppers in central business districts) emerge in the form of purely spontaneous and nonmarketable benefits. Goods and services that are characterized by any possible combination of these circumstances are not commodities. For brevity we refer to them as *noncommodities.*

This distinction between commodities and noncommodities has far-reaching analytical and, more importantly, policy-related implications. Noncommodities confront capitalist society with significant social, political, and administrative problems; for the production , exchange, and utilization of these noncommodities require modalities and norms of decision making that are quite alien to, and frequently irreconcilable with, the logic of capitalist society at large. Urban land is clearly a noncommodity in the sense that its intrinsic use value – differential locational advantage – is produced not by individual capitalists, but through the agency of the State and the *collective effects* of innumerable individual social and economic activities. Specifically, urban land is produced in a complex collective dynamic where the State provides major infrastructural services as well as various public goods which cannot be adequately produced in the commodity form; in addition, land finally becomes urban only when its private utilization consummates the 'useful effects of urban agglomeration.'[18] Like wheat, typewriters, or haircuts, urban land is essentially a human product; but, unlike them, it is a noncommodity. Admittedly, it is possible here and there (particularly in past historical periods) to find examples of the production of urban land in the commodity form (e.g. the development of new towns by private companies), though the total quantity of land produced in this way is evidently negligible and certainly destined to diminish even further, given the growing complexity and indivisibility of urban equipment. At the same time, while urban *land* is definitely a noncommodity, urban *floor space* is a true commodity in the strict sense, produced whenever private owners intensify the use of their land. The possibility of intensification, however, presupposes the *prior* existence of urban land. Intensification only gives access to an existing and socially given use value represented by urban land in the noncommodity form (i.e. a set of differential locational advantages). At the same time, and paradoxically, any increment in urban floor space in the commodity form inadvertently produces as an externality an increment in urban land in the noncommodity form. That is, any increase in the quantity of floor space generates for all *other* urban activities an increase in the supply of differential locational advantages.

Now, in contemporary North American society, the process of production of urban land occurs in two different phases involving two distinct and ultimately incompatible sets of social and property relations of production and exchange. The first phase is one in which the State provides (either directly or by subsidization) major infrastructural and other public facilities. The output of this phase consists of serviced, developable, but not yet developed land. So far as the use characteristics of urban land are concerned, this product is, as it were, unfinished; for the final spatial configuration of differential locational advantages is as yet undetermined. In the second phase, this serviced land is exchanged through a series of private market transactions which determine the geographical pattern and intensity of urban land uses. Only then does urban land, qua a system of differential locational advantages, acquire its final use characteristics. Of course, in reality, urban land is not then frozen in stasis, but is in a continual process of change and redevelopment.

The two phases of urban land development described above differ significantly. Whereas the first phase involves various forms of collective decision making concerning the quality, location, and timing of public facilities, the second phase is a purely anarchical process whose outcomes – despite a variety of land-use controls – are unplanned and undecidable at the outset. The first phase can, in principle, be undertaken according to the criteria of social costs and benefits. But the second phase can be undertaken only according to the criteria of purely private costs and benefits. In short, *while the outcomes of the first phase are objects of collective decisions, the outcomes of the second phase are not.* Consequently, since the second phase is a decisive element (both as cause and effect) in the production of urban land, it follows that the production process *as a whole* (as currently realized) is inevitably anarchical. Thus, under current social and property relations, the spatial configuration of differential locational advantages (and, hence, of land uses and land rents) is *not* an object of collective decisions. Nor can it ever be rationally planned in the absence of significant changes in existing social and property relations.

There is a further and equally important phenomenon that determines the nature of urban land. This is the slow convertibility of the built urban environment. This phenomenon is fairly obvious at the level of simple physical appearance. However, its analytical and policy-related implications are rather more subtle, requiring amplification.

In and of itself, slow convertibility has no special consequence in terms of the social and property relations that govern the production, exchange, and utilization of urban land. Its analytical and policy consequences derive from the relationship of slow convertibility to the non-producibility of urban land in the commodity form and to the collective undecidability of its use characteristics. In the first place, had it been possible to accommodate urban activities within nondurable, instantly convertible structures, the non-producibility of urban land in the commodity form would have resulted in markedly fewer urban problems than is currently the case. In these circumstances, any increase in the demand for urban land would have always called forth, and been met by, an immediate response on the part of private developers in the form of land-use intensifications giving increased accessibility to existing urban land. Had this been the case, we would have undoubtedly witnessed a vastly denser development of urban activities than is now the rule. Similarly, had urban infrastructural facilities been sufficiently easy to implant and renew, then the process of private land-use intensification would have resulted in fewer infrastructural shortages and bottlenecks. Under conditions such as these, urban land would have resembled one of those theoretical agricultural landscapes *à* Von Thünen, whose productivity, use characteristics, and geographical articulation can be instantly adjusted to meet every possible contingency. In the second place, had urban structure and infrastructure been swiftly convertible, the undecidability of the use characteristics of urban land would have posed only negligible difficulties. Collective irrationalities and negative external effects due to this undecidability would have been purely ephemeral. Moreover, every perturbation in the existing spatial configuration of differential locational advantages would have instantly produced a counter-response in the form of an overall conversion of preexisting structures to correspond to the changed configuration of differential locational advantages. Under these hypothetical conditions, the vast majority of the urban land problems that we discussed above would have remained only temporary and self-correcting, minor inconveniences.

In spite of the patent unreality of those analytical assaults on the urban land question that suppose, as a point of departure, that urban land uses are indeed perfectly fluid, such analyses seem to be the rule rather than the exception. Small wonder, then, that mainstream urban theory so frequently ends up producing comforting fictions about the processes of urban growth and development, together with mystified policy prescriptions built around the notion of the market as universal panacea. Hence, the land market is seen as: an auto-regulating phenomenon; an efficient allocator of scarce land units among competing users; a maximizer of collective benefits; a rational sorter and arranger of land uses; an operationalization of consumers' tastes and preferences; and so on. Quite apart from our earlier criticisms of mainstream urban theory, the evident fact of the slow convertibility of urban land itself poses a central dilemma for that theory. This is not, it should be noted, a dilemma that can be solved by simply rewriting the equations of spatial equilibrium to take account of certain inertial properties of the urban environment. It is, rather, a case of a fundamental, perennial, and intrinsic dissonance in capitalist cities: *the permanent mismatch of urban form, as the inert outcome of a private decisional calculus, and the functional imperatives of a transcendent social rationality.*

The urban land nexus and the genesis of urban land problems

In a society where urban land is produced, exchanged, and utilized in conformity with the social and property relations outlined above, certain immanent structural contradictions inevitably arise. These contradictions represent the mutual and dynamic opposition of antithetical component elements of the whole land development process; they are concretely manifested in the kinds of urban land problems discussed earlier. Attempts to deal with these manifest problems are likely to continue to fail unless the fundamental structural contradictions that underlie them are resolved. This can only occur when existing social and property relations are themselves fundamentally reformed.

In a capitalist system of production, the compelling force of competition turns every entrepreneur's technical *potential* for improved productivity into an *actuality*. Under the threat of ruin, individual firms are constantly forced to seek out their best possible (profit-maximizing) set of production relations. The consequence is a tendency to ever-increasing social productive capacity and, in theory at least, efficiency in resource allocation. By the term *efficiency*, we mean here exactly what the neoclassical economists mean: a given allocation is efficient if it is not possible to increase the output of some desired goods and services (urban accommodation, transport services, etc., in our case) except by decreasing the output of some other desired goods and services. It is our object here to show that the logic of land ownership and utilization in capitalist society can never lead to efficiency in this sense. That is, the allocation of urban land to competing uses under capitalism must persistently and inherently undermine whatever general tendencies may otherwise exist in the direction of efficiency.

At the outset, note that potential sporadic movements *away* from efficiency are endemic within contemporary cities. These movements will occur whenever there is a change in the spatial pattern of differential locational advantages (e.g. as a result of the servicing of peripheral urban land, the modification of existing infrastructure, or the private redevelopment of any plot of land, etc.). These events provoke derivative readjustments in land-use patterns. The question now remains: can a competitive land market, over time, restore equilibrium, given disturbances of this nature?

Neoclassical theory has a ready answer to this question: the private urban land market can indeed restore efficiency, *but only under circumstances of perfect constancy in all conditions exogenous to this land market.*[19] Let us throw the best light that we can on this assertion and see what it implies. Thus, all (exogenously given) changes in the geography of differential locational advantages must take place in an episodic temporal pattern, in a way such that any initial set of changes is followed by a prolonged period of quiescence (in exogenous conditions). This period of quiescence must be *at least* as long as the longest economic life of all preexisting urban structures whose location is made problematical by the initial change. Such structures will be those where land rent and the marginal productivity of the land are no longer equal, thus necessitating land-use conversions if efficiency is to be restored. Clearly, however, such quiescence is the exception, rather than the rule, in contemporary North American cities; indeed, the spatial configuration of differential locational advantages is in a state of constant flux. Thus, even assuming that there is a convergent process of readjustment as the urban system seeks to accommodate itself to some specific exogenous impulse, long before even a first round of readjustments has been accomplished a further exogenous impulse will have pushed the urban system once more away from equilibrium.[20] And even if a first round of readjustments *were* possible, this in and of itself would change the existing pattern of differential locational advantages, which in turn would require yet further rounds of readjustments in order to restore efficiency, and so on. We have shown, then, that on two counts the time period theoretically necessary for a free land market to restore efficiency once it has been lost is unrealistically long. At the same time, not only are land-use readjustments to *past events* rarely socially efficient, they are also rarely efficient with respect to potential *future events*. As Teitz has very convincingly demonstrated, locational decisions that are made in conformity with purely private costs and benefits tend to be intrinsically 'myopic' in the sense that they tend to preempt collectively rational options in the future.[21] In these various circumstances, only an administrative arrangement that transcends the market could harmonize and synchronize all decisions, inducing a condition of social efficiency, both spatially and temporally.

It is obvious that the state of affairs described above imposes various sorts of *collective* penalties on urban society. In addition, this inherent market inefficiency has deleterious effects on *individual* commodity producers and workers. Thus, a firm may make what initially appears to be an optimal (profit-maximizing) locational decision from its own point of view. Then, in the course of time, it is faced with unforeseen changes in the configuration of differential locational advantages that either vastly truncate or vastly augment its production possibilities. In either case, the firm's original decision is now revealed to be suboptimal in its own purely private terms. Moreover, the firm's committed fixed capital costs prevent it from readjusting to a fully optimal level, at least in the short run (though, of course, the firm will still seek to optimize whatever choice variables remain within its command in the interim). The more dramatic the changes in the configuration of differential locational advantages (e.g. the massive shift from rail to road transport some time around the 1940s), the more pervasive such dysfunctionalities. For similar reasons, urban residents and workers have frequently incurred heavy penalties as a consequence of urban change, especially as a consequence of the rapidly shifting location of work places relative to residential areas. This has led to many cases of social conflict and unrest, to frequent and recurrent demands for higher wages.

From all of this, it follows that capitalist social and property relations create two major contradictory tendencies around the issue of urban land. On the one hand, the logic of commodity production and the private appropriation of profit call for functionally efficient urban land-use patterns. On the other hand, the private ownership and control of urban land lead to a tendency away from such efficiency.

If this is true (that is, if market allocation of urban land gives rise to a tendency away from efficiency), then what prevents cities from falling progressively into massive disarray? In fact, land-use allocations in contemporary cities are not determined purely by a market, but are the outcome of a peculiar partnership: a *market* allocation process that leads away from efficiency and a *political* allocation process that unceasingly attempts to rectify this inefficiency (while simultaneously producing yet other inefficiencies by altering the fundamental distribution of differential locational advantages). It is not pure, idle speculation to suggest that widespread urban disarray would probably have been the rule in the absence of ever-increasing political intervention. The history of cities under early laissez-faire capitalism lends some credence to this assertion. A cursory recollection of the empirical circumstances preceding the enactment of urban hygiene laws in the nineteenth century, the municipalization of urban utilities, the application of zoning and subdivision controls, the introduction of the various planning enabling acts, etc., provides further strong support in favor of this notion.[22]

More generally, it seems evident that state intervention in commodity production is a historical necessity imposed by the developmental logic of commodity production itself and made irrevocable by the innate self-destructive tendencies of commodity production. As a consequence, the logic of this intervention derives directly from the structural conditions of commodity production, and it changes historically as these structural conditions change. In the domain of urban land development, and given the basic contradiction outlined immediately above, *state intervention derives its basic character from the necessity to mediate this contradiction as and when the socio-economic outcomes of the contradiction begin to undermine the viability of commodity production.* Given this historically specific task, and given that urban land problems tend to emerge slowly and gradually, rather than suddenly and cataclysmically, the State understandably tends to search for policies which promise maximal effectiveness, yet which require minimal changes in existing social and property relations. This double-edged strategy contributes to the emergence of a number of dilemmas and predicaments, whereby in seeking to resolve urban problems, the State finds itself participating in the creation of new problems. To complete our argument, we deal here briefly with four illustrative dilemmas.

DILEMMA1   In their perennial search for profits, firms seek out even more technically efficient production processes. In short, over time they tend to intensify their use of capital. One manifestation of this process is the continual intensification of land uses at certain locations. In an earlier paper it was shown that a profit-maximizing logic induces a pronounced process of land-use intensification at locations where land rents are already high; i.e. at accessible, centralized, and polarized locations.[23] However, as intensification proceeds, so these locations became progressively more congested and overloaded, and their transport access, progressively less adequate. This calls forth a response (i.e. reactive planning) on the part of the State, which intervenes to correct these problems through new investments in basic infrastructure. This immediately increases the differential locational advantages of any set of locations thus treated and sets in train a new round of land-use intensifications. The State is obliged once more to intervene, and so on.

DILEMMA2   Individual locators actively seek out beneficial urban agglomeration effects. Because these effects cannot be produced as commodities, locators are thus impelled to drift towards urban districts, where such beneficial effects have *already* emerged in definite and irreversible form. Conversely, locators tend to shun those districts where agglomeration effects are negative. This, however, leads to a self-perpetuating, vicious circle: Where agglomeration effects exist (due to historical circumstances), they grow; where they do not exist, they have little chance of appearing spontaneously. This process is further accentuated by the very same phenomenon of reactive planning and land-use intensification that we noted in the last paragraph. In this way, then, the capitalist logic of urban land development is self-constricting, for it discourages the very social outcomes (the emergence of new poles of aggregate urban activity with beneficial agglomeration effects) which might further its ends.

DILEMMA 3   Rapid peripheral urban expansion has been a characteristic feature of urban growth for many decades. It is at once a solution to certain problems, and yet itself creates new problems. Because of the distortions and disequilibria induced by past economic growth, the State has tended to encourage peripheral expansion as a way of relieving pressures in central cities. This encourages out-migration of urban activities. And where peripheral expansion has been especially aggressive (as in certain US cities), this leads in the course of time to moribundity of the central core. The State typically responds by treating only the symptoms of this process; i.e. by urban renewal. Urban renewal itself then produces a typical sequence of problems, especially the destruction of low-income housing.

DILEMMA 4   Throughout much of North America, large metropolitan regions are characteristically split up among semi-independent municipalities. Each municipality competes with other municipalities in seeking to maximize its own tax revenues. In this process, municipalities adopt zoning laws and fiscal arrangements that compromise the global efficiency of the whole metropolitan region. This leads typically to central city decay (and fiscal crisis), while the suburban communities enjoy an affluence that is bought at the expense of this decay.

In these four illustrative dilemmas we have tried to show that, given the fundamental social and property relations of commodity-producing society, the State becomes a part of the urban problem in the very process of seeking solutions.

In the light of the entire preceding argument, we shall now attempt to embed the urban land question more thoroughly in a map of contemporary social and political relations.

The historical process today

Current developments in state strategies

From the time of its inception, capital has never been able to provide for itself the social preconditions of its own existence. That is to say, the social underpinnings of the processes of the production and appropriation of surplus value have never been producible by capitalist means. For this purpose, capitalism has always required a State which confronts individual units of capital and labor as an agent of general collective capitalist imperatives.

Even so, the extent of state intervention in capitalist society has never been constant but has displayed a very pronounced historical tendency to grow. Under early capitalism, the State adopted the simple, yet effective, strategy of laissez-faire while guaranteeing the specific social and property relations necessary in a regime of free competitive enterprise. However, the historical trajectory of this system revealed its tendency to self-paralysis, a paralysis that culminated in the Great Depression of 1873 and the later collapse of 1929. These events ushered in the end of a laissez-faire strategy. This strategy was supplanted in the 1930s by a form of interventionism (justified theoretically by Keynes) that blossomed in due course into a full-blown Welfare-Statism. The State found itself compelled now to intervene directly in the processes themselves of production and reproduction. It had to create conditions for the more effective deployment of underutilized capital and labor, to underwrite the social costs of capitalist production, to regulate the economic cycle, and so on. Moreover, this new interventionism was visibly reflected in, and indeed largely effectuated through, *urban* policies and programs. In particular, throughout the 1950s and the 1960s, the State committed massive and ever-increasing quantities of public funds to interurban and intraurban transport systems, to urban renewal, to the provision of public housing, to new towns programs, and the rest. This aggressive interventionism had two major effects. First, by making elaborate networks of *collective* goods available for *private* use, the State helped raise labor productivity and, hence, the general profitability of capital. Second, while dealing fairly effectively with the urban question in the short run, this interventionism set in motion certain secondary circuits of appropriation that have contributed significantly to the growing problems of the welfare state. This latter point will be taken up again below.

Welfare-Statism was predicated upon the Keynesian faith that an aggressive state effort to underpin a system of propulsive economic activities (that is, to influence microeconomic decisions by the manipulation of macroeconomic variables) could secure continued smooth economic growth and social stability. The problem was that, for a time at least, the faith worked only too well. By the end of the 1960s, Welfare-Statism itself (and that structure of commodity production that it had helped to engender) began to evince signs of self-paralysis. The propulsive branches of economic activity were rapidly monopolized by a small number of firms, and this had deleterious effects on weaker and less technically advanced branches of production. Unable to withstand the growing economic power of the monopolies, the nonmonopolized branches of production have required increasing state support. Backward regions, heavily subsidized, failed to 'take-off,' continuing to absorb growing streams of public funds. Welfare programs turned more and more unemployed into permanent welfare recipients. Large metropolitan regions began to develop the first signs of a mounting fiscal crisis, and so on. Most severe of all, the State has been caught in a predicament where it cannot stimulate new growth without producing severe inflation; yet, if it does not stimulate new growth, high unemployment continues unabated. The result is an endemic condition of 'stagflation.' In brief, it is because Welfare-Statism was so successful in the 1950s and 1960s that it began to encounter growing difficulties by the beginning of the 1970s. Welfare-Statism has gradually eliminated precisely those structural conditions which made it workable at one time.

State strategy is now at a crossroads, and some new strategic directions are clearly imperative if major economic dislocations and concomitant social unrest are to be averted. In any case, a simple return to a full-blown welfare-state strategy seems scarcely to be practicable. Much less, of course, is there any possibility of a return to anything even approaching laissez-faire. In a definite sense it is apparent that yet *more* collective action is called for. At the same time, given the failures, inadequacies, and dangers of Keynesian strategies, future policy formulations in this domain are likely to be cautious and highly selective. There remains, however, a further domain of interventionist possibilities that has hitherto been explored, in practice, in only the most superficial and tangential ways, and yet would appear to offer some significant solutions to current policy dilemmas. This is the domain of *collective management,* in the sense of bureaucratic control, planning, and harmonization over wide areas of economic and social activity. Recent experiences with wage and price controls in many of the advanced capitalist countries would seem to represent an incipient phase of this new strategy that is quite evidently emerging.[24]

What these observations imply for the urban land question is that the palpable geometric increase in urban planning activities in recent decades is almost certain to continue unabated, but with some altered modulations. On the one hand, the State will undoubtedly continue to provide massive injections of public funds into physical urban infrastructure, though in a more highly discriminating and rationalized manner than has been the case in the recent past. On the other hand, the State will, to an increasing degree, impose administrative rules whereby the urban game is played. Zoning regulations and urban general plans, for example, are only the mildest apprehensions of this development. Thus we might expect yet more control of the housing market, yet more collective decision making about the development and redevelopment of urban land, yet more government direction of industrial locational decisions, yet more attempts to rationalize municipal finance, and so on. The compelling motivation of this activity, of course, is that it responds to urgent social and economic problems with little risk of triggering new rounds of inflation and/or unemployment. But it also means that urban life, in general, must become increasingly mediated by the State and, hence, directly politicized. We pursue this question further in the succeeding argument.

Current developments in class relationships

As we have shown in Figure 6.2, the total net national income generated by capitalist commodity production divides without residue into profits, wages, and rents. These abstract *claims* on net national income are appropriated by specific concrete *claimants,* so that they represent not simply analytical economic categories but, more importantly, real human interests. These interests, in fact, are the outward manifestation of a set of class relationships that are themselves determined by an underlying pattern of social and property relations. Moreover, because profits, wages, and rents are mutually exclusive quantities, the human interests that attach to them collide in various ways and thus have a definite political dimension.

In the course of the evolution of capitalist society, the abstract analytical categories – profits, wages, and rents – have remained as perennially unchanging and unambiguous claims on net national income. The claimants to profits, wages, and rents, however, have been markedly more mutable, and the class relations of late capitalist society are quite definitely different from those that characterized early capitalist society. This change has been made all the more complex by the fact that the primary distribution of net national income into profits, wages, and rents via the development of competitive markets is now complemented by (among other things) a complex series of secondary circuits of appropriation. Of all these circuits, those that emanate from the process of urbanization and urban planning are of particular significance and interest.

Let us now ponder the detailed relevance of these general matters for an extended analysis of the urban land question.

The structure of early capitalist society was fairly straightforward. It consisted of: capitalists who owned the means of production and appropriated a profit; workers who owned labor power that they sold for a wage; and landlords who owned land and appropriated a rent. These social groups, of course, were not perfectly internally homogeneous, nor were other kinds of social fractions absent. Even so, this tripartite social structure was the dominant historical (and classical) situation in England towards the end of the eighteenth century and the beginning of the nineteenth century. As a corollary, conflicts over income shares were visibly expressed in the form of an ongoing series of social and political struggles among these three different groups. These conflicts were, however, uneven. In the early stages of the development of capitalist production, capitalists fought a long and ultimately successful battle against the entrenched power and privileges of landed property (the hereditary aristocracy). Once this battle was won, decisively with the repeal of the Corn Laws in the early nineteenth century, capitalists and landlords tended to form an uneasy alliance against labor as an emerging social force on the basis of their shared material interest in the institution and ideology of private property.

Late capitalist society represents a significant departure from this early social map. In particular, the progressive nineteenth-century trend to a deliquescence of a specific landlord *class* has been definitely resolved, and this class has now more or less disappeared as a distinctive element of capitalist society. This is not to say that land as a factor of production has ceased to have significance or that land rent has ceased to exist as a specific share of net national income. On the contrary, they remain as significant and as real as ever. However, the class identity of owners of land (hence claimants to rent) has become quite diffuse. This diffuseness itself results from the fact that land in late capitalist society has been totally converted into an alienable and commercial value on a par with stocks, bonds, productive equipment, and the like. In this manner, land ownership has lost its former social specificity. Land is now owned severally (by commodity producers, financial institutions, a handful of professional *rentiers*, and by workers themselves) and variously applied to the purposes of production, residential activity, and simple speculative investment. However, whereas *ownership* of land has become diffuse, *private control* over land devlopment has become gradually more concentrated. For one thing, the tendency towards technical amelioration that is a persistent element of all capitalist branches of production has led to the emergence of a progressively rationalized and vertically integrated property development branch. For another, the increasing lumpiness of infrastructural artifacts has induced large-scale operations in this branch by permitting the internalization of significant external economies. The end result has been that this specific branch of capital generally, in pursuing a perfectly normal capitalistic logic, has come to be in dominant control of the whole land development process.

If the theses advanced above are correct, then the ways in which urban land problems translate themselves into specific social and political issues must have become markedly different from what was formerly the case. We shall take up this matter again briefly in the concluding paragraphs of this paper. For the moment, two points must be stressed.

First, with the general diffusion of land ownership, the prediction (frequently made by many Marxian theorists) of an irrevocable and impending confrontation between capitalists and landlords is quite certainly without foundation. No doubt this prediction had a certain validity in the context of the kind of social conditions that existed in England at the end of the eighteenth century. But in the context of modern capitalist society, the socio-economic function of land rent has been transformed, just as a landlord class has been largely dissolved. More particularly, land rent is no longer a kind of exogenously imposed tax on economic progress, a tax that is then dissipated by the extravagances of a parasitic landowning class. Rather, *as soon as it is appropriated, rent enters into the stream of new investment capital at large.*

Second, with the concentration of control over private land development into a functionally specific branch of production, conflicts over the urban land question have been absorbed (though with some ambiguity) into the conflict between capital and labor, generally. But, in addition, this same concentration has meant that some of these conflicts have taken on the form of specific conflicts among different branches of capital. As we have stressed in an earlier section, there is nothing unique about the general structure and rationality of the property development branch that causes it to be different in kind from other capitalist branches of production. What is unique, as our analysis suggests, is the final product: urban land itself. Moreover, precisely because this uniqueness resides, at least in large part, in the character of urban land as a rather curious phenomenon that is collectively produced but privately appropriated, exchanged, and utilized, conflicts over land issues have resulted in the ever-escalating intermediation of the State.

The State itself, then, has become intimately bound up with the urban land question. Furthermore, the State, through its planning apparatus, has become the nexus of a series of secondary circuits of appropriation of net national income. More particularly, profits, wages, and rents are now in part a direct function of state planning and control.

Thus, on the one hand, the State levies differential taxes on profits, wages, and rents, thus diminishing these quantities. The revenues raised in this way are then in part devoted to the multifarious material and immaterial infrastructures that are typically supplied by the State in modern urban centers: the construction and coordination of transport linkages, public housing projects, the establishment of recreational facilities, the administration of construction and zoning regulations, the subsidization of industrial enterprises, and so on. These benefits are then recaptured in the form of various increments to profits, wages, and rents. In the first place, these benefits increase *profits* by providing a variety of subsidized inputs to commodity production. In the second place, they increase *wages* (in the sense of total final consumption) both by direct subsidization and by augmenting the total range of social consumption and amenities. In the third place, they increase *land rents,* since all spatially differential improvements in the quality of the urban environment are ultimately capitalized in the price of land. Indeed, it appears more than likely that the predominant share of net national income that is redistributed via urbanization is finally captured in land rent.

Whatever its exact quantitative realization, this process of redistribution of net national income via state intervention in urban development has become the focus of an important secondary manifestation of the primary conflict described above over the distribution of net income shares to capital, labor, and land. This conflict is made politically tangible in the efforts and aspirations of various social groups, including those that have a purely territorial identification to ensure that public decisions on matters of urban development and control are dominantly favorable to them. But, more importantly, the emergence of these new circuits of appropriation has, to some degree, repoliticized and sharpened social and political conflicts. What was once left, for better or for worse, to the agency of a 'naturalistic' and, hence, depoliticized market is now the consequence (and is *seen* to be the consequence) of political and administrative decisions. In view of this development, we can now understand how it is that urban planning, which (at least in its institutionalized versions) began its history as a rather innocent and negligible undertaking, has in recent years been moving inexorably into a focal position of controversy and social significance. Concomitantly, it seems that urban planning is also deeply implicated in the 'legitimation crisis' of the modern State that Habermas has recently so cogently described.[25]

Prospects for the urban land question

If our assessment of the historical process today has any validity, then certain major developments (of far-reaching importance for the urban land question) are likely to become increasingly evident.

Thus, to the extent that the State itself becomes predominantly instrumental in controlling the distribution of the economic surplus, then to the same extent is the locus of socio-economic conflicts likely to shift from direct confrontations between workers and firms to indirect confrontations between workers and firms *via* clashes with the state apparatus. For example, in Canada today, much of the conflict over wage and price controls takes the form of unions and management independently pressuring and lobbying government agencies. In addition, recent events in many of the advanced capitalist countries suggest that the State is more and more likely to mediate subsidiary conflicts among various fractions of the citizenry at large. As it matures, this structural shift in the nature of social conflicts (causing the State to move ever-increasingly into a pivotal situation with respect to those conflicts) is likely to have a decisive impact on political consciousness and, hence, on political practice. It means that all social relations must now become more and more openly politicized. At the very least, this phenomenon is likely to repoliticize the workers' movement, encouraging it to go far beyond simplistic forms of trade unionism to rediscover its own origins in the search for human emancipation and self-determination.

As a corollary, that search, in all probability, will be nowhere more visible than in the general domain of urban life. Hitherto, this domain has tended to be splintered in a way such that workers' movements as such have focused on wage demands, while citizens' movements as such have focused on attempts at improving various aspects of the quality of the urban environment. Yet, for all the reasons adduced above, it is especially in the matter of urban development that we would expect the hand of the State to be evident in the near future. This emerging tendency may well finally reveal the structural connections (rooted as they are in a single, global process of appropriation of net income) between exploitation in the work place and disparities and dislocations in the community. In this way, there would seem to be real possibilities for progressive and mutually beneficial coalitions between workers' and citizens' organizations. Paradoxically, the great danger is that this same increased politicization of urban issues will have precisely an opposite effect, and that augmented doses of 'planning' will merely intensify intergroup and intercommunity antagonisms. This is surely an area in which meaningful political education is a real possibility and, indeed in part, a decisive element in the future that we choose for ourselves.

Clearly, prospective developments in state strategies, class relations, and political practices are likely both to undergo significant transformations in the future and to open up definite possibilities for political action around the urban land question.

In conclusion, it needs to be stressed that while we have predicted a number of future developments, we by no means assert the inevitability of those developments; they seem to us to be implicit in the current conjuncture, though they are definitely not certainties. Nevertheless, our analysis has revealed what is surely an irreversible historical process at the core of the urban land question; namely, the social momentum that emanates from the contradiction between the socialized production of urban land in its totality, on the one hand, and the privatization of concomitant benefits, on the other. For structural reasons, this momentum evolves progressively in the direction of yet more socializations but less privatization of the urban land nexus. Given this trend, conflicts surrounding the issue of urban land will likely (but not necessarily) lead to the gradual maturation of a specifically socialist political praxis. In the end, the fact that we refrain from any easy and rhetorical conclusions is only a recognition of the evident condition that the extent to which human beings *act* is contingent upon the degree of their political organization, education, and determination. These are matters that in part depend upon an adequate prior theoretical clarity, but they also crucially depend upon a real engagement in ongoing political practice.

Notes

1 Cf. Edward C. Banfield (1973) *The Unheavenly City.* Boston, Mass.: Little, Brown and Co. (1977).

2 For a very thorough review (from within) of this epistemology, see Harry W. Richardson (1977) *The New Urban Economics: And Alternatives.* London: Pion.

3 Cf. Stephen Gale and Eric G. Moore (eds.) (1975) *The Manipulated City.* Chicago: Maaroufa Press and London: Methuen.

4 Consider, for example, Graham Barker, Jennifer Penny and Wallace Seccombe (1973) *High-Rise and Superprofits.* Kitchener, Ontario: Dumont Press Graphix; also James Lorimer and E. Ross (eds) (1976) *The City Book.* Toronto: James Lorimer and Co.

5 David Harvey (1974) 'Class-monopoly rent, finance capital, and the urban revolution.' University of Toronto Department of Urban and Regional Planning, Papers on Planning and Design no. 4, Toronto. While the notion of class-monopoly rents is evidently Harvey's, it would, nevertheless, be unfair to treat Harvey as a simple manipulated city theorist. His work, in fact, has been consistently of a rigorousness and sophistication that sets it far above the easy assertions of those theorists.

6 J. S. Mill (1975) *Principles of Political Economy.* New York: Kelley.

7 Note, however, that these prices are not determined through the interplay of market forces, but through managerial control.

8 J. D. Herbert and Benjamin H. Stevens (1960) 'A model for the distribution of residential activity in urban areas.' *Journal of Regional Science 2:21–36.*

9 A philosophy that is criticized from within, as it were, on two fronts: the Romantic (cf. Jane Jacobs (1961) *The Death and Life of Great American Cities.* New York: Random House); and the reactionary (cf. Banfield, *Unheavenly City*).

10 To dramatize the point: if we could somehow return to the Middle Ages and take a sample of consumer preferences, it is even remotely conceivable that we might discover hidden, however deeply, in the psyche of some serf, a primeval dream of a suburban bungalow neatly set within a rectangle of lawn? Ultimately, the doctrine of consumer sovereignty bears a disconcerting resemblance to

Hegel's World Spirit. In this regard, Preteceille calls the doctrine *'idealist'* in the sense that it presupposes that objective socio-historical conditions are simple matters of subjective decidability; whereas, in fact, human subjectivity acquires a real (as opposed to a metaphysical) existence only in the context of a *given,* concrete historical reality. Cf. Edmond Preteceille (1975) 'Besoins sociaux et socialisation de la consommation.' *La Pensee* 180:22–60.

11 William Alonso (1965), *Location and Land Use.* Cambridge, Mass.: Harvard University Press. Richard F. Muth (1969) *Cities and Housing.* Chicago: University of Chicago Press. Robert M. Solow (1973) 'On equilibrium models of urban location' in M. Parkin (ed.) (1973) *Essays in Modern Economics.* London: Longman, pp. 2–16.

12 Friedrich Engels (1970) *The Housing Question.* Moscow: Progress Publishers.

13 The credulity of the reformist position is succinctly typified in a recent statement by Lorimer:

'. . . city governments in Canada are strongly and directly tied to the property investment and land development industry, with the strongest links being the arrangement which puts a hard core of small-time property industry people like contractors, real estate agents, architects, developers, and real estate lawyers on city councils. These politicians with property industry connections form the centre of a majority voting bloc which implements policies protecting and promoting the interests of developers, property investors, and other industry members.' (Cf. James Lorimer, 'Canada's urban experts: smoking out the Liberals,' in Lorimer and Ross, *City Book,* p. 98.)

Lorimer goes on with the well-meaning, but finally vapid, suggestion that 'The alternative to property industry domination of city hall is a radical alliance . . . [which could] . . . end the exploitation of city residents and the city itself by the property industry.'

14 Cf. the recent symposium (1975) *Rent Control: A Popular Paradox.* Vancouver: The Fraser Institute. Admittedly, this is a conservative organization and one less than adequate in its prescriptions, though devastatingly accurate in its analysis of the deficiencies of rent control policies.

15 No statement as to the general nature of commodity production can be complete without a concomitant theory of value (hence, a theory of profit). However, this is a complex, subtle and controversial question. Since we cannot do justice to the question in present context, we reluctantly adopt the *pis aller* of neglecting it.

16 Walter Isard (1956) *Location and Space-Economy.* Cambridge, Mass.: M.I.T. Press.

17 See, in particular, Piero Sraffa (1960) *Production of Commodities by Means of Commodities.* Cambridge: Cambridge University Press.

18 C. Topalov (1973) *Capital et propriété foncière.* Paris: Centre de Sociologie Urbaine.

19 Note, however, that the early work of Koopmans and Beckmann has seemed to show that, even under these circumstances, competitive locational systems are intrinsically unstable and divergent from equilibrium. This work has always had something of a maverick position in mainstream urban and spatial economic theory. We suggest that it may well prove to be of central importance in the light of our own discussion. Cf. Tjalling C. Koopmans and Martin Beckmann

(1957) 'Assignment problems and the location of economic activity.' *Econometrica* 25:53–76.

20 In seeking to maximize their private gains, individuals will, of course, undertake land-use conversions only when the private opportunity costs of conversion exceed the actual private costs of conversion; that is, these conversions are always delayed by a factor that is a function of the privately incurred fixed costs sunk into existing structures. In practice, these fixed costs are invariably heavy. The conversion process will, therefore, be far from instantaneous.

21 Michael B. Teitz (1968) 'Toward a theory of urban public facility location.' *Papers of the Regional Science Association* 21:35–51.

22 See, for example, the two outstanding historical studies: Leonardo Benevolo (1967) *The Origins of Modern Town Planning*. Cambridge, Mass.: M.I.T. Press; and Karl Polanyi (1944) *The Great Transformation*. Boston, Mass.: Beacon Press. A theoretical overview is presented in Shoukry T. Roweis (1975) *Urban Planning in Early and Late Capitalist Societies*. University of Toronto Department of Urban and Regional Planning. Papers on Planning and Design no. 7, Toronto.

23 Allen J. Scott (1976) 'Land Use and Commodity Production.' *Regional Science and Urban Economics* 6:147–60.

24 A strategy that Offe and Ronge have also recently identified as a strategy of so-called 'administrative recommodification.' See Claus Offe and V. Ronge (1975) 'Theses on the theory of the State.' *New German Critique* 6:137–45.

25 Jurgen Habermas (1973) *Legitimation Crisis*. Boston, Mass.: Beacon Press.

7 Urban planning in early and late capitalist societies: outline of a theoretical perspective

Shoukry T. Roweis

Current formulations in the field of 'planning theory' tend to be purely descriptive or purely normative. Little is done in the way of *analytical/ historical* formulations concerning the essence, roles and 'logic' of urban planning in capitalist societies.

As a consequence, planners tend to lack a systematic understanding of their discipline/profession; of the real constraints imposed on, and the objective opportunities open to, their practice. Lacking, as it were, a map of social reality in which they can situate themselves and their practice, planners tend to vacillate between utopianism and technical pragmatism. The former robs their practice of its potential effectiveness and their theory of its practical significance. The latter reduces practice to an aimless management of day-to-day bottlenecks and theory to a technical instrumentality of shortsighted 'problem-solving'.

This paper is an attempt to fill this gap. The purpose here is to situate urban planning in the concrete economic, social, and political context of capitalism. The paper deals with questions like: is urban planning exclusively a domain of activity of urban planners? Are urban planners, in other words, the only (or the most important) agents who attempt to influence urban growth and development? Who else does? Under which conditions does capitalist society witness serious intervention in the processes of urban growth and for what purposes? Under what conditions does society retrench from such attempts? What forces, in capitalist society, limit the capacity of urban interventions to reform and improve urban conditions? What forces call for urban reforms? And so on.

In what follows, I shall attempt to sketch the main outlines of a theoretical perspective capable (I hope) of answering such questions. The argument is presented as a series of interrelated propositions. Little empirical substantiation is attempted. The propositions, however, are testable (or can be made so). The argument is developed in three steps. First, I present some methodological propositions. Then I move on to sketch a theory of early and late capitalism. Finally, I deal with urban planning in these two historical contexts.

Methodological considerations

1 A plausible analysis of urban planning should start by *rejecting* even the possibility of carving out an area of activity and trying to *analyze* it in isolation from the overall social/historical context in which it occurs. Unless the analysis embraces the *totality* of society, we are doomed to produce distortions and invalid views. We will explain little.

2 Social analysis begins with the first awareness that what social agents *think* they are doing might not coincide with what in fact they *are* doing. So when we come to study planning we know that we cannot rely on urban planning literature as the sole, or even the most important, source of information. We should seek a variety of information sources and particularly historical documentation and 'raw' descriptions of what happened at the different time periods under investigation.

3 Urban planning is above all else a *social* activity. Even the dreamer who contemplates a utopian city presupposes a group of people *collectively* engaged in social life. An analysis of urban planning must therefore focus on understanding the *social relations* planning sustains, changes, or creates anew.

4 Unlike so-called 'autonomous spheres of social action', e.g. the market exchange of commodities, the 'private' management of family affairs, etc., urban planning since antiquity requires some measure of *deliberate collective action* (for an elaboration refer to proposition 21 below). Citizens of medieval cities would not have built defense walls had they not had some form of political decision-making mechanism allowing them to undertake collectively the financing and probably the construction of such walls. Nor would Torontonians have built their contemporary transit system short of such collective action. *At the heart of the activities we call urban development and urban planning is the need to institutionalize some form of collective action.*

5 I do not mean by propositions 3 and 4 above to suggest that individual actions and their social consequences are irrelevant or do not influence urban development and urban planning. Suburban expansion, for example, has rarely been a consciously deliberate *collective* action. Yet, as a result of hundreds of thousands of individual actions, suburbanization has obviously changed the entire urban landscape and had an enormous impact on urban planning.

6 Propositions 3 and 4 above suggest the importance, in an analysis of urban planning, of focusing on the investigation of *social* (as opposed to individual) decision-making processes. This requires an understanding of the structure and functions of the *State* (government, parliament, parties, etc.). Proposition 5 above suggests the need to give attention to the ongoing (and institutionalized) social and property relations (basically the relations of possession and non-possession of the various means of production; relations governing access of non-possessors to the means possessed by others, e.g. wage labour/capital relations; the distribution of social agents in the hierarchy of the division of labour; and relations governing the distribution of the social wealth thus produced). This requires an understanding of the structure and dynamics of the *civil society*: the patterns of 'autonomous' and institutionalized social relations. More importantly, propositions 3, 4 and 5 above taken together suggest that a plausible urban planning analysis must focus not only on the investigation of the State on the one hand and the civil society on the other, but on the *interface* of these two. We have to examine how urban development and urban planning actions *grow out of* this constant action and reaction of civil society and the State.

7 The structural characteristics of the State and of civil society change dramatically from one historical period to another. In nineteenth-century competitive capitalism, the State was an arena of fundamental political struggles regarding the class structuring of society and the relative power of the various classes. But its direct intervention in the day-to-day operations of civil society was limited. In the late Welfare State capitalism of the 1970s, there is hardly any arena of social activity which the State does not regulate (fiscal regulations, monetary regulations, foreign trade regulations, licensing, zoning, development controls, taxes, subsidies, income transfers, labour-capital arbitration, to mention just a few).

The same is true of civil society. In nineteenth-century civil society, small entrepreneurs competed 'freely'; landowners bought and sold, used and reused land in virtually any way they desired; conflicts between labour and capital were left to resolve themselves (by the mediation of Adam Smith's invisible hand) in 'natural' wages, 'natural' profits and 'natural' prices; and so on. In today's civil society multinational firms control supply and demand and prices; giant financial organizations control the investment process; labour unions organize the confrontations with capital; urban development corporations build entire suburbs and towns; and so on. Moreover, the structure of classes in today's civil society can no longer simply be characterized as a dual structure of capitalists and proletarians. Corporate managers do not own the capital they control; the working class has now encompassed an increasingly important segment of educated labour, technicians, professionals, technocrats, bureaucrats, social engineers, etc. A new analysis of contemporary civil society is desperately needed.

A reasonable analysis of urban planning must recognize these fundamental changes. We can no longer proceed along general, abstract and formalistic lines to discuss urban planning as if it existed in a stable, immutable or universal context. Hence, the title of this document: urban planning under *early* and *late capitalist societies.* More precisely, such an analysis must isolate *distinct phases* in the development of capitalism and attempt to understand the main characteristics of these phases. Only then can we proceed to analyze the nature of urban planning in these concrete social/historical contexts.

To simplify matters, I shall assume (at the risk of deemphasizing significant subtleties) that there are only two important phases: *early* and *late* capitalism.

Propositions concerning capitalist society (early vs. late capitalism)

Early capitalism (mid-seventeenth to late nineteenth century)

8 Early capitalist society can be described in terms of a number of essential socio-economic institutions. These institutions, which governed the basic social and property relations in civil society were maintained, perpetuated, and modified through the ongoing struggle between antagonistic classes and fractions thereof.

9 The basic socio-economic institutions were:

    i *private ownership* of capital (private ownership of land came later and even then some land remained inalienable, e.g. Crown land);

    ii *a market in labour* in which labour-power was sold and bought in a 'free' market, its price – wages – being determined by the prices of the essential goods and services necessary for the life-maintenance of the wage earners;

    iii a 'free' *competitive market* in goods and services following the principle of 'exchange of equivalents', in which what gets produced, how much gets produced, and for how much it gets sold, are all determined by the free play of competition; and,

    iv *incomes derived primarily by the sale of individually owned means of production,* and at the ongoing market prices. (If you own land, you get rent; if you own capital, you get profit; if you own labour-power, you get wages; if you own nothing . . . !)

Such 'arrangements' or social 'rules of the game' are referred to as *institutions* simply because they attain their dominance as 'rules of the game' only after a long and tedious process of institutionalization – a process which took Europe at least three hundred years to 'complete'.

Such was the structure of civil society under early capitalism. The property relations that followed determined the basic class structure (landlords/capitalists/workers). It was an explosive class structure and it required the intermediation of certain state apparatuses to preserve the 'tense balance' (to borrow a bit of jargon from H. Kissinger) between the classes.

10 The fundamental duties of the early capitalist State were:

i assuring the observance of the rules of the game. In other words, maintaining and perpetuating the essential socio-economic institutions of civil society;

ii securing internal social stability by invoking legal rights and, if necessary, by using force. The State has the monopoly over the legitimate use of force;

iii preserving and defending national sovereignty; and

iv administering a minimum of public works programmes which meant levying taxes and distributing subsidies.

With few exceptions, the role of the State, in this early phase, was limited. This was particularly true in the area of urban development and urban planning. The way things developed was not so much a consequence of deliberate state intervention as it was a result of the day-to-day operation of the essential socio-economic institutions in the civil society.

11 Although civil society under early capitalism constituted an 'autonomous' sphere, and although the State's role was limited, fundamental interactions and reactions took place between civil society and the State. The historical development of the one influenced, and was constantly influenced by, the development of the other.

An example: urban landowners in the nineteenth century, given the institution of private property in land, and given the leasehold system (whereby some land could not be sold but only leased for thirty, forty or, more frequently, ninety-nine years) would seek to develop their land in 'building estates'. They would lease the land to a contractor/capitalist (a leaseholder) who would, in turn, build working-class houses and rent them (the 'high classes' built on their own land or rented directly from landowners). The leaseholder, knowing that when the lease expires, he has to give all the houses to the landowner – which was the rule of the game then – tries to get as much profit as possible by charging high rents. The industrialists felt the pressure, for the higher house rents go, the more wages workers would demand. Repeated worker rebellions, work stoppages, and machine sabotages, led the industrialists to build housing for 'their' workers. They built cheap, unsanitary, small houses, and this led to substantial health and crime problems. State agencies, in an attempt to maintain civil order, passed legislation calling for minimum space, sanitary and construction standards. Such regulations imposed additional costs on the industrialists. They reacted by shifting working-class housing to cheaper peripheral sites far away from the factories at the centre. The high fares and long trips to and from work spurred the workers into fresh struggles with the owners. The State intervened, using its licensing powers to pressure railroad (and streetcar) companies to lower fares and extend more 'suburban lines' to the new housing quarters. A qualitatively different urban form began to emerge; *an urban form that nobody anticipated or tried consciously to bring about.*

Many more examples can be cited. But the point is this: the dialectical relationship between the civil society and the State has the following characteristics:

  i actions – or moves, to use the language of chess – in the civil society follow logically *not only* from the rules of its own socio-economic institutions, but also from the altered conditions brought about by state actions; and

  ii actions by the State not only reflect the differential political power of social classes and fractions thereof, but also emerge in response to pressures and explosive contradictions created by the day-to-day operations of the socio-economic institutions of civil society.

Late capitalism

12 The internal logic of early capitalist society carried in it the necessary and sufficient conditions for a qualitative transformation of laissez-faire into monopoly/welfare state capitalism. At the heart of this metamorphosis is the indisputable trend towards expanded and intensified state intervention into vital economic and social processes. Two important questions arise here:

  i what historical dynamics led to increased state intervention? and

  ii given this trend, is it still possible to characterize capitalist society along the lines presented in propositions 8 to 11 above?

13 To answer the first question, we should chart out the concrete forces *calling for* more and more state intervention, as well as those forces *resisting* such intervention. The following are the principal state functions that have emerged since the nineteenth century. They show why state intervention had to be expanded and intensified:

  i to redress grievances and inequalities emerging as a matter of course in early capitalist society, hence to stabilize the social order and prevent class conflicts from exploding into disruptive confrontations;

  ii to contain the crises caused by irrationalities embedded in the socio-economic institutions of early capitalist society; crises which affect not only the working class but the owning classes as well. The fundamental irrationalities are:

  – production for exchange, and not to meet social needs, which leads to overproduction, the need to expand markets, and finally economic and military warfare (Napoleonic wars, colonialism, First and Second World Wars, to mention just a few);

  – anarchic production decisions leading to booms and busts; and

  – market competition leading to the extermination of small businesses and the emergence of trusts, cartels and finally multinational firms capable of setting prices and curtailing production leading to contrived shortages and bottle-necks (e.g. corn prices in the 1830s);

  iii to organize the provision of public goods, such as roads, railroads, ports, power stations, schools, parks, research and development, etc., whose production is unprofitable and hence impossible in the civil society, but whose provision is imperative for the continued profitability of private production.

All three functions could hardly be fulfilled by the relatively weak and archaic organization of political governance (the State); and all three called for more and more state intervention.

There are forces, however, which have resisted (and continue to resist) the expansion and intensification of state intervention. These are:

i the institution of private property and its legal expression in laws protecting property rights and the owning classes; and

ii the capitalist mode of production: commodity production for exchange and its legal expression in laws protecting contractual rights, freedom of investment, 'autonomy' in setting production priorities, expansion decisions, prices, etc.;

iii the ambiguities and relative opaqueness of the functions performed by the technocratic and bureacratic 'strata' in state agencies and hence the reluctance (on the part of virtually all social classes) to grant them more powers and jurisdictions (see proposition 17 below).

The nature of the late capitalist State can be understood as embodying those structural characteristics which result from the balance of these two groups of forces (for and against state intervention) at any given time.

14 The second question (i.e. is it still possible to characterize late capitalist society along the same lines as we did with early capitalism?) is the subject of an ongoing debate among social theorists – Marxians and non-Marxians alike. My own view on this (admittedly tentative) is as follows.

In early capitalism, when the State performed limited functions (as an arena of class struggles) and when the civil society was rather autonomous, the social relations created and maintained by the socio-economic institutions were quite *objective*, objective in the sense that they could not be changed at will (i.e. by the sheer subjective intentionality of individuals or even large groups either in the civil society or in the State). The actions of the State only reinforced and helped maintain these objective social relations and their outcomes. The classical social scientists recognized this. Hence they saw the early capitalist economy as a *natural* economy immune from any arbitrary subjective political interference. As long as that was the case, it was indeed possible and valid to characterize society in terms of a *material base* (i.e. a domain of objective social relations) and a *political superstructure* (i.e. a domain of subjective intentions, legal codes, ideology, etc.). It was equally valid to see the latter as a mere reflection of the dictates of the former.

But in an era of pervasive state mediation and intervention, one can no longer reasonably speak of spheres of social relations which are free of state intervention (i.e. of political/subjective influence) and which constitute a domain of objective social relations. No longer is it possible, in other words, to rely on the earlier scheme of material base and political superstructure. Now both shade into one another, and the concepts have, as a result, lost much of their analytical value. A different theory of society is now necessary.

This, however, does not mean that we can no longer speak of a civil society and a State. *Nor* is it to suggest a pluralistic theory of society, which depicts late capitalism as a society made up of competing 'interest groups' presided over by a 'neutral' government seeking the welfare of society as a whole. *Nor* is it to espouse a 'power élite' theory of society, which posits a ruling minority of powerful economic, technocratic, and military interests and a ruled majority of powerless groups whose interests are repressed and thwarted.

All my argument suggests is this: that we have to *redefine* what is meant by the civil society and what is meant by the State in view of the significant changes that have taken place in both since the nineteenth century. To this I now turn.

15 In contrast to the characteristics of civil society under early capitalism, one can no longer speak of fully institutionalized (and hence 'autonomous') rules of the social game. *The principal institutions are still there, no doubt, but they can now be characterized as being under continuing political review.* In other words, the earlier core institutions, while preserved, are subjected, through the channels of the State, to a continuing process of amendment, suspension and modification. The 'map' of civil society now looks something like this:

  i private property is subject to *politically decided* limitations and con-
    straints (eminent domain, zoning, subdivision controls, regional plan-
    ning regulations, etc. in the case of land; fiscal, monetary, tariff and
    price regulations, etc. in the case of capital);
 ii a market in labour subject to *politically decided* provisos (legal rights
    to organize, strike, etc. on the part of labour; minimum wages, social
    security, health and safety regulations on the part of capital).
      Moreover, the State can now legally interfere to arbitrate labour-
    capital disputes whenever the institution of the labour market fails
    to 'resolve' such disputes;
iii a market in goods and services that is no longer 'free' nor competitive
    in the traditional sense. The market is now subjected to a variety
    of political regulations (anti-trust regulations, price controls, foreign
    trade regulations, incentive policies, subsidies, etc.). What is kept
    intact is *commodity production* (production for exchange; reliance
    on the market-price system; private appropriation of surplus); and
 iv incomes are no longer derived essentially by selling owned means
    of production, but to an increasing degree of direct or indirect political
    decisions (welfare programs, unemployment benefits, minimum in-
    come guarantees, inter-governmental transfers, regional equalization
    policies, subsidized social services, etc.).

The new 'map' of the civil society is obviously less 'autonomous' and more 'politicized'. That is, it is now more subject to *political and adminis-*

trative, as opposed to purely market and 'economic' considerations than was its predecessor in early capitalism. *But* the new civil society is never fully 'neutralized' to the extent of losing its capacity to shape the structure and functions of the State. *In the final analysis, the dictates of capitalist commodity production and exchange are still at the core of the civil society and, as such, continue to be the ultimate regulators of state activities and decision-processes.*

16 How can we characterize the State in late capitalist societies? The historical forces which have led to expanded and intensified state mediation and intervention, have also shaped the specific ways in which state apparatuses function. The 'new' State has emerged to deal with three paramount areas of societal problems:

i Problems bearing on economic stability such as hedging against recessions and depressions; fighting inflation and unemployment; easing localized shortages and bottlenecks in key supplies; and – most importantly for an analysis of urban planning – providing, expanding and rationalizing urban, regional and national infrastructure and social overhead facilities commensurate with, and necessary for, continued capitalist expansion. *From its former status as a realm 'beyond' the domain of commodity production and exchange, the State has now become the prime mediator of this process.*

ii Problems of foreign policy and foreign trade. This domain deals with imbalances and conflicts generated by commodity production and exchange at the supranational level of organization. Faced with external threats to the national stability (created directly by neocolonialist and imperialist claims to, and disputes over, geographic spheres of influence, e.g. European Common Market vs. US, US vs. Japan, US vs. Canada, Imperialists of all descriptions vs. the Third World, etc.), problems of foreign policy and foreign trade became as vital to the continued national stability as the economic problems described above. This domain deals with agreements, alliances, developmental policies, protectionist policies, and monetary policies.

iii Problems of mass loyalty. No longer is the problem one of legitimacy of the social order. Legitimation was necessary in the early days of capitalism. The emerging bourgeoisie were then still fighting – at the ideological level – the old feudal ideology (and its use in day-to-day reality) which insisted that feudalism was the only legitimate social order. After all, the Catholic Church, with all its entrenched legitimacy then, used to argue that God meant some to be landed aristocrats and some to be serfs. How could God be found wrong and the bourgeoisie right? This was the battle of legitimation *then*. All this is history by now. What is now important is *mass loyalty*. This does not depend as much on whether the mass of people believe in the legitimacy of late capitalism. What is at issue is whether their concrete claims and demands can be minimally responded to. The prerequisites of mass loyalty are pragmatic: sustained economic prosperity, a modicum of social and economic equality, and a minimally democratic formula of political governance. State apparatuses must now see to it that these prerequisites are fulfilled.

17 On the basis of propositions 12 to 15 above, and given the three problem areas outlined in 16, I can now formulate a basic proposition regarding the mode of operation of state apparatuses in late capitalist societies.

By contrast to their typical functions and modes of operation in the nineteenth century, state apparatuses have come to rely heavily on what can be called *the pre-politics 'processing' of political information*. This has become not only a prime function of virtually all state apparatuses, but a characteristic mode of their operation as well.

Pre-politics 'processing' of political information refers to the wide range of preparatory legwork performed by the complex (and growing) technocratic bureaus of the State in anticipation of, in response to, or in attempt to forestall political disputes or confrontations. The main purpose of this legwork is to set the stage, as it were, for maximally predictable and controllable politics. This requires, among other things, compiling and 'processing' politically relevant information and arriving at tenable conclusions regarding:

    i the relevant classes or fractions of classes in a given political issue or dispute;

    ii the nature and extent of their respective interests and claims in the dispute;

    iii their respective political power (or powerlessness);

    iv the likelihoods of alliances forming among some of these classes or fractions and the anticipated power of such alliances;

    v the degree of disruption likely to result if the interests/claims of some of these fractions are ignored or upset;

    vi the degree of disruption likely to result if the whole issue or dispute is sidestepped or postponed;

    vii the various ways in which fractions can be neutralized, immobilized, bought off, or otherwise dealt with;

    viii the range of politically, economically, and fiscally feasible 'truces', deals, or compromises which are likely to quell the situation;

    ix the respective costs (fiscal and political) of each of these feasible deals; and

    x the respective prospects of these deals (in terms of durability, future resurfacing of discontent, effects on fractions outside the particular issue or dispute, and so on).

18 The degree to which the mode of operation of state apparatuses has come to depend on pre-politics 'processing' of political information can be easily gleaned from the growing reliance on Royal Commissions, committees of inquiry, task-force reports, White Papers, Green Papers and so on. In conjunction with this, a growing number of state agencies have come to establish their in-house technocratic bureaus to undertake similar, if less spectacular, 'processing' functions.

In contrast to the 'direct politics' characteristic of early capitalism, pre-politics 'processing' of political information enables state agencies to act as buffers between the combating classes or fractions thereof: to handle otherwise volatile political situations; to contrive workable political 'truces', deals or compromises before it is too late; and hence to contain and temper the outcomes of the class struggle.

19 If we can no longer characterize the mode of operation of the State in terms of direct politics or its outcomes in terms of an uncritical bias towards the interests of the owning classes, then what are the main features of the new mode of operation? The following are some strikingly stable features:

i The greatest share of political attention, i.e. the highest priorities of political/administrative intervention, will be given to the socio-economic claims (demands) of those classes, fractions, groups or organizations who are most able to contribute effectively towards easing fundamental crises or reducing the risk of aggravating such crises. (The emerging farm land preservation policies in Ontario give high priority to the claims/demands of small farmers in the hope of easing regional unemployment problems, declining farm incomes, shrinking regional purchasing power and the unmanageable pace of rural-urban migration. This, despite the obvious fact that these policies run against the interests of the powerful, city-based, real estate and financial concerns.)

ii Social problems whose consequences do not have pervasive ramifications on the stability of the social order (at large) would tend to receive little, if any, attention. Housing, for instance, receives little more than lip service and mild policy attention as long as the housing problem remains localized and without threatening consequences on economic stability or mass loyalty. In other words, social demands that cannot present a strong case for the dangerous consequences that would ensue (or that they could precipitate) if they were ignored, will lie neglected on the periphery of the sphere of state action.

iii Since 'overdoses' of state interventions generally tend to give rise to problems in other areas, a minimum of regulation necessary for stability will tend not to be exceeded. (Land banking, or for that matter, the nationalization of urban land, might ease a number of persistent urban problems. In the absence of substantial crises that can be managed *only* by such actions, the attention will not go beyond simple slogans.)

iv The disparity between the attention given to problem areas with high crisis potential and those with low potential widens as the State's available resources and policy tools are needed more and more urgently for the range of problems related to continued prospects for profitable capital expansion. These are: securing sufficient effective demand, maintaining foreign trade relations and containing domestic conflicts. The widening disparity between the attention given to problem areas with high crisis potential and those with low potential is reflected in the discrepancy between the most advanced production and military apparatus and the stagnating organization of housing, transportation, health, education, daycare, etc.; also, the notorious contradictions between rational planning and regulation of fiscal, monetary and tariff policies and the anarchic, virtually unplanned development of cities and regions.

20 Propositions 16 to 19 above point to the emergence of a *new technist pragmatic mode of political mediation* in the advanced capitalist State. This new mode of political governance no longer seeks social reform as defined by the early liberal tradition (equality, liberty and progress, etc.), but only the *conservation* of those social relations whose capacity to avert, repress, or postpone fundamental crises is their sole justification. Intentions having genuinely democratic or even liberal origins and aims must rebound without much effect from the political administrative system because the *issue is no longer reform (ideological) but system stabilization (technical).* Only when the total social order is in jeopardy would such reforms be implemented, and only to the extent necessary to insure stabilization. This proposition has important implications with respect to the development of urban planning, both in theory and in practice. (See propositions 26, 27, 31 and 32 below.)

Propositions concerning urban planning under capitalism

General propositions

21 Urban planning in capitalism, both in theory and practice, and whether intentionally or unknowingly, attempts to grapple with a basic question: how can *collective action* (pertinent to decisions concerning the social utilization of urban land) be made possible under capitalism?

Collective action here means social activities *whose consequences are objects of political decision.* Collective action, as used in this paper, is devoid of connotations of literally 'getting together and doing things'. Collective action may, for example, be achieved by traditional/cultural consensus, by elected government through taxes, subsidies and public spending, or by dictatorial decisions. Business decisions of real estate corporations, for example, do not constitute collective action (although the aggregate consequences of such disjointed corporate actions are of major social importance). Such actions do not constitute collective action since their aggregate consequences (total number of housing starts; aggregate composition of dwelling types, sizes, price/rents, densities; annual staging of residential growth over urban space; etc.) are not themselves objects of political decision. The construction of a subway system in a metropolitan area, in contrast, is indeed a collective action.

The history and development of urban planning under capitalism can thus be seen as the history and development of modes of operation allowing for some measure of collective action (that affects decisions concerning the social utilization of urban land); modes of operation which must be feasible in a society whose basic social and property relations (or institutions) resist such action. (For the build up to this statement, refer to propositions 4, 8, 9, 11, 15, 16 and 19 above.)

22 Given the characteristics of civil society and the State in both early and late capitalism, urban planning has not been fully institutionalized in the civil society, nor has it been fully integrated in the operations of the State. To elaborate:

  i *Urban planning and the civil society:* The 'rules of the game' of the early civil society contained no provisions allowing for deliberate collective action in general and urban planning action in particular. Private property rights over land, private commodity production and exchange, and the one-to-one contractual nature of the labour market, not only omitted any provisions for collective action, but actively discouraged it. In practice, collective action was antithetical to bourgeois freedom and liberty. And the dynamics of the civil society tended to nullify the effectiveness of deliberate collective action, if and when taken (refer to example, proposition 11 above).

    Today's civil society with its institutions under political review (proposition 15 above) can no longer repress collective action as vigorously as it once did. But the fact still remains that the basic social and property relations of capitalist commodity production and exchange, as an intact core of civil society, are antithetical to collective action. Private ownership of land and capital is the cornerstone of capitalist commodity production and exchange. It presupposes anarchic decision-making (respectably known as liberty) which contradicts collective action. Commodity production for the realization of profit and not for the fulfillment of social needs, attempts to *privatize consumption*: cars instead of buses, TV instead of movies, records and stereos instead of concerts, individual washers and dryers instead of collective laundry facilities, and on and on. The reason is simple. Private consumption means private buyers and means easy and guaranteed realization of profit. Collective consumption creates the problem of excluding the non-payers and obstructs the realization of private profit. When combined, as indeed they are, private ownership and commodity exchange tend to negate any real possibilities of collective action. As a result, urban planning remains a 'foreign body' to the civil society. The structure of civil society, in other words, still tends to nullify urban planning activities even when forced on the civil society through the mediation of state agencies.

  ii *Urban planning and the State:* Given the weakness of the early capitalist State, urban planning, and the disciplined collective action it requires if it is to bear fruit, remained sporadic, *ad hoc,* remedial and, in the end, thwarted. But by no means was collective action ruled out. Nor could it be. The continued survival of capitalism *required* some measure of collective action (e.g. national defense and offense, taxation, provision of essential public goods and, most fundamentally, the enforcement of property rights).

In the 'new' capitalist State, things are not qualitatively different. The characteristic mode of operation of state apparatuses (proposition 17 above) and the various structural characteristics of state actions flowing from it (propositions 19 and 20) still preclude the full integration of urban planning in the activities of state apparatuses. *Urban planning remains an instrumentality picked up or neglected depending on whether, under the circumstances, it can be effectively used to stabilize the economy and/or to maintain a reasonable threshold of required mass loyalty.* (Societies with an established tradition of social democracy may exhibit different patterns.)

In summary, then, we have this key contradiction. Capitalism was and still is antithetical to that which is necessary for its survival: collective action. This contradiction helps explain the difficulties, hesitations, inconsistencies, compromises, and ineffectiveness of urban planning. But at the same time, it helps uncover the real historical essence of urban planning as well as its latent potential.

23 This objective contradiction is expressed in urban planning as a *contradiction between theory and practice. In theory,* urban planners deplore the urban 'ills' resulting from anarchic urban development and the lack of collective action. They never tire of attempting to show how wonderful urban life can be *if* capitalism and collective action can be reconciled.

In practice, however, urban planners are forced to recognize the near impossibility of such a reconciliation. They come to realize that if they are to be effective at all they have to learn how to exercise the 'art of the possible'. But to the extent they do exercise it, their practice diverges considerably from their theory.

As the gulf widens between urban planning theory and practice, the choices open to urban planners begin to crystallize. Either they must stick to a 'tough line' of critique of given urban realities and of advocacy of futuristic and more socially rational urban arrangements and forego any serious hopes for practical effectiveness; *or,* they must develop a 'softer' pragmatic, technist and less theoretical line, more in tune with what is feasible in practice, and forego any serious hopes of producing significant structural impacts on urban development and growth.

24 From propositions 22 and 23 above, it can now be seen that it is not what happens in the minds of urban planners (i.e. ideas, advocacies, self-images, professional ethos, theories, etc.) that primarily shapes the nature and development of urban planning practice (although it may have a negative influence), but rather the exact opposite. *It is the strains put on urban planning in practice* (i.e. the concrete manifestations of the irreconcilability of capitalism and collective action) *that constantly shape and reshape the theoretical superstructure of the field* (refer to propositions 1 and 2 above).

Hence, a critical analysis of urban planning must reject the inane notion that the successes or failures of urban planning practice are due to 'good' or 'bad' urban planning theory. It should rather seek to understand these successes and failures by analyzing the concrete socio-economic political conditions under which urban planning is practiced and by looking specifically, in every case, for the objective circumstances allowing (or demanding) the undertaking of more (or less) collective action in the domain of urban development.

25 In view of this last statement, it is necessary to make the distinction between periods during which capitalist urbanization passes through major territorial instability, dislocation and disarray (typically accompanied by social and political turmoil), and periods during which it experiences relative territorial stability, constancy and orderliness (typically accompanied by relative social and political tranquillity). The degree to which collective action is allowed, or called for, is directly related to those periods. During periods of urban territorial instability and turmoil, capitalism allows and indeed requires large 'doses' of collective action. During periods of relative territorial stability and tranquillity, it allows and can do with less. We shall, therefore, look at what happens to urban planning during each of these types of periods.

Urban planning in periods of territorial instability and social turmoil

26 The contradictions between urban planning theory and practice diminish considerably during periods of territorial instability and social turmoil. The bitter irony is that urban planning achieves most when capitalist urbanization is most troubled. It is in such periods that urban planning acquires relatively high social significance and is given the opportunity to influence the course of urban development most pronouncedly. Why?

27 Periods of territorial instability and disarray – typically occurring in the wake of economic crises and resulting from sudden, unplanned and drastic technological transformations in the methods of production, transportation, communication and construction – are characterized primarily by massive dislocations in the organization of urban land uses.

Because of the anarchic and unregulated utilization of urban land and infrastructure, these dislocations tend to precipitate particularly severe hardships such as housing crises, overcrowding, congestion, disruption of existing residential quarters, speculation, and the like. Land developers and builders rush to cater for the rapidly expanding needs of the booming businesses and the wealthy while working-class and lower-middle-class families are left to bear the full brunt of the disarray.

Such compounded hardships, particularly when concentrated on the working classes, and when they cease to be localized in few urban areas but reach national proportions, tend to intensify class conflict. They threaten to produce unmanageable class confrontations and social disruption.

It is only under such threats that the capitalist State faces the imperative of implementing significant urban reforms and hence intervenes to make possible sufficiently aggressive and disciplined collective action. It is during such troubled times that most significant urban reform legislations are passed, most extensive (and sensible) additions to urban infrastructure and social overhead facilities are constructed, and most meaningful urban development controls are enforced. (Refer to proposition 20 above.)

28 It is also during, or shortly after such troubled times that urban planning *theory* advances most visibly and enters into a more meaningful relationship with planning practices; a relationship where theory absorbs the lessons of practice and turns back to guide such practice. These are also the times in which urban planning theory goes beyond being a mere apology to the technist pragmatic interests of state officials and business and real estate concerns.

29 Concurrent with such liberal progressive theoretical 'leaps', however, a utopian romantic line tends to emerge (or re-emerge) in urban planning 'theory'. Vehement opponents to urban expansion, back-to-land 'decentralists', utopians and visionaries, all tend to make their voices heard. Disconnected as they usually are from practice (and hence incapable of guiding such practice) they tend unwittingly to play a significant role nonetheless. They tend to stir public opinion, articulate mass grievances and, hence, do most of the propaganda and public relations legwork needed for practicing urban planners to tap the full potential of the opportunities temporarily open for urban reform.

30 But even under such 'favourable' conditions, urban planning remains locked within definite constraints, and its achievements, in terms of social rationalization of urban life, cannot stretch beyond the limits imposed by the capitalist order of which urban planning is an integral part.

Urban planning in periods of relative territorial stability and social tranquillity

31 As capitalist urbanization steers away from troubled times and begins to enter a period of relative territorial stability and social tranquillity the exceptional tolerances for disciplined collective action begin to be curtailed. The 'rules' of the civil society take over once again, and urban development begins (although admittedly on a more rational plane) to proceed along its 'normal' anarchic/unregulated lines. This capitulation, however, takes some time. The progressive momentum gained during the crisis tends to carry on for a short period after the crisis. Urban planning practice tends to lose its power gradually. In such periods, capitalism lays the foundation for its next crisis.

32 The gulf widens again between urban planning theory and practice, and urban planners divide into two camps: (a) intellectual reformers and social critics, with decreasing abilities to influence the trends of urban development; and (b) urban managers, technocrats and social engineers

who see their task as cooperation with the inevitable: capital accumulation, commercial expansion and real-estate speculation. Individual planners quite frequently combine these two inconsistent postures. The former enhances the planner's self-image (as an agent of progressive social change) while the latter delusively reassures her/him of practical effectiveness. (This paper can properly be seen as a statement by a planner who has become conscious of the internal contradictions of his consciousness *as a planner.*)

33 Ironically again, during periods of relative territorial stability and social tranquillity urban planning theory stagnates, turns its attention to utopian visions and/or technical trivia, loses sight of the predicaments of practice, and gradually reaches an impasse. The field ceases to attract able and serious minds, precisely when the potential emerges for the orderly guidance of urban development.

The predicament carries within it its own resolution

34 The problems, predicaments, and contradictions in urban planning, not unlike those in capitalism itself, are pregnant with their own resolutions. There are two interrelated dimensions to this assertion: an objective and a subjective dimension.

35 If we focus on the dynamics of change in the civil society, we find a pronounced shift in the locus of class conflicts accompanying the shift from early to late capitalism. The main problems in early capitalism were problems of production (i.e. problems of insufficient aggregate supply, hence relatively scarcity). Under such conditions, class conflicts were over the distribution of wealth. Concretely, the conflicts were strictly labour/capital conflicts centred in the work place and focusing on wage/profit disputes.

In today's capitalism, this is no longer strictly true. *Problems of production gave way to problems of overproduction.* With capitalist commodity production and exchange still the intact core of capitalist society, there is always a tendency to expand production. But sooner or later a problem arises: where shall additional buyers come from? Goods which do not find buyers not only do not realize 'their' profit, but do not even return the initial capital invested in their production. The tighter the markets become (because everybody is overproducing), the more aggressive (monopolistic and oligopolistic) producers become in seeking higher labour productivity (to reduce costs of production and increase retained earnings). But this can only be accomplished through massive investments in technological improvements. Such massive investments produce substantial pressures on the money markets, set in motion a spiral of inflation, and yet do not *necessarily* contribute to the expansion of output or employment. Major sources of tax revenue begin to stagnate or decline. Fiscal deficits reach crisis magnitudes and state agencies begin to cut back on public expenditures. *In short, what labour takes with one hand (in the workplace struggle) it gives away with the other (in the urban living place).*

Labour unions, the traditional engine of class confrontation, gradually lose their effectiveness. With the intensified and expanded state intervention, the struggles over wages lose some of their meaning, and the struggle over political/administrative power begins to impose itself as a crucial, but yet unpursued struggle. At the same time, with the ever increasing urbanization of the population, most of these struggles acquire a definite urban character. *The struggle, in brief, is shifting from the sphere of production (of commodities and services) to the sphere of reproduction (i.e. the maintenance of a stable, if not an improving quality of urban living).*

In view of all of this, urban planning, as a domain of social action dealing with the social rationalization of urban life, traditionally marginal and 'foreign' to capitalism, now finds itself in the centre of the emerging mode of struggle for a better society. *This potential exists objectively, and irrespective of the will of urban planners.* It is concretely what is meant by: 'the predicament carries within it its own resolution'.

36 The subjective dimension of these new developments is to be found in the dynamics of planning education (the process by which planners' consciousness is shaped). The contradictions between urban planning theory and practice (refer to propositions 23 and 31 to 33 above) are directly reflected in planning education. Prospective planners are presented with 'bodies of knowledge', 'theories', 'methods', etc. which have one of two dominant characteristics.

On the one hand, there are those 'bodies of knowledge', etc. which bear little relationship to the realities of planning practice. Here, we find idealized notions about what planning *ought* to be or do; about how to go about planning in a society or a community *that wants to be planned for* (i.e. whose institutions do not resist planning); about the kinds of methods to use in the process of planning for a society *that has reached agreements about what it wants to achieve* (i.e. that is not divided by irreconcilable class interests), etc. The hallmark of this type of 'knowledge' is that it might be valuable in guiding planning practice *somewhere but definitely not in our current society* (as characterized on pp. 162–70 above). This insurmountable gap between this type of 'knowledge' and the day-to-day reality it purports to address (and change), makes the 'knowledge' incredible and unbelievable. At best, this type of 'knowledge' leaves seriously committed minds frustrated, and determined to find and develop more plausible knowledge that is indeed capable of guiding progressive practice here and now. Planning education, in this way, produces what it least seeks to produce: politicized/radicalized prospective planners. At worst, this type of 'knowledge' provides charlatans and neo-Machiavellians with a handy collection of slogans and smoke screens useful only for purposes of apology and justification.

On the other hand, prospective planners are presented with a range of pragmatic/technical tools which are directly applicable in practice, but which are utterly incapable of changing the *status quo* or of contributing to meaningful urban reform. (These amount, in most cases, to no more than *implicit and unselfconscious* manuals of how to undertake pre-politics 'processing' of political information regarding land-use conflicts, but with no awareness of the implications of this activity. See proposition 17 above.) Not unlike the former type of knowledge, the latter type unwittingly intensifies the process of politicization and radicalization of some planners. Here again, with respect to the subjective dimension of current urban planning, the predicament carries within it its own resolution. The more diligently society tries to produce (and does produce) planners capable of practising under these contradictory conditions, the more it breeds minds intent on, and capable of, transcending the contradictions. These remain a minority, but a growing one.

Yet it is crucial to realize that this transcendence, imminent as it is, is *not* going to happen while we sit and watch. The question now is this: will planners and *prospective* planners *act* to realize the potentials of urban planning? And the other question is: what if they do not act?

8 Notes on comparative urban research*

Michael Harloe

'But "general" arguments about imperialism which ignore or put into the background the fundamental difference of social-economic formations inevitably degenerate into empty banalities.'[1]

Introduction

Lenin's sharp reminder of the need to pay attention to the variety of national experiences in the analysis of imperialism applies equally strongly to the new theoretical approaches to the study of urban development which have been emerging in the past few years and which are represented in this collection. As yet, despite the growing mass of empirical and theoretical literature in this subject area and the stimulus that this has given to international scholarly discussion, there has been little consideration given to systematic comparative work although its importance is generally recognized. Far more needs to be done if the validity of the claims that patterns of urban development are essentially linked to characteristics of basic modes of production and of their associated class conflicts is to be firmly established and if these theories are to be further developed.

This short paper does not attempt to do much more than raise and

---

* This paper was completed in 1978. It does not therefore reflect recent important developments in the views of the political theorists discussed here, nor in my own views on the problems of comparative research and the prospects for such work.

illustrate some of the issues involved in the search for better comparative studies, in the hope that more attention will be paid to the matters which it raises in future. Firstly it demonstrates, with reference to recent influential work by Lojkine and Castells, the problems that generalizing from a single national example can cause for the analysis of urban policies and development in capitalist societies; secondly, it suggests that comparisons between capitalist and socialist societies (using these terms in the conventional sense) are of importance even if the main objective of most of the current work is a theory of capitalist urbanization; and thirdly it considers whether there are nevertheless some limitations to the usefulness of the comparative approach and how, if this is so, such problems might be resolved.

Lojkine and Castells – towards a general theory?

Both Jean Lojkine and Manuel Castells have attempted to establish some general propositions about the nature of urban development in advanced capitalist societies. In doing so they have made crucial contributions to the evolving body of urban research and theory despite, and even because of, the critical scrutiny that their ideas have been subjected to. Both of them accept the need for comparative work; indeed Castells has recently begun to develop his own work on French urban politics in a far more historical, and therefore nationally specific, direction than before.[2] Both of them would now accept that their earlier work suffered from over-generalization from the French experience; nevertheless, at least until these second thoughts find published expression, an examination of some of the problems raised by this earlier work is a useful way of illustrating the difficulties caused by the lack of a comparative perspective.

It is not necessary for the purposes of this paper to present a comprehensive critique of these authors' work, this has been done at length elsewhere.[3] Moreover, Castells's general approach is clearly outlined in his own contribution to this volume and in his book *The Urban Question* which has been available in a revised English edition for some time now.[4] On the other hand Lojkine's major theoretical work *Le Marxisme, l'État et la Question Urbaine* has still to be published in English and has been less widely discussed. So in this section of the paper some of the arguments in this latter work will be examined in order to show the difficulties that the lack of a sufficiently rigorous comparative approach causes. Then some observations follow on the ways in which particular aspects of French politics and urban development have influenced both theorists. The purpose of this is not to suggest that such specific concerns should not be of crucial importance to French urban research, but merely to point out that what is politically important in France may not be so elsewhere.

Although Lojkine refers at the beginning of his book to the need for historical and comparative studies of urban development in order to distinguish broader trends from more specifically French features, and even draws attention to the existence in France of a powerful and centralized State which developed historically far earlier there than in some other capitalist countries, it seems that he subsequently ignores the specificity of the political system caused by this factor. In recent years there have been very close links between advanced capital and the French State, which have been especially evident in the field of urban policy, and the French Communist Party has sought to show that there is a 'fusion' of monopoly capital and the State. Such analyses provided the theoretical backing for the French Communist Party (PCF) contention that conditions are ripe for the formation of an anti-monopoly alliance of all those social groups, ranging from workers to petty capitalists, who have been losing out in the face of the new ruling coalition. Lojkine's book is an attempt to demonstrate the applicability of this theory and of its political conclusions to urban development, not just in France but elsewhere, and this objective seems to lead him away from his initial concern with paying careful attention to national differences during the course of his analysis.[5]

He states that there are three obstacles to the new type of urban development which is occurring under the influence of the needs of advanced capitalism. All three derive from the contradiction between, on the one hand, the private ownership of capital and, on the other hand, the 'technical' necessity for the progressive socialization of production in order to maintain the extraction of surplus value in the face of growing competition. These three obstacles are, firstly the need for planning and coordination but also the necessity for the continuance of anarchic capitalist competition; secondly the hindrance to 'rational' urban development caused by the persistence of individual landownership – this he suggests is declining due to the appropriation of such ownership by the financial monopolies; and thirdly the difficulties caused by the need to finance necessary but unproductive means of collective consumption such as schools, mass transport and some housing. None of these problems can be wholly solved by individual capital, hence the necessity for state intervention and the consequent growth of urban politics centred on such issues.

Lojkine's concern to show that the monopoly/non-monopoly division is of fundamental importance then leads him to try and show that urban policies are now dominated by the interests of monopoly capital and that any measures which seem to favour dominated groups are, on closer inspection, of marginal value. But the way in which he goes about doing this is highly dubious, firstly for establishing the validity of his thesis for France, and secondly for extending it to other advanced capitalist countries, as he wishes to do. In order to establish the dominance of monopoly capital in the field of urban policy, if he had taken his own earlier warnings about the need for careful historical analysis specific to the country in question seriously, he would have had to examine who gained and who lost from the working of a wide range of urban policies. However, he chooses instead to select only four cases, where 'concessions' were made to dominated groups, for examination.[6] In each of these cases he demonstrates that the 'concessions' were not what they seemed and that, in reality, they were motivated by, and in the interests of, advanced capital, conferring little or no advantage on the dominated groups. Even if this is so a mere four examples cannot suffice to establish his general thesis, especially as the principles guiding this selection are not stated. Are those examples truly representative of the whole? The evidence on which a judgement may be made is simply not presented.

Having established, to his own satisfaction at least, the dominance of the monopolies in French urban politics, Lojkine argues that this is also the case in other advanced capitalist countries but he presents little convincing evidence for this. Thus, after discussing the way in which French local government (which in recent years has come under communist or socialist control in many cities) has come under stronger and more effective central government restrictions and hence under the tutelage of monopoly capital, he extends this to Britain by reference to the 1972 local government reform (involving the amalgamation of smaller into larger units) and the earlier establishment of Regional Economic Planning Boards (which are in fact largely powerless, although he does not seem to realise this).[7] But the extension of his general thesis to Britain on the basis of these two cases alone is hardly satisfactory and even though it can be accepted, on the basis of better evidence than Lojkine provides, that central control is increasing in Britain, it is still necessary to analyse whose interests this development serves. And this question can only be answered by detailed empirical studies of the nature and output of specific policies.[8]

Elsewhere too, Lojkine uses rather flimsy arguments to support the generality of his conclusions. Thus, reflecting a contemporary concern of the Left in France – excluded from national but not local government, as mentioned above – he stresses the severe limits to the power of left-wing local councils to carry out progressive policies in opposition to the central power. Using several examples, he argues that the increasing dominance of central government by the monopolies has resulted in funding for municipal programmes being cut, especially sharply when these are controlled by the Left. The evidence that he presents does indeed suggest that this has occurred in France but, quite insupportably, he then extends this conclusion to Britain by stating that, in the latter country, central government contributions covered 36.9 per cent of total local authority expenditures in 1963/64 but that in 1974 only 10 per cent of the budget of the biggest local authority, the Greater London Council, came from central government![9] Apart from the fact that he is not comparing like with like, the GLC figures are incorrect, referring only to a part of its budget. In reality, central government contributions to local government have risen rather steadily and now cover about 60 per cent of local expenditure.

This example shows that Lojkine does not merely wish to argue for a similar distribution of power in the urban sphere in Britain and in France, but that he also claims that the mechanisms by which this power is exercised are similar. Yet, if the variety of national experiences is important, as it empirically seems to be, it is surely rather unlikely that these mechanisms will be similar. The important task is to demonstrate that, underlying differing mechanisms of power, the basic nature of capitalist societies ensures rather similar policy effects or, if this is not so, to relate differences which do occur to features of the particular course of development taken by capitalism in the country concerned. Claiming that even the mechanics of power are similar in the two countries, when this is patently not so, is only likely to undermine the credibility of the broader and more important claim that there is a similarity between the fundamental nature of the two economic and political systems and in the State's role in urban development.

Both Castells and Lojkine, despite their differences, aim to define, in general terms, the content of the Marxist analysis of urban politics, seeking to establish that it is dominated by the interests of monopoly capital and that the contradictions which arise have far-reaching implications.[10] This, they argue, is because certain types of urban development are now indispensable to monopoly capital if it is to prosper and because a point has been reached where the future evolution of capitalism is seriously in doubt. In this crisis, protest movements which arise in opposition to urban policies can play a vital role, transcending the immediate issues with which they are concerned.

Both writers are trying to establish a special theoretical content to their studies, involving the formulation of universally applicable propositions about the role of urban development in modern capitalism. *If* such propositions exist beyond the most general level (and it is an important issue to discuss whether this is so or whether the whole enterprise might be misconceived, at least in the terms in which these theorists pose it – but one which will not be analysed here) then the best way to establish them would be by means of a careful comparative analysis such as that initially proposed by Lojkine, or, failing this, by a fairly comprehensive examination of French urban development which might then suggest directions for further comparative research. Yet neither *The Urban Question* or *Le Marxisme, l'État et la Question Urbaine* contain anything like this. In both cases the presentation and argument contains very little history and rather too many sweeping generalizations, perhaps even more so in Castells's writing than in Lojkine's book.[11] Despite this, it seems as if both of them have been influenced by French experience and history, abstracting elements which are specific to this country and incorporating them at an unjustifiable level of generality in their theories. One example has already been given: the references by Lojkine to reductions in central government contributions to local expenditures. Another example is Castells's suggestion that there is a new petty-bourgeois revolt based on an environmentalist counter-culture.[12] This is a popular political issue in France (and in West Germany), provoked by the rapid economic development in recent years.[13] But, despite particular protests such as the anti-nuclear and anti-motorway campaigns, it is hardly a key political issue in British urban politics – for the moment at least.

More importantly, the emphasis that both theories place on the State's closeness to the interests of monopoly capital, its centralized control and the relative weakness of countervailing pressures on policy making are understandable given:

a The historic centralization of the French State and the weakness of local government.[14]

b The close links between civil servants and advanced sections of capital which characterize the economic planning strategies of the last twenty years in France. In particular the Sixth Plan was primarily aimed at modernizing industry and building up its ability to meet foreign competition. In line with this policy urban investment was designed to support industrial expansion above all else. Furthermore, the close link between urban developers and the government during the Pompidou regime is generally acknowledged.[15] Such connections were becoming evident during the years when Castells and Lojkine were doing much of their original work on urban politics.

c The long period of domination of central government, and directly or indirectly of local government, by the Gaullists and the exclusion of the Left, especially the PCF.[16]

Other, less general, features of French development may also have over-influenced Castells's and Lojkine's conclusions. For example, in 1963 the French government decided to expand Paris and develop it as a tertiary centre. The subsequent urban renewal involved the removal of much of the working class to the periphery of the metropolitan area.[17] Has this influenced more general references in Castells's book to the functionality of such areas for monopoly capital and similar references in Lojkine's writing to the existence of a new segregation of space, contrasting metropolitan areas occupied by such capital with peripheral towns accommodating the working class and the lower orders of capital? If Castells and Lojkine had been living in London, where, if anything, the reverse has tended to occur, their conclusions might have been rather different.[18] To consider just one more example, Lojkine's stress on the destruction of small and medium capital in the face of the monopolies – hence deriving the possibility of an anti-monopoly alliance – apart from ignoring the question of whether the workers and former petty capitalists are likely to perceive common interests, seems to be heavily influenced by the nature of the French economy.[19] Up until the immediate post-war period this economy was largely based on smaller industrial and agricultural producers but, since then, it has rapidly been 'rationalized'. Moreover, there has been some quite significant political resistence to these changes from small business. Of course small business has also declined in Britain but not at the same rate and with the same consequences, in part at least because British industry was far more heavily concentrated already.[20] Examples such as these suggest that well-based urban theories need more detailed historical comparisons, or failing this, more restricted application than either Castells or Lojkine provide.

One final possible explanation of the limited nature of these two theories deserves mention. The connection between political strategy and analysis in Lojkine's case has already been noted. But did the aftermath of the 1968 struggles in France, with the return to a lower level of militancy in industrial relations combined with the increase which occurred then in urban research funds, stimulate radical sociologists to examine urban politics as a possible basis for mass political mobilization? Pickvance has suggested that this may, initially at least, have been so.[21] More recently this has been confirmed by Castells in the introduction to a new collection of his essays. He writes, 'The large social movement of revolt during that period [1968], created conditions which were extremely favourable to the development of a new type of urban research . . .'.[22] He continues by enumerating these conditions: they included the growth of radical ideas in the academic world; the progressive thinking of some civil servants leading to the provision of research funds; the reaction of ruling groups to the 1968 events – they transformed the social crisis into an urban crisis (in thought) and commissioned research; and the recognition by the Left that urban issues were important factors around which to organize. Given these factors, extensive state backing by officials concerned with problems which were seen in national terms, and the Left's concern with immediate and often very localized struggles, the lack of a comprehensive and comparative consideration of the politics of urban policy in advanced capitalism is understandable (together with other features such as the focus on 'urban social movements' rather than wider arenas of class struggle relevant to urban development). The lesson of this experience is surely that comparative studies are necessary, not just to ensure that national and international characteristics of urban development are not confused, but also to enable a conscious recognition to be made by researchers of the limitations that the very conditions which surround their activity impose on the content of their work, conditions which are likely to be highlighted by contrast with circumstances elsewhere.

Are East-West comparisons useful?

Given that comparative work is necessary, one of the key questions to be answered is: what selection of countries should be made? Most writers appear to think that direct comparisons between advanced Western nations and emergent capitalist countries at a lower level of development are of little use, although studies of the two groupings, which explore the relations of domination and dependence between them and its consequences for urban development, are being carried out.[23] Should comparative work be therefore mainly confined to examining urban policies in the advanced capitalist bloc? In recent discussions on this issue three different positions have been taken. Briefly stated these are:

1 The argument, strongly advanced by Pahl, that comparisons of urban policy and development in Eastern and Western (capitalist and

socialist) industrial societies are needed. For Pahl the common feature of East and West which should be stressed is the appropriation of an economic surplus (be it by private capital or by state enterprise) and the role of the State and its bureaucracy in applying a part of this surplus to urban policies which seem, in both systems, to be concerned with stimulating further economic development and with questions of social control. Although he admits that there are important differences between the two types of society, Pahl contends that there are also sufficient similarities between them, both in terms of the distribution of power (e.g. the role of state bureaucrats) and of urban inequalities (e.g. patterns of residential segregation), for comparisons to be valuable and indeed essential. He is arguing that a common concern with technology and rationality in both East and West has led to some rather similar urban outcomes, despite great differences in the mode of operation of the former factors in each case. In Szelenyi's words he regards the level rather than the mode of production as the more important determinant of urban development.[24]

2 In opposition to this Castells has argued as follows, '. . . the comparison between capitalism and socialism [i.e. East and West] can be misleading or impossible. Capitalism is a social system which functions according to specific rules . . . if we can understand these rules and show how they generate the contradictions which exist within capitalism, we have an analysis of capitalism which is valid without the need for any comparisons with socialist systems . . . we have to look at the variation between urban policies in different countries showing generally similar systems.'[25] In so far as urban inequalities are in fact rather similar in the two blocs, Castells explains that this is due to the persistence of elements of the capitalist mode of development in the socialist states.[26]

3 Szelenyi also disagrees with Pahl's views but, in his case, suggests that the crucial distinction is not between the urban effects of capitalism in the West and socialism in the East but that between the former factor in the West and a mode of production in the East based on an authoritarian State and bureaucracy.[27] So, despite limited agreement at least with Pahl's contention that there are similarities between urban policies and problems in East and West, he would agree with Castells that these are no more than formal correspondences which can only be explained in terms of different modes of production. In fact he goes further than Castells with this contention, rejecting the view that such similarities as do exist are due to the persistence of vestiges of the capitalist mode of production in the East.

As yet there has been little systematic discussion of these propositions, probably because attention is still concentrated on the effort to improve research and theory within national contexts and because of the obvious difficulties of access to data from Eastern Europe. It is not possible in this paper to attempt an exhaustive examination of these different views but merely to mention some of the problems which they raise and some of the implications for comparative studies.

The first point is that each writer admits to the existence of some urban policies and patterns of development which are similar in East and West. Pahl suggests that these are quite numerous; Castells seems to be less sure of this (although he has never really elaborated on his general acceptance of some similarities) and Szelenyi regards the similarities as even fewer and less significant. Clearly, therefore, despite Castells's reservations, there would be some purpose in comparative studies of urban policies and development in East and West in order to establish some empirical basis for the three writers' assertions, or at least to discover whether Pahl has exaggerated the similarities or whether he is correct. Such work seems an important first step and is also probably more practicable than theoretical studies, as most of the urban sociologists in Eastern Europe are involved in very practical work aiding planners and policy makers and have a good deal of empirical material as a by-product of this work. If the similarities do not appear to be nearly as prevalent as Pahl thinks they are, this would obviously tend to add support to the contention of Szelenyi and Castells that mode of production is more important than level of production.

Secondly, if there are substantial similarities between East and West, while this would tend to reinforce Pahl's views, it would certainly not substantiate them. In order to go further it would be essential to study urban developments in societies with rather differing levels of production to see whether the latter did in fact have a powerful influence on the former factors. In the light of this it does seem rather surprising that, as mentioned above, the main focus in the debate so far has been on comparisons between advanced industrial societies only. Of course to carry out any of these comparisons far more needs to be done to clarify the concepts involved, for while mode of production has a fairly well defined meaning, level of production – at least if it refers to more than a series of empirical facts such as GNP or capital/labour ratios – has not. In his writings Pahl seems to be principally concerned with factors such as bureaucratization, planning and technological developments. However, the first of these at least has characterized several very different types of society throughout history, as Weber showed, and, more generally, these three aspects of level of production need to be linked together into a coherent definition of the concept which they are intended to typify. Only when this is done will it be possible to solve some of the other problems that comparisons of urban development and levels of production will face; for example, the problem of what is to count as a similarity and what is to count as a difference. Thus one can even now show that access to urban facilities in both East and West is often stratified according to income. However, the differentials are normally far less in the East than the West. Is this a significant difference in theoretical terms or is it not? Furthermore is it significant that capitalist bureaucracy, technology and planning have tended to create decaying inner-city areas, with the poor trapped in the centre and the better-off at the periphery, whereas in Eastern Europe the reverse is often the case? Analyses of the inner-city problem in Britain have begun to relate this phenomenon to aspects of capitalist competition, referring to theories of the capitalist mode of production and the capitalist State; but to examine whether the level of production is a more useful theoretical approach, when the concept is so unclear, is difficult.[28]

This last point applies equally to Castells's and Szelenyi's contentions. Both of them believe that there is a different mode of production in the East. Castells's view is that there is an emergent socialist mode, dominant but coexisting with a vestigial capitalist mode. Szelenyi believes that capitalism has wholly vanished in the East and that a new state socialist mode is developing there. In both cases their references to a new mode of production are about as sketchy and unelaborated as Pahl's references to bureaucracy, technology and planning. This reflects the lack of attention paid by most Marxists to the concept of the socialist mode of production and/or to analysing the basic nature of the Eastern European societies. However, some work is now beginning to appear and its relevance to the explanation of patterns of urban development in these societies should be examined. Also, if Castells's contention that the socialist and capitalist modes coexist in Eastern Europe is to have any analytical value, some attempt must be made to describe the articulation of these two modes before making more detailed studies of matters such as urban development. If this is not done there is a danger that urban inequalities which seem inconsistent with the assumed socialist mode of production will be automatically assigned to the influence of vestigial capitalism, a procedure more suited to apologists rather than sociologists.

Enough has been said in this section to indicate that there are some interesting questions concerning the nature of the relationship of urban development to production which East/West comparisons, as well as comparisons between advanced and less advanced economies, would illuminate. Pahl's insistence that such research is worthwhile seems justified; Castells's wish to confine comparative work to countries with similar modes of production seems too restrictive. On the other hand, the section has also shown that the practical and theoretical problems involved are formidable (and these have only been sketched in here). It seems as if further progress can most usefully be made in two, for the moment, rather separate ways. Firstly, by attempting to collaborate with sociologists in Eastern Europe on building up empirical knowledge of, and comparisons between, urban patterns and policies in East and West without, at this stage, any very elaborate theoretical objectives. Secondly, by more discussion and elaboration of the concepts of level of production, of socialist or state socialist mode of production and of the articulation of coexisting modes of production in the East.

The implication of all this must be that no early or easy answer is likely to be found to Pahl's challenge to the Marxist theories of urban development, unless the empirical work referred to above shows that the starting point of his argument – that there are substantial similarities in urban outcomes in the two blocs – is false. For those who find the theoretical basis for Pahl's views ill defined, and/or who begin from the assumption that the capitalist mode of production is closely linked to urban development in the West, having as their main task the elaboration of the nature of the links between the two phenomena (and most Marxist urban studies are in this category), the outcome of this debate will be of little interest. On the other hand, at the very least, the three alternative views of the relationship of urban and economic development expressed by Pahl, Castells and Szelenyi, one from a Weberian and two from a Marxist perspective, have illustrated how much still needs to be done if the new urban social theories which have been developed in the last few years are to overcome what was one of the principal weaknesses of earlier urban sociology – its limited interest in anything other than conditions in Western capitalist societies.[29]

A limit to the value of comparisons?

Having argued for the value of comparative work, both between capitalist countries and between capitalist and socialist countries, this section considers whether there is nevertheless an important limitation to what such studies can achieve. In a recent paper Offe has discussed a number of different approaches to the analysis of capitalist States.[30] He points out that, in order to establish the necessity of the domination of the State by class-based interests, one must show how the 'systematic exclusion of all opposing interests come about'. (Note that, although Offe is only concerned to discuss capitalist States, a similar proviso would apply to the analysis of other types of class-based society including socialist ones.) He suggests that studies of the rules of exclusion, conceptualizing state activity as a 'selective event generating sets of rules . . . a sorting process', are required.

But the only way to establish the existence and nature of this selectivity is by empirical studies (for example of urban policies). However, this gives rise to what he calls an 'empirical methodological' dilemma. He writes, 'in order to gain an idea of the exclusion mechanisms and their affinity to class interests it is necessary to have at one's disposal a concept of what possibilities are negated by these mechanisms . . . but how can evidence of what is non-existent, the very thing that is excluded, be established sociologically?' This question has certainly not been faced by the type of urban research discussed in this paper, and it is interesting to note that Offe discusses and criticises a number of approaches which are inherent in this and other policy-oriented work. In particular he refers to the position

> 'which bases its critique on a deductively obtained concept of the revolutionary class and its "objective" interests. This type of procedure, which only supposedly follows Marxist "orthodoxy", runs the risk of raising the still-to-be-proved class character of the State to the theoretical premise, and at the same time degrading to triviality the historical pecularities of the selectiveness of a

concrete system of institutions – whether it can be brought into line with the dogmatically preconceived class concept or not.'[31]

However, Offe also raises an objection to the use of comparative analysis as a means for discovering the 'sorting process'. Comparisons, by their very nature, will tend to highlight differences in rules of selectivity between differing systems and, he adds, 'those types of factors common to the systems under comparison do not come to light'.[32] The relevance of this to the subject matter of this paper is obvious but, it is suggested, more because it points to some important considerations which comparative studies must take into account than because it rules them out altogether.

Firstly, it is worth noting that, at the very least, such studies should be able to identify important differences in the rules of operation of the system under comparison; and hence begin to generate explanations which do not ignore the importance of historical differences between societies and states which at a certain level are held to be formally similar. This is the point which was stressed in the second section of this paper. Secondly, Offe's criticism seems based on the idea that excluded action is 'non-existent'. In that something does not happen this is of course true, but one cannot necessarily conclude that exclusion leaves no 'clue' which can provide reasonable evidence of excluded alternatives and a basis for studying the ways in which their exclusion occurred. Studies of the history of urban policies and ideologies will reveal the existence of many proposals about the nature and form that urban development might take which were not adopted by the State. Also, especially when their proponents tried to take practical steps to achieve their goals, the resistance that they faced from the established political system and ideas may be quite revealing.[33]

However, the fact remains that not all possible demands on a political system which are excluded may be clearly enough articulated to be revealed by comparative studies. Indeed, the demands may be pathologically altered in the process of exclusion and lose all resemblance to a coherent alternative. Is, for example, the growth of personal violence in many urban areas in the West of this nature? At best there is an element of intelligent guesswork in the views of those who give it a politico-economic explanation. Most importantly however, difficulties arise because the State's role as a stabilizer of the social order (or, if one prefers, its self-interest) results in a tendency to use ideological explanations (in particular) in order to conceal, rather than reveal, the exclusion of possibilities which are regarded as destablizing (or not in its self-interest). Of course, these attempts to, for example, represent certain courses of action as in the 'general interest' do not always succeed, especially at times when conditions are favourable for the vigorous expression of opposing interests. It is at such times, times of crisis for the State, that the nature and limits of its power become most clear. Offe's view is that these are the only times when the rules of exclusion can be clearly studied. This is, of course, not the view taken in this paper but the importance of focusing attention on crisis periods can usefully be combined with a comparative approach. The example Offe gives is of the so-called full employment policy in the post-war years.

Until the events of the last few years, the view that full employment and growth could more or less be permanently achieved in advanced capitalist societies by state action was virtually unquestioned. Since this has been shown by events to be an illusion, the spotlight has been turned much more clearly, and not just by Marxists, on these States' dependence on economic interests and the political constraints which result from this situation. (Conversely it could be said that the attack on liberal welfare policies which has been going on sharpens our knowledge of which interests are excluded, especially when groups beyond the very poorest – who are usually excluded – begin to be affected.)[34] Offe's example refers to the present and to processes which are fundamental to contemporary Western society but the injunction to pay particular attention to times of crisis can equally well be applied to less fundamental areas of policy and to the crises which they face from time to time, crises which may or may not be connected to issues of concern for society as a whole on the scale of those now taking place.[35]

Conclusions

Despite the many criticisms of Castells's *The Urban Question,* his analysis of the faults of much previous urban sociology presented in the first part of his book remain persuasive. The crux of this critique was that most of the classic work remained bound within the time and the social system in which it was written, and – worse still – was oblivous of these facts. It would be ironic if the new urban research, despite its sensitivity to the importance of the link between urban development and the broader economic and political structure, fell into a similar trap and remained bound within national or supranational contexts. This paper has suggested that this has already occurred to an extent and that one of the ways in which such dangers could be avoided would be by paying far greater attention than hitherto to the development of comparative studies. It has also tried to show that there are a number of important theoretical and methodological issues which need to be resolved in order for such studies to be effective.

Apart from these difficulties, comparative studies are likely to be beset with practical problems which in themselves may often seem insurmountable. Firstly, most urban research is either directly sponsored by the State or by organizations which share its policy-oriented focus. Apart from the difficulty of obtaining support for critical research from such sources – post-1968 France excepted – experience suggests that many sponsors will not be very interested in international studies, beyond those which consist of fairly superficial accounts of aspects of current policy thought to be of direct and immediate relevance to their own national situations. Secondly, full-scale studies are likely to involve close collaboration between researchers in each of the countries concerned. A great deal of time and effort has to be put into identifying such people, finding a means for effective coordination of their efforts and clarifying the methods and theories to be employed. All this adds to research budgets which are likely to be high already, given the scale of the work, and to the difficulties of obtaining funding. Thirdly, access to data and historical material may be uneven and this may be a severe limitation on what can be achieved. Problems may also arise in standardizing data, and research which depends on elaborate data collection is almost certainly doomed. Finally, for the reasons already mentioned, it may be difficult to get common funding and be necessary to rely on separate national sources, all of whom may have rather different reasons for giving support and look for different results. This will put the researchers under conflicting pressures, adding to those already present as a result of the normal intellectual and cultural differences which are likely to typify an international collaborative project.

This all seems to amount to a rather formidable list of difficulties and it is hardly surprising that there has not yet been a very well developed body of comparative urban research or indeed of comparative social research generally. But there are some hopeful signs – such as the growing number of international meetings and publications contributed to by urban researchers – and the growing realization that the types of theories being discussed in these fora require testing and developing in a cross-national comparative perspective.

Notes

1 Quoted in Slater, D. (1975) 'Political economy of urbanisation in peripheral capitalist societies.' *International Journal of Urban and Regional Research* 2, 1 (March): 43.

2 Although Castells has recently provided a brief and interesting critical account of post-war French urban development and policies. Castells, M. (1978) 'Urban crisis, state policies and the crisis of the State: the French experience' in M. Castells, *City, Class and Power.* London: Macmillan, pp. 37–61.

3 An expanded version of much in this section of the paper is in Harloe, M. H. (1979) 'Marxism, the State and the Urban Question: Critical Notes on Two Recent French Theories' in C. Crouch (ed.) *State and Economy in Contemporary Capitalism.* London: Croom Helm.

4 Castells, M. (1977) *The Urban Question.* London: Edward Arnold. Lojkine, J. (1977) *Le Marxisme, l'État et la Questione Urbaine.* Paris: PUF.

5 Some of the theoretical aspects of Lojkine's position are discussed in the Introduction to Harloe, M. (ed.) (1977) *Captive Cities.* London and New York: John Wiley, pp. 25–9.

6 The cases discussed are hypermarket development, urban renewal, property compensation and local government autonomy.

7 Lojkine 1977: 290.

8 An example of a preliminary attempt to do this is contained in Harloe, M. 'Housing and the State: Recent British Developments.' *International Social Science Journal* 30, 3.

9 Lojkine 1977: 310.

10 See Castells (1978).

11 But in his latest collection of essays Castells's remarks that a better starting point for theory might have been historical analysis rather than the more abstracted methods of *The Urban Question*. See reference 2 above.

12 Castells 1977: 464.

13 Ardagh, J. (1977) *The New France*. Harmondsworth: Penguin Books, pp. 243–51.

14 The influence of this historic factor is commented on by many analyses of the modern French state and society. See, for example, Ardagh (1977) and in a comparative perspective, Shonfield, A. (1965) *Modern Capitalism*. London: OUP; Denton, G., Forsyth, M. and Maclennan, M. (1968) *Economic Planning and Policies in Britain, France and Germany*. London: Allen and Unwin; Hayward, J. and Watson, M. (1975) *Planning, Politics and Public Policy. The British, French and Italian Experience*. London: CUP.

15 On the relations between civil servants and big industry see the works cited above, also d'Archy and Jobert in particular refer to the explicit link between urban investment and the focus on developing advanced industry, the main priority of the Sixth French Plan (1970–5) viz., ' . . . it must be emphasized that, taking into account the priority given to industralization by the Sixth Plan, the Towns Commission placed much greater emphasis on investment in infrastructure which was necessary for industry, such as transport and telecommunications, while the preceding plans had stressed the necessity for developing public spending destined directly for the benefit of the inhabitants', in Hayward and Watson (1975: 305).

When discussing property development in Paris in the late 1960s and early 1970s, Ardagh refers to the positive climate towards such development from the government and 'the well-known links between President Pompidou and the big banks' (Ardagh 1977: 298, 301, 663).

16 Note also the weakness of French unions. For example many employers refused to bargain collectively before 1968, by the end of the 1960s only about 22 per cent of the workforces was unionized (42 per cent in Britain) and the movement was not unified. (See Hayward, J. (1975) 'Planning and the French labour market: incomes and industrial training, and "Comparative Conclusions" ' in Hayward and Watson (1975)).

17 The Fifth Plan (1966–70) aimed 'to modernise the Paris region rather than halt its development. The Plan accepts the fact that the growth of the national economy must involve at least a moderate expansion of industry and employment in the Paris region' (Denton *et al.* 1968: 324). Also Ardagh remarks on the policy of developing Paris as an office centre with some new housing and adds 'with land costs so high, the new housing is mostly expensive and middle class, while the former slum dwellers have been evicted to new houses in the other suburbs. Thus inner Paris has become increasingly a bourgeois "ghetto" – to the fury of the Left' (Ardagh 1977: 304).

18 Thus in London housing politics have revolved round the successful resistance of the outer boroughs to providing substantial rehousing for working class inner Londoners. For an account see Young, K. and Kramer, J. (1978) *Strategy and Conflict in Metropolitan Housing*. London: Heinemann; and Harloe, M., Issacharoff, R. and Minns, R. (1974) *The Organisation of Housing. Public and Private Enterprise in London*. London: Heinemann.

[19] Thus Watson notes the support of small business in France for moves to control trade union activity. ('Planning in the liberal democratic State' in Hayward and Watson (1975: 471).) Runciman's findings on relative deprivation might also tend to suggest that failed capitalists would be highly unlikely to feel much affinity for the working class they were joining. See Runciman, W. G. (1966) *Relative Deprivation and Social Justice.* London: Routledge and Kegan Paul.

[20] Ardagh refers to two economies 'uneasily coexisting' in France: a modern one of big firms and state enterprises, much of it implanted since the war and below 'an old creaking infrastructure based on artisanship, low turnover with high profits, and the ideal of the small family business'. He suggests that this situation has only really rapidly altered since the 1950s (Ardagh 1977: 24). He also quotes the fact that in 1964 there were sixty-four firms in the world with a turnover of over $1,000m including forty-nine American, five German, four British, and none French, and over 50 per cent of French workers were in firms with less than 200 employees (Ardagh 1977: 62). Evidence presented by Paris suggests a far greater preponderance of large firms in Britain than in France and a far less significant sector of small business. Thus, in 1963, 2.1 per cent of total manufacturing employment in the UK was in firms employing under ten persons. The figure for France was 10.8 per cent. See Prais, S. J. (1976) *The Evolution of Giant Firms in Britain.* Cambridge: CUP, Chapter 6 and Table 6.4, p. 160.

[21] Pickvance, C. G. (ed.) (1976) *Urban Sociology: Critical Essays.* London: Methuen, pp. 1–2; see also his comment in Harloe, M. (ed.) (1978) 'Urban Change and Conflict' in Proceedings of the Second York Conference. London: CES.

[22] Castells 1978: 9–10 ('Urban crisis, political process and urban theory').

[23] See for example, Walton, J. and Lubeck, P., 'Urban class conflict in Africa and Latin America: comparative analyses from a world systems perspective.' *International Journal of Urban and Regional Research,* forthcoming.

[24] See Pahl, R. E. (1977) 'Managers, technical experts and the State', in M. H. Harloe (ed.) (1977: 49–60); and also Szelenyi, I. 'The relative autonomy of the State or state mode of production?' Paper presented at a session of the Research Committee on the Sociology of Urban and Regional Development at the 9th World Congress of Sociology, Uppsala, Sweden, August 1978 (mimeo), p. 2.

[25] Remarks reported in Harloe, M. (ed.) (1975) *Proceedings of the Urban Change and Conflict Conference.* London: CES, pp. 17–19.

[26] See Castells (1977: 64–72).

[27] This position has been developed by Szelenyi in an, as yet unpublished, series of papers. See also reference 24 above.

[28] The most important analyses of British inner cities problems from a Marxist perspective are contained in the reports of the Community Development Project Collective. See, for example, National Community Development Project (1976 and 1977) *Gilding the Ghetto: The State and the Poverty Experiments and the Costs of Industrial Change.* London: Home Office.

[29] See the critique in Castells (1977) parts I–III.

[30] Offe, G. (1974) 'Structural problems of the capitalist State', in K. van Beyme (ed.) *German Political Studies.* London: Sage, Vol. 1: 31–57.

[31] Offe 1974: 43.

32 Offe 1974: 44.

33 See, for example, Fishman's account of how Ebenezer Howard's scheme for deurbanization based on a cooperative socialist system was taken up and promoted by businessmen and others and transmuted in the process. Fishman, F. (1977) *Urban Utopias in the Twenthieth Century.* New York: Basic Books, pp. 23–88.

34 For discussion of some of these issues see the symposium on economies in crisis, especially the contributions by Miller and Mingione (1978) in *International Journal of Urban and Regional Research* 2, 2 (June): 201–22.

35 See Harloe (1978) for an example of an analysis of this nature regarding British housing policy.

IV Commodity production and urban development

9 The UK electrical engineering and electronics industries: the implications of the crisis for the restructuring of capital and locational change*

Doreen Massey

Introduction

This paper presents some of the results of an empirical investigation of the spatial effects of certain aspects of the present British economic situation. The material is thus concerned with locational implications of retrenchment, rationalization and aggregate employment decline, and it indicates that the present economic situation in the United Kingdom may be having certain distinctive spatial/regional repercussions. But the analysis is also presented in order to make a number of more theoretical points. Firstly, it serves to emphasize the link between national economic developments and intra-national spatial employment patterns – to emphasize that regional employment changes must be understood in terms of an analysis of the economy as a whole. Secondly, the results indicate that a number of presently well-recognized spatial changes are themselves closely bound up with the effects of the present industrial retrenchment. The paper examines in particular the changing distribution of employment between the depressed and the relatively prosperous regions of the country, and the continuing decline, as centres of manufacturing, of the major conurbations of the country and particularly their inner areas. It is shown that both the intersectoral reorganization of production (the reallocation of capital between branches), and the differential impact of intra-sectoral attempts to reduce the cost of variable capital and increase the rate of surplus

---

* This is a revised version of a paper presented to the 23rd Annual Meeting of the North American Regional Science Association, in Toronto. See page 230 for a postscript to the text.

value, are contributing significantly to these spatial changes. Some of the political implications of this are mentioned at the end of the paper.

The approach taken here is to examine the response of individual firms to the particular economic problems facing them over the period 1968–72. This derives from one of the aims of the original research which was to explain the locational behaviour of individual firms in relation to the development of the economy of which they are part. In the established approach of industrial location theory, based on neo-classical economics, this relationship is rarely treated. At most, the structure within which the firms are operating is included merely as some vaguely conceptualized 'context'; and certainly not as an integral part of the explanatory framework.

But if neo-classical location theory has difficulty in moving from the individual firm to the structure, Marxist work, though correctly starting from the level of the structure as a whole, and from the overall process of accumulation, sometimes appears to neglect examination of the form taken by those structural movements at a more disaggregated level. The research presented here has attempted firstly to look at the very different ways in which the crisis may be articulated in different sectors of the economy, and secondly to analyse, in relation to this, the response of individual capitals.[1] This process of moving beyond the level of industrial capital as a whole is essential to any political understanding of the present situation.

Finally, examining spatial changes in this manner enables one to go beyond the frequent rather evocative references to 'uneven development' (references which, while recognizing the fact of regional differentiation, remain at the level of description) to analyse the operation of certain of the mechanisms producing that differentiation.

British capital was hit early and severely by the world economic crisis. By the late 1950s the rate of profit was already beginning to decline steeply. In the face of this, the general response of UK capital had to be to attempt to raise the rate of exploitation, in order to raise its rate of profit and to stave off realization problems by increasing its share of declining world markets. Since the mid-1960s the combined effects of the problem and of its attempted resolution have produced – among other things – a considerable reorganization of British capital, a dramatic upward shift in the national level of unemployment, and a fall in the number of people employed in manufacturing. However (and unsurprisingly since most capitalist countries are pursuing the same course), neither of these twin and related aims, of increasing the rate of profit and increasing the degree of international competitiveness, has been achieved. The processes discussed in this paper are likely to continue for some time yet.

One of the forms of response to this situation, and in many cases a necessary condition for the reorganization of productive capital, has been restructuring. By 'restructuring' is meant the reorganization of the ownership of capital, primarily through processes of centralization (see also on this, e.g. Fine and Harris 1975). Such processes – as Marx pointed out – are typically reinforced during periods of crisis. The concern of this paper is to follow up this crucial element of retrenchment, to analyse the forms taken by the restructuring process at the level of individual capitals and, the central theme, to assess its impact in terms of the spatial distribution of employment.

A further significant feature of the crisis has been the important role played by the State. The Industrial Reorganization Corporation (IRC), which lasted from 1966 to 1970, was one arm of this general intervention, with the explicit aim of restructuring crucial sectors of the economy. Some aspects of its interpretation of the term 'crucial' will emerge in the discussion in this paper. Certainly, potential for internationalization and for increases in the rate of exploitation were central.[2] The cases of restructuring examined here were all carried out in a general sense under this arm of Government policy. They included mergers of whole companies, the reorganization of branches of production through the amalgamation into new companies of overlapping product-ranges from a number of individual capitals, and the provision by the IRC of additional money capital for expansion of production. The IRC itself selected major branches of industry on which to concentrate, and the research presented here has focused on one of its early-established priority areas – electrical engineering and electronics.[3] In 1966, there were 1,911,000 people employed in the sector, representing about 14 per cent of all the workforce in all manufacturing industries (data from the Department of Employment Gazette, October 1975). According to the Census of Production, the sector accounted in 1968 for 10 per cent of net manufacturing output.

The decade of crisis in the UK has also been marked by significant changes in the geographical distribution of economic activity, changes which are parallelled to differing degrees in a number of countries of Western Europe and in Canada and the US. Figure 9.1 indicates the basic economic geography of the UK which is relevant to these changes. Firstly, the major conurbations marked have all been losing manufacturing employment at a rapid rate throughout the period. Secondly, the map indicates a 'policy division' of the United Kingdom into Development Areas and non-Development Areas. Development Areas are those which were assisted under the regional policies of the period.[4] In them, grants are available to manufacturing industry both on new capital investment and on all employment. It should be stressed that, in contrast perhaps to the sunbelt states of the US, or the predominantly agricultural periphery of France and Italy, these assisted areas are not on the whole non-industrialized. Indeed, as the map indicates, they include some of the major industrial conurbations. The Development Areas are defined primarily in terms of their unemployment rates, though this criterion is used in a fairly broad way. Since the war, the difference in unemployment rate between the present Development Areas and the rest of the country has fluctuated, but it has remained. This leads us to the second change over the last decade in the geography of UK economic activity. Between 1966 and 1976, a shift in this pattern became apparent, with the manufacturing employment shares of assisted areas noticeably increasing, and those of the non-assisted areas declining. Table 9.1, from Keeble (1977) gives the

Figure 9.1 Development Areas in the UK (1966)

Glasgow

Newcastle

Liverpool

Manchester

Birmingham

London

Development Areas

Table 9.1 Change in manufacturing employment share, 1965–75, in percentage points

		%
South East	non-assisted	−2.37
West Midlands	areas	−0.21
Scotland	Development	+0.04
Wales	Areas	+0.60
North		+0.70

figures for those regions for which data is available: This 'convergence' is a new phenomenon, and specifically it is one which runs counter to the common assumption that in periods when the national economy is 'in difficulty' regional differentials widen.

The results presented in this paper indicate some of the mechanisms through which the restructuring of capital in the face of international crisis conditions is contributing to both these aspects of geographical change.[5]

Data was collected both from published sources and from a long series of detailed interviews[6] with the firms concerned. The firms in the survey (which included all those in the above sectors which operated in any way under the aegis of the IRC) accounted for about 20 per cent of the sectoral employment at national level. This coverage was unevenly distributed, however, many specific product groups within these sectors having played little part in the processes we are examining. In those which are included, therefore, coverage is well above this figure, and in many product-groups is 100 per cent.

The research was structured around a series of questions, the sequence of which reflected the main directions of causality postulated and which incorporated, in a way traditional theory does not (indeed cannot), both the link between individual firm behaviour and the wider economic structure, and the centrality of the process of production. After the characterization of the general economic situation, the next stage analysed the specific way in which the crisis was articulated in each case; this gave us, therefore, the particular reasons for the restructuring which did take place. The third stage of the analysis examined the forms of reorganization of the production and labour processes which were consequent upon/enabled by the restructuring. Finally, the last stage was concerned with the spatial impact of this reorganization primarily in terms of broad patterns of employment and unemployment.[7]

The analysis revealed the existence of three distinct Groups amongst the cases of restructuring we examined – Groups that were distinct in terms of each of the stages of analysis – that is in terms of their relation to the overall crisis of the economy, their forms of production reorganization, and the spatial impact of this reorganization. It should be stressed, however, that although the locational changes introduced by the firms in the three Groups were, on aggregate, significantly different from each other, the point of the grouping is not to introduce some new form of simple correlation (restructuring of type x implies locational change of type y). The objective is to emphasize (a) that the overall crisis hits sectors in different ways, and (b) that locational change can only be understood as part and parcel of an economic analysis which takes account of such structural determinants.

To allow the overall perspective to be borne in mind, it is worth indicating in advance the content of the classification into these Groups. This classification does not simply follow a sectoral pattern (partly because of differences within sectors in the effects of the crisis, partly because of the different situations of specific individual capitals); however, the breakdown is roughly on the following lines:

> *Group 1* Heavy electrical engineering, supertension cables, aerospace equipment.
> *Group 2* Electronic capital goods.
> *Group 3* Individual capitals from all sectors, but involved in specific conditions in relation to the international market.

This classification thus does not cover all the industries initially listed; telecommunications, for instance, is missing. The reason for this is that most of the firms in these sectors span a wide range of product groups. In this paper, the next three sections (pp. 205–16) deal with the reason for restructuring, and the form it took, in terms of the sector or product group which was dominant (because of the nature of the effects of the crisis on it) in stimulating that restructuring. The identification of the dominant sectors was a significant part of the research and has been confirmed in all cases by published data and interviews. However, mergers of major conglomerates may have repercussions on all their divisions and not just on the sectors which provided the rationale for the restructuring. The aggregate results, which are presented on pages 216–27, therefore include all effects, in both 'dominant' sectors and in 'secondary' sectors.

The treatment of spatial implications is divided between all four of the following sections of the paper. In the next three sections (pp. 205–16) each of the three Groups is taken in turn. These sections concentrate on establishing the particular form of the effects of the crisis in each case. It is this discussion which makes the link between the crisis, as an 'a-spatial' phenomenon, and its effects on the geographical patterns of economic activity. At the end of each section, the spatial implications specific to each form of restructuring are very briefly indicated. The final section (pp. 216–27) then attempts to put all these together to assess their combined spatial implications. Thus the next three sections are accounts of the way in which spatial employment changes are produced at the level of the individual firm. The final section is an analysis of the aggregate effect of those individual changes.

Centralization in the face of surplus productive capital

The effect of the UK economic situation on the first Group of cases was dominated by the problem of surplus productive capital. There was simply too much capacity for it all to be employed at an adequate rate of profit. This situation typified a considerable range of the industries examined in the study, and clearly was the major factor behind the restructuring in supertension cables and aerospace, and most of the mergers in the heavy electrical machinery sector. The context of the problem was the slackening in the rate of growth of demand specifically for products of the electrical sector, and the exacerbation of this situation by the turndown in worldwide industrial activity. In all cases, previously secure domestic markets were both declining in size and being increasingly invaded by foreign capital, itself under similar pressures. In terms of potential exports, the problems were just as great. The Commonwealth was no longer such a protected, nor such a lucrative, market; export outlets had to be found elsewhere, but in a situation where international competition was becoming increasingly severe.

In most parts of these sectors the problem was expressed on the ground by the existence of actually unused fixed and variable capital. Some capacity was simply standing idle; orders were spread amongst the rest leaving even this capacity underused. The problem both for the individual capitals and in the sector as a whole was therefore to produce existing and projected output from a smaller amount of constant and variable capital, from which a higher rate of profit would consequently be appropriated. Without financial restructuring the situation was therefore a competitive one of which individual capitals would disappear and which would remain. The process of centralization of capital which took place allowed the solution to this problem by enabling a coordinated (instead of a competitive) reduction in capacity whereby no individual capital went bankrupt. Instead, these capitals merely withdrew from the sector involved, allowing the money capital representing its previous interests to be reabsorbed into more profitable areas of production.

The problems took slightly different forms in each of the product groups. In supertension cables imports were, and have remained, negligible. The basic problem was a decline in home (UK) demand, for which exports at that time were unable to compensate. Increasing amounts certainly were being sold to the non-Commonwealth Third World (the only remaining growth market), but margins were low and competition, especially from Japan, was severe. Moreover, the existence of overcapacity was made more severe as a problem for individual firms because of this industry's high degree of capital-intensity. The supertension cables sector falls within MLH 362, which is by most measures the most capital-intensive of the industries being examined here. Taking only the fixed element of capital costs, calculations from the Census of Production give a figure of £235 of capital expenditure per capita in MLH 362 as a whole, higher than for any other of the MLHs being studied except 366 (electronic computers).[8] In particular, the industry uses aluminum extruders, each costing around £700,000 (or probably about £400,000 in 1968). The financial implications of leaving such fixed capital investment idle are considerable. In fact, in order to appropriate an adequate return on such an outlay, three-shift working (i.e. twenty-four-hour use of the machinery) is necessary (interview).

The heavy electrical machinery sector produces power-generation equipment, such as turbine-generators, switchgear and transformers. In the early stages of this industry, British companies built up good market positions, which continued until the mid-1950s with a strong performance in both UK and export markets reflecting a high growth rate in the demand for electrical energy (see Scott 1974). The subsequent fall-off in that growth rate has resulted in a problem of overcapacity emerging at a world scale. 'Prices in export markets came under acute pressure. The result was a drastic reduction in profitability and almost all the subsidiary companies involved in turbo-generators, power transformers, heavy switchgear and power-cables experienced losses at some point' (Scott 1974: 34). British companies were particularly hard hit by the diminishing importance of Commonwealth markets. Most of our interviewees referred to this. In particular, it was stressed that the gradual decline in the market available to British companies resulted from their relatively greater losses in the Commonwealth, with the ending of Commonwealth preference and the beginning of home manufacture of the smaller end of the range of products. This latter development has meant some replacement of exports by overseas production at the lighter end. For the bigger machines, international tendering still operates, but in a situation of low profits and overcapacity.[9] The capacity problem was exacerbated in this sector too by technological changes – in this case the increasing size of the commodity being produced. This meant that production was increasingly based on a small number of large, discrete orders, frequently individually specified. Maintaining an even flow of capacity-utilization, and therefore rate of profit, becomes very difficult under such circumstances. Moreover, even without any reductions in aggregate sector output, orders become fewer, and competition for them consequently fiercer. The increasing size of individual products also increases the capital it is necessary to invest in production – for instance in testing facilities – producing some of the problems already outlined for supertension cables.

For both the sectors discussed so far it has been stated that declining UK demand was an important element in the changing situation. A few words more should be said about this, since the problem is popularly (and that includes most economists) ascribed to the drastic fall-off in investment, and consequently of ordering for capital goods, by the nationalized industries, most particularly the Central Electricity Generating Board (CEGB). This explanation, however, gives the situation an *ad hoc* status, and fails to relate it to the overall economy. Certainly the basic domestic market for these commodities is provided by the CEGB; and it is also clear that the major immediate reason for overcapacity was below-forecast demand from the CEGB. But this needs to be set in a broader context. Firstly, the decline was one from levels predicted in the National Plan of 1965, and this was, of course, itself an earlier response to the problems of the British economy. Secondly, such forecasts of CEGB demand are themselves overwhelmingly based on forecasts of demand by domestic industry as a whole. The new private investment forecast by the Plan did not materialize; little was yet being done to restructure industrial capital towards a higher rate of profit. Thirdly, in an economy where the bulk of production is in private hands, such 'plans' amount to no more than forecasts. The fact that nationalized industries are the only individual capitals to publish their investment and supply intentions on the basis of these forecasts is a result of their position within the economy – the fact, simply, that they are nationalized.

The problem was, of course, exacerbated by other uses made of nationalized industries in national economic planning. Most of our interviews in this Group produced comments on this. In particular, it was pointed out that the increasing emphasis on 'making nationalized industries pay their own way' (itself a demand of private capital) was having contradictory repercussions on specific sectors. Thus, for instance, changed policies towards nationalized industries' pricing and investment strategies meant that not only were some of the inputs to electricity generation (e.g., coal) becoming more expensive, but also that the increased cost of electricity itself was contributing to the declining demand. Finally, the situation was aggravated by the completion of the major Supergrid investment project. The view that the problem facing these sectors was a sudden one, and one solely attributable to mis-forecasting by a nationalized industry, is, therefore, inadequate. The fall-off in investment by nationalized industries was by no means an *ad hoc* phenomenon, but yet another reflection of the overall crisis.

The problem in the aerospace industry had many of the same characteristics. In the mid-1960s, all the four still-existing major companies were facing increasingly severe competition from abroad (primarily from the US, but also from parts of Europe), 'and this was the key to the ensuing mergers' (interview). Added to this was the collapse of the immediate post-Second World War sellers' market, where selling had been characterized by 'cost-plus, easy money, and defence contracts'. The ending of the Korean War, and defence cancellations in the UK (TSR2, 1154 and 681), severely increased competition in both domestic and world markets. Profitability was declining and the associated problem of 'overcapacity' became more acute. The collapse of a major customer, Rolls-Royce (after the mergers, but before the formation of Lucas Aerospace) added to already-existing sectoral problems.

In all of these industries, then, it was necessary to cut productive capital (and thus costs of production) in relation to the (almost unchanged) amount of commodities being produced, thereby increasing both the rate of surplus value and the rate of profit. The IRC encouraged an orchestrated process of further centralization which allowed major amounts of unused and underused capacity to be taken out of production in the sector without any individual firm going out of business. Indeed, it allowed these firms to re-allocate money capital into more profitable lines of production, and to lessen their involvement in these declining industries. The mergers which took place involved all the major firms in each of the product groups, and were as follows:

Industry	Acquiring firms	Acquired firms
Supertension cables	BICC	AEI (interests)
	PIRELLI	Enfield Cables Ltd (interests)
Heavy electrical machinery	GEC	English Electric and AEI
	REYROLLE	C.A. Parsons Ltd and Bruce Peebles Ltd.
Aerospace equipment	LUCAS AEROSPACE	English Electric (interests)

The capacity cuts enabled by these mergers went as high as 50 per cent in a number of product groups. But those firms which remained needed also to increase their share of the world market. Pressure was therefore directed towards further increasing productivity. A number of measures were adopted, with different emphases in different sectors. The measures can be broadly divided into two groups – those more or less directly associated with the reductions in capacity, and those relating to the reorganization of the labour process.

Profitability was of course increased simply by the abandonment of unused capacity; the rate of surplus value was increased by the organization of that abandonment. Firstly, the State was clearly not supporting all individual capitals; those which survived the restructuring process were those with the greatest potential for internationalization (see also e.g. Fine and Harris 1975). Secondly, it was the production processes with the highest labour productivity and technical composition of capital which were retained. These measures dominated supertension cables and were important in the heavy electrical sectors.

The second group of strategies designed to increase the rate of surplus value involved changes in the labour process. The simplest of these was the process ideologically known as 'the reduction of overmanning'. This involved increasing the intensity of labour by reducing the number of workers required to produce a given output, with no other changes in the production process necessitated. Frequently, for instance, it involved sacking indirect workers (mates, etc.). More complex reorganizations of the labour process took place as a result of the introduction of new techniques. These, too, took a number of forms. In the heavy end of the power-generation product groups, the large size, small number, and individual specification of many of the products makes mass production impossible. Here productivity was increased by standardization[10] of as many components and characteristics as possible (e.g. in switchgear), and by the introduction of conveyor production-line systems and automation to those processes amenable to such forms of organization. In aerospace, the small-batch, custom-built nature of much of the production meant again that automatic transfer and mass-production were not possible. On the other hand it was possible in such cases to introduce numerically-controlled machine tools which allowed both the number of workers and their average wages (because of the associated deskilling) to be lowered. At the lighter end of the electrical machinery sector, finally, where the number of products produced is larger, and their size smaller (e.g. distribution transformers), full standardization and automation has been possible. All of these changes are, at least in part, 'competitive' in nature at the level of the individual firm, and enable a relative increase in the rate of surplus value. They all also increase the technical composition of capital.

So what was the locational impact of all this? First of all, two basic facts should be established: that all of these industries are located primarily in the major industrial conurbations of England, and mainly outside the Development Areas (the major exceptions to this being Merseyside and the area of Newcastle-on-Tyne); and secondly that all were sectors which employed relatively high proportions of skilled craft labour.

The changes introduced into the production and labour processes can be characterized as:

1 capacity cutting/closures;
2 reduction of labour force without closure;
3 partial standardization and automation;
4 introduction of NC (numerically controlled machine tools);
5 full standardization and automation.

In terms of labour requirements, all are characterized by cuts in amount. The last three, moreover, are also characterized by changes in the *type* of labour required. Partial standardization, simply by reducing the individually-specified nature of the product, requires a less adaptable (to different specifications – i.e. less skilled) workforce (interview). Full standardization and automation reduces most of the individual labour processes involved to assembly-work. The skill-requirements are again lowered, and cheaper labour-power, including that of women, can be introduced. The introduction of NC has a similar impact, both reducing the number of workers employed and deskilling the requirements of the remainder (see, for instance, Palloix 1976).

Moreover, because of the nature of the changes, all measures 1 to 4 take place within the existing geographical confines of the industry concerned. Thus the conurbations of London, Manchester, Birmingham and Liverpool (three of which are outside the Development Areas) have seen a massive decline in both the absolute level of employment in these industries and, further, in the level of skill required of the (previously highly-skilled) remaining labour force. Where actual closures have taken place, these have frequently been in the inner parts of major cities, where plants are older, individual labour productivity therefore lower, and unionization frequently stronger.

In only one of the types of change in this Group, type 5, is any major relocation of fixed capital normally produced. Cases where full standardization and automation processes can be introduced more frequently merit investment in brand-new plant. It is common, in such cases, for a locational change to be required (see later) and the new lack of ties to skilled labour, combined with regional controls and incentives, mean that the new location is most often outside the major conurbations but within a Development Area. The original location will therefore lose jobs, and a Development Area may gain – though fewer jobs, and of lower skill requirements than the ones lost.

Problems of technological change

The second Group of industries was involved in financial restructuring as a result of pressures very different from those affecting the first Group. This is reflected in the forms of reorganization of the production and labour processes, and in their spatial repercussions. All the cases in this Group fall within 'electronics', the specific industries being manufacturing of computers, numerically controlled machine tools and industrial instruments. Two very different factors appear to have conditioned the restructuring process, and state involvement in that process, in these sectors. Firstly, at the level of the economy as a whole, there was a need both to increase and to cheapen the output of these industries, since this output constituted an important means of increasing productivity in the rest of the economy. Secondly, and located more at the level of the individual capitals involved, there was continued pressure to keep up with international competition.

The first of these pressures is one which can only be explicitly responded to at the level of the State, since the aim goes beyond the competitiveness of the individual capitals involved. At this level, however, these advanced Department I sectors are crucially important. Increased application of their output in the production processes of other industries will both increase the international competitiveness of those other industries and, when the effect has fed through to Department II sectors, and consequently to the value embodied in the necessary wage, will contribute to an increase in the rate of surplus value over the economy as a whole. The IRC was quite clear about the importance of restructuring advanced sections of the capital goods sector (see, for instance, Villiers 1969), and efforts to increase productivity in general, and to increase the application of new technology in particular, were major planks of state policy during this period (see, for example, Bailey 1968; Young and Lowe 1974; Massey and Meegan 1979). The results of this particular pressure for restructuring, then, were on the one hand growth of output in these sectors and on the other hand the cheapening of that output through changes in the process of production.

The second pressure for restructuring resulted from the need of the individual capitals to keep up with international competition. Even here, however, the form of this competition, and the consequent implications for the production process and spatial organization, were quite distinct from those in the electrical sector. During the whole of this period, that group of industries which is mainly in electronics maintained a momentum of growth far faster than that of the electrical industries, in terms both of rate of growth and of absolute growth in the amount of production. Here too, however, market and production possibilities are exhaustible. But while the major expansion has already taken place, technical innovation remains important in the electronics sector in providing the dynamic of inter-firm competition (see, for instance, Meek 1973).[11]

Again in contrast to the electrical industry, the market in electronics has from its inception been an internationally competitive one. Thus, while the import share of the UK market may be higher for some products than it is in the electrical sector, this does not represent a new departure; nor does it take place against a background of decline in that home market overall. The share of the UK domestic market taken by non-UK capital is also considerably higher than indicated by the share of imports, since significant shares of home production are accounted for by foreign-owed firms. Although there are obviously considerable variations between product markets, sales by foreign-owned enterprises as a percentage of total sales are on the whole higher in electronics than in the electrical industry. This distinction is particularly true of the specific products on which attention is focussed in this study.[12]

These different conditions in the electronics sectors mean that the economic crisis gives rise to different forms of problem there. Although particular periods of recession may now lead to closures, the main problem is one of competing on an international market in a highly-technological and fast-changing industry. The resultant characteristics of the industry may make increasing firm size advantageous within the framework of capitalist competition. The most obvious of these characteristics is the importance of research and development. A number of different effects of this should be distinguished at the level of the firm. The most obvious is the amount of expenditure devoted to R and D. The National Economic Development Office (NEDO) (1972) divides total investment into gross fixed-capital formation and current (only) R and D. Rough percentage estimates for the UK are as in Table 9.2.

There are a number of ways in which the higher proportions of expenditure devoted to R and D in electronics benefit from a restructuring of

Table 9.2 The division of total investment

	All manufacture	*Electronics*
	%	%
Gross fixed capital formation	74	31
Current research and development	26	69
Total	100	100

Source: NEDO 1972.

capital into larger units. Firstly, these numbers are presented as proportions, but it is not the *proportion* which is critical to technical advance; it is the absolute size of the R and D programme which is crucial to technical competitive strength. The point of a merger may well be to *reduce* the proportion of funds which must be devoted to R and D, while enabling retention of the same, or even an increased, absolute level of expenditure. The second point is related to this. It is frequently argued that innovation occurs more often in small firms than large. The evidence is ambiguous and the arguments somewhat *ad hoc.* Be that as it may, what *is* clear is that the post-innovation *development* stage requires the backing of large amounts of capital. The Electronics Economic Development Committee of NEDO divides current R and D expenditure into that for basic research (defined as work undertaken for the advancement of knowledge without specific commercial objectives), applied research (research undertaken with either a general or a particular object in view) and development (research directed to the introduction or improvement of specific products or applications, up to and including the prototype or pilot plant stage). Table 9.3 gives a breakdown of expenditure on R and D on electronics products carried out within private industry, for 1969/70.

Table 9.3 The importance of development in electronics R and D

			%	£000s	
Capital expenditure	—	total	5	6,096	
Current expenditure	—	total	95	124,219	
Basic research			2	2,795	sub-totals
Applied research			10	13,595	
Development			83	107,829	
Total			100	130,315	

Source: National Economic Development Office 1972

Thirdly the rate of technical change may itself demand the concentration of capitals into larger units. The rate of technological change is fast; equipment may frequently become technically obsolete before it is anything like physically worn out. The period of time over which its cost can be depreciated is thus reduced, and the fixed capital element of production will increase proportionately unless it can to some extent be offset by increasing intensity of use. Finally, the NEDO Industrial Review to 1977 (National Economic Development Office 1973) adds another aspect of the problem, and one which again reinforces the advantages of size. The report notes that 'the risks involved in developing new products in this industry, with its rapid rate of technological advance, mean that it needs to finance more of its capital investment (and virtually all of its private venture R and D) out of its own resources, than do other industries' (p. 47). Thus, in contrast to the electrical sector, where financial restructuring took place primarily in order to reduce competition, it was clearly the need for an absolute increase in the size of individual capitals which dominated the process in electronics.

The precise form of these pressures obviously varied from case to case. In the takeover of Cambridge Instruments by George Kent it was hoped to integrate scientific instrument development into that of industrial instruments, and also through increasing size to build up a 'total systems capability'. In the formation of Plessey Numerical Control, through the takeover of the NC interests of Ferranti and of Airmec-AEI, the main scale consideration was the necessary high absolute level of spending on R and D (see, for instance, Layton 1972), and it was this factor which dominated the takeover of Elliott Automation by English Electric in automation and industrial control systems. Finally, the integration of technological development programmes, and the presupposition of consequently greater development potential, was a major factor contributing to the formation of International Computers Ltd, through the merger of ICT and English Electric (and previously Elliott) interests.

The combination of all these pressures produced the following effects on the production and labour processes:

1 consolidation of R and D into a smaller number of larger groupings;
2 the reduction of labour content in the products, as part of the process of cheapening the output of this sector. (This, of course, is an ongoing change, and also does not apply equally to all products. But each generation of computers requires only one tenth of the labour amount of the previous generation; and the labour content of the present (sixth) generation of NC is down to 5 per cent of costs (interviews));
3 reduction in the level of skill required by the production labour-force, as a result of the process noted under 2. Production in many electronics industries is now largely a matter of semi-skilled assembly and simple testing. In electronics, the resultant dichotomization of the labour-force in terms of skill, between production and R and D and control, epitomizes that developing within the economy as a whole;
4 overall sectoral growth (in both output and employment).

This in turn produced the following clearly distinguishable spatial effects:

1 the growth and concentration of industrial research and development activity primarily within the south-east of England. This entails also, of course, the concentration in that region of the highly-qualified labour-force necessary for these activities;
2 the establishment of locational hierarchies of production. Such hierarchies have been discussed at an international level (e.g. by Hymer 1972), and at an intra-national level (e.g. by Buswell and Lewis 1970). Our own work has confirmed the existence of this tendency under conditions such as those typified by the sectors in this Group. By

such a locational hierarchy is meant a spatial division of labour between, for instance, research, development and initial production, and fully finalized mass-production. In this way individual firms take advantage of regional uneven development. Financial restructuring aided in a number of ways the full establishment of such hierarchies. Firstly, size in itself is necessary in order to maintain a viable development programme at all; secondly, greater size increases the feasibility of separable locations. The process again relates to the dichotomization of skills in these industries, with the dichotomization increasingly reflected in locational patterns. In a number of interviews it was stressed that research 'must' be carried on in the south-east of England (occasionally it was allowed that Edinburgh was also a feasible location). But the 'next stage' can rarely be full mass production, a period of developmental production often being necessary to iron out snags. It was argued in interview, therefore, that this stage should be located in close proximity to the research labs. Indeed this was felt to be sufficiently important to warrant making out a lengthy case for avoiding the locational requirements of regional policy. It was stressed a number of times that under present conditions 'one can train testers in Development Areas, but not designers';

3 the long-term growth nature of most of these industries meant that new capital investment dominated closures (i.e. the opposite of the situation in Group 1). Most of the new production facilities were established in Development Areas, as a result either of regional policy or of relaxed labour requirements (or, more usually, as a mutually-reinforcing combination of the two). In these areas they employed semi-skilled labour;

4 on-going increases in technical composition in the processes of production of some (not all) of these industries led to reduced demand for labour, and a downgrading of its required skill, at existing plants. These existing plants are in both assisted and non-assisted areas, but differently distributed from similar losses in Group 1 in being far less concentrated in 'older industrial areas', and specifically in their relatively rare occurrence in inner city areas.

Market power

The financial restructuring which took place in Group 3 was stimulated overwhelmingly by the need to strengthen international market position and power by additions to the sheer size of the acquiring company. That is, in this Group the advantages derived from financial restructuring are largely those of muscle and market share. They do not require, in order that advantage be taken of them, that the financial reorganization is accompanied by any major reorganization of production. There is consequently little effect on the labour process. Although the IRC tended to refer to potential economies of scale, the reality of which the relevant companies in interview subsequently denied, it was also quite clear that sheer size could be an advantage in its own right – 'international economic competition today involves a great deal of horse-trading and arm-twisting: bargaining situations and influence processes in which total size of resources deployable, rather than unit costs or rate of return on capital, may be the critical factor' (McClelland 1972: 25).

The process is best illustrated by reference to some of the individual mergers which come into this category. Racal's takeover of Controls and Communications Ltd (CCL) was concerned with military radio communications, and specifically military manpacks, with instrumentation playing a subsidiary role. Both sectors, especially the former, were growth areas, with increases occurring in both output and employment. The two firms were both small, but the sectors in which they were primarily involved were themselves not large. In 1975, the combined Racal group was producing 51 per cent of UK total production of ground radio communications equipment (Annual Report 1975). The IRC's expression of support was given simply in terms of 'the development of [an] outstanding company'. It was, significantly, the increase in competition abroad between the two companies which precipitated the merger. Their activities had been largely complementary until Racal entered the military manpack market. The new group was immediately one of the world's leading manufacturers of military radio equipment for manpack and mobile use. Finally, the company itself confirmed that this was the correct interpretation of the merger.

The second case was the acquisition of Isotope Developments Ltd and the nucleonic instrument business of EMI by Nuclear Enterprises Ltd. Once again, the rationale was that of the development of individual companies, of 'backing winners'. Hogg, the IRC representative in this instance, was quoted as saying 'We are taking a couple of entrepreneurs and backing them' (*Financial Times* 6 December 1967). It is, in fact, in Group 3 that the ideological content of the role of the IRC, in backing specific types of management, is most evident. This is, in turn, presumably a function of the nature of the stimulus to the mergers, being less of a response to the exigencies of production. Like Racal, although Nuclear Enterprises was 'small', it was important in its field. It was already the largest British company in the nucleonics field, accounting for at least 50 per cent of total exports (*The Times* and *Financial Times* 6 December 1967). Other cases in this Group were of a similar nature, and were in the sectors of software and of electric motors (the latter being a defensive response to the growth of GEC through mergers also covered in the survey). Finally, the takeover by GEC of AEI must also be classified here. Certainly many sectors in the combined company were considerably *affected by* the merger (telecommunications and traction are the best-known examples), but the problems and opportunities to which this rationalization was the response were not themselves the reason for the financial restructuring. This conclusion was confirmed in a large number of interviews. Specific possibilities (such as economies of scale in electronics, and the problems of technological change in telecommunications) were rejected as in any way having been the reason for the merger. The only sector where problems did play a part in promoting the acquisition was power-engineering. An extended quotation from Young and Lowe (1974) elaborates the argument. Having mentioned the possibility of the potential for the merger having repercussions through the whole sector of power-engineering, and a degree of duplication in a number of other sectors, they write:

'Thirdly, the industrial arguments in favour of the links were strengthened by the prospect of Arnold Weinstock presiding over the merger rationalisation and future development of the company. In order to appreciate the importance of this when looking back, it must be remembered just how high Weinstock's reputation stood in 1967, five years after he had become Managing Director of GEC. The company's record for the three years preceding the bid was way ahead of AEI's on the various indicators of company efficiency. Wilsher and Johnson (1967) summed up his work as "probably the most remarkable managerial transformation ever brought about in a British company". Probably of even more importance to the IRC Board than his record, were his international ambitions: to win the firm a secure and profitable future he and his top managers argued that they needed to greatly increase (sic) GEC's size and move into the international field away from the home market and government orders. AEI offered a solution in terms of assets, and from the IRC's view-point this expansion would be apt to spur Weinstock into maximising the chances the merger offered, as well as promising greatly increased exports. This was the first of several cases where the IRC sought, as Villiers later confirmed, to rationalise an industrial sector by supporting a management team that had been successful' (p. 64).

In spatial-locational terms, this Group may be quickly dealt with. As already outlined, the financial restructuring was primarily aimed at increasing market standing through sheer size and/or combination with potential competition. This was not the 'rationalization route' to greater international competitiveness. There were, therefore, virtually no effects on the spatial organisation of production, although reorganization of marketing and overseas supply functions was significant.

Interregional relations

This final section will draw together some threads from the previous separate discussions of restructuring processes, and will attempt to assess their combined spatial impact.

First of all, it is useful to examine the aggregate job-changes involved at the national level. These are presented below in terms of the three' Groupings. It should be pointed out that the figures include, not only those specific employment results of the mergers themselves, but also the effects of on-going changes (discussed in the previous section) which are integral to the overall reorganization of the production process which took place over the period. There is also another distinction, however, which was introduced in the first section (pp. 199–204): that between dominant and secondary sectors. So far, the discussion in this paper has concerned itself entirely with those sectors which predominated as stimuli for the financial restructuring. In the cases of large, multidivisional companies, however, such restructuring will also involve other sectors. Indeed, the merger once having taken place, these secondary sectors may be subject even to drastic rationalization. These effects are listed here for completeness, since they were an integral part of the employment consequences of financial restructuring. The sectors involved were telecommunications, traction and a variety of electronics product-groups. In the full research report, the production of spatial employment change within these branches is analysed in the same way as is that of dominant sectors. In the present context, there is space only to present the results. The national-level figures, then, are shown in Table 9.4.

**Table 9.4 Major national-level employment changes
as a result of restructuring processes**

	Absolute loss	Absolute gain	Locational shift	Total
Group 1	−24,113	+100	(966)	−24,013
Group 2	−4,006	+1,750	(703)	−2,256
Group 3	−210	+120	(150)	−90
Secondary sectors	−9,657	+0	(2,676)	−9,657
Total	−37,986	+1,970	(4,495)	−36,016

It is immediately clear that, unsurprisingly, the results were dominated by employment loss. It is also abundantly clear that, as would be expected from the analysis so far, the three different forms of financial restructuring contributed differentially to this loss, with Group 1 dominating and Group 3 producing relatively little job-loss. Again as would be expected, it was Group 1 which produced least in the way of newly generated employment. Such results confirm the importance of national and international economic developments as a dominant determinant of locational change. In the table, employment changes which represented a gain or loss 'to the country as a whole' are called 'absolute changes'. Those employment changes which, in contrast, result from geographical movement within the country are listed under 'locational shift'. It should also be noted that the figures in the 'locational shift' column represent the employment which was created in a new location after movement from a previous plant. This number is far smaller than the loss recorded at the original factories. Job movement, in other words, has frequently been either part of a process of overall cutbacks or has been the occasion for cutback. In the first case, overall cuts in capacity often entail concentrating the work of smaller factories on a reduced number of larger ones. Such moves are frequently announced as transfers, and indeed some production may well be moved. They do not, however, represent a transfer of all the jobs at the previous location. In the second case, locational shift may be the occasion for major changes

Table 9.5 Jobs lost during locational shift of production

	Transferred jobs	Jobs lost in transit[a]
Group 1	966	5,505
Group 2	703	641
Group 3	150	190
Secondary sectors	2,676	4,909
Total	4,495	11,245

[a] Included in *absolute loss* column in previous table.

in production technology (see Group 1, measure 5), again leading to a reduced workforce in the recipient region. The locational shift may be brought about because the nature of the technological change demands either new fixed capital or a new workforce. In the first case it may be necessary, in the second prudent, to move, thus avoiding conflict with the unions. The job-losses associated with transferred jobs are given in Table 9.5.

How, then, is all this employment change expressed at regional level, and what are its implications for interregional relations? The first question to be addressed is whether or not there was a differential impact, in terms of employment numbers, on Development Areas and non-Development Areas.

Table 9.6 Aggregate results for assisted and non-assisted areas

	Non-assisted Areas		Development Areas		
Locational gain	270		2,452		
Absolute gain	210	+	1,760	=	1,970
Locational loss[a]	(2,452)		(270)		
Absolute loss	(27,748)	+	(10,238)	=	(37,986)
Net total	(29,720)	+	(6,296)	=	(36,016)

[a] Excluding all moves *within* a regional category.

Table 9.6 gives the aggregate results by category of employment change for Development Areas and non-Development Areas. It is immediately clear that in these simple numerical terms, the Development Areas have fared far less badly from the processes analysed than have the non-Development Areas. This is true both of the net total change of employment (where Development Areas lost less than one-quarter the number of jobs lost in the rest of the country), and of each gross component of this change. In all categories the Development Areas record a better result.

They have (i) more locational gain; (ii) more absolute gain; (iii) less locational loss; (iv) less absolute loss.

Table 9.7 below confirms this bias in a comparison of the percentage of initial total employment in each of the two types of region with their respective percentages of the components of change:

Table 9.7 The major regions as a percentage of the national results, by component of employment change

	Development Areas	Non-Development Areas
	%	%
Initial employment:		
⌈manufacturing	25	75
total employment	27	73
⌊in survey[a]	26	74
Results/employment change:		
⌈absolute loss	27	73
absolute gain	89	11
⌊locational gain[b]	90	10
Total net change (loss)	17	83

[a] to avoid confusion in the discussion which follows it should be stressed that this percentage distribution is for the survey as a whole. Individual industries in the survey had initial distributions which differed from this.

[b] calculated as the percentage of jobs gained out of all those which shifted locations within the UK.

The percentage distributions of absolute gain, locational gain, and total net loss confirm unambiguously a relative improvement in the position of the Development Areas. Indeed, this is true also for the component 'absolute loss' in the sense that, as already pointed out, proportional equality of loss in a period of overall decline in employment is an improvement in the Development Areas' normal pattern of performance since the war. This point will be taken up later in the discussion. Certainly at this superficial level, the processes of restructuring analysed here do seem to be contributing to the interregional convergence of employment availability. It is necessary, however, to examine the results in more detail.

First of all, what is the relationship of the processes of restructuring to this regionally-differentiated impact? If a quick run-down is made of the locational implications listed in the last section, four major components of change stand out by which the Development Areas may be gaining in employment terms relative to the non-Development Areas. These are:

1 that they are losing less drastically from the capacity cuts resulting from the restructuring in Group 1;

  2 that they are gaining new plant from production relocated after major technical change;

  3 that they are receiving the bulk of new investment in the production end of the growth in Group 2;

  4 that they are losing less from the labour-saving technical change within electronics sectors in Group 2.

The importance of the first of these components is confirmed by the distribution of absolute losses in Group 1, presented in Table 9.8.

Table 9.8 Absolute losses in Group 1: distribution between Development and non-Development Areas

	MLH					
	354	*361*	*362*	*369*	*383*	*Total*
Development Areas	0	5,100	0	0	510	6,610
Non-Development Areas	450	13,795	1,128	470	1,660	17,503
Total	450	19,895	1,128	470	2,170	24,113

It is clear that in absolute terms the losses in non-assisted areas were far greater than those in Development Areas.[13] In part, of course, this is a structural phenomenon – indeed predominantly so. That is to say that as representative of an overall sectoral decline it is dependent on the initial distribution of those sectors. (It was pointed out in the second section (pp. 209–10) that all but one of the changes in these sectors would take place within the existing geographical distribution.) The non-assisted areas are therefore suffering absolute losses from restructuring in this Group to the same extent as are Development Areas. In broad historical terms, this result, an effect of the equal representation of a declining sector, constitutes a worsening of the relative situation of such areas. If such decline is becoming a more important phenomenon in non-Development Areas this is significant. The existing assisted areas owe a large measure of their historical employment problems to such structural changes. MLH 362 here represents supertension cables, where both the initial distribution, and the losses, were entirely confined to the South East. This was a simple case of plant closures. The other two sectors are more complicated. The losses in 383 in Development Areas were entirely in Liverpool, and again in this sector there was no relocation, for instance as a result of technical change. All losses took place on existing sites.

    This was not, however, the case with MLH 361, and here the second of the factors aiding the Development Areas begins to emerge. The power-engineering losses in MLH 361 were of course also heavily dominated by closures and on-site losses. But in this sector, not only the absolute but the percentage loss (of initial total employment in this industry), is lower in Development Areas. There is, in other words, not just a structural, but also a differential, element to this relative gain.[14] This does not necessarily mean, however, as is frequently interpreted that the differential gainer is in some sense the more 'competitive' region – that it has a locational advantage for any specified production process. Certainly this may be part of the explanation in this and other sectors where this phenomenon occurs. However, there is a further factor – that associated with intra-sectoral differentiation and technical change. To the extent that full automation is possible in this Group (measure 5), new plants are being established. With the labour requirements of the new techniques being limited in terms of skill, a non-conurbation Development Area location can now suffice where previously the craft skills of an old industrial area were necessary. A similar phenomenon appears to be occurring in telecommunications.

The third factor working in favour of the Development Areas concerned Group 2. Here the impetus to increase output led to new investment. The distribution of employment growth by absolute gain in this Group is presented in Table 9.9.

Table 9.9 Group 2: distribution of absolute gains

	Development Area	Non-Development Area	Total
MLH 366	850	–	850
MLH 367	800	100	900
Total	1,650	100	1,750

Again the predominance of Development Areas is apparent (they gained 1,650 jobs as against 100 in non-Development Areas). More significantly, however, and confirming the identification of the process at work, *all* the gains registered in this Group in Development Areas took place in new plant. One must be careful not to exaggerate the generalizability of this conclusion, since all these new plants, save one, were in a single Development Area – Scotland – renowned for its presence of externally-controlled electronics facilities. Nonetheless, the evidence is indicative.

Fourthly and finally, the labour-shedding resulting from increasing capital-intensity in this Group was concentrated in the non-Development Areas. Again there was a mixture of structural and differential effects. Thus considerable losses were associated with the running down of facilities at the 'older end' of the electronics spectrum. This applied particularly in instruments (MLH 354) and computers (MLH 366). In the latter, much of the large (3,550) absolute loss in non-Development Areas was due to the closure of electro-mechanical plants. Electro-mechanical equipment is an early, half-way stage to full electronics, and is considerably more labour-intensive. Moreover, as part of the initial development of the industry, it is more concentrated in the south-east of England, which region consequently suffered the bulk of the losses of this type. Within the fully-electronic parts of this Group, job-loss due to increasing capital-intensity was more dispersed but still overwhelmingly within the non-Development Areas.

The more detailed results therefore fully confirm the fact that the restructuring processes had differential effects on the depressed areas of the country on the one hand and on the previously relatively prosperous regions on the other, and that this differential worked in favour of the depressed areas. Such results are of immediate importance for two reasons. First, they are clear warning against easy (and apparently 'radical') assumptions that the aspatial centralization of capital is always mirrored in increased locational centralization of production. They also warn against ahistorical assumptions that regional policies of employment dispersal are necessarily contradictory to the process of accumulation at the national level.

Such tendencies, however, do not mean the end of a spatial division of labour, a homogenization of the whole country. They imply, rather, a different form of integration and new forms of differentiation. The most likely development suggested by the present investigation is related to the increasing dichotomization of labour skills, between a small and highly-qualified personnel and a semi-skilled/unskilled majority (a process which is in itself important in reducing the costs of reproduction of the mass of the labour force, and therefore potentially further increasing the rate of surplus value over the economy as a whole). There are a number of aspects to this.

The first point is that the centralization of capital involved in the cases studied did contribute to the further spatial centralization of control functions. Control over investment and production was increasingly withdrawn from most areas of the country (save perhaps the West Midlands), and concentrated in the south-east of England. This process occurred at the level of divisional control as well as at that of the whole firm. This is unsurprising, and corroborates other evidence, such as that of Parsons (1972) and Westaway (1974). It has not been dealt with in detail in this research because its aggregate effects on employment are not large. Nonetheless, it is an important aspect of interregional relations, and should therefore be noted.

The second aspect of changing interregional relations which emerged was the evidence of the hierarchical spatial separation of research/development and mass-production. This is an intra-product hierarchy closely resembling the various formulations based on the concept of the product-cycle (see, for instance, Vernon 1971). In the cases studied in the present research, this spatial separation was primarily between the south-east of England, together with some parts of central England, and the rest of Great Britain. This hierarchy mirrored, and therefore reinforced, that stemming from the spatial centralization of control functions already mentioned. It entailed not merely the build up of skilled scientific and managerial workers in the South East, but also the loss of some employment of this type in the rest of the country.[15] In this it was part of a cumulative process, the firms' response to location factors (locating research facilities in the South East because of the existing concentration there of suitable labour) being precisely to reinforce the differential nature of their distribution.

Lastly, there was a territorial division of labour within production at the *inter*-commodity level. Thus within a number of broad product-groups, the production of commodities requiring more standardized and less-skilled processes was concentrated in Development Areas, with the production of commodities requiring more skilled labour more frequently found outside of the Development Areas.[16] Evidence of the formation of this pattern was identified in a number of MLHs spreading right across the sectors investigated. It was usually associated with new investment in the non-conurbation parts of the Development Areas; this new investment was itself the result of two distinct processes. On the one hand some such investment (and consequently the establishment of such a hierarchy), resulted from major technical change in an existing production process. It was therefore accompanied by closure in the area of initial distribution. On the other hand, some of the new investment resulted simply from the extensive growth of a particular sector, the point here being that the production process in the expanding sectors under study tended to require semi-skilled assemblers rather than skilled craft workers.

This last 'hierarchy' (that between commodities and within production) was, however, unlike the other two, clearly not increasing in intensity and distinctness. On the contrary, in this case the evidence was for a weakening of spatial employment differentiation. For at the same time as the new investment in production was taking place, changes were also occurring in the nature of labour demand at the already existing production sites – which were more dominantly located outside of the Development Areas, and certainly outside of the non-conurbation parts of the Development Areas. Evidence has already been presented of, on the one hand, the run-down of the skilled-craft basis of electrical engineering in many of the older industrial areas of England, and, on the other hand, the declining aggregate labour-demand and increasing dichotomization of skills (predominantly a process of de-skilling) in the initial areas of distribution of the production of many commodities within electronics (1 and 4 in the list at the start of this section). Thus, outside of the south-east of England, and a small number of favoured locations in central England, not only unemployment levels, but also the demand for labour skills on any scale, may be tending to converge. The on-going changes in production (as opposed to the absolute decline of craft-based sectors such as power-engineering), themselves a result of the need to increase the rate of exploitation, are thus releasing certain branches of industry from their former locational dependence on reserves of skilled labour. Such changes are making feasible locations outside of the major industrial conurbations but within the depressed areas of the country.

The locations from which these industries are being released are also those in which the major absolute declines in skilled employment have taken place. These areas are of course the major industrial conurbations;

and any consideration of the evolving spatial division of labour must take account of the major changes taking place in the cities. The results of the present research were therefore disaggregated to examine the effects on four such conurbations: Manchester, London, Liverpool and Birmingham.[17] Table 9.10 gives the aggregate results for these cities and the disaggregation of those results by the forms of restructuring identified.

Table 9.10 Employment changes in the conurbations

	Absolute loss			Locational loss	Absolute gain	Locational gain
	In situ	In transit	Total			
Group 1	15,528	4,980	20,508	606	0	30
2	1,350	136	1,486	0	0	20
3	0	100	100	100	0	0
Secondary sectors	600	4,419	5,019	2,546	0	0
Total	17,478	9,635	27,113	3,252*	0	50
Net loss to cities	30,315					

*Excludes 270 intra-city job moves
            130 inter-city job moves

It is immediately evident that these processes were an important component of the total survey employment change. The cities suffered a far higher than proportionate absolute and net loss, and the results were also consistent across the individual conurbations. The lowest percentage loss (in Birmingham) was still far higher than for any other spatial division. The performance of the cities also differed from that of the other spatial categories in that their percentage locational loss was rather higher and their share of total absolute gain was zero. The explanation for this very specific pattern of results is once again found by looking back at the individual processes of restructuring.

As in the overall results, Group 1 accounted for the bulk of the absolute loss, and of the net change. Indeed, it was in the cities that the proportional effects of this form of restructuring was at a maximum. In the first place, the high level of initial presence of Group 1 plants in the cities meant that these areas were vulnerable to the employment cutbacks taking place as a result both of capacity-cutting *per se,* and of the various forms of technical changes outlined on pages 209–10. Secondly, the fact that the inner cities accounted for a higher-than-average proportion of older more labour-intensive plant, meant that they were more liable to both forms of employment reduction and change. On the one hand, selective capacity-cutting was primarily aimed at such factories. On the other hand, it was in the production and labour-process in these factories that the greatest technical changes could be made to save on labour costs. The result of all this was both a massive loss to the cities of jobs in skilled-craft occupations, and a reduction in the skill required in the employment which remained. Thirdly, none of the small amount of new investment and of associated absolute gain of employment, which was generated in Group 1, was located in the cities. It should be noted that this applied – as indeed did all these conclusions – as much to the Development Area conurbation (Liverpool) as to those outside Development Areas. Finally the locational losses in Group 1 were the result of two forces. On the one hand, the product-line reorganization within power-generation occasioned some 500 job losses within the cities. This was the result firstly of an overall programme of concentration on a few, major, differentiated sites, of which only one was in a conurbation. Secondly where, within this overall programme, explicit choices were available about the location of lay-offs or closures, it was often thought easier for the company if these latter were in conurbations, where their individual effect would be muted by the context of a large and varied job pattern, rather than in towns where the firm concerned dominated the labour market. Such a strategy was mentioned in a number of interviews.

The cities also suffered particularly severely from the changes in Group 2. This may seem strange since their initial presence in the industries constituting this Group was not dominant. It results from a combination of causes. Firstly, the intensification of the process of technical change in this Group produced labour-loss. Much of this took place in production processes within electronics and located primarily outside the cities. There was, however, also a more dramatic technological change in progress, such as that from electro-mechanical to electronic in MLH 366, where the older production processes *had* been located in the cities. It was of course also these processes which were both larger employers of labour, and which were closed rather than undergoing technical change on-site. The resultant losses to the cities were, therefore, again both absolutely and proportionately high. Secondly, the cities did not gain in employment terms from the consolidation of R and D activities in this Group. There *is* a presence of such activities, and the possibilities of gain should not be ruled out, but the numbers involved would be small, and the jobs would be located outside of the inner areas of the cities. Thirdly and finally, the more considerable new capital investment in this Group was located entirely outside the cities, whether or not these latter were in Development Areas. This is evidently the other aspect of the process already referred to by which a change in labour requirements may increase capital's locational flexibility. But there is a vast reservoir of unemployed unskilled labour[18] in these same cities that is being deserted. Why, then, the moves to areas outside conurbations? A number of indications (and they are no more than that) began to emerge from the present study. In the first place, if it was a case of major technical change and a complete switch of workforce (and consequently probably also of union) a locational shift might be used to avoid conflict with the unions. Secondly, the reserve of labour in the cities was seen as being more expensive, more highly unionized and more militant. It is labour already well-integrated into the labour force. So, instead of using this labour, capital is beginning to spread out. Small towns are now favoured locations. From being centres of skill the cities are becoming a location for the continually reconstituted reserve army of the unemployed.

What conclusions, then, can be drawn from these results? First, it is perhaps worth stressing again the importance of establishing the relation between locational change, and therefore the differential economic fortunes of different areas, and the overall process of accumulation. Present attempts in the UK to divert and disconcert working-class action over unemployment focus particularly on fostering the inter-area competitive approach to industrial location. The most obvious case of this is in the debate over inner cities where protest has to some extent been played off against the problems of Development Areas. Such tactics produce a tendency to see the causes of economic decline within the area itself. The Home Office's attitude to the Community Development Projects it had sponsored in cities grew increasingly hostile as the latter linked the problems of their areas to the problems of the national economy rather than to the skills and psychological propensities of inner-city dwellers (see, for instance, Community Development Project 1977). The problems of particular areas result from the particular form of use by capital, at any given period, of spatial differentiation. It has been the major concern of this paper firmly to establish the link between locational behaviour, changing patterns of employment, and developments at the level of the process of accumulation.

Second, in terms of the empirical results concerning locational behaviour, the present study indicates that one effect of restructuring during the period has been the emergence of a change in the form of territorial inequality. This change in form has been in terms of both its geographical pattern and its nature. The general disparities between Development Areas and non-Development Areas may be becoming less marked, in terms of employment numbers, while that between conurbations and non-conurbation areas seems likely to become more so. While the pattern of unemployment differentials may be evening-out between assisted and non-assisted areas, the dichotomization of skill-levels within manufacturing between the metropolis and the rest of the country may be being reinforced.

Such changing forms of inequality give a hint, also, of the reasons behind these locational shifts. As numerous examples in the previous pages have indicated, certain forms of technical change are increasingly freeing parts of capital from its previous ties to the skilled labour of the cities. In this context, capital is freer to search for labour which is less unionized, less militant. Such changes go hand in hand with regional policy, and the process of restructuring and the policy of regional location appear to have been mutually reinforcing. This exodus has been reinforced in its effect on the cities by the disproportionate declines there, in these sectors, of industries suffering problems of over-capacity.

Thirdly and finally, what of the implications of these shifts for the strength and the strategy of labour? In certain ways, the interregional evening-out of unemployment rates may make the management of high national-level unemployment easier for the State to handle. The problem is spatially diffused. On the other hand, this very diffusion at the same time should make clear the nature of the real issue – declining *national* employment opportunities, not the simple gain of one region at the expense of others. Inevitably, in a period of such high national unemployment, some areas are particularly hard hit. But the spatial pattern is changing – the 'black spots' now vary from Skelmersdale to Coventry. Most particularly, of course, the effect is being felt in the cities, where grass-roots activity is becoming increasingly important – through Trades Councils, in action groups for particular areas, such as the Docklands in London, and through the work of the Community Development Projects. The old industrial cities have for long been the bases of some of the strongest union power. The present severe decline of industries within them therefore undermines that strength. At the same time, for those industries which are not simply declining but which are moving out of the cities, decreasing demand for skilled labour means that considerations such as differential labour militancy and organization may increase in importance as location factors.

Notes

[1] It is the first of these which is given most attention in the present paper. The second is dealt with more fully, e.g. in Massey and Meegan (forthcoming).

[2] The nature of the IRC and of its intervention are analysed more fully in Massey and Meegan (forthcoming), which also contains more analysis of restructuring, and statistical information, etc., on the sectors analysed. In the present paper, the IRC's importance is as a means of identifying sectors and firms in which the restructuring process was significant during the late 1960s.

[3] The industries covered are the following (the numbers refer to the Minimum List Heading (MLH) of the British Standard Industrial Classification):

    354 Scientific and industrial instruments and systems
    361 Electrical machinery
    362 Insulated wires and cables
    363 Telegraph and telephone apparatus and equipment
    364 Radio and electronic components
    365 Broadcast, receiving and sound-reproducing equipment
    366 Electronic computers
    367 Radio, radar and electronic capital goods
    368 Electrical appliances primarily for domestic use
    369 Other electrical machinery
    383 Aerospace equipment.

[4] In fact the pattern of regional policy assistance is rather more complicated than this dichotomy, but it was this division which was dominant during this period.

[5] UK regional policy was implemented particularly strongly through this period, and the 'accepted' explanation of the convergance is that it was simply the result of state policies for geographical redistribution. This issue is not addressed

directly here (for a detailed argument, see Massey 1979). The aim of the present paper is simply to point out how restructuring such as that analysed here could also be a significant component of this convergence.

[6] Points of discussion or information in the text which are based specifically on material derived from those interviews are indicated '(interview)'. We should like to thank the interviewees for their very helpful contribution, though responsibility for the conclusions drawn is entirely our own.

[7] In the rather simplified version of the framework presented here, it appears that 'space' is merely a passive surface. In fact, of course, it is itself a social product and has its own effects.

[8] It should be noted that this figure is for capital *expenditure* per annum and not stock of capital. The correct measure would be total capital employed, but this is not available. Wood (1975) argues that expenditure is an adequate surrogate for the derivation of indicators of capital-intensity. Comparing MLHs 362 and 366, however, one a growing sector, the other stable, is bound to lead to bias. In this instance, the bias will only reinforce the point being made.

[9] Kravis and Lipsey (1971) are quoted in Glyn and Sutcliffe (1972) as recording that the British share of the world market in electrical machinery fell from 28 per cent to just over 13 per cent from the late 1950s to 1970.

[10] Standardization was a process related to internationalization directly, as well as through increasing productivity, since it involved also alignment of British and international specifications (see National Plan, HMSO 1965:105).

[11] It is nonetheless increasingly in the spread of existing applications that the industries must look for extending their markets. In turn, this means that after its long period of uninterrupted growth, the industries are becoming increasingly sensitive to the fluctuations of capital investment programmes.

[12] It should be pointed out here that, although the major part of IRC activity was directly to aid British-owned firms, and although the strength of productive capacity of foreign-owned firms within the UK may considerably have increased the degree of competition, and therefore the problems, of those British-owned firms, it does not necessarily follow that any state action will be directed *against* the foreign capital operating on UK soil (see, e.g. National Economic Development Office 1973). It is not possible simply to equate 'national economy' with 'national capital'.

[13] This conclusion is greatly heightened when it is noted that of the total DA loss, 4,910 jobs were from the Merseyside conurbation – see later.

[14] This dichotomization of employment decline should perhaps be clarified. It refers to a simple statistical disaggregation of employment change in a region within a nation into (a) that component of change which could be expected on the basis of regional *structure* (e.g. a regional employment decline representing the presence of a nationally declining industry), and (b) a *differential* (sometimes called a 'locational') component which indicates the differential performance within a region of any given industry (in the case of decline, the fact that an industry is declining faster in the region than it is nationally).

[15] It should be stressed that the spatial hierarchies under discussion were identified as general tendencies, and not as iron laws. Some of the firms surveyed are still entirely located with Development Areas.

[16] The old industrial conurbations of England which are within Development

Areas – Liverpool and Newcastle upon Tyne – also retain much of this production. The more detailed spatial pattern is discussed later in this section.

[17] The geographical areas used were as follows:

London: Greater London Council Area
Manchester: Manchester County Borough
Liverpool: Liverpool County Borough
Birmingham: Birmingham County Borough

The results of this aspect of the research are presented in Massey and Meegan (1978).

[18] Unskilled as unemployed labour seeking new jobs. This does not preclude the possibility that their previous employment was in fact highly skilled, but in crafts now no longer demanded by capital.

References

Bailey, R. (1968) *Managing the British Economy. A Guide to Economic Planning in Britain Since 1962.* London: Hutchinson and Co. Ltd.
Buswell, R. J. and Lewis, E. W. (1970) 'The geographical distribution of industrial research activity in the United Kingdom.' *Regional Studies* 4: 297–306.
Community Development Project (1977) *The costs of industrial change.* London: CDP.
Department of Economic Affairs (1968) 'Reorgcorp, London.' Progress Reports (May). London: DEA.
Economist (1970) 'Gunning for the IRC.' *Economist,* 12 February 1970: 61–2.
Fine, B. and Harris, L. (1975) 'The British Economy since March 1974.' *Bulletin of the Conference of Socialist Economists* IV, 3 (October): 12.
Glyn, A. and Sutcliffe, B. (1972) *British Capitalism, Workers and the Profits Squeeze.* Penguin Special. Harmondsworth: Penguin Books.
Her Majesty's Stationery Office (1975) *National Plan,* Cmnd. 2764. London: HMSO.
Hymer, S. (1972) 'The multinational corporation and the law of uneven development', in J. Bhagwati (ed.) *Economics and World Order from the 1970s to the 1990s.* New York: Collier Macmillan, pp. 113–40.
Keeble, D. (1977) *Industrial Location and Planning in the United Kingdom.* London: Methuen.
Kravis, I. and Lipsey, R. (1971) *Price Competitiveness in World Trade.* New York: National Bureau for Economic Research.
Layton, C. and de Hoghton, C. (1972) *Ten Innovations.* Political and Economic Planning. London: George Allen and Unwin.
Mandel, E. (1975) *Late Capitalism.* London: New Left Books.
Massey, D. B. (1979) 'In what sense a regional problem?' *Regional Studies* 13: 233–44.
Massey, D. B. and Meegan, R. A. (1978) 'Industrial restructuring versus the cities.' *Urban Studies* 15, 3 (October): 273–88.
—— (1979) 'The geography of industrial reorganization.' *Progress in Planning* 10, 3.
McClelland, W. G. (1972) 'The IRC 1966–1971: an experimental prod.' *Three Banks Review* (June).
Melk, E. G. (1973) *The Electronics Industry: an annual review of the industry and its prospects.* London: Hoare and Co. Govett, Investment Research Ltd.
National Economic Development Office (1972) *Annual Statistical Survey of the Electronics Industry.* Electronics Economic Development Committee, London: HMSO.
—— (1973) *Industrial Review to 1977.* London: HMSO.
Palloix, C. (1976) 'The labour process: from Fordism to neo-Fordism', in *The Labour Process and Class Struggle,* Conference of Socialist Economists, Pamphlet No. 1, Stage 1.
Parsons, G. F. (1972) 'The giant manufacturing corporations and balanced regional growth in Britain.' *Area* 4, 2: 99–103.

Scott, P. A. (1974) *'Electronics and Electrical Engineering.'* A survey by Joseph Sebag and Co., London.
Villiers, C. H. (1969) 'The Industrial Reorganisation Corporation.' *Investment Analyst* (December).
Westaway, J. (1974) 'The spatial hierarchy of business organisations and its implications for the British urban system.' *Regional Studies* 8: 145–55.
Wilsher and Johnson (1967) *Sunday Times,* 1 October.
Wood, E. G. (1975) *Comparative Performance of British Industries.* London: Graham and Trotman Ltd.
Young, S. and Lowe, A. V. (1974) *Intervention in the Mixed Economy.* London: Croom Helm.

Postscript

The changes that have taken place since this study was completed have made even clearer the relation between the general condition of accumulation and spatial unevenness. With changing degrees of relative importance, many of the processes discussed here continued through the 1970s. The variations in their strength and form have depended upon the combination of the overall economic situation and the dominant political strategy. In different periods, different policy responses have advanced particular aspects of the long-term shifts. The present situation is one of international recession and a dominant political strategy in Britain of non-intervention. In such a context, it is decline and de-industrialization that is the dominant characteristic of regional change. The 'regional problem' continues, and indeed looks set to intensify.

10 Policies as chameleons: an interpretation of regional policy and office policy in Britain

C. G. Pickvance

The aim of this paper is to put forward alternative interpretations of British regional and office policies.[1] The interpretations proposed will be supported by reference to empirical data on the evolution of these policies since their inception and by the theoretical framework for policy analysis set out in the introduction and discussed again in the conclusion.

Introduction: a framework for the analysis of state policy

The theoretical framework to be used in the analysis of regional and office policy will be briefly outlined here. It applies to the policies of capitalist States only.

There are two major distinctions between the liberal and Marxist conceptions of the State and state policy. (By State I mean central and local government, the judiciary, police and armed services.) For the former conception society consists of competing groups whose differences are not fundamentally opposed, and the State is an open and neutral institution which reflects the balance of these conflicting pressures. Its policies are thus in principle capable of ensuring harmony between the competing groups. For the Marxist conception society is riven into two major classes, capitalists and proletariat, whose interests are fundamentally opposed in the long term, and the State has a built-in link to the capitalist class. The range of variation of its policies is thus in principle limited by the need to secure the continued domination of the capitalist class. It follows that many problems will have a recurrent character, e.g. the housing question – whereas Engels (1969) showed there were definite limits to state policies, even if these limits have changed somewhat now.

These generalities need to be made more precise to develop our analytical framework. In particular we must emphasize the division of the proletariat into strata, of which the 'middle class' is one; and the division of the capitalist class into 'fractions' according to sphere of operation, e.g. industrial, banking, or property capitalists, and into 'monopoly' and small or 'competitive' sectors in each sphere according to their economic dominance and, empirically, geographical scale of operation. The conflicts between capitalists in these different spheres and sectors are important and limit the ability of any state policy to respond to any single capitalist fraction's interests. However the different fractions of capital share a long-term interest in the preservation of capitalist relations of production, e.g. the right of an industrial capitalist to decide how much capital to invest, what to produce and how to produce it, and how to dispose of the resulting surplus.

The fundamental opposition between capitalists and proletariat occurs because production is interdependent, or social, but appropriation (the power to dispose of the surplus) is private. The role of the State and nature of its policies can best be understood in terms of the evolution of this contradiction.

Competition at the world level forces the introduction of new technologies and extends the chain of interdependence in the production process. The colossal size of investments and the scale of the attendant risk forces the State to take on new functions *vis-à-vis* monopoly industrial capital since the alternative is the investment of capital in another nation. The range of 'general conditions of production' which the State is thus required to provide continually increases (Mandel 1975: 484). (General conditions of production are those prerequisites of production which individual capitalists cannot provide themselves. They vary between country but in most cases include roads, railways, telecommunications, power, etc., as well as a legal framework, a currency, an education system and so on.)

However the State's capacity to provide these general conditions is limited (Lojkine 1976). In particular a major limitation is financial.[2] This is visible in the increasing 'socialization' of production costs (i.e. bearing of these costs by the State) necessary to enable 'profits' to be made on the non-state-borne outlays. This gives rise to what O'Connor (1973) describes as the 'fiscal crisis of the State'. The built-in link of the capitalist class to the State forces it to give priority to the new exigencies of social production and private profit. To meet these demands the State is obliged either to print money (and cause inflation), increase corporate taxes (which at most it will do selectively since its aim is to facilitate corporate profits), increase personal taxes (which is what is happening in most capitalist countries),[3] or selectively reduce state expenditures (as is also happening widely). In the latter case it is often 'social' expenditures which are cut to make way for aid to industry. (An apparent fourth alternative, borrowing, affects the amount and time structure of state expenditure but does not remove the pressure.)

Which option is pursued depends upon the degree of political mobilization in each case. By nature the State will seek to avoid increasing corporate taxes. The degree to which it does increase them reflects the strength of political action among workers as taxpayers, state workers whose jobs are threatened by service cuts (e.g. teachers, social workers) and workers as consumers of education, medical care, etc. Thus the initial fiscal crisis may persist as a fiscal crisis or be transformed into inflation, the acceptance of new levels of taxation, or a service crisis, according to which options are open because of lack of mobilization.

This leads directly to the second aim of state policy: its concern to preserve capitalist relations of production, either by repression, or by the integration of the proletariat. To refer back to the evolution of the contradiction between social production and private appropriation, this contradiction finds political expression in various levels of class struggle organized by trade unions, political parties, etc. If the deepening objective contradiction is exploited by political groups, new demands will be expressed and a whole range of state policies emerges, with a contradictory character.

On the one hand they arise because of working-class demands (in the given conditions of production) and to that extent are 'concessions' required to preserve social stability. On the other hand they are compatible with renewed capitalist production, but within new constraints. (This is well demonstrated by Marx's analysis of the Factory Acts (1970: 298–302) cf. Lojkine 1976.) In this way state policies may have a contradictory character because of their two-fold role in the development of the contradiction between social production and private appropriation. (This development process is not however unilinear. In periods of economic crisis it may be reversed.)

To sum up, state policies can be viewed from the point of view of the general conditions of production and/or from the point of view of the preservation of capitalist domination. These two standpoints are impossible to separate because of their roots in the evolving contradiction between social production and private appropriation. It is the role of state policies in developing this contradiction which gives them their contradictory character. In the Conclusion we shall take up the question of whether *all* state policies have this contradictory character, which has been deliberately left unanswered here.

Regional policy

In this section we shall discuss the origin and evolution of policies which are explicitly titled 'regional' policy. This restriction is necessary because a large number of government policies have a differential regional impact without being called 'regional' policies. This is true for example of policies designed to assist specific industries which are regionally concentrated (e.g. textiles or shipbuilding), and of infrastructure provision (e.g. motorway construction). In addition, New Town policies which in the 1960s were seen as promoting depressed regions, will be excluded. Our discussion will be divided into three sections. We shall first describe the origins of regional policy, then its evolution since the war, and finally propose an interpretation of it.

Origins of regional policy

The precursors of regional policy in Britain start in the 1930s. Unemployment reached unprecedented levels in the early 1930s; the 1931 Census, for example, showed there were 3,289,000 unemployed, or 15 per cent of the workforce (Glyn and Oxborrow 1976: 148), and the rate rose further in 1932. At the same time the concentration of unemployment in the traditional industries (shipbuilding, iron and steel, mining, textiles, etc.) meant that areas based on these industries experienced rates very much above the average. A 'regional problem' was recognized in the face of average unemployment rates of 25 – 35 per cent in 1932 in Scotland, Wales and the North, with specific towns having rates well above 50 per cent.[4]

Although the 1930s were in general a time of moderate trade unionism,

> 'the Distressed Areas had a strong tradition of trade unionism and were rich in local organizations of many kinds. This led to a degree of solidarity and organization among the unemployed [in those areas] unparalleled before or since, made possible by the concentration of unemployment and the fact that local leaders of the labour and trade union movement were out of work along with the rest' (Branson and Heinemann 1973: 67).

Political response by the unemployed was led by the National Unemployed Workers Movement which organized the hunger marches of 1932, 1934, and 1936.

A month after the second march, an investigation was launched which led to the designation of Special Areas. The powers and funds granted to the Special Areas Commissioners however were quite inadequate. They were expanded in 1937 to allow loans and tax incentives to firms in Special Areas, the building of trading estates and the assistance of labour migration from the depressed areas. Although these powers were in force too short a time to have large results, their scale is significant. According to McCrone 'they embraced a wider range [of measures] than regional policy was to acquire again until the 1960s' (1969: 99).

Rearmament temporarily abolished the regional problem as the government directed armaments factories into the depressed regions for strategic reasons, and the centrally planned war economy not only reduced absolute levels of unemployment to a minimum but also removed the regional differential through the increased demand for ships, iron and steel, etc.

In 1938, however, a Royal Commission was set up to consider the possible future recurrence of regional inequalities, mass migration from the depressed areas and congestion in the growing areas. The report of this Commission, the 'Barlow Report', appeared in 1940 and although its members were divided regarding remedies, all agreed that the government should have the power to refuse a firm the right to expand in the congested areas of the country. No power of this kind had existed prior to the war.

Although negative controls on the location of industry were only introduced after the war, together with a whole host of other reforms, their
origins lie in the experience of war and the social and political changes
it entailed.

A recent historian of this period, Paul Addison, has argued that the
Coalition government 'proved to be the greatest reforming administration
since 1905–14' so that the 1945 Labour government's strategy 'was essentially a conservative one . . . to consolidate and extend the consensus
achieved under the coalition' (1977: 14, 261).

In terms of class relations the war brought about some changes in structures (e.g. the share of income from wages rose 18 per cent between 1938
and 1947 while the share from property fell 15 per cent (Addison 1977:
130)), but more significant perhaps were changes in consciousness. Egalitarian conditions became normal, and 'the social pressure for "equality of
sacrifice" was consistent and powerful' (Addison 1977: 131). New standards
of comparison emerged: if sacrifice was equal in war time why should
rewards not be equal too? 'The prevailing assumption was that the war
was being fought for the benefit of the common people' (Addison 1977:
131).

These changing expectations were reflected politically in a shift away
from the Conservative party. The Coalition government had been working
on a vast set of reforms to be implemented after the war, and which
corresponded to popular expectations. But although there existed a minority of Tory reformers, the party as a whole had 'lost touch both with the
progressive trend of government policy, and with the movement of popular opinion' (Addison 1977: 229). Thus the Labour government elected in
1945 was in a strong position to defend and extend the degree of state
economic intervention.

The regional policy which was introduced in 1945 had two arms. Firstly,
the Board of Trade (a central government department) was empowered
to offer incentives to firms to locate in the newly designated Development
Areas (e.g. grants and loans), and build factories in these areas itself or
via industrial estate companies. Secondly, and this was the new power
recommended by the Barlow Report, it was given the right to restrict
the building of new factories or extension of existing ones above a specified
size, in specified regions (viz. the 'congested' Midlands and South East),
by requiring them to obtain Industrial Development Certificates (IDCs)
in addition to planning permission granted by local authorities. However,
and this is a fundamental limit of the policy – given by the capitalist
nature of the economy (Castells 1976d: 166–7) – these negative powers
stopped short of state 'direction of industry'.

The policy, then, appeared as a decisive measure capable of eliminating
regional inequalities in unemployment via central control of industrial
location. The logic of the policy was that of 'stick' and 'carrot'. By preventing firms from expanding in the congested areas, and providing incentives
for them to do so in the depressed areas, the movement of jobs from
the former to the latter regions would be brought about and unemployment
rates equalized in the process.

Straightaway the ambiguity of the policy can be seen. Should the 'anti-market' interference with industrial location decisions be seen as anti-capitalist, i.e. as substituting alternative aims for the aims of individual capitalists, or for the collective aims of a capitalist State? Or should the policy be seen as a rationalizing measure either for the benefit of individual capitalists (e.g. by regulating competition between them – as in the case of the building of three aluminium smelters in Britain in the late 1960s (HC 347–I: 284–301) – or for the benefit of the capitalist State and indirect benefit of capitalists generally? An example of the latter would be a policy which takes social costs (borne at least partly by government) into account, and prevents the individual industrialist from carrying out *his* preferred location decision.

The ambiguity of the policy is a necessary condition of its subsequent chameleon-like evolution.

Evolution of regional policy since the war

Before describing how regional policy has evolved since 1945, we need to note its limits. Firstly the IDC control bears on the building of factories and thus only indirectly on the level of manufacturing employment. It does not affect the level of employment in existing factories either in the congested regions or in the depressed regions. (It also excludes factory buildings below a certain limit.)

Secondly the effectiveness of regional incentives depends on the number of firms a) planning to invest in new plant and b) whose factories are potentially mobile between regions. The former factor depends crucially on the economic climate. It thus follows that the scope of the policy is greatest during booms and least during slumps or crises.

These then are the limitations in principle of the policy's effect on regional unemployment. However, the actual effect depends also on the scale of incentives and severity of application of the controls in the 'congested' regions. These two factors have changed considerably over the post-war period and will be discussed in detail below.

Before describing these variations, however, it is important to emphasize the overall looseness of application of IDC controls. At no time have they represented a total prohibition of factory building in the Midlands and South East, or anything remotely approaching it. As can be seen from Table 10.1 the refusal rate for expansions or new factories above the threshold[5] in the South East and Midlands has generally ranged between 5 and 25 per cent. Even these figures exaggerate the real restrictive effect since firms can re-apply for IDCs after a refusal.

Since the data on IDC refusals in the Midlands and South East are important to our argument, it is worth saying a word about them. On the face of it these refusal rates are a good indicator of the severity of the policy. However, it is sometimes argued that the very existence of the system is in itself a deterrent: i.e. 'most companies will not bother to apply if they know they are going to be turned down' (Hardie 1972:

Table 10.1 Industrial Development Certificate refusal rates
in the South East and Midlands

	%		%		%
1950	22.6	1959	13.7	1968	22.1
1951	13.2	1960	16.7	1969	16.0
1952	15.7	1961	19.1	1970	17.1
1953	7.7	1962	24.2	1971	9.5
1954	6.4	1963	21.6	1972	8.7
1955	6.3	1964	26.1	1973	11.4
1956	1.8	1965	23.7	1974	12.5
1957	2.1	1966	29.9	1975	11.6
1958	13.8	1967	22.6		

Source: 1950-71: Moore and Rhodes (1976). 1972-5 figures kindly supplied by J. Rhodes. (These figures differ from those in Moore and Rhodes (1973) probably due to being standardized at a different exemption limit.)

233). Against this, however, there are two arguments. Firstly, a Department of Environment survey in 1976 of firms in the South East failed to uncover a single case where the system had acted as a deterrent (1976c: 80). Secondly, as Hardie implies, the force of the deterrence argument depends on the rate of refusals being very high. But this is the reverse of the truth: as the Department of Environment states 'most applications for IDCs in the South East are in fact granted' (1976b: 11). We thus follow the practice of Moore *et al.* (1977) in treating IDC refusals in the Midlands and South East as a good indicator of policy severity.

Moore *et al.* (1977) distinguish four periods in the operation of regional policy. These are identified by the level of incentives available and the relative 'severity' of IDC refusals. We shall distinguish the same four periods, but our interpretations are not necessarily the same as those of Moore *et al.*[6]

1945–50 This was a period of 'active' regional policy. Government expenditure at current prices ranged between £6m and £13m per year over the period, and was primarily devoted to the building of 'advance factories' and industrial estates (McCrone 1969: 114). In addition the armaments factories which had been built in depressed areas were now made available. Negative controls in this period were fairly strict – though no precise data are available. As well as IDCs, building licences were necessary and the latter were particularly effective from 1948 (Dow 1964: 150; Luttrell 1962: 68). The actual effects to be attributed to these policies are, however, difficult to estimate since the shortage of premises and labour in the main cities tended to force firms wishing to expand to the development areas. This market pressure was thus facilitated by government policy (Luttrell 1962: 71).

However, the policy's effects were greater before 1948 than after. The series of sterling crises starting in 1947, and the constraints of Marshall

Aid, led to new priorities: if a firm produced goods for export its contribution to the balance of payments overrode the constraint against locating in congested regions. As a result the share of new industrial building in Development Areas fell from 50 per cent (1945–8) to 17 per cent (1948–50). (Randall 1973: 29).

1951–8   This was a period of 'passive' policy. Government expenditure at current prices fell to between £3m and £6m per year (McCrone 1969: 114). Most of this spending was on factories and industrial estates, but a little was on loans and grants. At the same time the IDC refusal rate in the Midlands and South East fell from around 14 per cent in 1951 and 1952 to 2 per cent in 1956 and 1957. In 1958 it rose to 14 per cent again. The 1950s are thus a period without any effective regional policy.

Why was this? The main explanation is that the period was one of full employment: the percentage unemployed varied within the range 1.1 to 2.1 per cent between 1951 and 1958 (Worswick and Ady 1962: 536). The regional problem had disappeared as the traditional industries of the depressed regions had benefited from the boom. And neither of the standard texts on the period (Dow 1964, Worswick and Ady 1962) contains any reference to regional inequalities of unemployment. In addition the build-up of the eight London New Towns in this period necessitated an increase in manufacturing employment in the South East.

This period presents the paradox that in boom conditions, that is to say those in which the potential for influencing industrial movement to improve regional inequalities is greatest, such a policy was not implemented. Instead, as Holmans (1964a) shows, a large increase in manufacturing employment in the South East occurred, and this increased the favourable industrial structure of the South East, as measured by its balance of growing/declining industries. It has also been shown that much of this increase occurred independently of the 'open' IDC control system. Holmans (1964b) and Gracey (1973) emphasize how manufacturing employment increases could occur outside the IDC controls – a reminder of their limited bearing on employment mentioned earlier.

1959–62   Moore *et al.* describe this as a transitional period between the 'passive' policy of the 1950s and the subsequent 'active' policy. The level of government expenditure rose from £4m in 1958/59 to £33m in 1961/62. Factory building continued but the main increase was in loans, whose terms were eased.

A significant change in this period was the priority given to the relief of unemployment in regional policy. Firstly, the 1960 Local Employment Act introduced loans and grants which were conditional on the level of employment provision – and free depreciation which was not (McCrone 1969: 131–2; HC 347: 51–2). Secondly, a new criterion was introduced for defining 'Development Districts' – a 4½ per cent level of unemployment – which led to the designation of new and scattered areas. This 'social' emphasis can be seen as a move in the direction of the anti-capitalist pole of the policy continuum, but not to an extreme position. At the same time controls on the issuing of IDCs were tightened. Refusal rates in the 'congested areas' rose from 2 per cent in 1957 to 14 per cent in 1958 and 24 per cent in 1962.

This rapid turn in regional policy, occurring as it did during a period of Conservative rule, should lead us to question the emphasis sometimes given to Labour governments as the natural progenitors of active regional policies (see Keeble (1976:212)). The policy change in this period is more readily explained by the sudden return to salience of the regional problem after twenty years of absence due to the end of the boom in the traditional industries and a rise in the general level of unemployment.

1963 TO DATE   The final period of regional policy is the period from 1963 to date. This is a period of active policy, according to Moore *et al.* We agree that it is 'active' in terms of the level of incentives offered, but not as far as IDC refusals are concerned. The absence of strict controls will lead us to a different interpretation.

In most respects the level of incentives has indeed increased dramatically since 1963. In 1963 the incentives available since 1960 were simplified but remained based on employment creation (with the same exceptions). However the policy was increasingly attacked on two fronts. Firstly, the use of the unemployment criterion meant that the designation of Development Districts was continually changing causing difficulties for investing firms. Secondly, the policy emphasis on relieving unemployment was criticised for giving the highest priority to the areas which had the least prospect of economic growth (McCrone 1969: 125). This is another way of saying that the policy needed to move back towards the pro-capitalist end of the continuum.

These views were shared by the new Labour government in 1964 and in 1966 the policy shifted away from its 'social' emphasis. It may be suggested that this shift of view was due to the fall in overall unemployment and hence reduced salience of the regional problem after 1963. (Overall unemployment percentage rates fell from 2.0 per cent in 1962, and 2.5 per cent in 1963 to 1.6 per cent in 1964, 1.4 per cent in 1965 and 1.5 per cent in 1966 (HC 85–I: 715) and regional rates fell correspondingly.)

In 1966 new and larger Development Areas were designated, no longer on a strict employment criterion, and offering a 'better basis for sound regional economic growth' (McCrone 1969: 120). At the same time most of the existing incentives were replaced by investment grants (40 per cent in Development Areas, 20 per cent higher than elsewhere) for plant and machinery. The employment criterion was no longer applied, and the abolition of free depreciation meant that firms with no profits taxable in the UK would benefit. International oil companies were a category which benefited in both respects (Hardie 1972: 228).

In 1967 a major innovation was introduced, a labour subsidy, Regional Employment Premium (REP), paid in respect of all manufacturing employees in Development Areas. This was seen as a response to critics who argued with reason that existing grants were encouraging capital-intensive industry in the depressed areas (Keeble 1976: 229). *It did not, however, eliminate the greater percentage subsidy for capital (due to investment grants)* and thus capital-intensive firms still got an appreciably better deal (Hardie 1972: 230). REP also represented a new approach to the regional problem. Rather than attempting to increase employment in Development Areas by incentives to moving firms, it was a blanket subsidy to all firms in those areas.

Also in 1967 Special Development Areas were created as 'a simple welfare response to acute unemployment problems in certain former coal-mining areas' (Keeble 1976: 229) and improved incentives were made available in them. In 1970 a third category 'Intermediate Areas' was created with fewer incentives than 'Development Areas' – collectively the three categories, which exist to this day, are described as 'Assisted Areas'.

By 1969–70 current expenditure on regional policy had reached £299m, compared with £16m in 1963–4. One third of this took the form of investment grants, and another third REP. Ignoring REP, investment grants had increased dramatically yet over the same period IDC controls had *not* become more severe in the South East and Midlands. The refusal rate had risen from 22 per cent in 1963 to 30 per cent in 1966, but had then fallen to 16 per cent in 1969. We will return to this failure of stick and carrot to move in support of each other below.

The Conservative government returned in 1970 determined to reverse Labour's policy of helping 'lame ducks' and replaced the investment *grants* for plant and machinery (available to firms irrespective of profitability) by 100 per cent *tax allowances* (compared with 60 per cent outside the Development Areas). These would only benefit profitable firms. Building grants however were retained and their rates raised. A year later the areas designated as Intermediate and Special Development Areas were increased in size.

In 1972 however the Conservatives did a *volte-face* and returned to investment *grants*. In 1972 Industry Act introduced two kinds of regional incentive: Regional Development Grants (RDGs) of 20–22 per cent on buildings, plant and machinery which are available to manufacturing firms in the designated areas automatically and without regard for the number of jobs created; and Selective Assistance to firms in the Assisted Areas, whether in manufacturing or services, of up to 20 per cent which is discretionary. This usually takes the form of interest relief grants and is available either for employment creation or modernization. (See 'Towards an alternative interpretation of British regional policy' below.) RDGs and Selective Assistance have continued after 1974 under the Labour government and are the main forms of incentive available today. (REP was doubled in value by Labour in 1974 but allowed to disappear at the end of 1976.)

By 1975/6 regional expenditure had increased to £691m – of which nearly half was RDG and one third REP. Thus in the post-1963 period this expenditure increased twenty-fold at current prices, or nearly ten-fold at constant prices. Over the same period however, the IDC refusal rate has declined from a peak of 30 per cent in 1966 to 17 per cent in 1970, and has varied between 10 and 13 per cent between 1971 and 1975.

This paradox will be the starting point for our interpretative discussion.

Towards an alternative interpretation of British regional policy

In the remaining part of this discussion we shall focus on regional policy since 1960 since by doing so we exclude very little. In the last section we left unresolved a puzzle. Before backing up my interpretation of regional policy with evidence I will state it briefly since it enables this puzzle to be resolved. The puzzle was that whereas the logic of the 'stick and carrot' approach to the regional problem requires that a severe policy be associated with large incentives, and a loose policy with small incentives, the evidence for the post-1963 period showed IDC refusals falling from a 1966 peak of 30 per cent to a 10–13 per cent level in the 1970s while the amount spent on investment grants (ignoring REP) *rose continuously*. This suggests to me that a different logic is in operation.

In a nutshell I will argue that state regional policy has become, since 1966 if not before, a means by which accumulation by industrial capitalists is bolstered in a period of falling profitability. In other words the 'regional' element of the policy is completely subsidiary, and is an example of the way that thinking in terms of spatial units can conceal the real social processes involved – see Castells (1976b, 1976c).

This hypothesis resolves the puzzle stated above by interpreting the increasing size of the carrot not as a greater incentive to move to an Assisted Area, as the term 'regional incentive' implies, but as a large subsidy to industrial capital to enable it to survive the profits squeeze. If regional incentives are not aimed to move firms to regions but to act as an operating subsidy to help industrial firms' balance sheets *tout court* then the puzzle disappears.

A perfectly justified reaction to this hypothesis is that it is unsupported. In the remainder of this section I shall offer empirical support for it, but first I will briefly emphasize its consistency with the theoretical analysis of state policy on pages 231–3.

The analysis of British state policy in recent years usually starts from two points. Firstly the profits squeeze, and secondly the conflict between industrial and banking fractions of capital. Glyn and Sutcliffe (1972) have set out the evidence for a profits squeeze in Britain and Beckerman (1972) has shown how banking capital (either in domestic or IMF form) has insisted on deflation and cuts in state 'social' expenditure in response to balance of payments crises. The role of the profits squeeze is controversial. One argument states that declining profitability has hampered new investment and helped reproduce the vicious circle of low investment, low productivity growth, declining competitiveness. The contrary argument, better supported by evidence, states that the reason for low investment is not the lack of profits, but a) a shift in investment to low wage areas of the Third World and/or b) the decline in incentive to invest in Britain, reasons which are mutually compatible.

Whichever is the case British governments since the 1960s have given high priority to the 'regeneration' of industry, subject to limits set by banking capital. In particular, since 1974 the Labour government has increased the supply of funds for investment, but has had less success in ensuring these funds actually went into investment. Firstly, a new system of tax allowances has cut effective corporate taxes drastically, countering the squeeze on pre-tax profits (Field *et al.* 1977: 21). Twelve of the largest twenty industrial firms in Britain paid no corporation tax in 1977 on their 1976 profits of £2765m due to the generosity of these allowances. (They included BP, RTZ, Esso, Ford, BL, Reed, GKN.) Secondly, an explicit strategy for the 'Regeneration of British Industry' (Cmnd. 5710) is in effect, which involves channelling state aid to promote industrial efficiency. This is a case of 'capital restructuring' with state aid (Fine and Harris 1976).

Thus our earlier theoretical analysis (see the Introduction, pp. 231–3), concerning the general conditions of production, would seem to be fully borne out by recent British experience. Colossal transfers of funds are going to the industrial sector, either in cash or in tax allowances. This is another way of saying that the industrial fraction of capital is being given policy priority. But these flows also reveal one of the limits of capitalist state policy. If investment decisions remain in capitalist hands what certainty is there that the funds provided will be invested in industry in Britain? One of the aims of the 1974 obligatory 'Planning Agreements' was to ensure this, but they were rejected out of hand by industrialists. It seems to me that the advantage of using regional policy and the National Enterprise Board to channel funds for industrial restructuring is that they can be directed to industrial firms which are investing in Britain (even if they do not affect the investment decision!).[7]

EMPIRICAL EVIDENCE: THE ABSENCE OF AN INCENTIVE TO EMPLOYMENT CREATION   So far we have indicated a theoretical rationale for our interpretation of regional policy. We shall now provide some empirical support for this view, and at the same time attack the conventional view that regional policy is primarily concerned with reducing the level of unemployment in Assisted Areas.

The first major argument in favour of our interpretation is that *regional policy incentives have not generally been conditional on the provision of a given level of employment, and have in fact systematically favoured capital intensive firms,* i.e. those most relevant to the restructuring of capital.

As we saw earlier, the RDGs introduced in 1972 and the investment grants they replaced, as well as the previous depreciation allowance system, were not dependent on the number of jobs created. The REP might appear to be a contrary case, but firstly, it was not dependent on the creation of additional employment; secondly, employment creation was only one of its possible effects; thirdly, its blanket character enables one to treat it primarily as a colossal subsidy to industrial capital in the Assisted Areas;

and, fourthly, its relatively low value (which was allowed to decline in real terms till 1974) left an incentive in favour of capital intensity.

The sole exceptions to our argument are the job-related incentives which operated from 1960–6, and the Selective Assistance scheme of 1972. As suggested earlier, the former do represent a deviation in a 'social' direction from the main trend of regional policy. In contrast Selective Assistance has run at about one-sixth the level of RDG since 1972 (HC 600–II: 9–10) and as we shall see later has only a limited relation to job creation.

The general absence of a close link between incentives and additional employment is a very remarkable feature for a policy whose explicit aim is the reduction of unemployment. Strangely, this feature has been little discussed by academic commentators. Keeble (1976: 236–7) makes an oblique reference to it when he notes that in the 1960s there was an awareness that a reliance on investment incentives seemed to be encouraging capital-intensive industry to move to the peripheral regions. (This was first reported by the Hunt Committee – Cmnd. 3998: 43–4.) However from the standpoint of conventional analyses emphasizing the 'social' aims of regional policy, the finding remains an anomaly.

By contrast the 1972 government report, *Public Money in the Private Sector,* prepared before the 1972 Industry Act came into force, gave full recognition to this aspect of regional policy. The evidence taken from Rio Tinto Zinc and British Aluminium, for example, regarding the limited employment benefits of the aluminium smelters they built in Wales and Scotland refers to this (HC 347–I: 291 and 299). However the Report (HC 347: 51–3) made no proposals to alter the policy. The non-employment justifications were seen as sufficiently strong.

To provide further support for our argument regarding the way incentives have favoured capital-intensity rather than labour-intensity we will refer to the most recent evidence. Firstly we shall examine policy regarding Regional Development Grants, and secondly, policy regarding Selective Assistance.

The starting point for our discussion of RDGs is the report of the Comptroller and Auditor-General (HC 92) and the subsequent report of the Committee of Public Accounts (HC 536). In the course of examining expenditure on regional policy the Comptroller and Auditor-General discovered that £160m of RDGs was due to be paid towards the building of four oil terminals (costing £806m). These reports raise two major issues and one minor issue.

The first major point is that oil terminals (in which water and unstable gases are removed from crude oil, and in which oil is stored) are by no stretch of the imagination 'mobile industry'. By definition a location close to the oilfield is essential. The grants paid towards them therefore vividly illustrated the fact, discussed below, that location decisions are frequently made for reasons unconnected with incentives. The second major point is that oil terminals are highly capital-intensive. For example the Sullom Voe terminal in the Shetlands will employ 300 people, which means that the RDG of £90m (on a cost of £450m) represents a cost per job of £300,000.[8]

The Department of Industry's response to these points is crucial to our argument and will be quoted in full:

'In November 1974 and December 1975 the Department reviewed their policy on regional development grants to capital-intensive projects in the mining, construction, oil and chemical industries where geographical and wider economic factors were the most important considerations in capital investment decisions and where the availability of grants was unlikely to make any appreciable difference either to the decision to invest or to locating the investment in the assisted areas. *They concluded that such projects should remain eligible because investment in growth industries accorded with the Government's industrial strategy and the assisted areas needed a mix of capital-intensive and labour-intensive projects to provide balanced and vigorous development'* (HC 92: ix, emphasis added).

This statement makes clear that the use of automatic regional incentives (RDG) in this way was a matter of policy. In other words the recommendation in 1973 (see below) that they should be used selectively, in the sense of being witheld in cases where they had no bearing on the investment or location decisions, was in fact rejected.

Moreover, it is not the case that the use of investment grants in this way was new in 1974, or that it was unintended. There is evidence that highly capital-intensive plant received these grants in the 1960s (Cmnd. 3998: 43–4; HC 536: 542), and that the 1972 Industry Act was deliberately intended to continue this policy (HC 536: 546 and 548).

In brief, regional development grants are now used to 'foster investment' and 'regenerate British industry' (HC 536: 547; HC 600–II: 3) *irrespective of the extra employment created,* but this is merely a new rhetoric for a much longer-standing policy.

Secondly we will discuss Selective Assistance to Industry in the Assisted Areas under Section 7 of the Industry Act. Although expenditure on this type of aid is much less than that on RDGs it is worth examination since it is both discretionary and is aimed to 'encourage sound projects which will improve employment opportunities in the Assisted Areas' (HC 545: 45). It should therefore reveal the extent to which the employment criterion enters regional policy currently.

Eligible projects under this Section may be 'new projects . . . which create additional employment' or 'projects, e.g. for modernization or rationalization, which do not provide extra jobs but maintain or safeguard existing employment' (HC 545: 45). Unfortunately no evidence is available regarding the balance between these two types of project. However the evidence given by the Department of Industry to the Expenditure Committee in March 1978 suggests that the latter type of project is predominant.[9]

The following quotations contain the gist of the Department's view:

'the jobs we are talking about under Section 7 are permanent jobs' (HC 600–II: 34).

'Unless industry modernises on a viable basis then there will not be any employment at all. We do even give assistance under Section 7 for modernisation purposes which will result in a certain amount of redundancies, provided the level of employment which is retained is there on a long term basis and on a commercial basis. So it does not always follow that Section 7 assistance leads to an increase in employment immediately or directly. . . . [Our policies] are not to be looked at in the same way as temporary employment subsidies . . . they are essentially permanent and long term' (HC 600–II: 45–6).

Thus it can been seen that although Selective Assistance under Section 7 is said to be 'related to the concept of sustaining or creating employment' (HC 600–II: 3) – unlike the RDGs which are automatic and independent of employment created – this awkward phrase conceals the fact that this type of incentive too, frequently – and possibly generally – is given for projects where very few if any new jobs are created. It must be concluded that this type of regional incentive is completely dissimilar to that operative for a short time after 1960 where a 'social' objective of new employment creation was primary.

To sum up, the general absence of an incentive to job creation, and the continuing incentive to capital-intensity throughout the period (including that when REP was in operation), clearly confirmed as a policy aim in the statements quoted, are incontrovertible evidence that the social wrapper of regional policy does not refer to its economic effect. These features are far better interpreted as means of aiding the restructuring of industrial capital in a period of profit squeeze.

EMPIRICAL EVIDENCE: THE UNSELECTIVE CHARACTER OF REGIONAL INCENTIVES The second major argument in favour of our interpretation of regional policy is that while the policy is explicitly aimed at encouraging firms to move to the Development Areas, the incentives offered for this purpose are given to firms which would have moved to these areas anyway, as well as to firms where the incentive was decisive. The result of this unselective policy is that for firms in the former category the incentives are an aid to profits with no bearing on the location decision.

The major finding of the 1973 report, *Regional Development Incentives,* was the considerable proportion of firms whose moves to the Development Areas were independent of the incentives available. This was true both of many major British firms (Unilever, ICI, Plessey, Thorn) and of the US firms interviewed (IBM, Honeywell, Univac, Burroughs). It was not true of Chrysler and British Leyland, however. In fact among firms surveyed in the 'biggest 500' category, only 15 per cent gave regional incentives as a major reason for their investment decision, though 51 per cent said it had influenced their location decision (HCP 85–I: 676).

This finding may seem inexplicable. However it quickly becomes understandable when it is realised that many of these large firms are multinationals for whom the major choice is between Britain and some other *country.* If as is frequently the case it is the lower wage costs[10] in Britain which are decisive, then capital incentives will be irrelevant (HCP 85–I: 685). (The still lower labour costs in the Third World help explain the de-industrialization of the advanced capitalist countries.) Secondly, capital incentives may be outweighed by gains from transfer pricing (HCP 85–I: 685). Thirdly, there are various other reasons why an assisted area location may be chosen regardless of incentives. For example, the capital-intensive firms referred to in the Hunt report (e.g. in oil, steel and chemicals) denied that it was government inducements which had attracted them there, and mentioned access to deep-water, effluent disposal facilities, availability of water, and market accessibility (Cmnd. 3998: 44).

The unimportance of incentives revealed in the *Regional Development Incentives* report would strike at the heart of a regional policy whose real aims were social. The Committee did in fact recommend a more selective use of incentives (HCP 85: 75). But this was spitting in the wind. It is clear from the government's response (Cmnd. 6058: 10) and subsequent experience that in practice the use of these incentives has not changed one iota to being more selective in the sense the Committee intended. RDGs are given automatically, and Selective Assistance is not limited to firms which might have located outside an Assisted Area but is an additional grant to firms *already getting RDGs*.

On my interpretation the continued unselective character of regional policy is best seen as a major subsidy to industrial capital. This was certainly the Confederation of British Industry's view when they came out in favour of maintaining REP (HC 327: 375) even though it was recognized to be 'mainly a windfall' (HC 85: 35), and although they had opposed it two years earlier as 'a wasteful form of regional assistance since it goes to both efficient and inefficient firms' (HC 347–I: 305).

Furthermore, though the reports we have referred to have emphasized in no uncertain terms their low opinion of regional policy the policy has continued without basic change.

> 'We consider that far too little information is available to Government on the effectiveness of regional incentives. Large sums are being spent with inadequate knowledge of how far they are contributing towards the desired objectives. . . . There must be few areas of Government expenditure in which so much is spent but so little known about the success of the policy' (HC 347: 84 and 57).

And less than two years later,

> 'We are far from satisfied that the continuing search for a viable regional policy has been backed by a critical economic apparatus capable of analysing results and proposing alternative courses. . . . Regional policy has been empiricism run mad, a game of hit-and-miss, played with more enthusiasm than success' (HC 85: 72).

On the interpretation proposed here these judgments can be seen as reactions to the contradiction between the explicit and implicit aims of regional policy which by the early 1970s were seriously diverging.

We can now draw some conclusions to this section.

We have shown firstly that regional policy originated, in part at least, as a government response to the change in class forces during the war, and acquired its 'social' image at this time. Secondly, it has been argued by appeal to both theoretical argument and extensive empirical evidence that in the 1960s regional policy had become a means of aiding industrial capital (mostly 'monopoly' capital) through the subsidization of capital-intensive investment in the Assisted Areas, which would mostly have occurred there anyway. This orientation was apparent in the remarkable lack of attention to employment creation, and the persistent incentive to capital-intensity (the only exception being in the early 1960s), the continuing unselective character of the policy, and the reaffirmation of the policy in departmental statements in the face of House of Commons Committees' criticisms of its workings.

Regional policy in Britain thus shows a chameleon-like shift away from a concern with regional unemployment and towards the assistance of capital-intensive investment which due to market forces is increasingly attracted to Assisted Area locations. The term 'regional' thus refers to a secondary aspect of this process. In the Conclusion we shall locate the reasons for this shift in the evolving contradiction between social production and private appropriation.

Office policy

One of the paradoxes of regional policy on the conventional interpretation is that it applies to manufacturing industry whose share of employment has remained stable around 44 per cent since the war. By contrast the services sector which grew from 47 per cent to 53 per cent between 1951 and 1971 (Unit for Manpower Studies 1974: 14) has remained outside its scope, with a minor exception.

Office employment, a category which cuts across the manufacturing/services distinction (Daniels 1975), has grown even more remarkably – from 16 per cent of the working population in England and Wales in 1951 to 24 per cent in 1971. This growth has been concentrated in the South East where by 1971 31 per cent of all occupations were office jobs, and specifically in London where this figure rose to 38 per cent. (Department of Environment 1976a, Tables 1 and 2.) It is hardly surprising then that it was the London situation which gave rise to office policy in Britain.

Our discussion will be divided into three parts: the origins and early effects; the policy's aims and evolution; and an interpretation.

The origins and early effects of office policy

The year 1964 is the decisive date in the development of office policy. November 1964 marked the launching of a policy known as the Office Development Permit system (ODP) which was modelled on the IDC system. The policy required all office building above a specified size limit, in certain specified areas, to have an ODP granted by central government, in addition to local authority planning permission. The significance of

ODP powers being centrally located is that this was the only means of curbing the enthusiasm of local authorities for attracting office development. (The same rationale applies to IDC powers.) But unlike regional policy, the ODP was *not* accompanied by any incentives.

The new policy was launched with great *éclat* by the incoming Labour government. Crossman, the Minister of Housing at the time, describes the announcement of the Bill as 'a terrific success. It gave a tremendous sense of the Hundred Days' (1975: 46). (The reference is to the Wilson government's wish to emulate President Kennedy's first hundred days.)

The ODP system was not the first response to the growth of office employment in London. Awareness of the problems it was bringing grew in the 1950s and in 1963 the Conservative government issued a White Paper on the subject (Cmnd. 1952). This led to the closing of the 'Third Schedule' loophole which had enabled large increases in floorspace when old office buildings were demolished and replaced, a commitment to the dispersal of government offices and the encouragement of peripheral office centres, and the establishment of the Location of Offices Bureau (LOB).

Although from 1960 the Conservatives were moving towards a more interventive government economic policy (Brittan 1969: 149) the LOB epitomized economic liberalism. Its aim was to provide firms with information on office operating costs outside London in order to promote the decentralization of office jobs from central London – a trend which had existed for at least ten years, stimulated solely by market forces. The LOB was kept in operation by Labour and exists to this day. When Labour adopted a licensing approach (ODPs) the next year they were taking up an instrument rejected in the Conservative White Paper – a break with Conservatism which was symbolically important at the time.

The explicit aim of the ODP control system was to limit the growth of office jobs in the South East, and especially London. The information available at the time suggested that office jobs in central London were increasing at a rate of 15,000 per year. If this trend continued it would impose intolerable pressures on housing (causing more commuting), on traffic congestion, and on public transport (investment, travel conditions). Already by 1961 increased congestion had caused Middlesex and Essex to reverse their previous pro-office policies (Cowan 1969: 170). In fact when the results of the 1961 Census became available in 1965 it turned out that the growth rate in office jobs was much less than had been thought (Evans 1967; Cowan 1971). The policy was thus partly based on false premises.

For the Labour government the measure had additional attractions. It promised to curb the fortunes made by property developers in the 'property boom' decade after building licences were removed in 1954. As Crossman says, 'Nobody in a Labour cabinet is going to object to an action which is extremely popular outside London and which will only ruin property speculators' (1975: 46). Secondly the controls were seen as freeing labour and capital in the building industry for house-building, a traditional Labour policy priority (Crossman 1975: 33).

The policy came into effect in November 1964 and immediately became known as the 'Brown ban' – after the Minister, George Brown, who announced it. Crossman conveys the same image of decisive action when he describes the policy as a ' "plan" for stopping all office building in London' (1975: 46). Yet from 1965–68 8 million sq ft of new office floorspace was completed in central London alone (Rhodes and Kan 1971: 74).

In the remainder of this section we shall examine how this amount of building came about despite the 'ban' and hence what effect the policy really had in its early years.

ODP POLICY IN THE EARLY YEARS   There are two views of the effects of ODP policy in the first four years. The first is that the policy restricted the supply of new office space at a time when after a decade of rapid expansion vacant office blocks were starting to appear, and the continued upward movement of rents was starting to look uncertain[11] (Marriott 1969: 199, 213). This view argues that the policy played into the hands of the developers and finds support among both those on the Right who see government intervention as unnecessary, and thus futile when it is tried, and those on the Left who see all government intervention as inevitably *aimed* at supporting a particular fraction of capital.

While this first view emphasizes the policy's *restrictive* effects on the supply of office space, the second view emphasizes its *lack of restrictive effect,* at least in its early years. The latter view is based on the fact that property developers had prior knowledge of the announcement of the policy and 'naturally enough, developers started to sign every possible contract with their building contractors' (Marriott 1969: 213). Such contracts enabled a large amount of development to escape the permit requirement.

The paradox is that these conflicting views are presented in successive paragraphs by the same author (Marriott) without any comment from him – or from later commentators – on their inconsistency. In fact the most serious study of the subject, Rhodes and Kan's *Office Dispersal and Regional Policy,* supports the second view as to the economic effects of the policy in the first four years.

Their analysis depends on distinguishing the rate at which planning permissions are granted, the rate at which ODPs are granted and the rate at which new office space becomes available. The fundamental point is that long lags can occur between the granting of planning permissions or ODPs and the completion of building. This is due to property developers' judgements about the demand for office space (e.g. once they have planning permission or an ODP they can delay the start of building) and to the length of time between the conception, start of construction and completion of a building.

Rhodes and Kan show that from 1948 to 1964 3.5 million sq ft of office space were granted planning permission in central London, but only 3 million sq ft of new premises were on average built (i.e. permissions were hoarded, or not used), while in the period after the ODP policy,

1965–8, 0.6 million sq ft were approved each year while 2 million sq ft were actually completed (Rhodes and Kan 1971: 73–4). Thus, during the early years of the policy developers went ahead with a backlog of work for which contracts had been signed and planning permissions obtained prior to the announcement of the policy. They were thus able to escape the effects of the policy.

The gap between the conception and completion of an office block thus means that the effect of any restrictive policy, measured in terms of completed office space, will be delayed, e.g. by three years at least. Rhodes and Kan argue that the restrictions imposed from 1965 onwards only become apparent after 1968. They could not, therefore, have had an impact on the supply of office space, or on rents, in 1965 as Marriott argues.[12]

So far we have argued that any effect of the ODP policy was delayed due to the pipeline of existing contracts. However, we have not clarified the extent of the restrictive character of the policy. Earlier we quoted Crossman referring to a plan to 'stop all office building in London'. In fact, as the experience of the IDC controls should lead us to expect, the policy has been very far from a complete ban on office building.

Table 10.2 ODP refusal rate in central London (by gross area)

	%		%
1965 (Aug) — 1966 (Mar)	74	1972–3	51
1966 (Apr) — 1967 (Mar)	18	1973–4	13
1967–8	32	1974–5	49
1968–9	12	1975–6	16
1969–70[a]	33	1976–7[b]	13
1970–1[b]	14		
1971–2	25		

Source: Department of Environment (1976a) for figures up to 1974–5; HC 537 for later figures.

[a] From 1969–70 the figures cease to include offices ancillary to industrial development which accounted for 12½ per cent of space permitted. This change would lead us to expect slightly higher refusal figures from 1969–70 onwards.

[b] In December 1970 the exemption limit was raised from 3,000 to 10,000 sq ft, in June 1976 to 15,000 sq ft, and in June 1977 to 30,000 sq ft. After each increase one would expect, *cet. par.*, a higher refusal rate for the larger office blocks still subject to the control, since the refusal rate is lower for smaller offices than large.

Table 10.2 shows the gross area of offices in central London for which ODPs were refused, as a percentage of the total gross area of offices for which ODPs were issued or refused. As in our discussion of IDCs, we shall use the ODP refusal rate as evidence of the looseness or severity of the policy.

What is immediately striking about Table 10.2 is the very high refusal rate (74 per cent) reached in 1965–6, and the much lower rates in subsequent years (between 12 and 33 per cent for the 1966–77 period with the exceptions of 1972–3 (51 per cent and 1974–5 (49 per cent)). Because of the changing exemption limits a given refusal rate in the early years indicates a more severe policy than the same rate in later years. The 1976–7 figure of 13 per cent thus shows an exceptionally loose policy. Moreover the refusal rate for a given developer is lower still than is suggested by the figures in Table 10.2 since developers may resubmit an application following a refusal.[13]

Our conclusion so far then is firstly that the initial impact of the ODP controls was delayed by several years due to the 'pipeline' effect, and that the application of the controls in subsequent years in central London has mostly been very loose.

Policy severity, policy aims and policy effects

In this section we shall examine the aims and effects of the policy given this evidence as to the looseness of its actual operation. It will be argued that of the two stated aims of the policy, one (inter-regional distribution of office jobs) is symbolic, and the other (intra-regional distribution of office jobs) is pursued only to a limited degree. It will be suggested that the most significant effect of the policy is the promotion of central London as an international business centre – which is not an explicit aim.

In the first five years of ODP policy under Labour (1965–70) the contribution of the policy to improving the *inter-regional distribution of office jobs* was continually emphasized. In my view this aim was primarily symbolic.

The 1965–70 period was one in which the 'regional question' was an important focus of government policy. This priority was manifested both in the second wave of New Towns with their explicit 'regional development' aim and in the more active regional policy we described on pages 239–41. However, no element in office policy enabled it to have an effect on the inter-regional distribution of office jobs. In empirical confirmation of this the Location of Offices Bureau regularly reports that an overwhelming proportion of the London office firms which approach it for information move *within* the South East. The economists Rhodes and Kan have provided a theoretical rationale for this finding: they show that the costs of running an office fall sharply outside central London thus providing an incentive to move *within 100 miles*, but they could not 'find any factor which strongly favours the Development Areas . . .' (1971: 67). Finally no incentives to locate in such areas were available in the 1965–70 period, and in fact limited incentives were only introduced in 1973. (These were increased in 1976 by which time emphasis (still largely symbolic in my view) was again being placed on the inter-regional balance aim of the policy.) The first policy aim then, inter-regional balance, would seem to be mainly symbolic in importance.

The second policy aim was to improve the *intra-regional balance of office jobs*. This aim has been a continuing aim of ODP policy right up until the present. (Since 1977, however, the achievement of balance has been seen as requiring more office jobs in the *central* area, whereas before this the reverse was true). On the face of it the ODP system would appear capable of contributing to this aim. By holding down office building in the centre and encouraging it in the periphery of the conurbation, a decentralization of office jobs in the aggregate sense (i.e. an increase in the ratio of the number in peripheral areas to the total) would surely be achieved.

However the limits of the policy in achieving this aim are several. Firstly, the aggregate degree of decentralization of office jobs depends not only on the number of jobs moved from central to peripheral areas, but also on the net change in office jobs in each zone due to other processes, e.g. the arrival of new office firms in the central zone from abroad. It is thus perfectly possible for the outward moves of central zone offices to be counterbalanced by office firms moving into the central zone from elsewhere. (The paucity of research on such counterbalancing moves is very striking and is probably a direct reflection of policy emphasis on outward moves.)[14]

To what extent can these counterbalancing moves be affected by ODP policy? The answer is, only in so far as they involve *new office buildings.* Moves into existing buildings, e.g. those vacated by departing offices, cannot be controlled via ODP policy. To control these would require a change in planning permission from office use to some other use, or control of letting. Either of these would involve considerable interference with property rights – and thus only be possible under different economic and political conditions.

The second limit of ODP policy in achieving intra-regional balance is that it applies to new office building and not to office jobs in general. Like IDCs, ODPs have only an indirect effect on employment. They do not determine the number of jobs in a new office building, or of course the number in existing office buildings. (ODPs also exclude offices below the threshold.)

The third limit of ODP policy in securing intra-regional balance has been touched on already. The policy has only a marginal bearing on the *letting* of new office blocks, i.e. who the tenants are. This issue became controversial before 1964 when local authorities in outer London found they could not prevent new blocks being let to office firms from outside the London conurbation, rather than to decentralizing firms from the central zone. After 1964 a partial attempt to overcome this problem was made with the introduction of ODPs which specified 'named tenants', i.e. the ODP was issued to a developer on condition that a specified tenant became the occupier. However the weakness of this constraint was the stated reason for the Conservatives' rejection of a licensing system of office control. In the 1963 White Paper it was argued that unlike the case of IDCs, where the firm which will occupy a factory is known and the occupier of a factory does not change often,

> 'New office blocks . . . are more often than not built for letting . . . [so that] when the developer seeks planning permission he may not know how many tenants he will have or who his tenants will be; and these tenants may change at frequent intervals' (Cmnd. 1952: 4).

The objection that a licensing system which does not control letting is ineffective would appear to retain its cogency. Firstly, developers are not required to have a contractual relation with potential tenants. Thus the notion of issuing an ODP to a developer with a named tenant in mind seems nugatory. Secondly, although the Department of the Environment insists that if a tenant is unable to become the occupier, the developer must obtain a fresh ODP (1976a: §6.43), commercial pressures are so strong that the Department would have no effective power to prevent the developer letting to any willing tenant. The very limited effective power of ODP policy over the letting of new office blocks is thus a further factor weakening its ability to improve the intra-regional balance of office jobs.

In passing it might be argued that if the intra-regional *balance* is so difficult to alter via ODP policy, the policy might at least have some effect in constraining the growth in *absolute number* of office jobs in a single zone, central London. There is some truth in this argument. Even if ODP policy is only an indirect control on employment, and only applies to new office building, and cannot prevent the re-letting of vacated offices, it does operate *at the margin* to restrict additional growth of office jobs. But the figures in Table 10.2 show the limited extent to which even this potentiality of ODP policy has been put into effect.

Finally we may turn to the empirical evidence regarding the degree of constraint in issuing ODPs. Is it the case that they have been severely controlled in central London, and more loosely controlled elsewhere in the South East region, in order to improve the intra-regional balance? If this were so the figure in the first column of Table 10.3 (refusal rate in central London) for each year, should be lower than the figures in the remaining three columns, which describe the refusal rates in successive zones within the South East.

The remarkable result which appears in this table however is that a) *in no year during the 1965–77 period was the refusal rate higher in central London than in the 'rest of Greater London',* b) in only two years out of twelve was the central London refusal rate higher than that in the 'rest of the Metropolitan Region' and c) in only two years out of twelve was the central London refusal rate higher than that in the 'rest of the south-east region'. Thus the policy was not operated in such a way as to achieve the limited contribution to intra-regional balance of which it was capable. The LOB was thus quite justified when it complained that ODP policy was being applied so as to limit the possibility of decentralizing office moves. We shall suggest an explanation of the apparently bizarre way in which the ODP policy was applied below.

To sum up our analysis of the contribution of ODP policy to the intra-regional balance of office jobs, we have shown that not only is this contribution limited in principle because of the small number of parameters over which it exerts any control, but it is also limited in practice because of the failure to use it to control central London office building more rigorously than office development in the rest of London and the south-east region.

Table 10.3 ODP refusal rate in central London, the rest of Greater London, the rest of the Metropolitan Region, and the rest of the South East planning region (by gross area)

	Central London	Rest of Greater London	Rest of Metropolitan Region	Rest of SE Planning Region
	%	%	%	%
1965-6	74	74	64	–
1966-7	18	53	25	18
1967-8	32	35	45	34
1968-9	12	38	33	37
1969-70[a]	33	44	45	38
1970-1[b]	14	28	38	27
1971-2	25	58	41	27
1972-3	51	56	48	30
1973-4	13	33	59	34
1974-5	49	59	51	41
1975-6	16	34	47	43
1976-7[b]	13	21	26	24

Source: Department of Environment (1976a) for figures up to 1974-5; HC 537 for later figures.

[a] From 1969-70 the figures cease to include offices ancillary to industrial development which accounted for 12½ percent of space permitted. This change would lead us to expect slightly higher refusal figures from 1969-70 onwards.

[b] In December 1970 the exemption limit was raised from 3,000 to 10,000 sq ft, in June 1976 to 15,000 sq ft, and in June 1977 to 30,000 sq ft. After each increase one would expect, *cet. par.*, a higher refusal rate for the larger office blocks still subject to the control, since the refusal rate is lower for smaller offices than large.

A recent government report confirms one aspect of our conclusion when it argues that the impact of licensing controls is often exaggerated and states that 'the objective of decentralizing office employment from central London is being achieved by market forces, notably by rent differentials between the centre and other locations' (Department of the Environment, 1976b: 11). Characteristically, however, this report is not concerned with the counter-flow of office jobs into central London.

So far we have argued that one of the stated aims of ODP policy is symbolic and the other marginal in its effects. Rather than conclude that the policy is empty of significance I will suggest that a third implicit aim can be identified: *the promotion of central London* as an international-level centre for prestige corporate headquarters, banking and financial institutions and political representation. This involves such activities as government, overseas representation, finance, commerce and trade, head-quarters of professions, communications, etc. – see Department of the Environment (1973: 764), Greater London Council (1969: 10, 21).

This aim must be seen in an international context. The French government has a clear strategy of promoting Paris as a European and world-scale corporate control centre (Lojkine 1972) and urban renewal in Paris has been analysed as facilitating this by evicting the working-class population (Castells 1972). The development of New York has been analysed in similar terms (Fitch 1977). This evidence regarding Paris and New York is of course no more than suggestive about London.

To support our argument we can examine evidence from each of the three periods 1965–70 (Labour), 1970–4 (Conservative) and 1974–8 (Labour) in turn. Except for 1965–6 when 74 per cent of ODP applications in central London were refused, the first Labour period was one of relaxed policy: the refusal rate in 1966–70 varied between 12 and 33 per cent. (The rapid change after 1965–6 is no doubt linked to the publication in 1965 of data which showed that the projections of central London employment on which the policy was premissed were largely incorrect.) Ministerial statements in 1967 and 1969 referred to the need 'to take into consideration the efficient use of labour by those office employers who are unable to move out of the area of control' (H.C. Debs 1969, Vol. 776, 372). This statement is consistent with the promotion of central London as an international business centre.

In the Conservative period, 1970–4, the refusal rates fell to the 13–25 per cent range, with the exception of 1972–3. Because of the raised exemption limit in 1970 these figures underestimate the easing of the policy. This greater flexibility reflected a new policy orientation – at least at the level of Ministerial statements. The December 1970 statement explicitly emphasized the 'need to ensure that a sufficiency of new office space is being provided for the essential needs of commerce' (H. C. Debs 1970, Vol. 808: 45).

The second Labour period, 1974–8, is reminiscent of the first. An initial year of severe control (49 per cent refusals in 1974–5) was succeeded by very relaxed control (16 per cent 1975–6, 13 per cent 1976–7). The post-1976 figure underestimates the looseness of the control since the exemption limit was raised in 1976.

Once more Ministerial statements enable us to infer the same policy aim. The policy review initiated in 1974 was completed in April 1976 when the *Office Location Review* was published. The *Review* doubted whether the ODP system was still necessary since the main reason for the policy, i.e. the increase in office employment in central London, no longer applied. The policy was however retained, but relaxations were announced. The exemption limit was raised to 15,000 sq ft in June 1976, and some speculative building was allowed again (after the 1974–5 curbs). A year later, in May 1977, further 'measured relaxations' to assist inner London were announced. By this time the 'inner city' had been defined as a problem and office development could be promoted as a partial solution to it. In addition the LOB was asked to go into reverse gear and attract office jobs *back* to central and inner London.

What is significant for our argument is that in his House of Commons statement, the Minister explicitly referred to the need 'to attract international concerns to establish themselves in Great Britain – whether that be in London or in other major cities' (H. C. Debs 1977, Vol. 932, col. 318). This makes clear the reality of the competition between countries for specialized office functions.

A final and in my view highly significant point is that, as we saw in our discussion of the intra-regional balance aim, at no point during the ODP policy's operation have central London ODPs been harder to obtain (as measured by the refusal rate) than ODPs in the rest of Greater London. In other words, contrary to the intra-regional balance aim, the policy has been operated so as to *facilitate the building of new offices in central London relative to the rest of the conurbation.* This paradox is resolved as soon as it is suggested that a third and effective aim of the policy is precisely to promote central London as an international business centre.

The relevance of the evidence regarding the three periods to the claim that the promotion of central London was the implicit office policy, can be seen in two stages. Firstly, irrespective of whether Ministerial statements mention it or not, the effect of allowing market forces to operate with little interference – which describes the way ODP policy has been applied in all but three years – is to facilitate a double process of *decentralization/ recentralization: the dispersal of routine office functions from central London and their replacement by more specialized national and international functions.* (And any differential effect the policy has had as between the central and other zones of London has facilitated this same process.)

Secondly, the varying attention given to the promotion of central London in policy statements since 1965 can in my view be explained as follows. The absence of *explicit* mention of the promotion of central London in the 1965–70 period is due, I suggest, to Labour's ambivalent attitude to the City of London and to the expansion of office employment. This was the period when a tax was introduced to encourage manufacturing jobs and discourage growth in the services sector (Selective Employment Tax, 1966); yet also, one in which the importance of 'invisible' earnings in the balance of payments was recognized. By contrast the Conservatives did not share this ambivalent attitude and hence were explicit in their endorsement of central office employment. Finally in the post-1974 period Labour responded to the worsening economic crisis (e.g. unemployment rose from under 500,000 in 1974 to 1,500,000 in 1977) *inter alia* by providing subsidies and other incentives for the creation of jobs, whether in offices or not, to counteract some of those lost by its simultaneous policy of capital restructuring – one instrument of which is regional policy. In this period the distinctiveness of Labour policies was reduced considerably and its previous ambivalent attitude to office employment became a luxury which could no longer be afforded.

We have thus identified three aims of office policy. The aim of *inter*-regional balance in office employment, was argued to be primarily symbolic. We have seen that the aim of *intra*-regional balance in office employment can only be influenced to a very limited extent by the ODP system, and that in practice the policy has had the reverse effect, i.e. it has held down the growth of outer offices *more* than it has restricted central office building.

This led us to suggest that the real, but usually implicit, effect of office policy was to promote central London as an international business centre.

Explaining the origins and evolution of office policy

We have now outlined the origins of office policy, its changing severity of application, and the explicit and implicit aims it has sought. We now turn to the question of explanation.

The radical difference between the initial conception of the policy as a 'ban' and its subsequent reality as a very loose control makes clear its chameleon-like character and the need to explain the origin and subsequent evolution of the policy separately. From the model outlined on pages 231–3 we would expect to seek an explanation of policy origin and evolution in changing class interests and changing government priorities.

POLICY ORIGINS   I shall distinguish two non-explanations and two real explanations for the origins of the policy. Firstly, it is simply incorrect to assert that ODP policy was the result of working-class pressure due to travel conditions or traffic congestion in central London. (In fact it is the absence of questions asked in Parliament prior to 1964 which is notable (Pinniger 1978).) At most it might be suggested that the policy was a precaution against anticipated popular reaction. Apart from the difficulties with this type of explanation (Pickvance 1977b) there is no evidence that it applies here.

Secondly, it is also incorrect to see ODP policy as aimed at aiding the property industry. Even if this were one of its *effects* – a quite separate point – it would imply an improbable aim for a government in a very critical economic period. In fact the main policy priorities in the 1960s were the regeneration of British industry – i.e. aid to *industrial* capital – and the improvement of the balance of payments. The promotion of central London was a means to the latter end (since it would raise the level of 'invisible' earnings), and it entailed the building of new office blocks. But although property capital benefited at this stage, ODP policy was not designed to help it as a primary aim.

The two explanations of the policy's origins which are in my view valid relate firstly to the nature of the capitalist State and secondly to party politics. The fact that both parties, Conservatives in 1963 and Labour in 1964, made initiatives on the office problem is to be explained by the State's function in seeking to reconcile the conflicting interests of different fractions of capital, and the contradiction between capital and labour. In this case these conflicts seem to have been expressed in *financial* terms. For example, *local* authorities in a number of areas were facing increasing burdens due to office growth – which would be reflected in greater financial demands on central government and/or ratepayers. And the predictions about central London employment growth suggested a large increase in commuters, and in investment in railway infrastructure. Since nationalized industries had been given market targets to meet in 1961 this investment would imply higher rail fares as well as a higher central subsidy. It was,

I suggest, because of these financial implications of office growth that both parties were obliged to take some initiative. Ultimately these financial effects would be reflected in pressures for taxes either on higher profits or (more likely) on personal incomes.[15]

The second explanation concerns the *form* of initiative taken by each government. Here it seems to me that the sharp contrast between the liberal philosophy implicit in the LOB, and the ODP licensing system with its overtones of central planning can be directly attributed to party-political differences. The variety of quasi-planning institutions launched by the 1964–70 Labour governments is well known (Smith 1975a). The more potent policy instrument introduced by Labour can be explained by the greater Labour sensitivity to the issues of property speculation and housing.

POLICY EVOLUTION   The explanation of the metamorphosis of ODP policy away from the 'total ban' conception has already been hinted at. The evolution of the policy reflects directly the contradictions facing British capitalism and the government's priorities in responding to them.

In brief, the post-1965 period has been characterized by a series of balance of payments crises to which the government has responded cautiously placing monetary stability before growth at the behest of domestic banking capital and the IMF.

In order to provide a secure basis for continued capital accumulation, state policy has been to modernize and regenerate British industry (one element of which is regional policy) and also to improve earnings from exports, including 'invisibles' to overcome the external constraint. The evolution of ODP policy towards a loose policy in which the promotion of central London as an international business centre became the prime object is thus entirely consistent with the government's response to the contradictions facing it.

Conclusion

In this final part we shall bring together our analytical conclusions and relate them to some issues identified in the Introduction.

Our analysis of regional policy and office policy reveals some striking similarities. Apart from the fact that both operate through a licensing system, both had an 'anti-market', if not anti-capitalist, character at their origin, and both have evolved chameleon-like away from their original character while retaining their original 'social' wrapper. Their evolution has led ODPs, and to a slightly lesser extent IDCs, to operate so as to support market trends: the double process of decentralization/recentralization of office jobs in central London (a trend which has existed for several decades), and the largely self-propelled movement of capital-intensive industry to the Assisted Areas.

We explained the origins of office policy in terms of the threatening

financial implications of office growth in central London, and the origins
of regional policy in terms of the changing class relations which occurred
during the war. The second of these explanations appeals to the State's
role in preserving capitalist relations of production. But it would be misleading to couch the analysis of both origins and evolution of the two
policies in terms of the alleged 'accumulation' and 'legitimation' functions
of state policy, e.g. by interpreting their 'social' image as a necessary
means of legitimating their accumulation role.

There are two reasons for avoiding this approach, one general, and
one specific to the policies in question. The general reason is that there
is no such thing as a *necessary* legitimation function for state policy. As
we saw earlier, one of the two necessary state functions is the preservation
of capitalist relations of production. This may involve repression, it may
involve legitimation, and it may involve neither when political opposition
is lacking for other reasons. As Mann (1970, 1975) has pointed out, intellectuals are particularly prone to emphasizing the State's need to legitimate
its policies because their job involves them in continually trying to understand these policies and because their Left position makes them conscious
of the 'irrationalities' of policy. By contrast, Mann argues that the 'man
in the street' lacks this totalizing interest in policy rationality and has a
very pragmatic orientation to the social order – which implies a lesser
'need' for legitimation functions.

This first point is a salutary warning about the alleged legitimation
function of state policy *in general*. Secondly, we will suggest why it is
an inappropriate mode of analysis in the case of the *evolution* of regional
and office policy, and specifically why the 'social' images of these policies
have not played a role in their legitimation. To attribute a 'legitimation'
function to a policy is to assert its efficacy in limiting, or preventing, political
mobilization over some state intervention. For example, it might be argued
that regional policy's social image diverted mobilization around aid for
capital restructuring since the 1960s.

Now I would not deny that the 'social' images of both regional and
office policy *can* have a mystifying effect. This would apply if intervention
into decisions on industrial location and office location was seen as *ipso
facto* anti-capitalist in character. This misunderstanding is I think quite
widespread – both among individual capitalists who resent some kinds
of state interference, and among commentators especially on the Left who
read these protests as implying an anti-capitalist policy. However if an
'anti-market' intervention is one which brings about a different result from
the market, this 'different result' could either be basically compatible with
capitalism (e.g. town planning controls which insure a landowner against
his neighbour building an abattoir) or basically incompatible (e.g. the zoning of a city centre as a public park).[16] It follows that 'anti-market' does
not necessarily mean 'anti-capitalist'.[17] Nevertheless the 'anti-market' image of office and regional policies may have led to this confusion.

However the crucial stage of the 'legitimation' argument is to say that
the policies' images diverted political mobilization around the issues in

question. I would deny this for one simple reason: it grossly exaggerates the potential for political mobilization around these issues. It assumes that anti-capitalist forces were strong and needed to be blunted, whereas I would argue that two other factors made such mobilisation improbable, and hence that the 'social' images of the policies themselves were irrelevant. The first is the character of the Labour party which is not an anti-capitalist party putting forward 'class' interpretations of society but a reformist party with a 'national' rather than a 'class' ideology. For example, restructuring aid is given a national rather than a class interpretation. The second is the importance of spatial units (e.g. cities) as arenas of everyday life (e.g. as labour markets) and as political units electing MPs who are judged by what they achieve for the area. Particularly in a period of high unemployment a policy which apparently 'brings jobs' (like regional policy) – even if it is an integral part of a policy of closing old plant and creating redundancies – is valid currency in the local political arena.

I would thus dispute the argument that the anti-market images of regional and office policies are to be seen as legitimizing their actual functions in the restructuring of industrial capital, and in promoting the decentralization/recentralization of London office jobs.[18]

Rather than analyse the evolution of regional and office policies by appealing to the 'legitimation function' of their anti-market images, we have sought to analyse it in terms of the evolving contradiction between social production and private appropriation, and the changing role of the State in the accumulation process this has brought about.

The increasingly international character of social production has meant that both regional policy and office policy have come to reflect new demands. In the latter case the original aim of checking office growth in central London has given way to the promotion of central London as an international business centre. This reflects an attempt by the State to encourage 'invisible' earnings and hence counteract the external constraint on economic growth. Regional policy has likewise undergone a chameleon-like change so that it now involves large-scale incentives to capital-intensive industry mostly in respect of decisions which would have been made. This new policy clearly shows the State's lack of leverage faced with internationally mobile investment by capitalist firms. (The use of Selective Assistance specifically responds to this new reality.) The British government's strategy is to hold on to a share of this internationally mobile investment by offering increasing levels of restructuring aid.

Our analysis thus leads to the conclusion that not all state policies have a contradictory character as described earlier. We saw there that legislation like the Victorian Factory Acts was on the one hand a concession to the working class, yet on the other hand established new conditions for capital accumulation, in that case concerning the level of absolute surplus value. However in the case of office and regional policy their current character does not represent a response forced by working-class pressure. It reflects rather the evolution of the contradiction between social production and private appropriation mediated by a State which has largely integrated the working class. This is not to deny that the amounts of money channelled through regional policy, or the changing functions of central London are without their effects, but rather to distinguish between the character of policies which result from working-class pressure, and those which are pursued by the State in the absence of such pressure. In my view each has a different potential effect on the conditions of capital accumulation.

Notes

1 This paper was stimulated by reading Castells (1976a), although the resemblance is very slight. Earlier versions of it have served as a passport to seminars in various places since February 1977. The discussion of town planning policy included in earlier versions has not been reworked for reasons of time, and space.

2 For reasons of space other important limitations will not be dealt with here. See Pickvance (1978).

3 The tax share borne by corporations is falling, and that by individuals is rising. See Field *et al.* (1977) for British data, and Wolfe (1977) for Canada.

4 Due to the scale of migration from depressed regions both in the 1930s and since the war regional unemployment *rates* considerably underestimate the extent of regional inequalities in employment opportunities. See Sant (1975).

5 The thresholds have varied over time as follows for the South East and Midlands: 1945 10,000 sq ft; 1947 5,000 sq ft; 1965 1,000 sq ft; 1966 3,000 sq ft; 1970 5,000 sq ft; 1972 10,000 sq ft (SE) and 15,000 (M); 1976 12,500 sq ft (SE) and 15,000 sq ft (M). It is not possible to describe these changes as 'relaxations' or 'tightenings' of policy without further information. For example if the threshold were raised from 3,000 to 5,000 sq ft this would only be an effective relaxation if factories in the 3,000–5,000 sq ft range previously had a non-zero refusal rate.

6 Where not otherwise acknowledged our descriptions of policy measures are drawn from McCrone (1969), Keeble (1977) and Moore *et al.* (1977). The latter article is an invaluable source of data on expenditure on regional policy.

7 I would like to acknowledge my debt to Dunford's (1977) paper in which he shows how Italian regional policy in the Mezzogiorno can be analysed as a means of assisting capital restructuring. The sole contribution of the remainder of this part of the paper is to show that British regional policy can be interpreted in the same way.

8 The minor issue was that eligibility for RDG depended on a majority of employees being engaged in 'qualifying activities'. Since dewatering and destabilizing alone were qualifying activities, employees on the oil storage side were disallowed. Thus the £3m RDG given to the Anglesey terminal in Wales was based on sixteen of the twenty-eight full-time employees being engaged in qualifying activities. Presumably two fewer employees would have prevented the allocation of the grant. The Department of Industry defended the headcount criterion because it was a 'simple and liberal' method as required by the Industry Act (HC 92: vi, viii).

9 The Committee did discover that the actual number of jobs created by firms in projects receiving this aid was only two-thirds of the figure they originally

estimated (HC 600–II: 4). The Department of Industry later tried to withdraw this estimate as dubious (HC 600, xx) but it was confirmed subsequently in a Public Accounts Committee report (*Guardian*, 5 Oct. 1978).

[10] The extent of Britain's low wage advantage can be seen from the following figures. If hourly labour costs including social charges in Britain in 1964 are equated to 100, the comparative figures are USA 268, Canada 192, Sweden 153, Norway 122, Germany 119, France 103, Italy 93, Japan 42. By 1974 (Britain = 100) Britain's relative cheapness 'advantage' had increased in all but the first two cases: USA 194, Canada 186, Sweden 208, Norway 189, Germany 185, France 118, Italy 122, Japan 105 (Counter Information Services 1977: 17).

[11] Rising rents are vital for the financial stability of the property industry since as long as they go on rising the value of properties as collateral for bank loans is protected, and new loans can be obtained (CIS 1973).

[12] Unfortunately however the figures Rhodes and Kan provide are not fully adequate to demonstrate their conclusions. It is worth noting that central London office rents accelerated in 1969 and continued to rise at the faster pace until 1974 (Department of Environment 1976a). If demand were constant, this rise would be consistent with ODP controls imposed in 1965/66 biting three to four years later. However it could also be due to an increased level of demand resulting from the promotion of London as an international centre of finance and corporate headquarters.

[13] An analysis of forty-one 'refusals' of large offices (over 50,000 sq ft) showed that twenty-four subsequently obtained an ODP (Department of Environment 1976a: §6.9).

[14] One of the rare articles on the subject (Hall 1970) argues that vacated offices were not re-occupied, but this ignores contrary evidence in the same article that 73 per cent of vacated offices *were* re-occupied. This inconsistency may be due to the author being an employee of the LOB – committed at that time to the decentralization of office jobs.

[15] The same theme underlies the reorganization of local government in London (1964) and the rest of England and Wales (1974) where priority was also given to the rationalizing concerns of central government – at the expense of 'local democracy' (Sharpe 1978).

[16] In the former case although *individual* capitalists may have their investment opportunities limited and may protest, state policy promotes the overall functioning of capitalism. In the latter case although some *individual* capitalists might benefit (e.g. developers of offices on the edge of the park) the policy would have an overall anti-capitalist character. There is an obvious gradation in the concept of 'anti-capitalist' and it is inevitable that in a capitalist society, some category of capitalist will benefit from any 'anti-capitalist' policy. But this does not invalidate the conceptual distinction. See Esping-Andersen *et al.* (1976).

[17] In an earlier article I used the terms 'trend' and 'interventive' planning as though the latter necessarily meant planning against *some* capitalist interest (Pickvance 1977a). Such a distinction makes clear that town planning may simply reproduce market trends but overlooks the fact that some degree of 'intervention' is not only compatible with, but also necessary for capitalist functioning.

[18] If the legitimation function of regional policy is unimportant today it might

be asked, why does industrial restructuring aid need a regional wrapper? On the one hand, as we have seen, regional incentives are *not* the only channel for such aid. On the other hand, as suggested earlier, regional incentives do at least ensure that the aid goes to industrial capitalists – unlike tax allowances, for example, which benefit corporations indiscriminately.

References

Parliamentary Publications (all published by HMSO, London)

House of Commons Debates (H. C. Debs) (Hansard).

Cmd. 4153 (1942) *Report of the Royal Commission on the Distribution of the Industrial Population* (the 'Barlow Report').
Cmnd. 1952 (1963) *London. Employment: Housing: Land.*
Cmnd. 3998 (1969) *The Intermediate Areas* (the 'Hunt Report').
Cmnd. 5710 (1974) *The Regeneration of British Industry.*
Cmnd. 6058 (1975) *Regional Development Incentives* (Government Observations on the Second Report of the Expenditure Committee).

Sixth Report from the Expenditure Committee. Public Money in the Private Sector (1972) HC 347, 347–I, 347–II (1971–2).
Second Report from the Expenditure Committee. Regional Development Incentives (1973) HC 327 (1972–3), HC 85 and 85–I (1973–4).
Appropriation Accounts. 1975–6. Volume II (1977) HC 92 (1976–7).
Tenth Report from the Committee of Public Acounts (1977) HC 536 (1976–7).
Annual Report by the Secretary of State for the Environment. Town and Country Planning Act 1971. Control of Office Development (1977) HC 537 (1976–7).
Annual Report by the Secretaries of State for Industry, Scotland and Wales. Industry Act 1972. (1977) HC 545 (1976–7).
Eighth Report from the Expenditure Committee. Selected Public Expenditure Programmes, and Chapter Two: Regional and Selective Assistance to Industry (1978) HC 600 and 600–II (1977–8).

Addison, P. (1977) *The Road to 1945.* London: Quartet.
Beckerman, W. (1972) 'Objectives and performance: an overall view', in W. Beckerman (ed.) *The Labour Government's Economic Record.* London: Duckworth.
Branson, N. and Heinemann, M. (1973) Britain in the Nineteen-thirties. London: Panther.
Brittan, S. (1969) *Steering the Economy.* London: Secker and Warburg.
Castells, M. (1972) 'Urban renewal and social conflict.' *Social Sciences Information* 11:93–124. (Reprinted in Castells 1978.)
—— (1976a) 'Crise de l'état, consommation collective et contradictions urbaines', in N. Poulantzas (ed.) *La crise de l'état.* Paris: PUF. (English translation in Castells 1978.)
—— (1976b) 'Is there an urban sociology?' in Pickvance (1976).
—— (1976c) 'Theory and ideology in urban sociology', in Pickvance (1976).
—— (1976d) 'Theoretical propositions for an experimental study of urban social movements', in Pickvance (1976).
—— (1978) *City, Class and Power.* London: Macmillan.
Counter Information Services (1973) *The Recurrent Crisis of London.* London: CIS.
—— (1977) *Paying for the Crisis.* London: CIS.
Cowan, P. (1969) *The office – a facet of urban growth.* London: Heinemann.
—— (1971) 'Employment and offices', in J. Hillman (ed.) *Planning for London.* Harmondsworth: Penguin Books.
Crossman, R. H. S. (1975) *The Diaries of a Cabinet Minister, Volume One.* London: Hamish Hamilton.
Daniels, P. W. (1975) *Office Location.* London: Bell.

Department of the Environment (1973) *Greater London Development Plan: Report of the Panel of Inquiry (2 vols).* London: HMSO.
—— (1976a) *Office Location Review.* London: Department of the Environment.
—— (1976b) *Strategy for the South East: 1976 Review.* London: HMSO.
—— (1976c) *Strategy for the South East: 1976 Review. Report of the Economy Group.* London: Department of the Environment.
Dow, J. C. R. (1964) *The Management of the British Economy 1945–1960.* Cambridge: Cambridge University Press.
Dunford, M. F. (1977) 'Regional policy and the restructuring of capital.' Working Paper in Urban and Regional Studies No. 4. Brighton: University of Sussex.
Engels, F. (1969) 'The housing question (1887)', in K. Marx and F. Engels *Selected Works* (in 3 volumes), *Vol. 2.* Moscow: Progress Publishers.
Esping-Andersen, G., Friedland, R. and Wright, E. O. (1976) 'Modes of class struggle and the capitalist state.' *Kapitalistate* 4–5: 186–220.
Evans, A. W. (1967) 'Myths about employment in central London.' *Journal of Transport Economics and Policy* 1: 214–25.
Field, F., Meacher, M. and Pond, C. (1977) *To Him who Hath: a study of poverty and taxation.* Harmondsworth: Penguin Books.
Fine, B. and Harris, L. (1976) 'The British economy: May 1975—January1976.' *Bulletin of the Conference of Socialist Economists* 14: 1–24.
Fitch, R. (1977) 'Planning New York', in R. E. Alcaly and D. Mermelstein (eds) *The Fiscal Crisis of American Cities.* New York: Vintage.
Glyn, A. and Sutcliffe, B. (1972) *British Capitalism, Workers and the Profits Squeeze.* Harmondsworth: Penguin Books.
Glynn, S. and Oxborrow, J. (1976) *Interwar Britain: A Social and Economic History.* London: George Allen and Unwin.
Gracey, H. (1973) 'The control of employment', in P. Hall *et al. The Containment of Urban England, Vol. 2.* London: George Allen and Unwin and Beverly Hills: Sage.
Greater London Council (1969) *Tomorrow's London.* London: GLC.
Hall, R. K. (1970) 'The vacated offices controversy.' *Journal of the Royal Town Planning Institute* 56: 298–300.
Hardie, J. (1972) 'Regional policy', in W. Beckerman (ed.) *The Labour Government's Economic Record.* London: Duckworth.
Holmans, A. (1964a) 'Restriction of industrial expansion in south-east England: a reappraisal.' *Oxford Economic Papers* 16: 235–61.
—— (1964b) 'Industrial development certificates and control of the growth of employment of south-east England.' *Urban Studies* 1: 138–52.
Keeble, D. (1977) *Industrial Location and Planning in the United Kingdom.* London: Methuen.
Lojkine, J. (1972) *La politique urbaine dans la région parisienne 1945–1971.* Paris: Mouton.
—— (1976) 'Contribution to a Marxist theory of capitalist urbanization', in Pickvance (1976).
Luttrell, W. F. (1962) *Factory Location and Industrial Movement* (2 vols.). London: National Institute of Economic and Social Research.
Mandel, E. (1975) *Late Capitalism.* London: New Left Books.
Mann, M. (1970) 'The social cohesion of liberal democracy.' *American Sociological Review* 35: 432–39.
—— (1975) 'The ideology of intellectuals and other people in the development of capitalism', in L. Lindberg *et al.* (eds) *Stress and Contradiction in Modern Capitalism.* Lexington: D.C. Heath.
Marriott, O. (1969) *The Property Boom.* London: Pan.
Marx, K. (1970) *Capital, Vol. 1.* London: Lawrence and Wishart.
McCrone, G. (1969) *Regional Policy in Britain.* London: George Allen and Unwin.
Moore, B. and Rhodes, J. (1973) 'Evaluating the effects of British regional economic policy.' *Economic Journal* 83: 87–110.
—— —— (1976) 'Regional economic policy and the movement of manufacturing firms to development areas.' *Economica* 43: 17–31.
Moore, B., Rhodes, J. and Tyler, P. (1977) 'The impact of regional policy in the 1970s.' *CES Review* 1: 67–77.

O'Connor, J. (1973) *The Fiscal Crisis of the State*. New York: St. Martin's Press.

Pickvance, C. G. (1976) *Urban Sociology: Critical Essays*. London: Tavistock and New York: St. Martin's Press.

—— (1977a) 'Physical planning and market forces in urban development.' *National Westminster Bank Quarterly Review* (August): 41–50.

—— (1977b) 'Marxist approaches to the study of urban politics: divergences among some recent French studies.' *International Journal of Urban and Regional Research* 1: 249–55. (Reprinted in Pickvance 1981.)

—— (1978) 'Explaining state intervention: some theoretical and empirical considerations', in M. Harloe (ed.) *Conference on Urban Change and Conflict 1977*. London: Centre for Environmental Studies. (Reprinted in Pickvance 1981.)

—— (1981) *New Directions in Urban Sociology*. London: Tavistock.

Pinniger, R. (1978) *Office Policy in Central London*. Unpublished M.Phil. Thesis, University of Kent at Canterbury.

Randall, P. J. (1973) 'The history of British regional policy', in G. Hallett (ed.) *Regional Policy for Ever?* London: Institute of Economic Affairs.

Rhodes, J. and Kan, A. (1971) *Office Dispersal and Regional Policy*. Cambridge: Cambridge University Press.

Sant, M. (1975) *Industrial Movement and Regional Development: the British Case*. Oxford: Pergamon.

Sharpe, L. J. (1978) ' "Reforming" the grass roots: an alternative analysis', in D. Butler and A. H. Halsey (eds) *Policy and Politics*. London: Macmillan.

Smith, T. (1975a) 'Industrial planning in Britain', in J. Hayward and M. Watson (eds) *Planning, Politics and Public Policy*. Cambridge University Press.

—— (1975b) 'Britain', in J. Hayward and M. Watson (eds) *Planning, Politics and Public Policy*. Cambridge: Cambridge University Press.

Unit for Manpower Studies (1976) *The Changing Structure of the Labour Force*. London: Department of Employment.

Wolfe, D. (1977) 'The state and economic policy', in L. Panitch (ed.) *The Canadian State*. Toronto: University of Toronto Press.

Worswick, G. D. N. and Ady, P. H. (1962) *The British Economy in the 1950s*. Oxford: Clarendon Press.

11 The property sector in late capitalism: the case of Britain*

Martin Boddy

Since the last war a major feature of the restructuring of space by capital, in Britain as elsewhere, has been the massive scale of commercial and industrial property development. Office development, shopping centres, town-centre redevelopment, warehouse and distribution centres and industrial estates have been the main components. This restructuring of the physical infrastructure reflects major changes in the processes of commodity production, circulation and consumption which it supports. At a general level these changes have included the concentration and centralization of capital (industrial, commercial and interest-bearing) within individual enterprises; the sectoral decline in importance of manufacturing industry and within this sector the increasing importance of light manufacturing industry; the growth of wholesale and retail activity; and finally the massive expansion of the office-based service sector.

Looking briefly at the main sectors of development activity, the great expansion of office-based administrative and organizational functions has resulted in part from the absolute and relative expansion of the service sector; but the increasing concentration and centralization of capital within industrial corporations and financial institutions, and the increasingly international character of capital has contributed to the demand for major

* This chapter is derived from a more general, exploratory paper 'Finance Capital, Commodity Production, and the Production of Urban Built Form', presented to the Conference of Socialist Economists Money Group, London, March 1978. Numerous people have provided detailed critical comments on that original paper which have been taken into account in preparing the work presented here.

office developments to house and facilitate functions previously split between many small enterprises and between different countries. Added to this has been the expansion of the administrative functions of the local and central State. Mollenkopf, for example, has observed that 'the economy of the largest central cities has shifted from a factory basis of organisation to one of office-based command and control functions'.[1] As Marriot has argued, relatively little speculative office development took place between the wars, what development there was being primarily for owner-occupation; demand for office space consequent upon growth and structural change in the economy and changing office practices could thus no longer be met by the increasingly obsolete legacy of nineteenth-century office accommodation.[2]

Retail development between the wars had mainly been confined to suburban locations following new housing development, the rise of multiple stores and changing retail practices in city centres generally being accommodated in existing premises. Suburban retail development continued after the war but the major post-war activity in this sector has been town-centre redevelopment, initially in bomb-damaged cities such as Bristol and Plymouth but continuing into the 1960s and early 1970s in many provincial cities. Demand for redevelopment reflected the increasing disjuncture between the existing physical fabric – small individual retail units with poor access and storage facilities – with the changing nature of the retail trade – the increasing concentration of turnover in multiples requiring large floorspace and easy access for bulk delivery, for example.

Warehousing and distribution has become increasingly road and motorway-intersection oriented, switching away from urban and particularly port and railway locations. Furthermore the nature of the goods handled has shifted away somewhat from bulk commodities needing specialized premises and handling equipment, and methods of handling – palletization and containerization – have become more standardized. These changes have increased the demand for standardized accomodation with access the major feature, entailing in most cases new development in suburban and non-urban greenfield locations.

Finally, the shift in emphasis within the industrial sector towards light manufacturing, assembly and technology-based activities in what Mandel terms the 'Third Technological Revolution'[3] rendered much of the existing 'factory' accomodation obsolete and created a massive demand for standardized single-storey premises oriented, as with warehousing, to the motorway network.

Two aspects of these various trends in property development and redevelopment represent particularly significant changes in the restructuring of space and have wider economic and political implications. First is the emergence of a relatively autonomous 'property sector' managing the development process and the production of physical infrastructure, and based on specialized property development and investment companies. In the office sector, for example, whereas before the war firms requiring accommodation would generally acquire a site and have premises built under contract for their own occupation, in the post-war period office accommodation is increasingly rented from property-development/investment companies. In the warehousing and industrial sectors changing practices reduced the need for specialized premises and for fixed capital investment tied to particular sites which had encouraged owner-occupation. This then opened the way for standard estate-type development by property companies. The fact that particular developments are not tied by their specialized nature to the fortunes of a particular enterprise or branch of activity also increased the possibility of attracting long-term property investment into the sector, though it is still the warehousing sector, the most standardized type of development, rather than premises for manufacturing industry where development and investment activity is concentrated.

Second is the increasing involvement of financial institutions – primarily insurance companies and pension funds – in funding property development and, more particularly, in long-term investment in land and property. Institutions have, furthermore, tended to become increasingly directly involved in the property sector in contrast to their traditionally more remote role as a source of loan-finance for property-development companies. This chapter focuses on the changing scale and nature of financial institutions' investment in property in Britain. The first section sets out, briefly, an analytic framework as the basis for examining at the empirical level the institutional agents involved in the development process and their particular forms of articulation in this. The second section explores the particular, historically determined and changing, articulation of agents at the empirical level. The third section, finally, draws out the implications of this empirical analysis for the nature of development and redevelopment; the nature and effects of state intervention in the property sector, planning strategy, policies for industrial and regional development, and inner-city regeneration; the relationship between different agents and fractions of capital at the economic and political levels both within the property sector and in its relation to the wider socioeconomic structure.

I Commodity production and forms of capital

Commercial and industrial property is normally produced as a commodity; it is a vehicle for the production of surplus value and capital accumulation. At the level of the economy as a whole, capitalist commodity production is described by Marx in the form of the money circuit of *industrial capital*:[4]

$$M - C \quad \begin{cases} LP \\ MP \end{cases} \quad \dots\dots\dots\dots P \dots\dots\dots\dots C' - M'$$

in which industrial capital passes through three forms, money capital (M), commodity capital (C) and productive capital (P). Money capital is exchanged for commodity inputs (C) including labour power (LP) and means of production (MP) which, after purchase, form productive capital (P). In production, labour acts upon other means of production to produce commodity outputs (C') converted to the final money form (M') by the act of sale or realization. The commodity outputs contain surplus value equal to (C' – C), in the money form (M' – M). The sphere of circulation consists of the states and exchanges (C') to (C). Surplus value is produced only in the sphere of production (P). The processes of circulation are necessary to the extended reproduction of industrial capital but unproductive of new value.

Marx distinguishes two separate circuits of circulating capital which develop as manifestations of the increasing division of labour in capitalist society. To quote Thompson, 'Particular functions become separated off and the prerogative of special groups of capitalists given the growth in complexity of capitalist society and with it the growth in the division of labour.'[5] These are described in the circuits of commercial capital and of interest-bearing (or money-dealing) capital.[6]

Commercial capital circulates in the form:

$$M - C - M'$$

Commodities (C) are purchased with money capital (M) and sold for (M') at a profit of (M' – M) in value terms. Commercial capital allows the realization of commodity capital in the circuit of industrial capital in the money form before the commodities have been sold to the final consumer. The articulation of the two circuits can be represented as:

$$M'_I \text{---} C'_I \ldots P_I \ldots C'_I \diagdown M'_I \qquad \textit{Industrial capital}$$
$$M'_c \diagup C'_c \text{---} M''_c \qquad \textit{Commercial capital}$$

in which the diagonals represent exchanges of commodities or money between the different circuits.

Interest-bearing capital circulates in the form:

$$M - M'$$

Money capital (M) is advanced and subsequently repaid with interest equal to (M' – M) in value terms. Interest-bearing capital may be advanced to provide money capital to initiate a circuit of industrial capital (a), of commercial capital (b) or, as in consumer credit, to allow industrial capital to realize in the money form the value of commodities produced but not yet sold (c).

Marx developed these schema with reference to the economy as a whole, expressed in value terms. They can also be employed as an analytic framework to examine a particular branch of production. Clearly there are, at a formal, theoretical level, many possible combinations of the three basic circuits; one can imagine cascades of interlocking, interdependent circuits. But, although constituted at the theoretical level, neither the general schema nor their application to a specific branch of production (here prop-

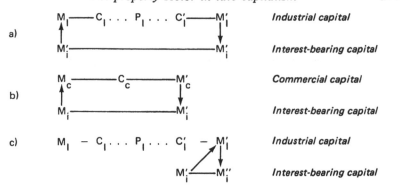

erty), represent an ahistorical abstract structure of categories. The form of the general schema at the level of the economy as a whole is historically determined having as its conditions of existence, first, the dominance of capitalist commodity production and, second, the differentiation of particular functions (industrial, commercial and interest-bearing capital) with the growth in complexity of capitalist society. Similarly the particular configuration of forms and circuits of capital representing property development, and the particular structure of institutional agents which constitute this configuration at the empirical level is historically determined (and therefore continually changing) within particular social formations. Of themselves therefore, the schema, and possible configurations tell us little about changing structures and processes but serve, rather, as an analytic framework through which to interrogate the empirical, observable level.

IIa Property development

Property is normally produced in the course of a circuit of industrial capital but with two specific conditions. First, land is an essential basis and is, through private control over rights of usage, monopolizable; this may enable the appropriation of excess profit earned by the users of the property produced as ground rent (see later).

Second, thinking in terms of major office developments, town-centre redevelopment or entire industrial estates, property is a very high-priced commodity requiring a long production period before a single saleable unit is produced; it thus requires that very large sums of capital be assembled in particular enterprises for considerable time periods before any of the industrial capital advanced can be realized in money form again. The corollary of this is that if the enterprise producing property is immediately to realize in full the capital invested in the production process, users of the property will similarly be required to devote a major sum of capital to purchasing the development outright, pre-empting investment in their specialist function such as manufacturing. In practice we find, however, a relatively high degree of functional differentiation between different capitals in the development process reflecting in part the growing complexity of capitalist society and in part the degree of centralization of capital which has developed over time within different forms of enterprise.

In a typical sequence of events in the development process a property company will obtain short-term development finance, purchase land, and arrange the construction of property thereon by a builder working under contract to the property company. The completed property is handed over to the property company which pays the builder for the work done. The core of the process is thus a circuit of industrial capital managed by the building contractor; the contractor will often borrow at least part of the building finance from the property company or possibly from a third party – usually a bank, the circuit of industrial capital being thus integrated with a circuit of interest-bearing capital.

The property company, in its turn, will usually have borrowed at least part of the necessary development finance covering land purchase, fees, payments to the contractor, etc.

The importance of different sources of development finance has varied considerably through time, according to their relative cost and availability. Mortgage-loans to property companies were the classic source of development finance in the 1950s and early 1960s. The normal loan was two-thirds of the market value of the *finished* property; the interest rate was generally tied to the yield on gilt-edge stock equivalent in length to the loan term, plus 1–2 per cent. Insurance companies and, to a lesser extent, deposit-taking banks and other institutions were the usual source of loans. Since the amount loaned was based on the estimated final market value and no interest was charged until completion of the development, developers could raise the entire cost of development with virtually no contribution of their own capital. However, in the latter half of the 1960s, declining market values and rising mortgage interest rates made this form of building-finance impossible and stimulated alternative forms of funding. The common feature of these alternatives was that institutions tended to take a more direct interest in the assets (property) created or the income (rent or share dividends) those assets generated, rather than simply providing mortgage-loans secured against the value of the development.

Two main alternative forms of funding developed. First, mortgage-loans or debenture (fixed-interest security) finance advanced to property companies at below the market interest rate in return for the right of the financing institution (e.g. an insurance company) to buy shares in the property company and so, in theory, benefit from the share dividends and increase in market value generated by the development. For example, in 1973 the Friends Provident Life Office loaned Haslemere Estates £10m for approved developments over five years at 6¾ per cent repayable after thirty years, with an option to subscribe up to 500,000 shares at 355p per share anytime before the end of 1978.[7] Second, the 'sale and leaseback' mechanism developed whereby a property company owning a piece of land would sell the freehold to a financial institution in return for, first, a loan to finance building offices on the land (interest being deferred until completion) and, second, agreement to lease the completed development back to the property company. The company pays rent to the institution providing the latter with its normal rate of return; the company then sublets the property to an occupier. Through sale and leaseback the financial institution acquires a stake (the freehold) in a *specific* development whereas through buying property company shares it acquires a stake in the *general* equity of the company.

In the early 1970s borrowing from banks became cheaper than debenture or other forms of finance for development and it rose substantially – from £362m in February 1971 to £2584m in February 1974 (amount outstanding).[8] This expansion was based on both the clearing banks and on fringe banks and finance houses; clearing-bank lending rose 470 per cent in 1970–5 and 'other' bank lending by 530 per cent. Bank finance was freely available up to 1974/5 and little short-term building finance was raised by any other method until the revival of the new share-issue market in 1975/6. Bank lending has been almost entirely short-term finance for development, with some longer-term lending in 1971/2, mostly recalled as interest rates rose. Like other financial institutions, banks generally attempt to match liabilities with assets of similar term or 'maturity' so that the general flow of withdrawals is covered by the flow of loans being repaid. Since the maturity of bank deposits ranges from zero to, at most, five to seven years they are reluctant to lend out money for longer periods, so while banks are an important source of short-term, building finance they are not drawn into long-term property *investment*.

The development process and finance for development is thus typically orchestrated and directed by the property company. Having completed the development the company may, if its financial resources allow, retain it as a long-term investment having paid off any short-term development finance raised earlier, selling the right to use the development to an occupier in return for a stream of rental payments. The company thus realises its investment over an extended consumption period. Alternatively the company can sell the completed development typically to an insurance company or pension fund seeking an outlet for investment funds (usually having assured a rental income by finding a tenant for the development) – the form of long-term investment is taken up again in the following sub-section.

In either case, the property company operates a circuit of commercial capital generally based, in part, on a ciruit of interest-bearing capital. In schematic form the development process may be summarized in the following form:

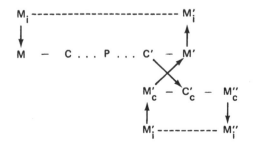

Interest-bearing capital
loaned to builder

Industrial capital
directed by builder

Commercial capital
directed by property company

Interest-bearing capital
loaned to property company

IIb Property investment

The dominant investors in commercial and industrial property are the insurance companies and pension funds; property unit trusts, property bonds and various charities account for a smaller though significant share. Institutions may invest *indirectly* by making long-term loans to, or purchasing shares in, property companies which themselves hold a portfolio of property as a long-term investment, or *directly* by themselves purchasing properties from property companies to hold as investments. The return on indirect investment takes the form of interest or dividends and does not involve any direct rent relationship; that on direct investment is received as rent and derives from ownership of land. Since the last war indirect investment has increasingly taken the form of share-purchase rather than loans so that institutions have become more directly tied to property companies, while direct investment has expanded in relation to indirect investment as a whole.

INDIRECT INVESTMENT   Short-term loans to property companies are generally made to finance the construction of particular buildings but institutions also provide a large proportion of property company funds in the form of long-term loans (Table 11.1). These may be tied to specific 'approved development' or may in effect contribute to property companies' general funds devoted to long-term investment as well as building and development.

Over half of property companies' funds are raised as long-term loans

Table 11.1 Sources of property company funds 1968/9—1972/3

	For nine major property companies, which accounted in 1970 for about 50 per cent of total property companies' assets		Estimate for all property companies, assuming the same structure for sources
	£m	% of total	£m
Internal:			
Retentions	14.4	0.9	26.6
Depreciation	5.5	0.3	10.1
Tax and dividends	23.3	1.4	43.0
External:			
Equity capital	195.7	11.9	361.0
Preference shares	−0.2	−	−0.4
Long-term loans	831.9	50.4	1,534.9
Short-term loans	280.2	17.0	517.0
Creditors	75.7	4.5	139.7
Other sources	224.1	13.6	414.5
Total	1,650.6	100	3,046.4

Source: Property, Hoare and Co. Govett, London (1972).

mainly from insurance companies.[9] Share-issues are a significant source of long-term funding but internal finance is relatively insignificant.

Turning to shareholding, a recent survey listed twenty-seven property companies where more than 10 per cent of shares were owned by financial institutions, in some cases several different institutions each holding over 10 per cent (see examples in Table 11.2). Many holdings arose when institutions took up options to buy shares given when building finance was loaned to property companies. Institutional shareholding in property companies appears as 'ordinary shares' and 'equities' in Tables 11.4 and 11.5, rather than (and therefore additional to) 'land, property and ground rents'.

Table 11.2 Institutional shareholding in property companies

Company	Institution	% of equity held
		%
Aquis Securities	Guardian Royal Exchange	64.5
Berkeley Hambro	Prudential Assurance	13.0
	Hambros	42.6
Capital and Counties	Union Corporation	24.6
Haslemere Estates	Hill Samuel	16.3
	Phoenix	12.2
	Throgmorton Securities	18.0
Sunley (Bernard) Invst.	Eagle Star	33.2

Source: Panmure Gordon and Co. (1976), see note 7.

Insurance companies in particular appear to have been particularly active since 1971, acquiring majority and 100 per cent interests in the shares of individual property companies.[10] Table 11.3, taken from a recent survey by Franklin, lists eleven property companies where insurance companies acquired 100 per cent of their shares from 1971–5. Acquisition on this scale is a relatively recent phenomenon and reached a peak in 1974/5. Share purchase by financial institutions merges the circuits of capital operated by property companies and the institutions into a single institutional capital.

DIRECT INVESTMENT   Institutions invest in freehold and leasehold property to obtain investment yield in the form of rental payment from the occupier. Yields will normally be compared with those on, for example, company shares and fixed interest securities (government stock, local authority bonds, etc.), the relative yield influencing the volume of new investment drawn into long-term property investment. The appreciating capital value of (freehold) land and property, generally maintaining the value of institutions' assets in the face of inflation, has also encouraged investment. Before the last war insurance companies and pension funds invested pri-

Table 11.3 Property companies acquired by insurance companies 1971–5

Date	Acquiring company	Acquired company	Purchase price £ m
July 1971	Commercial Union	Holloway Sackville Properties	20.4
July 1971	Guardian Royal Exchange	Metropolitan Railway Estates	5.3
November 1972	Commercial Union	West Bar (Leeds) Estate Co.	0.3
March 1973	Commercial Union	Weatherall Property and Services	2.2
March 1973	Guardian Royal Exchange	Tanway Properties	7.3
August 1973	Royal	Sterling Estates	13.1
September 1973	Prudential	Edger Investments	24.6
October 1973	Legal and General	Cavendish Land Co. Ltd.	45.1
1974	None		
March 1975	Eagle Star	Grovewood Securities	14.7
October 1975	General Accident	Brighton, Worthing and District Property and Investment Corpn (BPIC)	3.7
December 1975	Pearl Assurance	New London Properties	5.3
		Total	142.0
		Average	12.9

Source: Franklin (1976), see note 10.

marily in fixed-interest securities, both company and British government, etc., stock. After the war investment in shares increased as growth of dividend-income and capital values exceeded the return on fixed-interest securities. From the mid-1960s, however, much greater emphasis was placed on investment in commercial property where rental yields and capital values were rising rapidly, continuing until the 1974/5 slump. The increased importance of land and property in institutional portfolios since the mid-1960s is evident in Table 11.4. In purchasing completed developments from property companies, institutions realize the commercial capital invested by these companies in the development process and take upon themselves the extended sale/realization period by letting to the final occupier.

Pension funds and insurance companies meet all their liabilities from income and capital appreciation on their investments. The 'Pilcher Report' on *Commercial Property Development* describes their property investment as follows:

'The practice of the insurance companies and pension funds is to try to match liabilities with investments of a similar term. The average term of life insurance liabilities is long and the average length of liabilities of pension funds tends to be longer still. They need to be balanced by long-term assets. Before the last war, the assets required for this purpose were generally long-dated or irredeemable fixed interest securities. Since then, property, in the form of freeholds or very long leaseholds, has been in increasing demand from the

Table 11.4 Investment holdings of insurance companies and pension funds

		Land property, ground rents		Ordinary shares		British govt securities		Other[a]		Total
		£m	%	£m	%	£m	%	£m	%	£m
Insurance[b]	1964	851	9	1,933	21	2,189	24	4,159	46	9,132
companies	1974	3,904	16	5,051	21	4,877	20	10,525	43	24,357
Pension[c]	1964	176	4	1,964	41	979	20	1,735	36	4,854
funds	1974	1,594	12	3,396	37	1,213	13	3,066	33	9,269

Source: Financial Statistics (HMSO)

[a] Includes non-government public sector fixed interest securities and private sector securities other than ordinary shares.

[b] Book value.

[c] Market value.

institutions for this purpose because of its superior return and in most conditions the comparative ability of its income and capital value to keep pace with inflation. This consideration is of increasing importance as a result of the trend towards maintaining the value of pensions in real terms. Because returns have been maintained for many years in real terms and because of the steady annual growth of premium income and of contributions, it has been rarely, if ever, necessary for the institutions to realise property assets in order to meet liabilities. The institutions have therefore come to regard their property assets as in effect permanent and their energies have been devoted to maintaining and increasing the annual income from these assets in the long term.'[11]

It is this same steady growth of funds and lack of need to realize assets that has led the institutions into investing in agricultural land and the art/antique market. Figures for the net flow of investment by pension funds and insurance companies (Table 11.5) demonstrate the increased importance of land and property, but also indicate the considerable variations from year to year in the split of investment between the three main heads.

Direct investment has been encouraged by the tax position of financial institutions since the introduction of corporation tax in 1965. Property companies generally pay tax at 52 per cent and pay out share-dividends after tax, whereas the tax position of many institutions is such that they are not liable to corporation tax. Direct investment in property rather than indirect investment through purchasing property company shares is thus more efficient for the institutions. They are similarly more favourably placed through exemption from capital gains tax. Their advantage over property companies is reduced, however, in the case of companies with an active development programme since interest on money borrowed for development may be set against tax liability (i.e. they are not taxed on income used to pay loan-interest).

Table 11.5 Net annual investment, insurance companies and pension funds £m (percentage of total net investment)[a]

Year	Insurance companies[b]			Pension funds[d]		
	Fixed interest[c] securities	Equities	Land property, ground rents	Fixed interest[c] securities	Equities	Land, property, ground rents
1965	467 (72)	86 (13)	89 (14)	162 (59)	102 (37)	23 (8)
1970	292 (29)	271 (27)	198 (20)	24 (7)	165 (47)	70 (20)
1971	663 (53)	376 (30)	198 (16)	199 (54)	170 (46)	72 (20)
1972	510 (31)	710 (43)	131 (8)	−42 (−10)	276 (68)	99 (24)
1973	677 (40)	387 (23)	307 (18)	155 (28)	126 (23)	102 (18)
1974	346 (20)	37 (2)	353 (42)	78 (11)	93 (13)	113 (16)
1975	1,845 (74)	414 (17)	406 (16)	445 (45)	668 (67)	200 (20)
1976	1,855 (61)	306 (10)	449 (15)	572 (46)	409 (33)	248 (20)

Source: Financial Statistics (HMSO) and Business Monitor M5 (HMSO)

[a] Net, therefore not summing to 100 per cent. Note, the three headings do not sum to total investment.

[b] 1965 figures exclude Commonwealth Life Companies. Also, revised basis after 1974, therefore not strictly comparable.

[c] Includes British govt, local authority, overseas govt securities; debentures, preference shares, loans and mortgages.

[d] Private only.

Direct investment and increasingly direct involvement in development finance has been further encouraged by the desire to participate in the growth of market values reflected in share prices and book values of directly held property, and in rental growth as a source of income. In part this represents a response to the impact of rising inflation rates on institutions' asset values and income in real terms. More significantly, the increasingly direct participation of institutions in the property sector represents 'vertical integration' within the development process and growing centralization of capital under the control of the institutions. This enables institutions both to capture a larger share of 'developers' profit' and to control the nature and progress of development more closely.

IIc The function of commercial and industrial property in the economy

So far this account has stressed the way property functions as a financial asset and a source of income. Many 'radical' analyses have emphasized the speculative profit derived from offices and town-centre redevelopment, the escalating asset value of property bought by financial institutions, or drawn attention to the scandal of over-provision and empty office blocks – examples are Counter Information Services *The Current Crisis of London* and *Your Money and Your Life,* Marriot's *The Property Boom,* or the Labour Party National Executive Committee's analysis in *Banking and Finance.*[12] These analyses all focus on the circulation of commercial property and the sequence of exchanges in the property *market.* It is as if values are created out of thin air. According to the second CIS report: 'The gigantic increases in asset values are *totally dependent* on what the insurance companies and pension funds are prepared to pay for the properties' (my emphasis).[13] This is of course nonsense. For, though a certain amount of monetary inflation can push up property prices for a while, what the institutions are prepared to pay for the properties depends on the relative return yielded by that investment, that is, on the rent paid by the occupier of that property. Income from property and the market value of offices depends ultimately on rent levels, and empty office blocks make financial sense only under occasional particular circumstances. It is worth turning, briefly, to the role of property in the economy in relation to the actual consumers or occupiers of office space, and the origin of profit from property development and investment. This will clarify some of the implications of the empirical analysis which are raised in the final section.

The property sector has been analysed above in terms of the three basic forms of capital identified by Marx, industrial, commercial and interest-bearing. Lamarche has argued that 'like commercial and financial capital another specialized capital exists with the sole function of planning and equipping space in order to increase the efficiency of commercial, financial and administrative activities';[14] this he terms 'property capital'. The argument is based on the assertion that 'space is a more important factor in time and labour than the capital invested in circulation as a whole'.[15] Since there is no way in which the relative 'importance' of 'space' and the amount of capital 'invested' as factors in circulation can be measured, this argument appears ill founded. According to Lamarche property capital 'plays the same role at the level of property as does commercial capital at the level of movable goods: buying in order to sell at a higher price $(M - C - M')$. Its own commodity is floorspace . . .'.[16] But there is no justification for distinguishing commercial capital functioning to realize the commodity *property* from that realizing any other commodity such as shoes; the qualitative distinction made by Lamarche between 'property' and 'movable goods' is insufficient to establish 'property capital' as a separate category. However, Lamarche's analysis does pose the question of the origin of profit reaped by commercial capital in the property sector.

At first sight one might argue that commercial capital directed by property companies 'contributes to the speeding up of the circulation of the capital invested in the building industry and thereby benefits from a part of the surplus value created within that industry'.[17] Profit on commercial capital would thus be explained with reference to the industrial capital whose productivity it increases. Lamarche raises two objections to this.

First, capital invested in property construction is of relatively low productivity. It is fragmented into relatively small production units, lacks monopoly pricing, exhibits technical lag in relation to other sectors of manufacturing industry, and is fragmented between many different trades. The relatively low productivity of capital invested in property construction would thus not seem to justify the existence of commercial capital sharing in the mass of surplus produced. This argument is appealing though not totally convincing – 'productivity of capital' is impossible to measure directly; production of large office developments by major construction companies appears 'relatively efficient', subjectively speaking; finally, the emergence of a secondary circuit of commercial capital in any branch of production must depend not on whether industrial capital produces enough surplus to 'share' with commercial capital but on the increase in total surplus produced brought about by the intervention of commercial capital.

Second, the proposition implies that property is sold at a price based on its initial production cost and that profits originate solely from the surplus value produced at the time of construction. Against this Lamarche argues, correctly, that profit from property development and investment originate not so much from the construction process but from the difference between the construction cost and what is paid in rent for the use of the premises. Rents and prices (capitalized rents) of commercial and industrial property commonly bear little relation to the original construction cost. They are strongly determined by replacement cost and by supply and demand factors over the life of the property. Rental levels on a five-year-old office block may bear little relation to its original cost of production, let alone the case of a fifty-year-old property on which continuing rental payments have repaid the original capital cost of production several times over, where there have been frequent rent reviews, and which has been bought and sold between several different circuits of commercial capital. The extended and complex consumption stage acquires a large degree of autonomy from the original circuit of industrial capital and the surplus produced therein. Furthermore the commodity originally purchased by commercial capital continues to yield a profit apparently without reference to profit on the original industrial capital. Where, then, do these profits originate? The answer to this lies in the function of commercial and industrial property in the economic system as a whole.

At the level of the economy as a whole, costs incurred in the circulation of capital and the coordination of the economic system are reflected in the overall rate of profit. Property development and investment equips space and creates physical infrastructure so as to facilitate the circulation of capital in its various forms, the exchange of information, and the physical and legal transactions on which commercial, administrative, governmental and financial functions are predicated. The proximity and articulation in space of the unit of built form within which these functions are facilitated, speeds up the circulation of capital and facilitates the control and coordination of commodity production; it thereby lowers the costs of circulation and coordination, increases the total surplus produced and hence raises the average rate of profit in the economy as a whole. This is reflected anecdotally in the use of City of London office space shown in Table 11.6.

Table 11.6 Users of city office space (percentage of total)

	%		%
Banking	25	Printing and publishing	5
Insurance	15	Commodities	3
Public sector	9	Professional and scientific	
Stockbroking and other		services	7
financial services	7	Wholesaling	3
Shipping	7	Other	19

Wholesale and retail developments reduce the costs of circulation of commodities and maximize the turnover of consumer goods. Enterprises may benefit individually from the efficiency of scale and collective service provision in large-scale developments. Finally, the evolution of an autonomous property sector displacing owner-occupation of office, wholesale and retail or industrial premises is likely to generate benefits first, from functional differentiation and specialization of different capitals and second, by releasing occupiers' capital resources which would otherwise be tied up in the premises occupied, the occupier instead being able to pay for the use of the premises by rental payments over time. It is these various savings in costs, and increased efficiency for which occupiers pay and which determine the level of rent paid.

Rental payments relate to the excess profit which enterprises anticipate will be generated by occupying particular property developments in particular locations. And the major part of the profit on property development and investment originates as a share of this excess profit captured by virtue of property companies' ownership and control of physical infrastructure. The return on commercial capital in the property sector yielded by its ownership and control of space take the form of ground-rent (in the Marxian sense), 'that economic form in which landed property is realised'.[18] If, returning to Lamarche's earlier argument, there exists a specific 'property capital', its specificity is defined by the form of the return on such capital, namely ground rent. This remains, however, an inadequate basis on which to establish a fourth form of capital alongside industrial, commercial and interest bearing. These three forms are defined not by the nature of the return they receive but by the manner in which they circulate. As argued earlier, capital directed by property companies circulates as commercial capital; the fact that in the commodity form it takes the shape of an office block does not distinguish it from commercial capital directed by, for example, a shoe merchant.

III Implications and conclusions

In the previous sections the increasing scale and increasingly direct nature of investment by financial institutions has been set out, and the articulation of different agents in the process of property development and investment elaborated. This final section sets out a number of the economic and political implications of these structural changes in the property sector and in its relations to the wider economy.

1 The extent and increasingly more direct involvement of financial institutions in commercial property, coupled with the interlinkage of financial institutions as a whole, exposes the financial system to the effects of booms and slumps in property values and the interaction of demand and supply in the property market – which may themselves relate in turn to interest rates in the finance market and relative yields on different forms of investment. To quote *The Banker* commenting on the repercussions of the slump in property values in 1973/4:

> 'Virtually no corner of the City has emerged from the past twelve months unscathed. The secondary banks have been decimated, and the Bank of England and the clearing banks are still labouring under a rescue operation that has been transformed from the comparatively simple task of recycling liquid funds to the secondary banks to the far more onerous job of propping up fundamentally weak fringe institutions. At the same time the crisis of confidence and liquidity has spread far further than anyone could have guessed, to engulf some of the leading finance houses . . . Even the insurance industry, for so long a bastion of security, had its problems. Difficulties of Nation Life, Welfare Insurance and London Indemnity and General forced all investors to think again about just how safe their savings were.'[19]

The 'lifeboat' operation launched by the Bank of England, though relatively successful, required a massive injection of public money to stabilize the financial superstructure. Banks had to make substantial provision for writing-off losses due to default on loans to property companies, and £3,000m was estimated to be outstanding in bank loans to property companies in 1975.[20] Certain features of the early 1970s 'boom and bust', it may be argued, were unique. But these events undoubtedly drew attention to the critical nature of the links between the financial system and the property sector. Discussing insurance company ownership of property companies, Franklin has drawn attention in particular to 'the nature, number and extent of formal interdependencies existing and being forged between insurance and property companies (and between banks and insurance companies)' suggesting that:

'Economically, the considerable concentration of financial and real assets among few companies is a potentially worrying feature of the contemporary financial system . . . the more extensive the links, possibly the harder it will become to ensure the survival of the system in fact should another fundamental crisis of confidence occur.'[21]

In October 1977 the former president of the British Property Federation was again warning on the front page of *Estates Times* that recent trends in bank lending risked another property crash.

2 In 1975 the 'Pilcher Report' stated:

'Since the assumption is that most, if not all the funds for commercial development will in the foreseeable future be provided by the private sector, conditions must be secured in which these institutions are willing to supply the bulk of the monies needed.'[22]

As this report noted, this assumption embraces most of the commercial (and industrial) activities of private development companies and of local authorities and, to an increasing extent, new-town development corporations. Thus urban redevelopment, inner-city regeneration, regional policy and industrial strategies will be increasingly influenced in their feasibility and form by the investment criteria of the financial institutions to the extent that these policies and processes rely on institutional funding. The attraction of funding such investment (short or long term) will be assessed, and the terms on which money will be forthcoming determined in relation to gilt-edged stock, industrial company shares, agricultural land, etc. Thus for example office development or inner-city regeneration which fails to show an adequate return on strict financial criteria will be funded only if subsidized by public money. A more specific issue arose, for example, with the government's requirement in the Community Land Act[23] that local authorities sell land to the private sector leasehold and with a presumption for ninety-nine-year rather than longer leases – institutions require a higher percentage yield when investing in development on leasehold as opposed to freehold land and on shorter rather than longer leases – an institution might require that development on a ninety-nine-year lease yield 2 per cent more than equivalent development on freehold land and 1 per cent more than a 125-year leasehold development.

3 In *Banking and Finance* the Labour Party National Executive blamed Britain's poor industrial performance in part on the preference of financial institutions for 'non-productive' investment, particularly the upsurge in property investment in the early 1970s.[24] At the time, in fact, industrial companies were not particularly short of loan-finance and many were themselves investing in government stock, etc., rather than new productive capacity, given the low rate of return to industrial capital. But a boom in the industrial sector could bring the productive industrial sector into more direct competition for institutional finance with the unproductive (or 'indirectly' productive) commercial property.

4 Since rent paid to financial institutions investing in property directly, or indirectly via property companies, originates as surplus produced by industrial capital, financial landownership is increasingly placed in a contradictory relationship with industrial capital over the division of the total surplus. Increased rents may affect industrial companies directly if they rent office space and indirectly if they use the services of businesses occupying office space, and passing on in their charges the higher rents.[25] *Banking and Finance* raised the issue of investment 'diversion' which, though it may have implications for long-term industrial profitability does not pose an immediate contradiction between fractions of capital as does the division of excess profit into profit retained by industrial capital or captured as rent by the property sector.

5 The contradictions between public planning and private land-ownership and appropriation of ground-rent was demonstrated vividly by the effects of efforts to control office development in particular in London in the 1960s. The shortage of space relative to demand created by the restriction of new office building after 1964 through the use of Office Development Permits (the 'Brown Ban') ensured the massive profitability of those developments completed since the relaxation of the post-war restrictions.

6 The growth of financial institutions' interest in land and property has changed the structure of opposition to more fundamental moves designed to control development gain when planning permission is granted, to intervene in the development process (as in the Community Land Act) or, ultimately, to nationalize land. While such moves may not threaten the actual existence of the financial institutions (since they could theoretically switch to other sectors) it would remove or reduce the attraction of a sector on which they have depended heavily for assets and investment yield. The effect of such proposals on the flow of finance into urban development (long and short term) is an increasingly significant deterrent to radical measures – viz. the extent to which the Community Land and Development Land Tax Acts were watered down to become ineffectual at least in the short term. Financial institutions would also be affected by any such measures which affect property companies' development and investment activ-

ity which, as Massey and Catelano point out, are likely to be more severely affected due to their more direct and exclusive dependence on investment in landownership and property.[26] The changing structure of ownership and investment in the property sector is thus reflected in changing relations between groups of owners and investors, and the State.

Notes

[1] Mollenkopf, J. H. (1978) 'The Postwar Politics of Urban Development', in W. K. Tabb and L. Sawyer, *Marxism and the Metropolis.* Oxford: Oxford University Press, p. 121. See also, for a more general account of the underlying structural changes, E. Mandel (1975) *Late Capitalism.* London: New Left Books, Chapter 12, 'The expansion of the services sector, the "consumer society", and the realisation of surplus value'.

[2] Marriott, O. (1967) *The Property Boom.* London: Hamish Hamilton.

[3] Mandel (1975), Chapter 6, 'The specific nature of the third technological revolution', and Chapter 8, 'The acceleration of technological innovation'.

[4] See *Capital,* Vol. II, Chapter 1.

[5] Thompson, G. (1977) 'The relationship between the financial and industrial sectors in the United Kingdom economy.' *Economy and Society* 6, 3: 239–40.

[6] See *Capital,* Vol. III, Chapter XVI, and Vol. III, Chapters XIX and XXI respectively.

[7] Panmure Gordon and Co. (1976) *The Property Sector 1976–1977.* London: Panmure Gordon and Co., p. 41.

[8] Panmure Gordon and Co. 1976: 44.

[9] At the end of 1976, £3,157m were outstanding in loans and mortgages to insurance companies of which just under a half may have been loans to property companies *(Financial Statistics.* London: HMSO).

[10] Franklin, P. J. (1976) 'Insurance into property.' *The Banker* 126: 1127–9.

[11] Department of the Environment (1975) *Commercial Property Development.* London: HMSO, report of an advisory group chaired by Sir Denis Pilcher, p. 23.

[12] Labour Party National Executive Committee (1976) *Banking and Finance.* London: LPNEC. Counter Information Services (1973) *The Recurrent Crisis of London.* London: CIS. Counter Information Services (1974) *Your Money and Your Life.* London: CIS. Marriot 1967, note 2.

[13] CIS 1974: 11, note 12.

[14] Lamarche, F. 'Property development and the economic foundations of the urban question', in C. G. Pickvance (ed.) *Urban Sociology: Critical Essays.* London: Tavistock, p. 91.

[15] Pickvance (ed.) 1975: 91.

[16] Pickvance (ed.) 1976: 93.

[17] Pickvance (ed.) 1976: 93.

[18] *Capital,* Vol. III, Chapter XXXVII.

[19] *The Banker* (1975) 'The City – Annual Survey' (February).

[20] Massey, D. and Catelano, A. (1978) *Capital and Land.* London: Edward Arnold, p. 128.

[21] Franklin 1976, note 10.

[22] Department of the Environment 1975: 21, note 11.

[23] Legislation enacted in 1975 allowing local authorities to acquire development land in order to sell it on to the private sector intended to enable authorities to stimulate development through 'positive planning'.

[24] Labour Party 1976. Note 12.

[25] See Massey and Catelano 1978: 153, note 20 and Chapter 7.

[26] Massey and Catelano 1978, Chapter 8.

12 The new international division of labor, multinational corporations and urban hierarchy

R. B. Cohen

The international economic relationships which dominated the immediate post-war period have been overturned. While the most noticeable change in the 1970s has been the rising power and wealth of the OPEC nations, a more significant transformation has occurred through the elaboration of an international system for producing goods and services primarily accomplished by the largest multinational firms. A key element in this new order is the growing importance of the newly industralized nations, or NICs, as centers for productive activities. This essay will examine the new international division of labor and discuss how it has affected corporations, financial institutions, national economies (in both the developed and developing world) and the world's urban centers. By linking these international changes to domestic problems, this review should provide planners, geographers and urban economists with a framework with which to consider future world economic changes and their impact on cities.

The discussion which follows is divided into four sections. The first section defines the new international division of labor (NIDL) and delineates the difference between it and previous divisions of labor in the world economy. This part reviews the contradictions which gave rise to the NIDL, the contradictions inherent in the NIDL itself and the spatial ramifications of the NIDL. The second section reviews the changes which have taken place in the modern corporation. Many of these changes have been in response to the NIDL and the more complex economic environment it has created for corporations. This section focuses upon the new corporate demands for services from banks, law firms and accounting firms and links these new demands to both changes in the nature of institutions providing these services and to the emergence of a series of global cities. Such global cities act as centers of corporate control and coordination for the new international system. The third section explores the dimensions of the emerging world urban hierarchy of corporate headquarters and sophisticated corporate services, placing its main emphasis on the US, but drawing some parallels with developments in Western Europe. Finally, the fourth section explores a number of possible future trends in the new international division of labor and urban hierarchy.

The new international division of labor (NIDL)

Definition

What is the *new* international division of labor (NIDL) and what distinguishes it from past patterns of organization of the world economy? Until recently, the international division of labor reflected differences in *trade* between firms producing goods in different nations. In large part, the division of labor was founded *upon the sourcing of raw material* inputs largely from the southern hemispheres (Africa, Latin America, South-east Asia) and Mid East, for use by industrial companies in developed nations, mostly in Western Europe, Japan and the US. However, the *new* international division of labor reflects a number of transformations of the world economy. First, it is based mainly upon the *international spread of manufacturing* with trade occurring more frequently between subsidiaries or joint-ventures of global corporations producing goods in different parts of the world.[1] This has given rise to the *international spread of corporate-related services,* including multinational banks, law firms, accounting firms, advertising firms, and contracting firms. Perhaps one of the most important features of the present world economy, is the existence of a *system of international financial markets,* less subject to regulation by national-based central banks, but rather tied to the needs of major international firms and large multinational banks.[2] As currently constituted, these international financial markets are no longer closely linked to supporting trade between firms but have increasingly supported the expansion of the manufacturing systems of large multinational firms.

In essence, what the NIDL represents is a system for production on a world scale in which even greater numbers of people are integrated into activities carried on by large international producers of goods and by international firms which service these producers. Both the work process and the facilities used to produce goods and services are organized according to the demands of firms operating in a world market. In many cases, this integration of production and corporate services on a world scale has drawn increasing numbers of people into an industrial and service work-force, greatly diminishing the importance of agricultural and handicraft production.

But in perhaps its most profound impact, the NIDL has integrated qualitatively different types of laborers with very different levels of work experience, highly varied types of social backgrounds and vastly divergent histories of labor organization into corporate organizations which operate on a world level. Thus, the NIDL may be characterized as a complex hierarchical system which integrates different types of useful forms of labor of individual producers carried on under the aegis of large, highly-integrated international companies. This hierarchical system reflects vast changes that have occurred in the conditions of production and exchange on a world level, in particular, the integration of entirely new groups of laborers into a rapidly expanding world market characterized by *internalized systems of production and exchange* created by the largest multinational firms.

However, it is not simply the use of different types of laborers, receiving divergent levels of wages and located in various parts of the globe which define the NIDL. It is also the ability to produce semi-finished goods and to deliver different types of services in various parts of the world and to link them to an overall system of providing goods and services to a world market. Thus, we need to keep in mind the diverse ways by which multinational firms and service organizations participate in the new world economy. Multinational corporations take advantage not only of qualitative differences between laborers and differences in wages, but also utilize the benefits which they may obtain by operating in free trade zones (Volkswagen's Puebla plant in Mexico), participating in joint or multipartite ventures, operating via barter agreements (as in Coca Cola's vodka-cola swap with the Soviet Union) or utilizing management contracts.

Here, multinational corporations are reacting to both (i) the need to create new means to control more complex and farflung operations through new types of profitable arrangements; and (ii) the necessity to position themselves so they may react with greater flexibility to economic and political changes worldwide.[3] In some cases, such as in the petroleum industry, the new conditions have forced multinational firms to shift investment from certain initial stage processes to final stage refining, and in many cases, to make the more controllable final stages greater centers for profits than before. This latter type of behavior was adopted by most US oil companies following the OPEC nations' oil embargo.

In addition, international firms have become greater providers of services than ever before. Through management contracts, the construction of turn-key factories and barter arrangements, firms have placed greater reliance upon services in generating profits. In these ventures, the management of new plants or preexisting operations represents the major way in which multinational firms obtain profits. They are not responsible for the sale of goods in any particular market, but only in charge of managing specific facilities.

Besides the need for increasing their control and flexibility, multinational corporations also must participate in the restructuring of many of the industries in which they operate, particularly as the developing nations become major sites of industrial production and operations in developed nations become relatively less profitable.

Thus, multinationals have demanded more support from what I have called the advanced corporate services (primarily banks, investment banks, law firms and accounting firms). These services now help companies to develop their overseas operations, to acquire dynamic companies, to restructure their industries, and to adapt to political and economic change. Thus, these service firms enable corporations to position themselves to obtain profits even in a disorderly world. The ways in which corporations have transformed the role of advanced services is discussed in greater detail below. Thus, in large measure, corporate change has been propelled by the new international division of labor.

Many of the world's major corporate service organizations have responded to the NIDL by creating new structures which broaden their opportunities for new profits. In banking, what began as the Eurodollar market – a non US-based market for dollar lending, nearly unaffected by the regulations of domestic monetary authorities – has become a worldwide system of capital markets, largely serving the needs of major world corporations and major international banks. Singapore, the Bahamas, Panama, Bahrain and Luxembourg have become way-stations in the complicated financial transactions that are the lifeblood of the new world economy. This financial network facilitates the management and development of ever more farflung centers of operations by multinational corporations. It also provides them with greater flexibility in their adaptation to any new political and economic changes which occur in different parts of the world.

Contradictions and conflicts which gave rise to the NIDL

What forces led to the development of the new structure of the world economy? How did the NIDL respond to these forces, creating a very different hierarchical system from the one which had existed previously?

This growing movement internationally has resulted not merely from the desire of firms to utilize less-expensive sources of labor and more profitable situations for production. It has also been in response to a number of new trends in the international economy. First, this movement has resulted from the greater desire of corporations to obtain more flexibility and control over their operations, in light of the changing geopolitical situation. In the 1960s, nationalist movements expropriated numerous firms, disrupting production, lowering profits, and creating difficult situations for many companies. In the 1970s, firms responded by both seeking new centers of production with greater guarantees of stability and by utilizing new forms of foreign investment. Second, firms responded to the growing bargaining power of certain developing nations.[4] During the post-war period, negotiators for these nations had little idea of how to deal with international companies and even less knowledge of the impacts such firms might have on their economies. With the rise of OPEC and the formation of other commodity cartels, some of these developing countries demonstrated their ability to gain greater benefits from the operation of foreign firms. But this enhanced bargaining power, while stimulating the industralization of certain countries, also served to draw attention to the great differences among developing nations, particularly between oil-exporting nations and their non-oil-exporting counterparts. Third, firms went abroad to respond to the challenge of growing international competition. No major world firm could rest on profits garnered from its existing markets. Each challenged and counterchallenged its main rivals, often using diversification into highly lucrative fields, particularly via mergers and acquisition, to build further economic power and market positions. Fourth, firms moved overseas to counterpoise the challenge of well-organized labor and government regulations with much more highly profitable operations that were subject to far less government regulation and usually manned by laborers whose output and cost compared most favorably with the situation in developed nations. Such 'run-away' shops did much to weaken the power that existing labor organizations had over investment decisions by corporations in the context of national economies, and forced organized and unorganized labor to face a drastically altered economic world.

While most attention has focused upon the division of labor in the world market, little attention has been paid to internal shifts in firms. Yet it has been suggested that foreign direct investment is positively correlated to a firm's overall efficiency. In addition, studies have demonstrated that large firms specialize in activities where size is of great importance, and where competition is limited because of the considerable resources required. Hymer[5] in particular has emphasized that the internal division of labor in the firm has significant implications for the structure of the international economy. He lists a number of qualitative changes which are linked to the new hierarchy within the firm. These include: (i) the fact that a multinational corporation focuses on its *world market position* rather than its national market position;[6] (ii) the use of foreign investment as a means to restrain competition between firms from different nations; and (iii) the fact that the efficiency of the international corporation hinges on *the direction of change in the world economy,* rather than in the domestic economy.

These changes within the firm also gave rise to additional contradictions and conflicts which resulted in a NIDL. National policies in advanced nations, particularly nations where multinationals were domiciled, were unable to respond to the new transformations in both the world and domestic economies which arose largely as a result of the changes in the world market and changes in the corporation mentioned above. Thus, old, previously safe jobs were lost by large numbers of workers and the nation-state was unable to respond to the resulting 'job crisis,' and unable to create new jobs in traditional blue collar areas. Secondly, there were increasing pressures on the governments of developed nations to regulate the growing changes in the world economy and to preserve the living standards of large sectors of the population. Thirdly, there were increasing pressures on governments to increase protectionism and to control the flow of capital overseas, but in many cases, particularly in the case of capital movements, the new international market structures provided a large number of escape valves for international companies.

Besides the conflicts resulting from pressures upon national governments by labor, there were also important geopolitical factors. The growing flow of technology and productive facilities from advanced to developing nations and the enhanced bargaining power of developing nations led to the rise of new national centers of power. Such nations perceived that they could exert greater leverage over world markets during periods of economic crises. Such moves, while they did not throw the world into prolonged economic stagnation, could certainly be influential in restructuring the world economy. Thus, the rise of OPEC undermined some of the last structural arrangements which had existed since the end of the Second World War. Coming at a time when there had been a long decline in corporate profitability,[7] it stimulated moves aimed at the construction of a NIDL, or a new structure to give shape and stability to a highly reorganized world economy.

The contradictions and conflicts in the NIDL

What arose from the contradictions in the old international system has not been a stable international structure, but rather a NIDL fraught with many major contradictions which have not really been resolved. The present world economy is plagued by a number of key economic, social and political conflicts which will continue to have significant impacts on both international production, foreign trade, and upon domestic economies. These include:

1 The widening conflicts between developed nations (and firms in developed nations), marked by contradictory and often highly competitive approaches to resolving interregional or international problems.[8]
2 The emerging conflicts between policies of different groups of developing nations, not only between oil-producing countries and non-oil-producing nations, but also between different members of OPEC who face extremely diverse development and social welfare problems.[9,10]
3 The rising conflict between developed nations and developing nations, which is reflected not only in the ability of the Western nations to forestall redistribution schemes based upon calls for a new economic order (as seen in UNCTAD IV), but also in the growing impact which newly industrialized developing nations are having on the economies of more advanced nations.
4 Conflicts within the ranks of organized labor internationally, which is torn between pressures for protectionism, the improvement of working conditions in advanced nations and the relocation of productive facilities to developing areas which undercuts the power of organized labor in western nations. With the international spread of productive facilities, additional laborers will be organized and there may be more

cooperation between laborers internationally, posing a far greater threat to international firms than presently exists. But whether this potential new power is undermined by conflicts between workers' demands in developed nations and workers' demands in developing nations remains to be seen.

The spatial ramifications of the NIDL

The new changes which are part of the NIDL have reshaped the spatial hierarchy of economic activities. The three main dimensions of the new world system of production described earlier, namely, the international spread of productive facilities, the international spread of corporate-related services and the rise of a system of international capital markets, have created a new geographical differentiation between developed and developing nations.

In the developed nations, those international centers of business like New York, London, Frankfurt and Zurich have become more significant to the new world order. This is true, as we shall see later, not only because they serve as centers of corporations, but also because they are key centers of corporate-related services and part of the international network of financial centers. Thus, they play a role both as centers for individual corporate strategy formulation, and as centers where decisions which influence the restructuring of the world economy are made. This includes not only decisions about the relocation of activities from more-advanced nations to less-advanced ones, but also economic responses to the restructuring of production in advanced nations which are made necessary because of the declining international competitiveness of certain sectors of production. These latter decisions have become much more political and social in nature, since they mark a turning point in the evolution of industrial society.

In the developing nations, a major difference in spatial development is that while new centers of corporate growth or national development have emerged, few of these are also centers of corporate services and part of the new system of international financial markets. In fact, international financial centers are specialized havens from high corporate taxes, such as Panama, Bahrein, Luxembourg, Singapore and Hong Kong. In addition, although industrialization has advanced in some countries, a large part of the population derives little benefit from it. Often this gives rise to a highly dualistic economy, where most have little hope of escaping from poverty.

Thus, while centers of production have arisen in developing nations, centers of corporate strategy formulation and international finance have not. Particularly in light of the recent evolution of the corporation, this development bodes ill for the ability of the developing nations to control their own future. This is especially true in the area of finance, where private financial institutions have gained a new role as the main providers of development loans.

The changing corporation, corporate services and urban hierarchy

Changes in the modern corporation and its means of operating in the NIDL

During the last two decades, there has been a substantial change in the functioning of the large corporation. This change has been reflected in both the *internal* decision-making and strategy formulation of the corporation and in the external ties between the corporation and institutions which are critical to corporate viability, the banks, investment banks, law firms and accounting firms which I call the advanced corporate services.

INTERNAL CHANGES IN INTERNATIONAL CORPORATIONS
By the 1960s it was evident that corporations were operating in a new economic context – US and foreign businesses were demonstrably more tied to a world market than ever before. This new world market was their primary focus and they viewed corporate operations on a *world*, not a *national level*.[11]

In order to cope with the more complicated nature of worldwide operations, corporations reorganized their head-office structure (Gulf Oil and New York's Citibank are excellent examples of this) to provide better long-range planning and smoother internal management. The NIDL was a major factor in forcing corporations to restructure their internal operations. They not only had to face a larger world market, but also needed to respond to new economic and political challenges.[12] These challenges included the control of more complex operations, the need to obtain greater decision-making flexibility and the requirement that companies participate in the restructuring of industry on an international scale. Such new areas of corporate concern resulted in significant changes in corporate strategy and structure. They also transformed the relationship between the corporation and firms in the advanced corporate services.

With the growth of foreign affiliates, the management of large corporations had more problems with their traditional multidivisional structure. Not only was the international setting a challenge but the spectrum of management functions was also considerably wider than those of most parent concerns.[13] Subsidiaries forced home offices to contend with important new decisions, including:

> 'negotiations with foreign authorities; dealing with supranational bodies; and being able to adapt its objectives, policies, and methods to . . . a particular country . . [while creating] sufficient uniformity and continuity to enable effective coordination and control of the entire multinational structure'.[14]

Thus, as US firms expanded operations overseas, especially during the rapid growth of foreign business after World War II, they faced a set of vexing problems. Operating abroad meant that corporations had different needs than they faced in their rather diverse national operations. Some of the major differences included:

1 Corporations had to control a much more dispersed and diversified system of production.[15]
2 Control over adequate financial information was crucial; exchange-rate risks, interest-rate differences and price-trend diversities offered both 'new threats and new opportunities'.[16]
3 A world orientation had to be achieved among management; executives had to have broader horizons. As one corporate report noted:

> 'We had to shift from the corporate vision of the home country – the US – as our primary market, to a vision of the world as a total market, and all its people as in need of services. Then the world would become a single production/marketing area . . . Eaton would become a world citizen and operate in whatever part of the world is most conducive to our business.'[17]

4 Management often had to learn how to deal with fundamental shifts in strategy; in many cases, this was related to managers' efforts to stabilize the complex international environment in which they operated and the need to devise innovative strategies to meet international competition.[18]
5 New specialists had to be integrated into the corporation to help control more competitive and complex markets and the sophisticated knowledge required by global operations; some investigators have singled out the increased complexity of law, the more expensive cost financing and the instability of foreign currencies as major problems for foreign investors.[19] The need to make decisions about establishing overseas subsidiaries confronted management with the necessity to make judgments about problems which were unfamiliar and not as predictable as those they usually dealt with in the United States.[20]
6 A much more heterogeneous labor force had to be organized and coordinated, and major differences in the national systems of production had to be reconciled.
7 Complicated communications had to be streamlined and made available to those most in need of information. Large corporations operating on an international level had to reconcile the fact that different parts of their organizations retained distinguishably different goals with the need to find 'workable criteria' to evaluate the contribution of each part of the enterprise to the overall performance. In addition, there was usually an increase in intracorporate sales, which had to be managed and coordinated.[21]
8 Planning had to be highly coordinated and controlled in a sophisticated manner.[22]
9 Firms needed to obtain access to enormous amounts of capital.

How did firms adapt to their growing international operations? As Brooke and Remmers realized:

> 'reorganizations show the influence of the foreign operations on the total organization of the company. Once a company commits itself to [having] a considerable portion of its assets overseas, it commits itself to internal upheavals which do not affect a purely national company'.[23]

Thus firms reacted in a number of new ways:

1 market interactions were internalized within the firm, thereby achieving greater control over the business environment; further horizontal and vertical integration was the keystone to growth;
2 an enormous expansion of corporate cash flow was undertaken, which supported the further growth and consolidation of business;[24]
3 policy questions were centralized even further among top management; centralization was propelled by the process of integration;[25]
4 the administrative structure of the firm began to change, initially adding international activities as an appendage or reorganizing operations on a geographical or product basis; as global business grew, multidimensional structures – grids and matrices – came into use, because of the need to increase the coordination of corporate operations and rationalize the flow of information to top management. New groupings emerged within top management, such as the office of the chief executive, which attempted complex decision-making. A single corporate president could no longer command the global firm.[26]

These internal changes made the multinational corporation very different from the multidivisional corporation Alfred J. Chandler, Jr.[27] cites as the model for large corporations since the 1940s. As Peter Drucker has pointed out, General Motors no longer serves as the model for today's organizational needs for at least five reasons:

1 General Motors is essentially a single-product, single-technology, single-market business. Today's business is multiproduct, multitechnology and multimarket and faces the central problem of complexity and diversity.
2 General Motors is primarily a US operation, with appended international operations.
3 Information-handling is not a major organizational problem or concern.
4 General Motors employs 'yesterday's' labor force – manual production workers or clerks on routine tasks. Today's corporations are much more concerned with knowledge work and knowledge workers.
5 General Motors has been a 'managerial' rather than an 'entrepreneurial' business. Its strength lay in its ability to manage, and manage superbly, what was already there and known. Today's organizer is challenged by an increasing demand to organize entrepreneurship and innovation.[28]

The internal changes in the corporation thus facilitated operating under the conditions of greater complexity which characterize the NIDL. They provided for the integration of geographically dispersed operations and for the retention of the vertical hierarchy of command in the corporation. In addition, the internal changes allowed the corporation to become more selective in its strategy, choosing to emphasize control of productive operations where necessary (where high technology, entry into new markets or the development of new products was important), and deciding to operate more flexibly, more like a financial holding company, when investment meant facing considerable risks of losses, expropriation or probable disinvestment.

The shift in analyzing corporate behavior which I want to emphasize here links the strategy of the corporation to the need to adapt to a changing economic environment; it does not emphasize the differences among corporate executives as a means of explaining corporate evolution. Indeed, the survival of the corporation is now much more closely linked to rather different managerial skills than were dominant in the past. Before the 1950s, expanding production and entering new markets were the key elements of corporate success. Now mergers and acquisitions and the use of flexible investment strategies (i.e. seen in the increased use of management contracts, new forms of investment and service-delivery activities) have become central to corporate strategies.

The NIDL, in providing a new context for international business, has been a significant factor forcing corporations to change their strategy and structure. Because the NIDL will probably continue to pose additional challenges to corporations, the transformations and adaptations we have already witnessed will probably continue over the next decade.

CHANGES IN THE IMPORTANCE OF EXTERNAL INSTITUTIONS TO THE CORPORATION[29] Changes in corporate strategy and innovations in corporate structure have provided large firms with the ability to operate in the vastly different business world they face. In addition, these adjustments also have led corporations to change the demands they place on advanced corporate services. But the new demands have been tied to other innovations and changes in the corporation as well. With the rationalization of the firm's financial structure, the financial community – banks, insurance companies and investment banks – and accounting firms have become qualitatively more important. While they often provided support for business operations in the past, they have now become significant sources of information and intelligence that are directly tied to immediate decisions and strategic planning in the executive office. In a number of cases they had provided technical expertise for reorganizing the firm's financial structure (as in the case of accounting firms) or offered ways to manage funds more profitably on an international scale (in the case of banks).

Similar changes have occurred in corporate interactions with governments and other regulatory bodies. Since such dealings or negotiations have the potential to grow into confrontations which could have long-lasting implications for the firm, they are critical to the firm's development. Thus, major law firms, rather than acting as technical specialists handling problems which are ancillary to the main interests of the company, have taken a central role in solving key legal issues which might have significant positive and/or negative impacts on the entire corporation. As such problems have come to concern the broader *political* framework in which large businesses operate, key law firms have played an even more significant role for the corporations. Now, rather than limiting their role to legal interpretations, lawyers, many of whom have had experience in the diplomatic corps, act almost as private-sector diplomats in sensitive corporate negotiations. This is especially true of foreign problems, but lawyers sometimes play a similar role domestically. In many of these cases, the lawyers become political intermediaries who often have personal contacts and rapport with foreign government officials, unlike the corporate executives who hired them. Where time is short and matters are highly important, the experienced law firm partner can often not only mediate in problems, but also can provide executives with invaluable political sensitivity and intelligence.

As crises become commonplace in world events – the raising of oil prices by OPEC nations and the new demands by developing countries, serve as two of the most evident examples of the changing world – corporations have required much better information and have had to develop ways to operate more flexibly than in the past. To meet these needs, several advanced corporate services have been linked more closely to the corporate executive office.

The role of the advanced corporate services has expanded also because corporations have needed to respond to new initiatives taken by their competitors, both at home and abroad. The struggle for entry into new international markets or for major parts of established ones has pitted some of the largest multinationals against each other. While some of these firms gain advantages over their competitors because they have organized efficient global systems of production, others find that political influence is critical in establishing or consolidating a presence, particularly in some parts of the developing world. In other instances, as in Japan or Brazil, multinational corporations have faced competitors in a highly regulated market, where stringent requirements were often placed upon ownership patterns and corporate behavior. In these cases, while local expertise is sometimes sought, many corporations have turned to their legal firms or accounting firms to formulate the best strategy to enter such markets and the particular legal form of organization that would be most effective. This often has entailed negotiations between members of advanced service firms and representatives of non-US multinationals, in addition to foreign governments.

Thus, the transformed environment faced by large corporations has brought about vast changes in the role of the advanced corporate services. These services have been especially important in facilitating corporate responses to the NIDL, permitting corporations to gain greater returns under conditions of increasing instability than would have been possible had they relied solely upon skills present within corporate organizations.

New corporate demands for services and the changing structure of advanced corporate services

The new corporate demands were important in changing the advanced corporate services. Much of the change came because of the growing involvement of the largest advanced services firms in the international economy. Particularly where they took on roles which substantially enhanced the performance of their corporate clients, the larger service firms became even more influential than their smaller rivals. Even before the 1960s, a small group of firms dominated each of these services. In accounting, the eight largest firms, 'The Big Eight', employ roughly 10 per cent of all certified public accountants but are responsible for 90 per cent of the industry's billings. In banking, the top four banks account for nearly a third of the deposits of the top 300 commercial banks and much more of international banking business. Only in law has there not been the extreme concentration of business in a small number of firms. While concentration has remained stable or increased in the past decade, the reasons for concentration have shifted.

In banking, international loans and a broad range of financial services were demanded by corporate clients. Only a few of the top banks, particularly the heavily internationally involved Citibank, and its chief rivals, Bank of America and Chase Manhattan Bank, were able to develop the extensive foreign networks needed to serve the multinational corporations which were their clients. Their ability to provide loans at home and abroad, particularly in times of monetary stringency, and to provide international financial management services to major corporations led them to a dominant position among US banks. Due to these activities, the top three US banks accounted for much of the growth of foreign bank deposits and almost all of the growth assets of overseas branches of US banks in the early 1970s.[30] Between 1970 and 1975, 95 per cent of the growth in earnings of five of the top banks was due to their increase in international earnings.[31]

The eight largest accounting firms had 94 per cent of the *Fortune* 500 firms as clients in both 1964 and 1974.[32] There has been significant structural change in the profession. The recent expansion of management advisory services and international accounting services for corporate clients has led to major adjustments by the largest accounting firms. There has been an explosion of competition and a much more aggressive seeking of clients by the Big Eight firms which dominate accounting. Most visible among the changes in the hierarchy of the Big Eight has been the rise of the Chicago-based Arthur Andersen to a position second only to New York's Price Waterhouse. The major accounting firms have played an important role in mergers and acquisitions by major corporate clients and have diversified to provide more services to government and labor unions. They have also developed sophisticated accounting skills for the new needs of their multinational corporate clients, enabling them to cope with foreign exchange variations and inflation. In addition, because of their experience with corporate operations and corporate management, they can offer more sophisticated management services than any of their smaller competitors.

The hierarchy of firms in investment banking is closely related to the ability of investment houses to service their corporate clients. With the recent growth of competition between investment banks, price-cutting has been an important impetus to mergers, but a firm's ties to top corporations still determines its status on Wall Street. Certain firms, like Goldman Sachs and Merrill Lynch have made remarkable strides in status, both because they have taken part in a number of contested mergers on behalf of their victorious clients and because they have greatly expanded their capacity to finance operations on the Eurodollar market. In 1970, the top ten investment banks managed 70 per cent of all negotiated underwritings, but by 1976, their share had risen to 90 per cent.[33]

Particularly with their concentration on international aspects of business and the more specialized aspects of law which are of particular interest to large corporations, a number of the bigger law firms have been able to place themselves beyond much of the rest of the field in terms of competition for the business of the *Fortune* 500. They have a number of distinct advantages over smaller firms. They (i) have been able to develop the research capabilities which enable them to redefine the law in pathbreaking briefs;[34] (ii) have a 'flow of experience' with the latest and most complex legal problems which allows them to refine their skills in a way smaller firms cannot; (iii) are more likely to be knowledgeable in international law; and (iv) can also more easily develop skills in new areas of law.

Thus, while there have been major changes in the size of large law firms, particularly of those outside New York, it is not clear that the most rapidly growing firms have taken clients away from the large New York firms which have the best links to major corporations. The sixteen law firms with the greatest number of *Fortune* 500 clients now account for almost 30 per cent of the firms on the *Fortune* list.[35]

The emergence of global cities

Changes in the corporation and in the structure of the advanced corporate services have led to the emergence of a series of global cities which serve as international centers for business decision-making and corporate strategy formulation. In a broader sense, these places have emerged as cities for the coordination and control of the NIDL.

If one examines available data on international and domestic activities for the *Fortune* 500 firms and for a number of advanced services, it is apparent that only a few cities in the United States are vastly more important as centers of international business and corporate services than as centers of national business. Firms headquartered in New York account for 40 per cent of foreign sales of *Fortune* 500 companies, compared to 30 per cent of all sales of *Fortune* 500 firms. Banks in New York and San Francisco have 54 and 18 per cent of foreign bank deposits in US banks, compared to 31 and 13 per cent of total US bank deposits.

This concentration of *international* corporate decision-making and corporate services in a few US cities is tied to the emergence of the NIDL, the ensuing need for changes in corporate strategy and structure, and the transformation of the advanced corporate services. These changes and the increased centralization of corporate-linked functions have changed the dimensions of urban hierarchy in the United States.

The movement of US corporations overseas had several important impacts on the institutional structure of large US cities. International decision-making by major firms was largely concentrated in two cities, which were major centers of corporate headquarters and finance: New York and San Francisco. This trend was supported by major banks from these cities which developed an elaborate network of subsidiaries to service the foreign operations of major corporations. As a result, *cities which had been important centers of business in an earlier, more national-oriented phase of the economy began to lose economic stature to these global cities* because their firms were not as internationally oriented as others. Jobs related to international operations did not develop as extensively in places like Cleveland, St. Louis and Boston, as they did in New York and San Francisco. While the centralization of international activities in a few cities may be reversed somewhat in the future, as more US firms develop foreign operations, this trend will probably continue. This is especially true, since once firms lose time to their competitors on the international scene, it is often regained only with great difficulty.

In conjunction with the growth of international corporate activity, advanced corporate services, including banks, law firms, accounting firms and management consulting firms, expanded their international skills and their overseas operations. Yet, these, too, grew in but a few urban centers and thus drew even those firms with international operations which were headquartered in regional or national centers of business to more international centers. Thus, just as few banks with international expertise were located outside of New York, San Francisco or Chicago, few law firms with international competence were centered outside of Washington, Los Angeles or New York. Washington became especially significant as a center for international law because of the contacts which firms there had developed with the US State Department and with foreign governments. New York emerged as perhaps the single most important center of international accounting expertise in the nation, since even some of the largest accounting firms headquartered outside of the city do much of their international work there.

To quantify the present status of different US cities as centers of international business, I have constructed a 'multinational index' for cities (see Table 12.1). This index compares the percentage share of a city's *Fortune* 500 firms in total foreign sales to their percentage share of total sales. For instance, in the New York Standard Metropolitan Statistical Area (SMSA), *Fortune* 500 firms account for 40.5 per cent of all foreign sales by *Fortune* firms, and 30.3 per cent of total sales by *Fortune* firms. Dividing the first percentage by the second one gives us a multinational index of 1.34. Where most of the corporations domiciled in a city have extensive international operations, the city's score on the multinational index will be almost 1.0 or higher. This defines international cities. Those metropolitan areas with some firms having international subsidiaries score between 0.7 and 0.9, a lower score, and I have defined these as national cities. Finally, where there is little evidence that the firms in a city have any

Table 12.1 The urban distribution of *Fortune* 500 firms' total sales and foreign sales in 1974—selected SMSAs

SMSA	Fortune firms (1)	Percentage of total sales (2)	Percentage of foreign sales [a] (3)	Multinational index (3) ÷ (2)
New York	107	30.3	40.5	1.34
Los Angeles	21	4.6	3.8	.83
Chicago	48	7.3	4.6	.83
Philadelphia	15	1.6	1.0	.77
Detroit	12	9.1	8.8	.97
San Francisco	12	3.2	5.4	1.69

Sources: *Fortune Magazine* (May 1975); corporate annual reports, *Wall Street Transcript*, and Securities and Exchange Commission prospectuses.

[a] The foreign sales data were compiled for 321 of the largest firms. Data for the other, mostly smaller, firms was not available.

share of foreign corporate sales relative to their share of total sales, I define them as regional cities.

This index permits the assessment of a city's relative strength as a center of international business in relation to its strength as a center of national business. The index defines 'business strength' at three different levels based not on absolute share of foreign sales or total sales of the *Fortune* 500 firms, or on the size and industrial structure of a city's firms, but on the relative share of international business compared to all business of those firms headquartered in a city.

What is remarkable about the results obtained by using this index is that *several large cities which we usually think of as 'national' centers turn out to be very weak as centers of international business.* San Francisco, New York, Houston, Boston, Pittsburgh, Seattle, Rochester, Akron and Detroit are strong centers of international business based upon this index. But Chicago, St. Louis, Cleveland, Dallas and Milwaukee are merely 'regional' centers. They are cities with few businesses that have extensive international operations. Most other large cities are 'national', with corporations having somewhat smaller relative shares of foreign sales compared to total sales for the *Fortune* 500.

I would argue that merely examining the foreign and total sales of corporations headquartered in a city is not sufficient to gauge its importance as a center of international business. One has to know if the city is a strong center of international banking and strategic corporate services. Only a place with a wide range of international business institutions can be truly called a world city.

Thus, I devised a similar 'multinational banking index' which compares the share of all foreign bank deposits of the top 300 banks held by banks in one city to their share of all domestic deposits held by the same banks.[36] Using this index, several of the strong 'international' corporate centers drop to a 'national' or 'regional' classification. Pittsburgh and Boston only rank as 'national' banking centers, and Houston, Seattle and Detroit are 'regional' banking centers. Rochester and Akron do not even rank as regional centers. On the other hand, Chicago and Dallas move up to positions as global cities with international centers of banking, along with New York and San Francisco. Most of the other large cities – Los Angeles, Philadelphia, St. Louis, Atlanta, Cleveland, Minneapolis – are only 'regional' banking centers.

From these two indices it is clear that only two places achieve classification as global cities for both corporate and banking business: New York and San Francisco. Chicago and Houston may be moving to 'international' status, but the former has relatively few internationally oriented corporations while the latter has few banks with sizeable international operations.

The emerging world urban hierarchy

Industrial restructuring and urban hierarchy

Changes in the international 'competitiveness' of a number of world industries have been a major factor in the reshaping of urban hierarchy throughout the world. The restructuring of industry on an international scale (movements of plants from developed to 'developing' areas both within and between nations, closing of 'noncompetitive' plants in older, industrialized centers, and the technological improvement of industry to increase productivity) have all contributed to major shifts in employment and trade, with the greatest impacts being felt in both the urban centers of developed nations and the larger cities of developing countries.

What forces are behind this restructuring? The desire by multinational corporations to seek new markets and more profitable ways to organize production on a world scale,[37] national policies by developed nations to strengthen the future international competitive position of selected industries,[38] and national policies on the part of developing nations to stimulate the growth of export sectors, largely by attracting subsidiaries of multinational corporations.[39] These strategies of multinationals and policies of governments have resulted in a situation where the profitability of productive facilities in certain parts of the developed world relative to those in other parts of the world, particularly the 'newly industrializing countries' (NICs), has declined considerably. This has resulted in the inability of many older plants to obtain adequate profits, compared to those obtainable elsewhere.

In response to this situation, corporations in many of the developed nations have begun to restructure their operations. This process has included: relocating firms even more extensively to areas where less expensive conditions for production are available, changing the process of production by introducing new technology or altering the labor process, or by closing down existing plants. Both corporate restructuring and national policies, which have been focused upon a few specific industries, have resulted in dramatic changes in the structure of employment in both developed and developing nations, and have altered the pattern of trade between nations. Between 1970 and 1976, just fifty of 422 individual SITC commodity classifications accounted for 80 per cent of the fourfold, $23,000,000,000, increase in the total value of exports by NICs to developed market economies, with the top ten items accounting for 44 per cent of the increase.[40] While the direct effect of imports on employment is usually seen as less important than that due to technical progress in productivity, changes in the international competitive position of specific industries may have very deleterious affects on employment. Thus, Doreen Massey and Richard Meegan have found that declines in the international position of several British industries were responsible for nearly *70 per cent* of the job losses in Greater London, Manchester, Birmingham and Liverpool between 1966 and 1972.[41]

Such effects and international changes have had an important impact on corporations and governments. They have made it increasingly apparent that successful multinational companies will, in the future, not be those who exploit product life cycles, as much as they will be those who master the process of 'global scanning'[42] and are able to integrate their operations on a world scale. These changes have also resulted in increased intervention by governments in the economies of developed nations, responding both to the need to adjust to job losses in declining sectors and to support corporate restructuring strategies in more dynamic sectors. The result has been greater conflict and instability in the international economy. Governments and firms in certain nations have devised 'national industrial strategies' to enable the future expansion and adjustment of their economies to progress more successfully than their rivals, but how well the private goals of companies and the public goals of governments will be integrated remains to be seen. Other attempts to facilitate international adjustment to the new economic and political relationships of the 1980s, notably by the Trilateral Commission,[43] appear to have met with failure because of the intensity of the basic conflicts which must be overcome.

The urban impacts of these problems, which have been gathered under the rubric of 'structural adjustment problems', is often severe. In a number of cases, the very existence of medium or large-size cities has been threatened. In Japan, the imminent closing of one of Nippon Steel's oldest plants may have dire consequences for the entire town of Kamaishi.[44] In the Massey and Meegan study cited above, job losses due to declines in the international competitive position of British firms have had their greatest impact on skilled or semi-skilled workers, particularly on older, unionized workers located largely in declining urban centers. New employment in Britain appears to have focused upon less-skilled, lower-paid workers, leaving those displaced from jobs with little chance to maintain their economic status.[45] Even the most sanguine studies of future changes in the pattern of trade conclude that while total job losses in developed nations may not outnumber gains, employment dislocation in certain sectors may be severe.[46]

The United States' urban hierarchy

CENTERS OF CORPORATE HEADQUARTERS AND CORPO-
RATE SERVICES   As illustrated in the previous section, only a few
centers now stand out as places where key international functions are
agglomerated in the United States. This contrasts sharply with the more
regional hierarchy of national corporate functions, which is readily apparent from an examination of the location of *Fortune* 500 corporate headquarters. In the latter list, a number of regional centers stand out as important
centers of headquarters.

In the 1950s, the 'metropolis and region' structure of the United States,
which was fairly well established by 1913, was a rather accurate reflection
of urban hierarchy. Each regional metropolis was an important center
of corporate services. Cleveland, for example, had a large number of corporate law firms to complement its important group of corporations. Major
accounting firms were headquartered in New York, Chicago and Cleveland.
Important regional banks could be found in nearly all of the centers of
corporate head offices.

**Table 12.2 Urban centers with more than ten *Fortune* 500
corporate headquarters**

Fortune *500* Headquarters 1957		Fortune *500* Headquarters 1974	
New York	144	New York	107
Chicago	54	Chicago	48
Pittsburg	24	Los Angeles	21
Philadelphia	21	Cleveland	17
Detroit	17	Pittsburg	15
Cleveland	16	Philadelphia	15
Los Angeles	14	Detroit	12
San Francisco	13	San Francisco	12
St. Louis	13	Minneapolis	10
		St. Louis	10
		Milwaukee	10

By 1974, although the 'metropolis and region' network of corporate
head offices and corporate services had expanded (see Table 12.2), international business functions had become much more significant than before.
Thus, an analysis of the urban hierarchy in the US must take these new
and often more complex functions into account. If this is done, we find
that an extremely limited number of cities act as world centers of business
and corporate services. It should also be noted that as the top level cities,
primarily New York and San Francisco,[47] emerged as key international
centers, even the international activities of firms headquartered outside
these cities were increasingly linked to financial institutions and corporate
services located within them.[48]

THE EMERGING DIVISION OF LABOR IN THE US AND CONTRADICTIONS IN THE URBAN HIERARCHY   The new structure of urban hierarchy in the US is one dominated by two main international centers for corporate control and coordination. While nine or more regional corporate metropolises remain important as corporate centers (see Table 12.2), they appear to have decreased in importance if international corporate services are taken into account. In addition, with the evident centralization of the more sophisticated corporate services in just a few regional centers, places like Cleveland and St. Louis which have seen numerous blue-collar jobs leave the metropolitan area have not had a growing service sector to cushion part of the economic decline.

Indeed, the growing importance of some cities, even regional cities, as corporate centers reflects a number of contradictory forces in US urban development. First, certain places, by becoming centers of corporate operations and services, present less opportunities for blue-collar jobs and job mobility than they have traditionally done, and risk not only losing their middle class, but also 'marginalizing' the lower class which has traditionally found job mobility extremely difficult. Second, as corporations and the rest of the business community realize the significant role played by certain metropolises, there is more likelihood that they will shift from a laissez-faire attitude to urban economic development to more active support of planned urban development. But if such planning focuses primarily upon corporate-linked activities it also risks alienating the middle and lower classes. Third, the relative loss of power by national and regional cities may lead them to become the hinterland of new world metropolises. This would lead such cities to follow a pattern of development similar to the dependency relationship which has occurred between Canada and the US or what Kari Levitt has called 'branch plant development'. This results

> 'in the erosion of local enterprise, as local firms are bought out and potential local entrepreneurs become the salaried employees of the multinational corporation. Enterprises which remain locally owned tend to be marginal in the sense that they are small, or inefficient, or operate in industries which do not lend themselves to corporate organization.'[49]

To examine uneven development within the labor forces of different types of cities, I have quantified the differences between employment in the core and peripheral industries of these cities.[50] Core industries were defined by factor analysis and were characterized by high concentration ratios, large establishment size, high rates of unionization, capital intensity, few black workers and high male educational attainment. My analysis showed that the more important a city was as a national or regional center, the greater the share of its employment in dynamic, core industries. But once a city became a global city with a multinational index score of 1.0 or more, it lost employment in the core industries.[51] These results indicate that international administrative centers, like New York and San Francisco, may be qualitatively different from more national centers of corporate activities. In part, the difference may reflect higher prices for rents and services which result from the pronounced agglomeration of corporate and corporate-related activities in these centers. On the other hand, it may also reflect the fact that since the Second World War, such centers have grown largely by attracting a specialized range of activities, following a pattern that may have begun, as in the case of New York, as early as the 1900s.

The new hierarchy of world cities

THE NEW WORLD CITIES   The new hierarchy of world cities has a number of parallels with the changing hierarchy of US cities. For both groups, the rise of international corporate activities and of international corporate services has made a significant difference in the pattern of urban hierarchy.

To examine the world hierarchy of cities of the 1950s and early 1960s, we can rely upon the world cities selected by Peter Hall. These included the traditional large national centers of business and government: London, Paris, Amsterdam-Rotterdam, the Rhine-Ruhr complex, Tokyo and Moscow.[52] These places were certainly important for their concentrations of finance and corporate services, for as Hall pointed out, they attracted corporate-related jobs, 'for every producer of goods, more and more people are needed at office desks to achieve good design, to finance and plan production, to sell the goods, to promote efficient nationwide and world-wide distribution.'[53]

As the 1970s began, multinational business and international finance began to play a more dominant role in Europe and Asia. Traditional national centers which lacked concentrations of international corporations and banks declined in importance. If one had to assess the structure of urban hierarchy outside the US using the same approach which has been employed in studying the US urban hierarchy of corporate functions, a different group of cities might be called 'world cities' than those selected by Hall. According to Table 12.3, corporate head offices of non-US multi-

Table 12.3 Main locations of corporate headquarters, 198 largest non-US corporations, 1978

Metropolitan area	Number of headquarters
Tokyo	30
London	28
Osaka	13
Paris	12
Rhine-Ruhr	10.5[a]
Randstad Holland	3.5[a]

Source: United Nations *Transnational Corporations in World Development: A Re-examination*, Appendix IV, and *Jane's Major Companies of Europe, 1976.*

[a] When a firm has headquarters in two places, each locality is given 0.5 offices.

national corporations are most highly concentrated in London and Tokyo.[54] Only London, Frankfurt and Zurich have gained enhanced stature because they are centers of the Eurodollar market.

Thus, New York, Tokyo and London, are predominant as world centers of corporations and finance. Osaka, the Rhine-Ruhr, Chicago, Paris, Frankfurt and Zurich are second-level world cities as far as international corporate activities are concerned. There is at least anecdotal evidence that with the emergence of Frankfurt as a major international financial center, corporate service activities began to concentrate in the city.[55]

THE WORLD HIERARCHY OF CITIES AND ITS CONTRA-DICTIONS   The new urban hierarchy of corporate head office activities outside the US also reflects a new division of labor between centres of corporate control and coordination and more 'national-oriented' urban places.[56] The Amsterdam-Rotterdam and Rhine-Ruhr agglomerations and Paris, cited by Hall as world cities, are much less a part of the international system of corporate and financial activities than are the international financial centers in Europe. It is difficult to assess how the more 'national-oriented' European centres should be compared to financial centers, such as Singapore, Panama, Hong Kong and the Bahamas. But if we examine these latter centers as places for corporate decision-making, it is apparent that they serve primarily as centers for moving and mobilizing financial resources, rather than as centers of control.

As international boundaries become blurred by the increasingly global nature of corporations, numerous contradictions will arise within the world hierarchy of cities. First, there will be contradictions that will arise because private institutions, particularly large multinational corporations and banks, are able to undermine or contravene established government policy.[57] This contributes to the erosion of the position of certain traditional centers of government policy where corporate head offices or major financial institutions are not present in large numbers. Second, shifts in the importance of the Eurodollar market will probably create conflicts between new centers of finance and older, more national ones.

The NIDL will have a particularly strong impact on cities in the developing nations, particularly on those in the newly-industrializing countries, or NICs. Because multinational corporations, aided by international banks, will probably accelerate the establishment of foreign sibsidiaries in these nations over the next decade with the assistance and support of governments that view industrialization as a 'redemptive mystique,'[58] the perpetual crisis of cities in developing nations will almost certainly be exacerbated. In large part, this will be due to a phenomenon described by John Friedmann and Flora Sullivan. The labor force in cities of developing nations is divided into three sectors (a small corporate sector, a family-enterprise sector and an individual enterprise sector), according to their analysis, with the individual sector remaining structurally isolated from the others, thus countermanding the ordinary behavior of wages in an integrated labor market, which would tend to drop when there are more workers than jobs. Instead, wages in the individual sector are depressed to subsistence levels when additional job seekers enter this part of the labor market, but remain unaffected in other sectors. Friedmann and Sullivan claim this phenomenon may be explained by three effects.

First, jobs in the family-enterprise sector are destroyed due to an increase in competitive production in the corporate sector, contributing to labor absorption problems and a decline in job mobility. Second, since only wages in the individual sector are depressed by new entrants to the city's labor force, this sector is marked by prolonged unemployment, near subsistence wages and fragile support systems for job seekers whose kin are often unwilling or unable to provide them with support. Third, the relatively high wage costs and subsidized capital investment in the corporate sector lead to more capital-intensive development, decreasing the labor absorption capacity of this sector.[59] Friedmann and Sullivan note that the trichotomized structure of the labor force results in urban crisis even when a nation's gross national product is expanding because of the inability of manufacturing companies to create enough new jobs, the destruction of jobs in the family-enterprise sector and the accelerating flow of people into cities. What is needed, they argue, are development policies focused upon human potential rather than growth. This would include promoting greater national autonomy, industrial dualism (rather than import substitution), balanced rural-urban growth, greater equality, and the stabilization of the population.[60]

While this sentiment is admirable, it does not take into account the structural limitations of development in the Third World which make it extremely difficult for even those governments that want to change policies to implement them. This is certainly reflected in the restrictions placed upon many loans granted by the World Bank.[61] In more recent years, the emergence of commercial banks as major lenders to governments of developing nations has resulted in severe limitations on the direction of development policies. Two cases which have been studied, Peru and Chile, illustrate how commercial banks pursued lending policies which complemented the goals of multinational corporations. In the case of Peru,[62] government borrowing was restricted, in concert with the World Bank, until more traditional national economic policies were reinstated, following an attempt at income redistribution and nationalization of important sectors of the economy. In the case of Chile,[63] borrowing was reduced to almost zero following the rise of Allende and the nationalization of certain key industries. The lack of funds contributed to the economic problems which ensued, making it extremely difficult to pursue an alternative pattern of economic development.

In sum, the future of the city in developing nations must also be analyzed in relation to international and national development.[64] Certain forces will certainly continue to exacerbate the flow of people into the cities, such as the international sourcing of raw materials and the use of land for more export-related (largely agricultural) activities.[65] In addition, the need for entrepreneurial and technical skills, information from personal contacts, and the resolution of problems through extralegal procedures will contribute to the continued agglomeration of corporate manufacturing activities in the major cities of developing nations.[66] Yet what results from the pattern of development described here is an increasingly dualistic society, with the cities of developing nations being characterized by extremely high unemployment among relatively more educated, largely male, urban immigrants, who are usually sixteen to forty-five years of age.[67] The social and political turmoil which can result from the pursuit of such policies, even if they now result in profits for multinational companies and higher GNP per capita in developing nations, will certainly have a severe impact upon both the developed and developing nations in the 1980s and 1990s.

Future trends in the NIDL and urban hierarchy

A number of changes will probably occur over the next decade which will restructure the NIDL. These changes may have significant impact on both corporations and nation-states. They will both create more formidable problems for corporate managers by making the international economy more complex and more competitive, by eroding the traditional industrial structure of developed nations further, and by causing sweeping changes in the employment structure of developing nations.

These changes may be categorized into two groups: changes in corporate and financial institutions, and changes in developing nations. In the first group, the probable changes include: (i) a rapid expansion of the international activities of both US and non-US medium-sized corporations, including corporations from developing nations; (ii) a shift of the center of the Eurodollar market from London to New York; (iii) a continued growth in the dependency of US and foreign corporations on advanced corporate services; and (iv) a rapid growth of foreign investment in the United States. These changes would place further pressures upon management to become global in perspective, to increase its control over operations, to plan over longer time horizons, and to develop more flexible strategies.

The second group of changes would also increase pressures on corporations and financial institutions. They would probably include: (i) the rapid emergence of newly-industrializing nations as centers of production and as domiciles for new competitors for multinational firms; (ii) the assertion of power by OPEC-like commodity cartels for copper, aluminum, certain foodstuffs and other basic commodities; (iii) the erosion or enhancement of the power of OPEC by the development of current non-OPEC centers of oil reserves, Mexico being the prime example; (iv) a rapid growth of the debt burden of poorer developing nations, making them more vulnerable to potentially-disruptive economic transformations due to foreign investment and less able to purchase foreign goods; and (v) a growing political instability among the more developed of the developing nations.

These trends, since they place more pressure on corporations and financial institutions in developed nations, will probably tend to further concentrate corporate and financial decision-making in present world centers, drawing decision-making activities away from national or regional centers. This would exacerbate uneven urban development in developed nations, particularly since the continued movement of productive activities overseas would further erode urban blue-collar jobs.

Notes

[1] For a good review of these changes see United Nations, Economic and Social Council, Commission on Transnational Corporations (1978) *Transnational Corporations in World Development: A Re-examination.* New York: United Nations, particularly part 3, 'Patterns and Trends in Transnational Corporation Activities'.

[2] On this phenomenon, see the studies of the Eurodollar market, for instance those by G. Bell (1973) *The Eurodollar and the International Financial System.* London: Macmillan; R. F. Mikesell and M. H. Furth (1974) *Foreign Dollar Balances and the International Role of the Dollar.* New York: National Bureau of Economic Research; and the recent study by X. Gorostiaga (1978) *Los centros financieros internacionales en los paises subdesarrollados.* Mexico City: Instituto Latinoamericano de Estudios Transnacionales.

[3] A good example of the latter trend has been the growth of new forms of investment, such as management contracts, especially during the last five years. The major aim of such forms is to permit much less exposure of corporate capital to nationalization, to shifts in commodity markets, and to other threats which might lead to disinvestment. Management contracts usually allow a company to invest very little in productive facilities or to retain an important position in nationalized operations (OPEC oil fields) and to obtain profits from providing management skills for such operations.

[4] The enhanced bargaining power of developing nations has not been studied very extensively. Perhaps the best sources are three unpublished case studies of bargaining power in Brazil, Colombia and Mexico recently prepared by Ronald E. Müller and David Moore for the United Nations' Centre on Transnational Corporations.

[5] S. H. Hymer's paper (1979) 'The Multinational Corporation and the International Division of Labor', in S. H. Hymer *The Multinational Corporation: A Radical Approach.* Cambridge: Cambridge University Press, is one exception.

[6] Restraints which arise because of international competition require internal changes in structure. (See Hymer 1979.)

[7] The paper by J. R. Hiller (1978) 'Long-Run Profit Maximization: An Empirical Text.' *Kyklos* 31, 475–90, documents this secular decline in profits.

[8] See the negotiations for a European Monetary System, talks on trade restrictions between Japan and other advanced countries or attempts by nations to enter the China market for examples of such conflicts.

[9] Differences in the current account surplus of OPEC member countries are one indication of disparities, even among the relatively well-to-do developing nations. In 1977, four of the thirteen OPEC member countries – Algeria, Venezuela, Nigeria and Ecuador – registered total account deficits of $5,600,000,000. In

1978, the deficit of these nations plus Indonesia is expected to exceed $10,000,000,000.

10 'The OPEC Surplus Dwindles.' *Chase International Finance* (11 December 1978): 6–8.

11 From 1950 to 1960, foreign investment by US corporations rose from $12,000,000,000 to $33,000,000,000; by 1975, the total reached $110,000,000,000. Profits from overseas operations, which accounted for 7.3 per cent of all profits in 1950, reached 14.6 per cent in 1960 and 27.0 per cent of the total in 1972.[12] Nearly all of this international activity by US firms was accounted for by the 200 largest corporations.

12 T. Weisskopf (1975) 'American Economic Interests in Foreign Countries' and 'Global Crossroads.' *Wall Street Journal* (3 December): 24.

13 E. J. Kolde (1966) 'Business Enterprise in a Global Context.' *California Management Review* (Summer): 40.

14 Kolde 1966: 40.

15 William C. Goggin, Chairman and Chief Executive Officer of Dow Chemical notes that in his company, marketing managers did not know how much it cost to produce a product. See W. C. Goggin's article (1974) 'How the Multidimensional Structure Works at Dow Corning.' *Harvard Business Review* (January-February): 54

16 R. Vernon (1971) *Sovereignty at Bay*. New York and London: Basic Books, Inc., p. 131.

17 Eaton Corporation 'Eaton – A world corporation in a multination climate', offset, p. 2.

18 Vernon, 1971: 29–31.

19 See C.-A. Michalet and M. Delapierre (n.d.) *The Multinationalization of French Firms*. Chicago: Academy of International Business (translated by David Ashton), p. 45.

20 Vernon 1971: 115.

21 Vernon 1971: 116.

22 United States Senate, Committee on Finance (1973) *Implications of Multinational Firms for World Trade and Investment and for U.S. Trade and Labor*. Washington, DC: USGPO, p. 159.

23 Brooke, M. Z. and Remmers, H. L. (1970) *The Strategy of Multinational Enterprise*. London: Longman.

24 Charles Levinson explains this development in his book (1971) *Capital, Inflation and Multinationals*. London: George Allen and Unwin.

25 Several early studies of multinational corporations noted this trend, including Dunning, J. H. (1958) *American Investment in British Manufacturing Industry*. London: Ruskin House/George Allen and Unwin, p. 106–112 and Kolde, E. J. (1963) 'The functions of foreign business affiliates in the administrative structure of international business enterprise.' *University of Washington Business Review* (February). Both these articles are cited by Behrman, who elaborates on the causes of centralization in his (1970) *Some Patterns in the Rise of the Multinational Enterprise*. Chapel Hill, N.C.: Graduate School of Business Administration, University of North Carolina.

26 This trend has been suggested by many papers, including Kiefe, D. M. (1966) 'Reorganization for International Operations.' *Chemical and Engineering News,* (27 June).

27 See Chandler, A. D., Jr. (1962) *Strategy and Structure.* Cambridge, Mass.: MIT Press; and (1977) *The Visible Hand.* Cambridge, Mass. and London: Harvard University Press.

28 Drucker, P. F. (1974) 'New Templates for Today's Organizations.' *Harvard Business Review* (January-February): 47–9.

29 For a more detailed discussion of these changes, see Chapter 4 of my forthcoming book, *The Corporation and the City.* Cambridge: Cambridge University Press.

30 D'Arista, J. (1976) 'US Banks Abroad,' in US Congress, House *FINE: Financial Institutions and the Nation's Economy,* Book II, 94th Congress, 2nd Session, p. 813.

31 Mason, J. J. and Lissman, B. M. (1975) 'World trade trends and the major US multinational banks.' *Loeb Rhoades Bank Stock Overview,* Special Supplement (21 November): 18.

32 Montagna, P. (1974) *Certified Public Accounting: A Sociological View of a Profession in Change,* Appendix Table C-1; and special survey of corporate bond underwritings by the *Fortune* 500 done for my book, *The Corporation and the City.* Cambridge: Cambridge University Press, forthcoming.

33 'Where is the capital spending boom?' (1976) *Business Week,* (13 September): 66.

34 This aspect is mentioned by E. O. Smigel (1969) in *The Wall Street Lawyer,* p. 7, and has also been corroborated during a number of interviews with lawyers.

35 This data is derived from survey of corporate bond underwriting done for my book, *The Corporation and the City.* Cambridge: Cambridge University Press, (forthcoming).

36 See my unpublished paper, 'Urban effects of the internationalization of capital', for more detail.

37 For an interesting critical evaluation of these new strategies, see Hymer, S. H. (1972) 'The internationalization of capital.' *The Journal of Economic Issues* (March); and (1972) 'The United States multinational corporation and Japanese competition in the Pacific.' *Chuokoron-sha* (Spring). Both of these articles are reprinted in S. H. Hymer (1979) *The Multinational Corporation: A Radical Approach.* Cambridge: Cambridge University Press.

38 The Japanese Ministry of International Trade and Industry has been particularly active in aiding in the strengthening of Japanese industry. For some examples of its role, see Japan External Trade Organization (1973), *White Paper on International Trade,* especially Chapter 2, 'Changes in the international environment and Japan's external economic activities'.

39 'Those other Japans.' *The Economist* (10 June 1978): 84–5.

40 *The Economist* (10 June 1978): 84.

41 Massey, D. and Meegan, R. A. (1978) 'Industrial Restructuring Versus the Cities.' *Urban Studies* 15: 273–88.

42 For a discussion of this behavior, see R. Vernon (1971) *Sovereignty at Bay.* New York and London: Basic Books, Inc.., pp. 107–12.

[43] See the studies sponsored by the Trilateral Commission of the world economy, including: Gardner, R. N., Okita, S. and Udink, B. J. 'OPEC, the trilateral world and the developing countries: new arrangements for cooperation, 1976–1980'; Bergsten, C. F., Berthoin, G. and Mushakoji, K. 'The reform of international institutions'; and Cooper, R. N., Kaiser, K. and Kosaka, M. 'Towards a renovated international system.'

[44] Kanabayashi, M. (1979) 'Youngstown East: Japanese town faces an economic crisis as steel industry pushes rationalization.' *Wall Street Journal* (14 February).

[45] Massey and Meegan (1978): 279–80, 283–5, 287.

[46] See the following studies: Baldwin, R. E. (1976) 'Trade and employment effects in the United States of multilateral tariff reductions.' *American Economic Review* 66 (Papers and Proceedings) 1978 (May): 142–8; Commissariat Général du Plan (1978) *Rapport du groupe chargé d'étudier l'evolution des économies du Tiers-Monde et l'appareil productif francais.* Paris (January); and Schumacher, D. (1977) 'Increased trade with the Third World: German workers will have to switch jobs but not lose them.' *DIW Wochenbericht* 5 (February).

[47] Cheng, H. S. (1976) 'US west coast as an international financial center.' *Federal Reserve Bank of San Francisco Economic Review* (Spring): 9–19.

[48] Stodden, J. R. (1973) 'Their small size costs banks business of large companies.' *Federal Reserve Bank of Dallas Business Review* (October): 6–7.

[49] Levitt, K. (1971) *Silent Surrender: The American Economic Empire in Canada.* New York: Liveright, p. 104.

[50] This analysis follows the approach used by Robert Averitt in his book (1970) *The Dual Economy.* New York: Norton.

[51] These results are discussed in two unpublished papers which I have written. They are based upon industrial structure during the 1960s and have not been revised for the 1970s.

[52] Hall, P. (1971) *The World Cities.* New York and Toronto: McGraw-Hill Book Co.

[53] Hall 1971: 9.

[54] Westaway, J. (1974) 'The spatial hierarchy of business organizations and its implications for the British urban system.' *Regional Studies* 8: 145–55, discusses reasons for the concentration of corporate activities in London.

[55] 'Financial Times Survey—Frankfurt.' *Financial Times* (17 November 1978): 19–26.

[56] Here, I am drawing upon Christian Palloix's discussion of hierarchy between national economies which are dominant and less dominant internationally, i.e. the relative positions of Germany and France, for example, to set up an urban parallel between 'internationally-oriented' and 'nationally-oriented' cities. See Palloix, C. (1975) *L'internationalisation du Capital.* Paris: Maspero.

[57] Some examples of this are cited in Barnett, R. and Müller, R. (1975) *Global Reach.* New York: Simon and Schuster, pp. 254–302.

[58] Johnson, E. A. J. (1970) *The Organization of Space in Developing Countries.* Cambridge, Mass.: Harvard University Press.

[59] Friedmann, J. and Sullivan, F. (1975) 'The absorption of labor in the urban economy: the case of the developing countries', in J. Friedmann and W. Alonso

Regional Policy: Readings in Theory and Applications. Cambridge, Mass. and London: MIT Press, pp. 475–92.

60 Friedmann and Alonso 1975: 474, 493–9.

61 See Hayter, T. (1971) *Aid as Imperialism.* Harmondsworth: Penguin Books.

62 Stallings, B. (1979) 'Peru and the US banks: who has the upper hand?' in Fagen, R. (ed.) *Capitalism and the State in US-Latin American Relations.* Stanford: Stanford University Press.

63 Griffith-Jones, S. (1978) 'A Critical Evaluation of Popular Unity's Short-Term and Financial Policy.' *World Development* (July).

64 Friedmann and Sullivan (1975: 500) make a similar point, but limit their focus to national development processes.

65 See J. Collins and F. M. Lappe (1977) *Food First.* New York: Simon and Schuster, for a discussion of this problem in Northwest Mexico.

66 These points are made in W. Alonso (1975) 'Urban and Regional Imbalances in Economic Development', in Friedmann and Alonso, pp. 626–7. However, I would agree with Alonso's conclusion that efficiency in developing nations is best served by concentration of industrial activities. See Alonso 1975: 629.

67 Johnson 1970: 159–60.

V Reproduction and the dynamics of urban life

13 Community and accumulation

John Mollenkopf

The growth of large cities has constituted the most remarkable and defining part of our national experience over the last 150 years. Two themes recur in the literature which has grown up around this phenomenon. First, our cities have been described in terms of the dualism of contrasting and contending elements, whether order and disorder, opportunity and deprivation or growth and decay. Second, these oppositions mesh with a developmental cycle which leads from periods of organizational expansion into periods of conflict and breakdown which, in turn, generate new forms of expansion. These two themes are related; wealth and poverty, the cosmopolitan business center and the parochial 'urban village' and the central city and suburb do not stand in static opposition. Instead, in a manner resembling the old-fashioned pump-handled railroad cart, the contending elements of urban life act reciprocally on each other to propel urban development forward.

A proper theory of urban development must accurately capture these basic elements of the urban experience. It must account, on the one hand, for the elements or actors which stand in dynamic tension with each other, and it must also show how their interlocking sets of choices drive systemic development.[1] The following discussion outlines the basic building blocks for such a theory by way of introduction to the more detailed specification which must be worked out.[2]

Towards a theory of urban development

Essentially, cities concentrate and contain two kinds of relationships: those of production and economic accumulation and those of social interaction and community formation. By appreciating the strong but asymmetric and ultimately antagonistic interdependence between accumulation and community, we can clarify the duet between urban growth and crisis. Each of these aspects of city life presupposes the other, yet each operates by a distinct, unequal and ultimately opposing logic. The ensuing tension deeply permeates urban institutions, urban form and urban life.

The term 'accumulation' refers to how a society creates, expands, and distributes its means of well-being. The idea goes beyond the normal terms 'economics' or 'economic growth' because it takes in consumption patterns, the occupational order, the organization of work, and other things which are not always put under the heading of economics. The process of accumulation by definition assumes the existence of some form of society. In capitalist societies (and perhaps socialist ones as well), market values, particularly profit maximization, occupy an overriding place.

The term 'community' is more difficult to define because common language puts the word to so many different uses. Here, community may be thought of as a dimension or continuum. It refers to the bonds people build with one another which enable them to trust in and rely on each other. The sense of community becomes stronger as it is built up on a variety of foundations – for example, kinship, residence in the same neighborhood and shared occupation. Community does not exist in any *a priori* way. It is something which is slowly created over time, and may wax, wane, or be altogether absent. Nor do communities share only *a priori* common interest other than the stake participants have in being able to rely on their communities. Indeed, communities ordinarily find themselves divided by many conflicts of interest (for example, landlords versus tenants). For the most part, when the term 'community' is employed below, it refers to neighborhoods in which ethnicity, class, communal institutions and other building blocks have built up an overlapping set of bonds through the years. Unlike the accumulation process, market values play relatively little role in shaping community interactions. Instead, non-market values like reciprocity, mutual support and informal helping patterns are central. While communities could not exist without some form of accumulation, it is easy to see how the difference in their defining values could lead to conflict. Each depends to some degree on the other; yet each operates according to distinct, and, in the extreme case, perhaps incompatible values.[3]

Because this approach stresses the dominant role economic accumulation has played in shaping and constraining the choices made by urban actors, it differs from the social psychological theories which have traditionally characterized urban studies.[4] Historically, actors who have invented new and competitively successful methods for organizing the creation of wealth have launched social changes which conditioned choices made by everyone else. The industrial revolution undermined traditional settlement patterns and created cites as we know them today. Individuals made choices in response – whether to migrate from the country to the city or to join a trade union – but their choices were shaped and constrained by the evolving structural context.

Similarly, because the approach taken here attributes a strong, if ultimately subordinate, causal role to people's efforts to create community, it also differs substantially from the main lines of thinking emerging with neo-Marxian urban political economy.[5] Although these theorists differ as to the means by which dominance is achieved, they agree that the dominant interests within the system of accumulation (the capitalists) also successfully dominate the society as a whole, including such areas as government, urban form and even popular culture.[6] The argument presented here contrasts with this view in four basic respects.

First, nothing guarantees that either the laws of the market place or the power of economic élites will successfully dominate society. Before the emergence of capitalism, political and religious power obviously constrained economic actors and their market arenas; today economic actors must compete (along with everyone else) to define and preserve the rules of the game. Politics, therefore, counts heavily and can be ignored only with peril by those who are presently winning the game.

Second, to win at politics, those with a stake in the prevailing system of accumulation must create a positive following. They must motivate the balance of society to participate willingly. In other words, they must try to create constituencies which genuinely support the prevailing rules of the game. Simple coercion, manipulation and conquering by dividing cannot, by themselves, produce the desired result, at least over long periods of time. Rather, the players who want to keep on winning in the system of accumulation must create a mass constituency for it.

Third, because at bottom the accumulation process and community relations operate according to different values, the process of forging an accumulation-oriented sense of community is inherently fraught with difficulty. Within any society, and particularly within urban society, there are always people who want to place community life before economic growth. The increasingly inter-dependent nature of modern capitalism reinforces this tendency. The more economic actors try to negate 'economically irrational' communal responses, the more they run the danger of triggering political conflict.[7] Yet where traditional communal patterns impede the advance of accumulation, economic élites usually take the risk.

Finally, because Marxian theories stress the system of accumulation to the exclusion of community formation, they lack an adequate vocabulary for analyzing politics. Class analysis reveals much about the underlying distribution of interests within society, but politics has a frustrating way of not corresponding with underlying interests. Government, like cities themselves, is suspended between community and accumulation. In its political aspects, it is rooted primarily in the former rather than the latter. The basic building blocks of community – ethnic and kinship bonds, geographic propinquity, voluntary associations, shared political convictions – have far more to do with forms of political participation than does class. Even efforts to create a political consensus *favoring* the existing rules of accumulation must naturally have a multi-class, multi-interest base.

The theoretical framework for analyzing urban development presented in the following pages thus challenges both individualistic and structuralist approaches. It attempts to show that the path of urban development is charted through the interaction of two sets of forces and the choices of two sets of actors. Cities have concentrated both kinds of forces and they have proven useful (if not always successful) in containing their volatile interaction. As Lewis Mumford has written,

'The city was the container that brought about the concentrated development of new technology, and through its very form held together new forces, intensified their internal relations, and raised the whole level of achievement.'[8]

On the one hand, cities concentrated the capital, workers, production technology and marketing organization through which the industrial revolution triumphed over pre-industrial patterns. Historically, urbanization and industrialization have been highly correlated. The expanding process of accumulation necessitated the creation of ever-larger population concentrations. And the actors for whom accumulation was paramount sought to develop a sense of community among these people which included a commitment to the prevailing rules of the game.

The increasing velocity of human interaction and economic opportunity within cities drew migrants, who made cities the source of intellectual, personal, technical, cultural and political innovation. At one level, all sought to advance their own economic interests, and cities helped them to do so. But at the same time, these urban dwellers sought more than networks of streets, factories and other physical investments designed to maximize profit. They sought to turn urban environments towards communal as well as economic ends. Out of many fragments, and frequently in the face of great economic difficulty, the urban population built communities based to varying degrees on reciprocity rather than exchange and profit. Though subordinated to the patterns of economic growth and hardly opposed to the principle of material well-being, these emerging neighborhoods sought life beyond work. These communities typically became the main source of resistance, when economic élites felt that economic expansion could only be achieved by reorganizing social life. Thus, as cities concentrated relations both of accumulation and community, they intensified the potential for conflict. As we shall see, the development of various urban institutions, the urban political arena, and even the city itself as a social organization reflects this tension. Figure 13.1 depicts this asymmetric relationship in which the accumulation process lays the basis for expanding communal relationships which in turn tend to conflict with market values. Mediating institutions, including the city itself as a social invention, contain, moderate and reflect this basic tension. Urban political institutions play a key role in this process.

The next section applies this abstract set of distinctions to the development of US cities since the early nineteenth century in order to clarify the actors at work on each side, the strategies they adopted in pursuing

Figure 13.1 The asymmetry between accumulation and community in the urban system

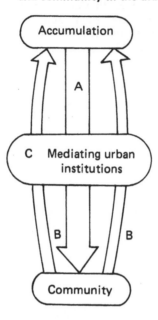

A   Accumulation's impact on community

- creates work, organizes working population
- allocates private resources according to profit criterion
- expands labor, land, consumption goods markets
- influences physical infrastructure, land use patterns

B   Community's impact on accumulation

- differentiates consumption classes, division of labor
- creates communal institutions seeking to maximize well-being of stable social units
- leads to political and other efforts to influence public and private resource allocation for non-market ends
- provides social basis for politics, government action

C   Urban institutions mediating interaction

- government structure, organization of local politics
- traditional organization of ethnically segmented division of labor (embodied in unions, professional groupings, etc.)
- Non-profit organizations of all types, ranging from religious institutions to education system to voluntary associations
- Spatial segmentation of residence patterns

their interests, and the resulting impact on the nature of our cities. It traces the attempt by economic élites at once to undermine ethnic working-class communities which threatened factory production and exercised costly political influence. It also traces community-based responses. The last section abstracts from this discussion the key dimensions on which future analysis should be based.

The contradiction between accumulation and community: phases of American urban development

Though mediating institutions enable cities to intensify relations of accumulation and community, what works in one period can frequently become an impediment in the next. These impediments must be broken apart and reshaped for a new stage of growth to occur. It is no surprise, then, that US cities have experienced a cycle of political and social crisis which parallels and meshes with the major phases of economic development. In US urban history, three such periods can be broadly, if schematically, outlined.

The industrial city

Between 1840 and about the turn of the century, the industrial revolution and the growth of the city-system nourished each other. Industrial capitalism amassed and organized large amounts of capital and people. Capital accumulated during this period both by expanding the scale of production and by intensifying work. As antagonisms between capital and labor mounted, capital responded with generous applications of factory power in its many forms: wage cuts, unemployment, Pinkertons, labor-saving technology, and in increasingly finely-graded division of labor.[9]

The growth of big cities catalyzed and multiplied this development. Chicago, a fifty-person trading outpost in 1830, had swelled to 1.1 million by 1890. Eighty per cent were immigrants, attracted by railroad construction, the stockyards, farm equipment production, and the burgeoning steel industry. They came from the US countryside, Ireland, Germany and later Italy and central Europe. While their heterogeneity and geographic mobility deterred any form of class-wide political action, the immigrant communities did challenge the authority of economic élites. Only the relentlessly commercial urban political machines deflected their efforts.

Indeed, if the urban political machines had not driven out both Yankee aristocratic rule and the fledgling American socialism, it is doubtful that the US industrial revolution could have occurred with such speed and effectiveness. Through them, the highest bidder reigned not only over individual lives, but also over the rate and type of public investments in roads, real estate, transportation, sewers, public utilities and, indeed, over the character of the urban political community.

At this point, urban growth stemmed from rising industrial production.

As Mumford said, 'The factory became the nucleus of the new urban organism. Every other detail of life was subordinate to it.'[10] As the modern business enterprise evolved, however, the biggest cities also developed the seeds of a new era. They became headquarters for increasingly far-flung industrial enterprises. Urban cores began to grow upwards as well as outwards. The modern skyscraper, invented in Chicago around the 1890s, spread rapidly to other cities.

Industrialization organized the size and overall form of cities, and dominated community formation and the integration of immigrant populations into the industrial structure. Machine politics regulated the relationship between accumulation and community, helping to instill a factory-like hierarchy within ethnic communities. It often controlled access to factory, as well as public, jobs. While stabilizing urban conflict, it thus also integrated ethnic élites into the accumulation process. The result was a period of tremendous growth in which cities like Chicago asserted themselves not only over the agrarian midwest, but also over the entire world.

The transition to modern corporate society

In the years after 1900, it became increasingly clear that these highly functional arrangements had begun to create difficulties. Intense competition and technology improvements led to chronic overproduction and economic instability. In this environment, only the moderate corporation, which integrated many production units with increasingly wide-scale marketing, could survive. In local politics during the Progressive era, the emerging managerial class sought to break the hold of ethnic politics on government by making it 'businesslike' and apolitical. Cities thus entered a second, transitional period in which economic élites fragmented and sanitized the centralized power of the ethnic blue-collar political machines.

As working-class ethnic communities built self-help institutions (like savings and burial societies), developed enclaves of small-scale property ownership, and became peacefully integrated into trades or occupations, the political machine became less critical as an engine of social peace. At the same time, the machine's venality and political inertia made life increasingly difficult for the emergent corporations. Even for businessmen with good political connections, the machine represented a politically unreliable monetary burden.[11]

Roughly speaking, the New Deal marked a turning point in the evolution from the factory city to the administrative city. The Depression enabled business élites to complete local government reforms, the brunt of which was to depoliticize and professionalize large areas of public policy, thus removing them from the vagaries of the political marketplace. In 1932, for example, 'the banks agreed to bail out New York, but only if the Mayor took marching orders from the banks and cut the budget drastically.'[12] Detroit, Fall River and a score more cities experienced municipal defaults during the 1930s, and all endured significant periods during which bankers ran their affairs.[13]

Even relatively sound cities found themselves governed by a growing number of semi-autonomous agencies and subjected to increasing state and federal regulation. Civil service, non-partisan at-large elections, and the other post-1900 local government reforms demobilized city politics. And the steady growth of suburbs, which since about 1920 effectively resisted central-city annexation attempts, helped diminish the influence of central-city political conflicts on regional growth.

With this shift in government structure came a noticeable, if at first modest, change in the location of urban economic activities. In a remarkable 1937 report of the National Resources Commission entitled *Our Cities: Their Role in the National Economy,* all the problems of the post-war 'urban crisis' were outlined: decentralization of industry and population, loss of central-city tax base, rapidly rising public-sector wage bill, growing slums, lack of comprehensive planning and the like. The report clearly analyzed these changes as a function of the shift from manufacturing to administration within central cities, which it attributed in turn to changes in the structure of the economy and to the desire of firms to escape 'the obnoxious aspects of urban life.'[14] By this it meant the urban community which nineteenth-century industrialization had created.

Our Cities conveys two overwhelming images of the city. The first is its deeply contradictory nature: cities are 'the focal point of much that is threatening and much that is promising,' one of the economy's 'fundamental supports but also one of the primary problems.'[15] The nation's cities were rapidly 'growing, but also wasting away, and traces of this deterioration are with us today in the form of many blighted neighborhoods.'[16] And the effect on community clearly followed: even though cities had grown rapidly and cities were economic generators, 'the various parts and participants of the urban economy are very highly specialized and the urban way of life is often socially disconnected though economically interdependent.' The result had been serious urban conflict during the previous decades:

'The city is both the great playground and the great battlefield of the Nation – at once the vibrant center of a world of hectic amusement lovers and also the dusty and sometimes smoldering and reddened arena of industrial conflict . . . It is the cities which must deal with the tragic border lines of order and justice in bitter industrial struggles. On these . . . alone many a "good government" has been wrecked.'[17]

In the report's view, a new political way to organize urban communal life was needed to bridge such conflicts.

'How to prevent these strains of separation from disrupting the whole city or its civic groups or even its families, how to weave these vivid and variegated cultures into a positive civic program of intercommunication and cooperation is one of the challenging problems of the coming decade.'

It then proceeds to call for public ownership of land-development rights, a substantial low-rent housing program, better trained municipal administrators, a coherent federal policy toward cities, and the institution of strong regional government. It was a blueprint for planned growth which could not have been more eloquently stated today.

The report's second theme concerns the incipient, but clear, sea-change about to occur in the urban economy. 'Far from being on the decline, the city gives evidence mainly of a new phase of its growth by emptying at the center and spilling over its own corporate boundaries.'[18] Manufacturing was relocating to suburban locations, or to new cities in the South or West, 'to gain competitive advantages derivable from such factors as lessened transport time and cost, freedom from collective bargaining and urban social control, decreased labor turnover, and lower land values and taxes'.[19] Instead, the biggest cities would rely in the future on commercial and service activities necessary for the new corporate economy. 'Thus, the city is both a product and a cause of the division of labor and of specialization.'[20] Unfortunately, gains from these trends might be 'offset by the disadvantages to the community and a loss to the central city.'[21] Hence, on this count strong public planning would be required to smooth the transition and spread the burdens.

New York City took the lead in responding to this need to coordinate the transition towards the post-industrial city. It did so, however, with a disregard for urban working-class neighborhoods which was to become characteristic of the planning profession for decades to come. In a powerful set of documents issued beginning in 1929, the Regional Plan Association (prominent executives backed by the city's major foundations) proposed a set of highway, mass transit, rail and port investments, together with zoning plans which would encourage highrise office building development while removing 'noxious' industrial uses. These documents constituted a blueprint for the next forty years of development. They are all the more remarkable because nearly every element of the proposed plan, save capital intensive suburban mass transit, was ultimately realized. The heart of the Plan was a series of radial and circumferential highways which would allow Manhattan to exert the lure of development within and control the flow of commerce through a fifty-mile radius, and in the process expel unsightly and noxious industries within the city.[22] Those who stress Robert Moses's agency in this change must concede that, regardless of his personality conflicts with the RPA, his highways, parks, bridges and public works followed the RPA's proposals down almost to the last foot.[23]

In sum, three factors set the stage for reorganizing urban form and the nature of urban communal relations between the two world wars. The changing nature of the modern business enterprise led to a centralized consolidation of administration and the increasing relocation of production units beyond the boundaries defined by the nineteenth century industrial city. Second, with corporate initiative, Progressive reforms broke the strong, if corrupted, hold that big city working-class voters held over political institutions. Once necessary to insure a modicum of social peace, the urban political machine had become so much wasteful overhead. And finally, the shifting patterns of investment held out the promise that old, outmoded communal forms could be undermined by the creation of new and more promising forms (whether based on centralized administrative activities or on decentralized production activities). Thus, the simmering class, ethnic and political conflicts characterizing big city life could be overcome.

The post-industrial city

In the years after the Second World War, US cities entered their third phase of development. The full development of the administrative city, and the attendant demise of industrial cities, rested on institutional foundations erected in the inter-war years. The outlines of post-war development have received considerable attention[24] and may be briefly summarized as follows:

- Production activities, including manufacturing and wholesaling, were decentralized to suburban areas and beyond; middle-class housing was almost exclusively located in suburban areas.

- Administration (both public and private), as well as high-level service activity (again, both public, non-profit and private) were largely centralized in the biggest central cities.

- Within the large metropolitan areas, a plethora of independent political jurisdictions were created, including central-city development institutions largely independent of traditional city-hall politics.

- The social base of politics, both in the central city and in the suburbs, was recycled; the traditional blue-collar base for big city politics was attenuated.

- Political authority shifted upward in the system to regional agencies, state government and the federal government; employment at all levels of government expanded substantially and new public services were initiated.

In part, these changes grew out of initiatives taken by economic élites; they were intended to undercut the power and stability of nineteenth-century communal patterns. To a great degree, they succeeded. The success of these initiatives should not, however, be taken as a sign that élites could thoroughly manipulate society against its interests. Rather, the success of these initiatives can in large part be attributed to the fact that they led to the development of new communities which large numbers of central-city residents readily embraced. For a time at least, these new communal forms proved more attractive, more supportive of the accumulation process and less politically difficult. The post-war, post-industrial city was an élite planning experiment which not only worked economically but which recruited a mass constituency.

But in spite of this constituency, the post-war development process was neither democratic nor supportive of existing communal patterns. New public agencies with little accountability (whether suburban governments or central city development agencies) mobilized tremendous amounts of capital to make this form of development possible. Extension of the urban periphery and more intense use of the central business district revolutionized land-use patterns. In the center, ethnic working-class government was undermined, while on the periphery government was characterized by apathy and depoliticization.

Once again, New York City provides an extreme but nonetheless illustrative example of the national development pattern. The Port Authority, the Triborough Bridge and Tunnel Authority, the city's urban renewal agency, the highway planning process, and literally dozens of other autonomous agencies (each with tax powers or dedicated funding streams) assiduously reshaped the city's social, economic, governmental and physical topography. In the sixties, the New York area sprawled outwards ('spread city') while Manhattan experienced an office-space construction boom ('slab city'). The combined public and private debt for the suburban housing tracts and shopping centers, public bridges, roads, tunnels, and transit systems, and central-city public works, office buildings, and other infrastructure rapidly grew to a staggering figure. (Nationally, such debt grew from $355,000,000,000 in 1947 to $1,780,000,000,000 in 1973 – a fivefold increase.) Investment in this new urban form constituted a major component of post-war economic growth. Though real-estate finance reached its apogee in New York City, tax loopholes, real-estate investment trusts, syndication, the secondary mortgage market, 'moral obligation' bonds and other novel forms of public and private financing became widespread. Since these devices depend on constantly rising land values, city governments and independent agencies attempted to produce increases as a matter of self-interest.[25]

Spending for government services rose in parallel with these physical investments. The educational, manpower, transportation, land-use and policy-planning needs generated by the 'post industrialization' of the urban form propelled the growth of city government budgets and increases in inter-governmental transfers. In New York, city government employment rose from 246,800 in 1960 to 371,000 in 1970 (the largest increases occurred in education, police and welfare agencies). The city's total spending doubled during the period with major increases in income transfers (480 per cent), health services (370 per cent) and education (270 per cent).[26] While some of this spending compensated central-city minorities who had been excluded from the benefits of metropolitan expansion, it is interesting to note that the bulk of the increases were directly related to expanding high-level services (health, education) in the central city. Doctors, hospitals, professors and educational institutions most directly benefited from the growth of government. In turn, they and their colleagues in other high-level services helped reshape New York's physical form and demographic composition.

The net result, in Theodore Lowi's terms, was that New York and the other large cities were 'well run but ungoverned'. Specialized, autonomous agencies like the Metropolitan Transit Agency or the urban renewal administration each promoted growth in their own way. No single authority oversaw the whole or could guarantee rational coordination. This arrangement stimulated growth while fragmenting the basis on which political opposition might arise, but it also compounded costs. Together with the damage such agencies inflicted on neighborhoods in New York it created troubles for the post-war balance between accumulation and community.

Dénouement of the post-industrial city

While the post-war strategy of urban development created and satisfied one set of constituencies, it soured others. Growth in government spending proved to be beyond the control of economic élites as well as the public at large. This troubled the private sector. Their dismay grew sharper as growth-oriented agencies appeared increasingly unable to pay their own way with real economic expansion. The devices which economic élites had invented to pull urban development out of its previous problems had, like the brooms which carried pails of water for the Sorcerer's Apprentice, created new problems through their very success. The ungoverned public sector thus lost an important part of its élite constituency because it was unclear whether the expansion of accumulation was being helped or hindered.[27]

Many of those living within the central cities suffered no such ambivalence. In the course of erecting the post-war pattern of urbanization, government agencies literally demolished many inner-city neighborhoods. What the bulldozers failed to reshape in the late 1950s and early 1960s, 'softer' programs of rehabilitation and upgrading accomplished in the late 1960s and 1970s. Even in areas where market demand would not respond to government programs, these programs hastened the withdrawal of equity and the collapse of communal relations. This assault on central-city neighborhoods produced a veritable 'community revolution'. While central-city residents, often but not always minority group members, had never strongly favored the development patterns set in motion after the Second World War, they reacted vigorously, sometimes violently, when they were selected to bear the costs. The post-industrial development strategy thus lost legitimacy from below as well as from above. From the 1960s to the present, these central-city neighborhoods developed a sense of community deeply suspicious of government authority in service to market forces.

Finally, there are growing signs that post-war urbanization is losing even the strongest of its friends from the 1950s. By urbanizing the suburbs, post-war development patterns created suburban variants of the big-city problems which capital had hoped to escape. Congested journeys to work, housing shortages and rising housing costs, difficulty in financing public services and the lack of public amenities have become fairly typical suburban problems, especially in inner-ring areas. Some suburbs face declining prestige, while others have become so expensive that the children who grew up in them cannot afford them. The urban malaise has thus affected suburbs as well as central cities.

Summation

Standing back from this historical sketch of the interplay between the expanding accumulation process and community formation, several general themes can be drawn out. Most evident is a cycle of growth and conflict in which the accumulation process leads to the formation of communities which it ultimately finds to be an impediment to further expansion. In the industrial period, highly concentrated working-class neighborhoods represented at first a direct threat and subsequently (with the development of machine politics) a heavy carrying cost. In the post-war period, post-industrial urbanization encountered trouble both from the rising tide of neighborhood and civil rights activism and from a decay in the growth 'yield' produced by government programs. In both cases, techniques for expanding accumulation which had great success in the beginning ended up at an impasse.

Second, leading economic actors have typically responded to such problems by undermining prevailing communal forms while simultaneously erecting new, more accommodating communal forms. Over the long term, economic actors have attempted to transcend the industrial city's generic problems by breaking the historic link between production and urbanization. By creating 'factories in the fields' on the one hand and a new central-city community of cosmopolitan professionals on the other, economic élites mostly succeeded. But the dissolution of the nineteenth-century urban industrial social fabric was accomplished at great and continuing cost. If those who seek to advance the process of accumulation beyond its current impasse adapt this traditional strategy to current conditions, they will probably have to pay (or exact) a correspondingly higher cost.

Third, the attempt to undermine some forms of community while creating others was inextricably tied up with changing the structure and accountability of government. The increasing size of government provided the resources and organizational capacity necessary to implement such unprecedented social change. But in order to make sure this power did not fall into the wrong hands, innovators built a lack of accountability and political responsiveness (to élites as well as to everyone else!) into the structure of government. As a result, the difficulties faced by those who seek to intervene in the structure of community formation have gone up a level of magnitude. Since much of the post-industrial city's fabric depends on the government, achieving greater responsiveness to the needs of the accumulation process may call for truly traumatic surgery.

Finally, throughout these phases, the communities under assault have shown an increased willingness to organize in their own behalf. The 1960s revolt of central-city minority neighborhoods invented a political vocabulary which has been embraced not only by service professionals choosing to reside in the central city, but by the suburban dwellers as well. This has induced government agencies and local political leaders to become much more neighborhood-oriented and participation-oriented at least in their rhetorical style. If the private sector directly challenges the public sector, it may reinforce this link.

Having sketched out the phases of development within US urban history over the last 140-odd years, and having drawn some elementary conclusions from this sketch, what can be said about the analytic dimensions which deserve more careful attention?

Levels of analysis in the study of urban development

The terms used in this sketch of the history and politics of urban development span an uncomfortably large gap between broad-scale structural movements of the economy and the specific nature of an individual's experience with urban life. At the structural, systematic and abstract level, notions such as accumulation and the progressively increasing role of government seem mutually exclusive with those at the individual level such as neighborhood, communal interaction, social networks and leadership patterns.

Debate within both orthodox and Marxian social science has been ambivalent, even confused, about which end of this continuum to use as the epistemological starting point.[28] The nub of the problem is that structural explanations suppress the issue of action and therefore tend towards the ahistorical, abstract and static. Reliance on individualism, however, continues an opposite weakness. Concentrating on individual actions tends to produce voluntarist, pluralist conceptions of how social systems operate. Epistemological debate has unfortunately called for choosing one approach over the other. The real problem is to construct theories which operate at *both* levels simultaneously and which isolate the key interactions among these levels. The following pages try to bring to the surface the dimensions of analysis implicit in the previous discussion.

Analytic levels and key variables

At the greatest level of abstraction, lies the (i) *process of accumulation* itself. The way a society produces and reproduces its means of sustenance has a central bearing on all its other aspects. Through this lens, the analyst can observe the formation of classes and occupational orders, their social, economic and political relationship to each other, different actors' market position, and the generic conflicts which result. At this level, historical change appears driven by external, impersonal and uncontrollable forces. The centuries-long process by which capitalism exploded and reorganized agrarian feudalism provides one example of this phenomenon. The twentieth-century domination a small number of multinational firms exercise over a highly integrated world economy is another. While individuals and organizations take actions which advance such changes, they do so because they are responding to structurally defined possibilities. In the urban context, this level of analysis requires us to decode the frameworks which define urban options. These include the economic roles played by different kinds of cities; the influence of wider markets on city-country relations and the inter-city competition; the broad organization of political authority; and the nature of cultural values.

The next lower level concerns the (ii) *institutional structures* within the economy and polity *which create the social capacity to act.* The city itself is a complex device built on a web of individual institutions designed for this purpose. These institutions assemble and transform streams of resources in order to achieve expanding production, exchange, and profit (if they are in the private sector), or useful but not necessarily marketable goods and services (if they are public or non-profit).

As Weber commented in connection with the origins of urban administration during the medieval twilight,

> 'Within a city the forms of technical administration available for regulating relations between the estates cannot be altered at will. The forms of political administration obey their own laws and are not to be interpreted merely as a superstructure resting on economic foundations.'[29]

'Merely' is a key to this sentence: while economic factors influence public and non-profit administrative devices, they are not completely determining. Theda Skoepel has stressed the need to consider this level of analysis in addition to higher and lower levels. ' "Ruling Class" explanations . . . tend to argue from class interests to intentions to political outcomes,' she writes, 'while missing the point that state structures mediate and modify the capacity of classes to attain their ends'.[30] In the previous discussion, both the changing nature of business enterprises and the changing structure of government require close attention. In particular, private actors apparently appealed to higher levels of government to help overcome their problems with urban politics. They also sought to reorganize local government with a good deal of success. These institutional changes clearly altered the relative power of different social groups to determine urban outcomes.

Just as the broader outlines of the accumulation process shape and circumscribe institutional structures, so the mosaic of such structures which make up the city sets the parameters for (iii) *political alliance formation and mobilization.* Political movements and alliances are rarely based on structural distinctions in any clear way. Instead, potential political alliances and the interests which undergird them arise through the operation of specific institutions such as the post-war development agencies. The source of inputs, the way institutions allocate them and the constituencies to which institutions respond each provide the basis for political mobilization and alliance formation.

In some important respects, political interests discussed previously were in fact defined by government (rather than the reverse which is traditionally assumed). Public housing tenants, defense contractors, municipal service recipients and urban property owners provide examples. These groups find their political vocabulary in the language of program design, forms of participation and constitutional authority. The interests and alliances so formed congeal over time into fixed political relations which may even become institutions. Such organizations exert their own peculiar impact:

entrepreneurship, brokering and negotiating, interest 'marketing' and organizational maintenance all contribute a special flavor to larger political
tensions. And at the same time they internalize the larger tensions to a
considerable degree. This is certainly true of the urban development agencies previously discussed.

Finally, we come to the most individual and subjective level: how people
interact with each other on a day-to-day basis. The (iv) *network of social
ties* is obviously shaped by all the previously mentioned levels, for people
interact on the basis of where they live, where they work, which social
stratum (with attendant culture) they belong to and which institutions
they join (whether by necessity or choice). In distinction to higher levels,
these attributes are not determined impersonally. People can and do take
a hand shaping their own ties. Society circumscribes this ability, but it
is perhaps mankind's most basic genius that it does not completely succeed.
Even in the worst of cases, new connections have always been possible.
If anything, the urban development patterns discussed above have given
wider scope to this human ability. As urbanization has broken down kinship
patterns, religious association and fixed status groups, it has opened the
way for self-created links, for intentional community.

Links among the levels of analysis

It should be apparent from this discussion that each level contributes
something special to the organization of urban politics and urban life.
Each has a distinct logic and scope of impact. As a result, the forces
operating on the city may best be understood as discrete but interdependent,
and as a result potentially at odds. No single level can displace the others
as an exclusive mode of explanation. Nor can the overall phenomenon
of urban life be understood without conceding that each level makes a
contribution.

So far, stress has been placed on how higher levels constrain lower
ones, and how historical trends circumscribe agents' ability to act and
define the ends which they are likely to seek. Seen this way, lower levels
of analysis appear to be nested within higher ones. The reverse is also
true, however. Lower levels, since they are the vehicles through which
higher level forces come to reality, exert imperatives upwards which may
break open the shell of constraint. Political movements, churches, government programs and firms rely, for example, on established patterns of
social ties. If these change rapidly, higher levels may begin to unravel.
History provides mankind not only with conditions to be accepted, but
also with conditions which provide the chance to succeed or fail at deliberate social change. The previous discussion hazards the conclusion that
inner-city neighborhoods share this opportunity (though their job is clearly
much more difficult) with economic élites.

The implications of multilevel analysis

Several important, if perhaps obvious, points follow from this view. First,
a satisfactory theory must examine the interactions among the levels as

they change over time. In particular, theories of urban development must look at how the structural context motivates individual action and how these actions in turn modify the content, the 'rules of the game'. The interplay between community and accumulation has been offered as a way of thinking about these issues.

It follows, secondly, that different historical periods will be characterized by distinct balances between levels of causation, that is to say, between context and actors. At some points, actors can line up forces at each level which unlock the way to major social changes. The ascendance of nineteenth-century factory production provides us with one example; post-war development provides another. These strategies for expanding urban development were worked out by many parties, contending in a hit-or-miss way, over many years. For a time they succeeded, but almost inevitably they generated new conflicts that required new alignments to be worked out. To date, such efforts have given cities their dual qualities of destruction and regeneration.

Thirdly, and finally, given the failure of past methods to continue smoothly and 'automatically' to generate urban development, politics nec-essarily comes to the fore. To overcome nineteenth-century problems, élites negotiated the expansion of government while reducing its accountability. This has increased the intervention required for setting any new direction by a considerable amount and has made urban government once more a focal point of important systemic conflict.

As government at all levels has expanded and multiplied, it has become less capable of absorbing and deflecting systemic conflict. Ironically, though large cities influence increasingly large areas because government plays such a large role in organizing urban life, the cities have become more, not less, subject to challenge.

The outcome is far from clear. Those who believe that community forms must once more be demolished and reorganized in order to promote re-newed accumulation may well be willing to demand the increasingly high price which will be required. Twenty-five years ago their counterparts were perfectly willing to obliterate 'outmoded' land-use patterns; many seem willing today to undermine 'outmoded' urban government, social services and community patterns. Because of past choices, they must pull down an even larger edifice than before and probably lose some of their own foundation in the process.

On the other hand, many of those who have experienced or observed the process by which communal values were undermined for the sake of market values have come to question the necessity as well as the wisdom of such actions. Speaking of the time when Robert Moses drove the Cross Bronx Expressway through 'the shoddy but vibrant world of a second generation Jewish neighborhood a generation ago', Marshall Berman has noted,

'The ironic fact is that the neighborhood people themselves lacked these ideas [the value of neighborhoods] when they were most in need of them; they did not have the vocabulary to defend their neighborhoods because, until the Sixties, that vocabulary simply did not exist.'[31]

Perhaps now that we have it, we will invent a way to advance urban development without sacrificing our communities.

Notes

[1] My epistemological starting point resembles Melville's discussion of the Loom of Time in *Moby Dick,* in which 'chance, free will, and necessity – no wise incompatible – all interweavingly work together'. Melville, H. (1950) *Moby Dick.* New York: Random House, p. 213.

[2] For further elaboration, see my *Building the Unfinished: The Politics of Urban Development and the Development of Urban Politics* (forthcoming). This paper is dedicated to the people of Bernal Heights, who are working to define their own sense of community. Thanks to Richard Ingersoll and Kathleen Gerson for editorial aid.

[3] Fischer, C. *et al.* (1978) *Networks and Places.* New York: Free Press, contains one of the best discussions of the meanings which the notion of community might take on.

[4] The Chicago school of sociology stresses an individualist, social psychological approach to the city. For a recent summary of this literature, see Fischer, C. (1976) *The Urban Experience.* New York: Harcourt Brace Jovanovich.

[5] Harvey, D. (1973) *Social Justice and the City.* Baltimore: Johns Hopkins; Castells, M. (1978) *The Urban Question.* Cambridge, Mass.: MIT Press.

[6] David Gordon has argued, for example, that urban development 'corresponds' to the stage of accumulation and was designed 'at least partly to reproduce capitalist control'. 'Capitalist development and the history of American cities', in W. Tabb and L. Sawers (1978) *Marxism and the Metropolis.* London: Oxford University Press, p. 27.

[7] Not all Marxian theorists have taken the route of economic determinism. For one thoughtful commentary on the problems of creating political community oriented toward accumulation, see Wolfe, A. (1978) *The Limits of Legitimacy.* New York: Free Press.

[8] Mumford, L. (1961) *The City in History.* New York: Harcourt Brace Jovanovich, p. 34.

[9] In general this discussion follows Braverman, H. (1974) *Labor and Monopoly Capital.* New York: Monthly Review Press; Stone, K. (1974) 'The origins of job structures in the steel industry.' *Review of Radical Political Economy* 6 (Summer); and Brody, D. (1962) *Steelworkers in America: The Non Union Era.* Cambridge, Mass.: Harvard University Press.

[10] Mumford 1961: 458.

[11] Hayes, S. (1964) 'The politics of reform in municipal government in the Progressive era.' *Pacific Northwest Quarterly* 55 (October): p. 157–69, remains the landmark study of reform in this era. See also Shafter, M. (1974) 'The emergence of the political machine, an alternative view', in M. Lipsky, W. Hawley *et al. New Theoretical Perspectives on Urban Politics.* New Jersey: Prentice-Hall, pp. 14–44.

12 'Experts fear growth in costly city debts.' *New York Times* (25 November 1974): 48. The article relates current problems to New York's Depression experience.

13 An extremely interesting series of case studies of Depression-era defaults is given in the Advisory Commission on Intergovernmental Relations (1973) report 'City financial emergencies: the intergovernmental dimension.' Washington, DC.

14 National Resources Committee Report of the Urbanism Committee (1937) *Our Cities: Their Role in the National Economy.* Washington, DC: US Government Printing Office, p. 35.

15 National Resources Committee 1937: v.

16 National Resources Committee 1937: vii.

17 National Resources Committee 1937: vii.

18 National Resources Committee 1937: 35.

19 National Resources Committee 1937: 38.

20 National Resources Committee 1937: 37.

21 National Resources Committee 1937: 38.

22 Regional Plan Association (1929) *Regional Plan of New York and Its Environs,* 2 vols. New York: Regional Plan Association.

23 Caro, R. (1974) *The Power Broker.* New York: Knopf, fails to mention RPA. In his one major battle with the RPA, over whether a bridge or tunnel should be built to get from Battery Park to Brooklyn, Moses lost. See Hays, F. (1965) *Community Leadership.* New York: Columbia University Press, for a revealing study of RPA. See also Fitch, R. (1978) 'Planning New York City,' in R. Alcaly and D. Mermelstein (eds) *The Fiscal Crisis of the Cities.* New York: Random House.

24 Wood, R. (1961) *1400 Governments.* Cambridge, Mass.: Harvard University Press, offers the classic description of metropolitan political fragmentation as a device for subordinating government to the market place. See also Hoover, E. and Vernon, R. (1962) *Anatomy of a Metropolis.* New York: Doubleday; and my own (1975) 'Postwar politics of urban development.' *Politics and Society* 5, 3.

25 Harvey, D. (1975) 'The political economy of urbanization in advanced capitalist countries – the case of the US', in G. Gappert and H. Rose (eds) *Urban Affairs Annual No. 9.* Beverly Hills: Sage Publications; and Stone, M. E. (1975) 'The housing crisis, mortgage lending, and class struggle.' *Antipode* 7 (September): 22–37, thoroughly analyzes the scope and import of this development.

26 Brecher, C. (1974) *Where Have All the Dollars Gone?* New York: Praeger, pp. 20–5.

27 *Business Week* (22 September 1975) noted that 'private investment cannot increase as a share of the GNP unless government spending declines' (115).

28 For one discussion, see Eulau, H. (1969) *Micro-Macro Political Analysis.* Chicago: Aldine.

29 Weber, M. (1958) *The City.* New York: Free Press, p. 166.

30 Skoepel, T. (1973) 'A critical review of Barrington Moore's Social Origins of Dictatorship and Democracy.' *Politics and Society* 4 (Fall).

31 Berman, M. (1975) 'Buildings Are Judgments.' *Ramparts* (March): 56.

14 Accumulation versus reproduction in the inner city: *The Recurrent Crisis of London* revisited*

Damaris Rose

'Public clearances have made less than no impression since the census of 1891 on the general want of housing . . . the deficit of sleeping room for the poor workers is worse now than it was then . . . the courts and streets where these poor folks lived are being destroyed, and soon will be replaced by new buildings, and put to commercial uses, which are infinitely more remunerative than housing the people . . . We find ourselves, then, at this pass. A multitude of poor workers' families must live near the centre; and the land near the centre has attained a value the equivalent of which they cannot conceivably pay, and it goes on increasing with fatal certainty. . . . To say that the community should buy up slum areas, and erect on them an adequate number of workmen's dwellings, under present conditions of cost, is appalling. . . . We must be prepared, if we go in for municipal housing of this kind, to face the fact that we will be sacrificing vast portions of the possible economic value of sites in the very centre of the world, and will be providing homes for certain needful sections of the community at wholly artificial rents. . . . But can we avoid this necessity in the long run, if we are really in earnest about the housing of the casual poor in London?' (B. F. C. Costelloe (1898–9) 'The Housing Problem.' *Trans. Manchester Statistical Society:* 32–4, quoted in Rubinstein 1974: 185–7.)

* This paper is based on Chapter 6 of an unpublished MA research paper (1978) 'Housing policy, urbanization and the reproduction of labour power in mid-to-late nineteenth-century Britain: a conceptual appraisal', University of Toronto. My thanks to the editors of the present volume, and to all who made helpful comments and criticisms of earlier versions – especially Jim Lemon, Shoukry Roweis, Andrew Sayer, Nathan Edelson and numerous other members of the Union of Socialist Geographers.

Introduction

The dilemma encapsulated in this extract from a speech by a Progressive member of the London County Council in 1898 has a familiar ring to it, both for academics studying 'inner-city problems' and for those engaged in urban community movements and struggles over renewal and redevelopment schemes in the 1970s. The persistence and apparent intractability of 'urban problems' in spite of a number of years of state attention to 'urban deprivation' and poverty (see Community Development Project 1977b), coupled with the failure of 'mainstream' urban geographical and sociological analysis to provide convincing accounts of the nature of the problems, led to the emergence in the mid-1970s of a body of 'radical' analysis. Although not homogeneous, this literature tended to discuss the conflicts in the city in terms of Property Developers versus the People (Ambrose and Colenutt 1975; Barker *et. al.* 1973), *Profits against Houses* (Community Development Project 1976) or the *Hegemony of Finance Capital over the Community* (Harvey 1974). In the British context, important questions were raised by such analyses, such as: why was central city land laid waste or used for commercial redevelopment when thousands of people were homeless or living in substandard conditions? why did the urban planning process consistently seem to create an inequitable distribution of costs and benefits in spite of the intentions of the planners? why, despite a century of state housing and planning policies, was there still a *Recurrent Crisis of London?*[1]

As the debate on these and related questions matured, it became increasingly apparent that 'urban problems', and the nature of state housing and planning policies, could only be adequately understood using historical materialist analysis, which saw the city neither as an 'independent variable' amenable to positivistic measurement techniques nor as the unfortunate object of cynical manipulation by unscrupulous individuals or institutions, but rather as the dynamic expression of a historically-determined *process* of urbanization – a process shaped in specific ways (which were yet to be specified) by the relations of production and reproduction of capitalist society.

The present paper came to be written in the context of this maturing debate and may best be seen as an intellectual 'staging post'. It focuses on London, and represents a preliminary attempt to analyse the historical development of the stubbornest of the city's problems from the mid-nineteenth century onwards, and to use this analysis to clarify the nature of the dilemmas associated with state housing and planning policies in the metropolitan city. The discussion centres on the chain of events leading to and following on from what became widely known as London's 'housing crisis' of the 1880s, using secondary sources illustrated by a small amount of primary data. By reassembling and reinterpreting existing scholarly research on urbanization, housing and the structure of labour markets in nineteenth-century London, and by briefly reviewing contemporary developments in the inner city, the paper attempts to cast some new light on the nature of the 'recurrent crisis'.

The paper is divided into four sections: the first, an account of the genesis of the crisis of the 1880s; the second, an analysis of the responses of central and local government to the crisis; the third, an assessment of the extent to which these responses constituted a 'solution' both in the short and the longer term, in the light of a reassessment of the nature of the conflicts and their relationship to two key 'structural' imperatives of capitalist development; and finally a re-evaluation of the nature of the questions we should be asking about 'urban problems'.

The genesis of the crisis

During the first thirty years or so of the nineteenth century the idea of 'housing policy' or 'urban policy' as an arena for active state intervention was not legitimate or even conceivable in Britain. The establishment of a 'free market' in labour, foodstuffs and other basic commodities was essential to the development of the emerging form of industrial capitalism into a continually-expanding process of capital accumulation. All able-bodied people had thus to be forced to sell their labour power for a wage with which to purchase their own food and shelter, without assistance from the State. This was largely achieved by the Poor Law Amendment Act of 1834, which abolished the 'Speenhamland system' of poor relief, restricting payments to those who were unemployed and prepared to stand the rigours of the 'workhouse test', and which thus led to an accelerated rate of rural-urban migration and the mass urbanization of the proletariat. (See e.g. Checkland and Checkland 1974: 9–47; Polanyi 1944: 77–85.)

By mid-century, however, the insanitary nature of the conditions in the newly-industrialized towns and cities, and the threats to public order and to the very survival of the industrial labour-force which they seemed to pose, had become so apparent that a limited form of state intervention in the city began to gain a reluctant acceptance, signalled by the passage of the first parliamentary public health legislation in 1848.[2] The 'urban question' was, however, viewed mainly as a question of health and sanitation. Legislation passed in the 1850s and 1860s was mainly concerned with the regulation of lodging houses and with rudimentary aspects of public health. Local authorities were given increased powers for entry to premises, for ordering that repairs to insanitary property be carried out, for prosecuting the proprietors of overcrowded dwellings and for enforcing certain standards of sewage disposal and ventilation (see Table 14.1, and Gauldie 1974; Tarn 1971; Wohl 1977). Yet by 1866 there were signs that the housing question was beginning to be viewed in broader terms than the physical (and moral) proximity of the occupants of individual dwellings; in the Sanitary Act of 1866 'the concept of overcrowding was taken somewhat outside of the context of sanitary conditions and [placed] within a context of the availability of accommodation', reflecting a new concern with the 'quantity as well as the quality of working-class houses' (Wohl 1977: 81). Yet, despite increasingly thorough research into housing and sanitary conditions by the expanding profession of public-health officers, there was relatively little action to curb overcrowding because no legislative procedures were available for demolition of insanitary houses or for rehousing the evicted. It was assumed by central government that the free market could provide for working-class housing, if given legislative encouragement in the form of liberalization of land-transaction laws and the provision of subsidized loans to Model Dwellings Companies[3] and other public and private bodies prepared to build low-cost housing meeting certain sanitary standards. This belief was institutionalized in the Labouring Classes Dwelling House Act, 1866 (Gauldie 1974: 261).

In response to this inaction, a bill was introduced into the Commons in 1866 by William Torrens, 'designed to establish local municipal government as a major agency of slum clearance and urban renewal' (Wohl 1977: 84). In its original form it emphasized overcrowding and the need for local vestries 'to demolish insanitary houses, purchase land for building purposes, and build and own dwellings. . . . Action could be demanded of the vestries by the Home Secretary' (Wohl 1977: 85). However, the Act which was finally passed, after many stormy debates, in 1868, was greatly emasculated: all references to rehousing were removed and the Act was limited to the compulsory purchase of single dwellings or small pockets of slums for demolition or improvement by the vestries (Wohl 1971: 85–92; Gauldie 1974: 267–72 give details).

The Torrens Act was little used. Dissatisfaction with its piecemeal nature and pressures for large-scale demolition and rehousing mounted in the early 1870s, leading to the introduction of a bill by Richard Cross (Gladstone's Home Secretary) in 1875. This became the Artisans and Labourers Dwellings Improvement Act, which gave powers to London's Metropolitan Board of Works to clear large tracts of insanitary housing. (It also applied to all towns with a population greater than 25,000.) The land thus cleared was to be either leased or sold at below its market value to individual private builders or to Model Dwellings Associations and other semi-philanthropic bodies, on the condition that they rehoused all those people displaced, on site, within three years of demolition.[4] Alternatively, local authorities could themselves build working-class housing on cleared sites, but they were not encouraged to do so. Rehousing powers could only be used at the discretion of the Home Secretary (or the Local Government Board outside of London) (Wohl 1977: 102) where private bodies did not take up the cleared land, and any houses built by the local authorities had to be sold within ten years.[5] Landlords had to be compensated for compulsory purchases, but the full costs of the land were not to be passed on to the rehousing agencies for fear of deterring them from taking up the land for redevelopment.

Different tiers of local government were involved in the operation of the Torrens and Cross Acts in London. This often led to conflicts between the Metropolitan Board of Works and the vestries (of which there were more than 200) as to which body should implement what clearance schemes and accordingly shoulder the attendant burden on the rates, which could be substantial (Wohl 1977: 128–9). Nevertheless, the powers contained in the Cross Act were used fairly extensively in London – especially in the early years of its operation. In 1877 alone, over 14,000 people were displaced from twenty-three acres of land (Wohl 1977: 135). This Act, combined with various other pieces of legislation clustered under an ideological umbrella of 'improvement', notably street-widening schemes (see below, also Wohl 1977: 26–9, 35; Jones 1976: 180–3; Weinberg 1974: 109–13; Smith 1977: 22–3) were used to raze to the ground extensive areas of the most unhealthy slums in London (Wohl 1977: 133); and, on a smaller scale, in other cities such as Liverpool (Taylor 1974: 75), dispersing some of the more notorious criminal haunts and making street policing easier in the process (Gauldie 1974: 267; Jones 1976: 166–7).

However, when it came to disposal of the cleared lands, local authorities encountered great difficulties in finding buyers or lessees willing to build working-class housing on the cleared sites for numbers of people equal to those displaced. As a contributor to the *Quarterly Review* in 1884 pointed out,

'The fact is that rebuilding has not kept pace with pulling down. Notwithstanding the enormous sacrifice at which the Metropolitan Board of Works is forced to offer sites for sale, purchasers cannot be found. Artisans' dwellings, built according to the Board's plans and under its supervision, are not a promising speculation. Even the great building companies are not attracted.' (Quoted in Rubinstein 1974: 160)[6]

In part this was due to technical problems, and hence high building costs, entailed in building high-density housing meeting the new standards of building design and sanitation, which led to lower rates of return for the Model Dwellings Companies (Gauldie 1974: 163, 225); similar problems were also faced by local authorities planning rehousing schemes (Taylor 1974: 75). Although this could have been solved by adopting techniques of multi-storey building on a much larger scale, as was done in some provincial cities and on the continent, the initial capital outlay that would have been necessary was beyond the reach of the small speculative builder who dominated the supply side of the market in working-class housing. Additionally, multi-family dwellings were not popular among those artisans and 'labour aristocrats' who could exercise some degree of choice about their living conditions. In the uncertain economic conditions which persisted even in the 'boom years' of mid-century (Church 1975: 72–3; Tholfsen 1976: 193), it was important for skilled workers to maintain their position and status by escaping identification with the 'residuum' or 'underclass' of casual workers and unemployed (Gray 1976: 88–9). An important means of achieving this differentiation was by living and bringing up one's children in the privacy of a 'respectable' single-family home, if possible in a residential neighbourhood away from the stigmatized slums (Davidoff *et al.* 1976: 153; Gray 1976: 98; Meacham 1977: 28–9).

From the speculative builder's viewpoint there were other more attractive alternatives for housing investment than building working-class housing in the inner city. In the lusher pastures of the new commuter suburbs land was relatively cheap and semi-detached 'villa' building for the better-off artisans and burgeoning middle-classes (often using capital borrowed from building societies, especially after the Building Societies Act, 1874) was becoming a fast-booming business (Dyos 1961: 82–3, 122–37; Wohl 1977: 228, 293–4; Gauldie 1974: 281). Working-class housing could not compete:

> 'Sites in the inner city were too expensive, and the inability of the masses to afford the rents which new buildings in the centre would have had to command to be profitable meant that private capital could hardly be attracted into the field, especially as capital already had so many other profitable outlets, including suburban development.' (Wohl 1977: 304)

The Metropolitan Board of Works was unwilling to build dwellings itself as land costs had become even more inflated by speculators as a result of the generous compensation given to slum landlords, although the rate of compensation was lowered in 1879 (Dewsnup 1907: 227; Wohl 1977: 105, 131–2). Meanwhile – and crucially – inner-city land was urgently required for the further expansion of commercial activities. (Earlier expansion had been facilitated by the construction of the London railway termini and lines in the mid-decades as well as by street improvements and other public works – we will return to this below.) This created a tremendous upward pressure on land values, the competitive market in land having been established by the operation of the leasehold system in London (Weinberg 1974: 56–7). Central government soon capitulated to this pressure after a Select Committee had reported in 1882 that the requirement to rehouse all those evicted in the same district was holding up both residential and commercial development. In 1879 local authorities had been given permission to rehouse the evicted in 'equally convenient accommodation' elsewhere in their own districts (Wohl 1977: 105–6) and in 1882 the obligation to rehouse was reduced to 50 per cent of those displaced (Gauldie 1974: 284–5), thus 'releasing the potential' of the cleared land, which local authorities were now explicitly permitted to re-let for commercial uses. They found no shortage of commercial takers (Gauldie 1974: 282). By 1890 local authorities were being encouraged to build working-class housing outside of their own inner-urban areas, with exceptions made for the rehousing of small groups of essential workers in the inner city. Thus, as early as the 1880s, a precedent was set for the allocation of land by the State to its 'highest and best' use rather than according to any criteria of 'social rationality', defined independently of capitalist production relations. This precedent had specific historical origins in the nineteenth-century capitalist urbanization process. Yet it has been enshrined as a *natural* and eternal 'law' in neo-classical urban land-rent theory[7] and incorporated into contemporary urban planning doctrine.[8]

It was consistently argued by Parliament and housing reformers in the

1870s that those displaced by clearance schemes were being adequately provided for in the suburbs by private enterprise construction. In this respect, Joseph Chamberlain, as radical Tory mayor of Birmingham, did much to bring his city's policies into the public eye. Birmingham Corporation had purchased inner-city sites – factories as well as houses – and had thus been able to decentralize jobs as well as people, making way for commercial development in the vacated central areas of the city at a substantial profit to the Corporation; this money was used for further street improvements and public works to enhance the city's attraction as a centre of commerce in the provinces. In contrast to the large sector of casualized labour in London, Birmingham had a substantial and prosperous 'core' of artisans who provided a secure market for suburban speculators (Gauldie 1974: 155); indeed, that city's freehold land and building societies were forerunners in the movement and enabled a substantial number to become owner-occupiers (Chapman and Bartlett 1971: 238). Birmingham's experience was widely discussed and admired but the significance of the differences in the social structure and the labour market for the construction of new housing was not yet fully appreciated (Wohl 1977: 229; Gauldie 1974: 279–80). Moreover, housing construction reached a national peak in 1876, but twenty years of continuous decline in the rate of building were to follow. Yet, as Gauldie (1974: 281) points out: 'It was not immediately apparent in the 1870s that those moving into the new houses were seldom those displaced from the old, and that a substantial section of the population was becoming unhousable by private enterprise.' Nor was it immediately realised that even when new housing was built in the inner cities,

'The houses that were created were too good for the class of tenants that we displaced in the lower part of the town; and the consequence was that the houses were never let; but when they were exposed for sale they were [sic] very eagerly looked after by artisans of a very superior class.' (Evidence given by Edinburgh town clerk to Royal Commission on the Housing of the Working Classes (Scotland) 1884–5, quoted in Gray 1976: 96)

Thus, the persistence of non-action by the central State in legislating for compulsory building of working-class houses, and the slowness of local authorities in implementing permissive legislation, appeared to be justified by the facts. A Select Committee, meeting to discuss the implementation of the Cross Act, was confident in its conclusion that private enterprise was still doing the job – despite hearing only incomplete evidence from just three cities (Gauldie 1974: 266). Interestingly, this was the very same committee which also argued that the requirements to 'rehouse on site' or in the immediate vicinity of clearances were holding up development.

There was a widespread belief that suburban development created vacancies in good-quality housing in the inner city for those working-class people displaced by clearances, so that, it was argued, building for the rich or the middle-class would eventually benefit all classes (Jones 1976: 206–8). This process became known as the 'filtering', 'turnover' or 'levelling-up' mechanism;[9] it was subsequently accorded academic respectability by the urban ecologists of the Chicago school in the 1920s, and later elevated to the status of an absolute truth used to legitimate the status quo in the urban land and housing system by their geographical disciples.[10] In the early 1880s these arguments continued to be used to justify the state-assisted process of intensification of central land uses, although by this time disturbing evidence was coming to light to belie the apparent smoothness of the suburbanization trend and to challenge the efficacy of the filtering mechanism in solving the 'overcrowding problems'.[11] The low incomes of the casual workers who moved into the inner-city houses vacated by the better-off forced them to share dwellings with other families because of ever-increasing ground rents, thus increasing overcrowding and accelerating the physical deterioration of the properties (Dyos and Reeder 1973: 361). Suburban development did not provide a solution to the crowding of casual workers. For even if they could have afforded suburban rents and train fares, the only chance of employment often necessitated their being on site in central London before the first trains arrived (Jones 1976: 83, 209; Dyos and Reeder 1973: 368–9). This was especially true for those who worked on the docks, who needed to arrive at 6 a.m. and fight their way to the front of the crowd in order to get any work (Clayre 1974: 172–3; Jones 1976: 82–3). So casual workers had to live within walking distance of their places of work. This was particularly important in the case of women and children, to give them better opportunities of a job without removing them too far from their additional responsibilities in the home. Further, their irregular earnings made proximity to local cheap food markets and neighbourhood pawnshops and grocery store credit particularly important – especially in immigrant areas such as London's East End, which supplied a large proportion of the casual workforce, and where inability to speak English increased reliance on local sources of credit (Jones 1976: 87–8).[12]

Despite the nominal acceptance by the State of the principle of rehousing the poor, its housing, sanitary and urban improvement legislation differed little in its *effects* on the poorest inner-city dwellers from the impacts of the previous forty years of privately-financed, but state-assisted, railway construction (Jones 1976: 204–5), in the course of which over 50,000 working-class people had been displaced in London, 37,000 of them in 1850–67 alone (Wohl 1977: 36), and thousands more in provincial cities such as Leicester, Newcastle, Liverpool, Birmingham and Nottingham (Pritchard 1976: 73; Simmons 1973: 297). Railway lines almost always cut through densely populated working-class zones, the lines of least resistance where little or no compensation had to be paid to tenants and where landowners were happy to sell out (Gauldie 1974: 248; Jones 1976: 161–3; Wohl 1977: 36–9).[13] It was not, however, until the Royal Commission on the Housing of the Working Classes made its report public in 1885 that the cumulative effects of the clearances and their exacerbation by state legislation began to be widely appreciated. Far from speeding up the decentralization of the slum-dwellers, the clearances had the effect of squeezing the casual poor out of the central city districts and into adjacent suburbs hitherto populated mainly by artisans – thus increasing overcrowding and exacerbating slum conditions to seemingly horrendous proportions (Gauldie 1974: 286–9; Jones 1976: 209–14; Taylor 1974: 48, 51 on Liverpool, 1851–71). The filtering mechanism had broken down because suburban rail fares were still too high to enable any but the middle class and best-off artisans to move out to the further suburbs and commute daily (Dyos 1961: 73, 75–6; Jones 1976: 213–4).[14]

The response to the crisis

Yet the concern about London expressed at the time by 'do-gooders', 'social investigators' and parliamentarians alike was related not so much to the physical condition of the new slums – which was generally improving, although at the same time overcrowding was increasing, as Shaftesbury pointed out in 1883 (Wohl 1977: 311) – as to the fact that class-mixing was going on. Concern was shifting away from questions of public health and sanitation to more profound questions of social stability. Anxiety was widespread that the respectable, thrifty artisans ran the risk of being contaminated by the 'moral degeneracy' of the casual poor, that 'pauperism'[15] would spread upwards, so that artisans would be discouraged from doing 'an honest day's work for an honest day's pay', from raising productivity levels and achieving satisfaction through financial gain which they would use to provide a stable existence and improved quality of life for self and family. Mellor (1977: 123) argues that

> 'The housing crisis was deemed an issue meriting departure from previous laissez-faire attitudes precisely because shortage threatened the established differentiation of residence, as in conditions of employment, of the skilled craftsmen and the unskilled labourers. Just as the former began to benefit from national prosperity (in that wages rose steadily through the last quarter of the nineteenth century), and their political maturity seemed assured, rising rents and enforced overcrowding in the inner districts, particularly of London, threatened living standards and social stability. Enfranchisement itself depended on an adequate supply of housing.'

Wohl (1977: 40) points out that before the Royal Commission it was believed that industrious and thrifty workers could always find physically and morally 'decent' accommodation, but the Commission's findings showed that 'even highly skilled artisans were living in overcrowded single-roomed flats and were often forced to share dwellings with the criminal poor'.[16]

No doubt the brief rise of Marxian socialism among sections of London's artisan population in the 1880s, and a series of militant demonstrations (Jones 1976: 290–6, 337–41) contributed to anxieties about public order, but what was really at stake was not just the question of public order but the very maintenance and continued growth of London as the capital city. London's supremacy, and indeed its very functioning, depended on a clearly-stratified labour force: sectors of casualized labour to build railways and buildings for industry and commerce, to work in the docks, to provide cheap consumer goods according to the dictates of seasonal trade by means of sweated female labour and outwork[17] and to finish consumer goods made elsewhere for distribution to the national market and – increasingly – the Empire; a sector of skilled craftspeople to provide specialized luxury goods for the aristocracy and the financial bourgeoisie; and a burgeoning sector of office workers and public servants to work in commerce and administration (national and imperial) – from which the emergent managerial and executive strata were largely drawn (Jones 1976: 19–51; Weinberg 1974: 13–14).

This increasingly sophisticated division of labour was associated with specific forms of urban residential differentiation. For the growth of socially and physically distinct residential areas, separated from each other as well as from the increasingly concentrated industrial and commercial ag-glomerations, and facilitated by suburban railway development, restricted people's contact with those of other strata in their social lives and cultural activities. And as the proportion of time spent in the workplace under the direct domination of capitalist production relations diminished, 'resi-dential environments', with differential access to schooling, leisure and various forms of social services and activities, came to play a key part in the reproduction of a labour force increasingly stratified into 'layers' with different skills, value systems and aspirations, reflecting and reinforc-ing their position in the social and technical division of labour. (See Castells 1977a, b; Gray 1976; Harvey 1975; Lojkine 1976; Meller 1976.) Whereas the urban structure of the 'coketowns' of the industrial north of England tended to mirror the relatively simple class division between worker and capitalist, the picture in the metropolis was far more complex, tending to reflect the fractional divisions within the bourgeoisie and within the working class characteristic of a major commercial city in the mid-to-late nineteenth century. (See Weinberg 1974: 29–30, 47–51 on London; Gray 1976: 97 on Edinburgh.)

However, the class-mixing which brought London's 'housing crisis' into sharp relief indicated that such an 'appropriate' pattern of housing types, infrastructure and residential neighbourhoods could not be maintained by the uncoordinated accumulations of decisions made by private interests and individuals. Although recognition by the State of the potential need to intervene in the process of residential social stratification may have dawned as early as 1851 (Rose 1978: 65–8), it took the crisis of the 1880s to prompt any definitive action, as we shall see below.

During the late 1870s and early 1880s charitable organizations and philanthropists had increasingly been attacked for perpetuating a state of pauperism among the casual poor and thus delaying the improvement of their housing conditions by 'self-help' (Jones 1976: 241–61, 271–80).[18] However, the discovery that artisans were being forced to live in cramped conditions 'was bound to shake up the accepted cause-and-effect pattern of character and environment' (Wohl 1977: 56). It was gradually realized, initially in the 1870s by the public health profession and more widely in the wake of the 'crisis', that overcrowding and poverty were closely linked; that the poverty of the casual poor was a *structural* problem related to the uneven nature of demand for casual labour, the need to provide cheap consumer goods in order to remain competitive, and the inevitable decline of old-established industries; and that these workers were being forced to remain in the central cities in increasingly crowded conditions because land rents were continually increasing (Jones 1976: 243; Wohl 1977: 310–15). Mellor (1977: 29) suggests that this realization was a long time in coming because in London the contradiction between labour and capital was in no way apparent as compared with the factory towns – for London 'was itself largely non-industrialized, and had at its heart the greatest pool of unemployed and underemployed in the nation'. Thus, middle-class reformers saw the poor as 'a problem of restricted consumption not a failure of the productive machine to provide jobs: with good reason – there was no evident productive machine to the metropolis'.[19] Not until the 1890s was the argument being made that the only lasting solution lay in 'increasing wages, so that people can pay a fair rent', in the words of one Medical Officer of Health (Wohl 1977: 311).

The recommendations of the Royal Commission of 1885, however, were quite restricted. One of its main proposals was the extension of the number of lines on which 'workmen's fares' were in force, so that the process of suburbanization of all those who did not have to live in the centre apart from the very rich, could be speeded up, reducing (in theory) the level of overcrowding among the casual workers who remained in the centre (Stewart 1900: 12–17). This recommendation was partially implemented with the passage of legislation to force the railway companies to extend the geographical coverage of workmen's fares. These fares had existed along some S.E. London lines since 1865 (Dyos 1961: 76–7), but the Great Eastern Railway had been the first to introduce them on a large scale in the 1880s following on the first Cheap Trains Act, 1883, which required the extension of cheap commuter services up to 8 a.m. along lines where some working-class dwellings already existed (Jackson 1973: 24; Stewart 1900: 93–111; Wohl 1977: 287). This led to 'a spectacular development of working-class suburbia' (Wohl 1977: 278; Dyos 1961: 76). By the close of the century the north-eastern suburbs of London were becoming increasingly overcrowded with a vast 'one-class' population – generating renewed fears of social unrest among members of the newly-created (1888–9) London County Council. Not until 1914 was a real geographical spread of workmen's fares on trains and trams established (Jackson 1973: 24–32; Wohl 1977: 290).[20]

Equally important for the growth of working-class suburbs were the tramway lines which the LCC began to municipalize in the 1890s, under pressure from its housing committee (Wohl 1977: 292) which realized that the private sector could not co-ordinate the building of new working-class suburbs with the provision of cheap transport for the workers. The Totterdown Fields estate in Tooting, South London, the LCC's first 'cottage' estate, was planned in conjunction with the opening by the LCC of a new commuter tram-line in 1903 (Jackson 1973: 54–6); this marked a new and important recognition, by the LCC, of the simultaneous needs to directly provide, or at least subsidize the costs of transportation to and from work for the working class, and to take responsibility for housing some sections of the working class. These needs, and others, were later to be co-ordinated in a planning process which involved the partial socialization of the means of reproduction of London's labour-power, both on a day-to-day and an extended basis. The necessity for such a centralized co-ordinating body was recognised by the central State in the Local Government Act of 1888, which abolished the Metropolitan Board of Works and created the London County Council, in addition to establishing the 'county council' system of elected representatives and appointed officers all over the country (Dewsnup 1907: 103–4; Gauldie 1974: 295). This was followed in 1899 by the London Government Act which abolished the vestries and district boards and created twenty-eight London boroughs – with elected members and paid officers, again – which were given powers to build housing in their own right (Dewsnup 1907: 112).

London's housing crisis, and its exposure by the Royal Commission, thus seemed to provoke an awareness that the suburbanization process, left in private hands, could not provide a smooth solution to the problems of crowding and class-mixing, and that substantial sections of the working class – not just the very poor – were now being left out in the cold by private enterprise. Combined with the attractiveness of alternative investments in suburban building speculation, foreign railway stock and government bonds,[21] 'the investment reluctance of private enterprise was a logical outcome of the operation of free market forces which had as their prime determinant the low level of wages in a glutted labour market' (Taylor 1974: 80). Taylor is writing in reference tò Liverpool, but his argument seems equally relevant to London, which in the late nineteenth century had a similarly glutted casual labour market, as employment in the docks became decasualized and the old sweated industries began to decline, not only with cyclical fluctuations in trade cycles but also irreversibly in the face of provincial and foreign competition in the consumer goods market (Jones 1976: 152–5).

High profits could still be extracted from the 'management' of old inner-city working-class houses occupied by large numbers of casual workers and their families, despite the low rent-paying ability of each individual family (Byrne and Damer 1978: 35–6; Checkland 1964: 240; Jones 1976: 206–14).[22] But *new* dwellings would have to comply with higher standards of sanitation and lower rates of occupancy, while ground landlords would extract rapidly increasing rents from the property owners as the pressure on inner-city sites increased and as the terms of old leases expired or were breached by lessees who built at higher densities or let at higher occupancy rates than the leases stipulated (Dyos 1961: 45–9). This combination of factors would reduce the rates of return on new properties, especially since the inner-city owner of working-class housing would have to rely increasingly on casual workers to provide his rental income as suburbanization of the better-off workers proceeded apace. Rent levels rose by from 14 to 30 per cent from 1880 to 1914 (Wohl 1977: 304) but this did nothing to encourage new construction for rent since the rent levels reflected the 'sum necessary to deny the space on the ground to some other use' (Dyos and Reeder 1973: 380) and thus benefited the ground landlord more than the leaseholder. Additionally, lessees would have to 'carry the can' for the sanitary condition of the properties; yet the relatively short leases (thirty-one to ninety-nine years) gave them little incentive to carry out improvements, the value of which would revert to the ground landlord. Hence the shoddy standards of building and frequent neglect of working-class leasehold properties (Backwell 1977: 4; Dyos 1961: 64–9; Jones 1976: 210).

The debate over the 1885 Housing Bill, introduced by Lord Salisbury, focused on London, on 'the great disproportion that existed between the incomes of the working classes and the rents which they were forced to pay in order to live near their work' (Wohl 1977: 246), and the problems of 'stimulating the provision of houses in places where work people have their work to do'. This debate indicated, for the first time, 'a new willingness to apply the resources of the State, however tentatively, to the supply of working-class accommodation', resulting from the realization that 'the laws of supply and demand in central London had broken down' (Wohl 1977: 246) and the Bill went swiftly through Parliament (Gauldie 1974: 290). The 'pauperism' argument, criticized in the late 1870s by public health officials, could no longer be used to legitimate legislative inaction, for not only was Parliament confronted with London's crisis in its political as well as its economic manifestations, but the *ideological credibility* of the argument had been undermined by the gradual insinuating acceptance of the principle of the need for state intervention in the reproduction of labour-power. It became clear that, de facto, laissez-faire was dead. For, as one Medical Officer of Health argued, perceptively, to the National Association for the Promotion of Social Science, as early as 1878:

> 'When what is called public opinion has been sufficiently worked up in any particular subject we find that Parliament is not only willing to "pauperize" to an indefinite extent, but will even go out of its way to declare that for the particular purpose in hand pauperism shall not be pauperism! I am referring of course to the Elementary Education Act 1870 in which exactly this course has been pursued, and certainly it does not seem easy to show a reason in the nature of the case why a man should be held to be pauperized by receiving state aid in order to house his children decently, when he is not held to be so pauperized by receiving the same aid in order to educate them.' (Quoted in Rubinstein 1974: 188–90.)[23]

The now urgent imperative of intervention in the reproduction of London's labour-power through the housing process finally forced central government to abandon all pretences that the policy of laissez-faire was still operative and still viable. From the beginning of the 1890s housing policy thus reflected an ideological shift as well as a new political commitment.

The Housing of the Working Classes Act, 1890, and its extension in 1900, which consolidated previous legislation and streamlined the procedure for local authority house-building (Gauldie 1974: 293–4), was implemented with an avidity hitherto unseen (Dewsnup 1907: 219 ff.). The London County Council's own house-building programme started in the 1890s, followed by some of the London boroughs. Authorities up and down the country began to borrow money from the Public Works Loan Board for their housing programmes – especially after 1900 – which could by this time include building on new sites as well as rehousing in redevelopment schemes. Between 1890 and 1904 £4–5 million had been borrowed by eighty towns (Gauldie 1974: 294; figures for London in Wohl 1977: 362–7; details of borrowing by English provincial authorities in Dewsnup 1907). It seemed that a commitment to 'council housing' had finally been made – although the requirement that local authorities offer their housing for sale to the private sector was not lifted until 1909 (Gauldie 1974: 305) and parliamentary wrangling over housing finance continued until the First World War.[24]

I have examined some of the factors which led to a commitment to building council housing, in conjunction with ensuring cheap transport for working people in London, and I have noted that authorities in other parts of the country seemed to follow London's example after the LCC's programme got under way in the 1890s. However, these factors were not only historically but also locationally specific. London's special status as the commercial capital not only of Britain but of the Empire, and the particular stratification of the labour market and the urban structure consequent on this function, created specific problems in the reproduction of labour-power and social relations which were not necessarily found elsewhere in the country. To a certain extent, Liverpool and Glasgow experienced similar problems. They were also centres of commerce with a large casualized sector which glutted the labour market (swelled by Irish immigrants fleeing the great famine to take employment in the docks and sweated trades) in the 1840s and 1850s. There too, overcrowding and the decline of the privately rented housing market became critical issues, though twenty years earlier than in London, provoking the earliest local government intervention in direct house-building in Britain – St. Martin's Cottages, built in Liverpool in 1869 (Tarn 1968; Taylor 1974). We cannot draw firm conclusions about the impact of policies in Liverpool and Glasgow on national legislation, although the Torrens and Cross Acts were modelled on standards built into local Acts in Liverpool and the major Scottish cities; yet as Wohl (1977: 107) points out, 'London was the imperial centre and what was permitted in provincial towns under local acts took on a new ideological significance when applied to the capital'. Mellor (1977: 28) argues that London's financial bourgeoisie became the ideologically and politically dominant fraction of British capital in the late nineteenth century, with the relative decline of British manufacturing industry, based in the provinces, and Britain's increasing dependence on imperial supremacy to guarantee an outlet for British goods, which required that there be no obstacles to maintaining London as the headquarters of the money market and of international trade.

> 'Whereas in the early years of the century it was to the new cities of the provinces that they had responded and recoiled, "everywhere barbarous indifference, hard egoism on the one hand, and nameless misery on the other", later it was the metropolis which governed the attitudes of the professional classes from whom the intelligentsia was drawn. . . . London, with its tightly packed alleyways and courts, its crowds and its unknown poor, was the source of moral conscience and political turmoil and no longer the factories and working people of the industrial towns.' (Mellor 1977: 28)

In order fully to understand regional variations in the adoption of housing legislation it would be necessary to undertake comparative studies of the economic bases of different cities and regions – and in particular to compare areas whose industry was oriented toward export production with those whose production was oriented toward goods for the home market; and to compare areas where new technologically advanced industry was developing with those areas of low investment and declining industries dependent on cheap labour for their survival. It would then be necessary to analyse the local political and ideological dominance of particular fractions of the bourgeoisie controlling the various sectors of industry and commerce, and the strength of local working-class organization. These are important and relatively unexplored areas of study which are unfortunately beyond the scope of the present paper.[25]

Accumulation versus reproduction: toward a reappraisal of the 'recurrent crisis'

By the last decade of the nineteenth century, then, council housing was established in practice as well as in principle. In addition, a new shoot was emerging in the State's 'solution' to the 'housing question' with the beginnings of active state encouragement to owner-occupation as a desirable tenure form.[26] Building societies, which had originally sprung up as 'terminating' societies for groups of working-class people, who saved money together until all had purchased their own homes, were established as reliable savings institutions for the small-to-medium investor (regardless of whether he wanted to buy a house) by the Building Societies Act, 1874 (Gauldie 1974: 200–1; Weinberg 1974: 74–80). Thus the principle of mortgage finance was established, while in 1899 the system of 'council' mortgages for better-off working-class people was inaugurated with the passage of the Small Dwellings (Acquisitions) Act, which allowed local authorities to lend money to individuals for house purchase (Cullingworth 1966: 233; Dewsnup 1907: 109–11). Some years earlier, in 1879 and 1881, two small Acts had been passed which were also 'designed to assist the working man in buying his own house' (Wohl 1977: 106).

However, this two-pronged strategy did nothing positive for the casual poor in the centre of London, despite Lord Salisbury's intention, expressed in 1885, 'to benefit the most needy classes of society' (quoted in Wohl

1977: 247) by building houses close to their place of work. Even after the exodus of those sections of the working class with steady employment who could afford to commute from suburbia, conditions did not improve in the inner city for those who had to remain there. Rents and overcrowding continued to increase in the period from 1901 to 1914[27] as, under the insistent pressure for land-use change, thousands of working-class rooms were demolished and replaced by commercial developments and new streets. The rate of new building declined dramatically in most of the London Boroughs (see Table 14.2) and wages for casual workers in the glutted labour market remained low. It was not until well after the First World War that overcrowding in inner London began to decline steadily. The war brought an absolute reduction in the size of the male labour-force, while post-war restructuring of London's industrial base with the growth of new light industries in some of the more outlying suburbs absorbed many ex-casual workers (Jones 1976: 348–9). Additionally, renewed suburbanization and the growth of home-ownership, combined with increased male wages in real terms, greatly reduced the number of women in the labour market and re-established the attractiveness of their role as guardians of the 'Ideal Home' (Davidoff *et al.* 1976: 172).

Yet for the 'residual' population of essential service workers, immigrants brought in in times of labour shortage to do the 'dirty' jobs, and those marginalized in other ways, the only state housing policies were palliative ones. Rent controls (first introduced in 1915 and maintained, to a varying extent, ever since) cushioned these groups somewhat from the impact of continuing pressure on inner-city land values on their housing rents. Rent controls used the small private landlord as a scapegoat, who could often no longer afford to maintain deteriorating properties, and effectively subsidized employers, especially those in the trades employing casual and low-skill labour, by enabling them to keep their wage bills down and thus maintain profitability. As Fremantle (1927: 21) explains, in the years before the First World War, 'persons of the higher wage-earning or salaried class' would often buy a few houses for renting out, financed by a building society mortgage:

> 'It is these small owners – or rather their several successors in course of time – whose property is most likely to fall slowly behind in decorations and repair; for it often provides an essential part of an inadequate income for an old pensioner, retired tradesman, invalid or widow without other resources. But for the poorest and most casual classes it has always been and will always be impossible to provide new houses at an economic rent; they naturally succeed to the oldest and worst available. So as the years go by, the fashionable quarters go downhill in the social scale; Aldgate moves to Bloomsbury and Bloomsbury to Mayfair. The wealthy merchant's house in Aldgate of the eighteenth century holds in the twentieth half a dozen families and in its old age is the centre of a slum.'

It is this group – which has refused to disappear despite continued attempts by planners and others to 'assume it away' – that in the 1960s and early 1970s came to be most closely and most intractably associated with the 'housing problem' and 'urban deprivation', and which has become the subject of a large number of investigations (mainly of the 'moralistic' variety) and the target of government policies to combat the so-called 'cycle of deprivation' in parts of London and in other cities whose industrial base is declining, such as Glasgow and Liverpool.[28]

The connection between nineteenth-century housing problems and the 'hard core of the housing problem' today seems to be a strong one. But it would be trivial to blame inept legislation, lack of foresight, unwillingness to reject laissez-faire, etc., for reproducing these problems in the twentieth century. Worse, laying such 'blame' casts an ideological smokescreen by distracting from questions about the very *nature* of state housing policy. In order to *understand* our *present* housing situation it is necessary to reject such explanations[29] and to look more systematically at the structural forces shaping previous iterations of state policies, and their articulation in the social and spatial fabric of late nineteenth-century London – a fabric which is, in large measure, still with us, in its effects if not entirely in its visible form. I shall now, therefore, try to re-examine how the structure of late nineteenth-century London was shaped by the dual imperatives of maintaining the necessary conditions for capital accumulation and reproducing the labour-power and social relations appropriate to London's role as a centre of accumulation.

The mid-nineteenth century saw the first attempts to 'enhance London's capital accumulation potential'. As Weinberg (1974: 82–3) explains, this meant increasing

'its economic efficiency as a centre for production, exchange and distribution. Under capitalist market institutions and in an era prior to the widespread diffusion of electronic communications technologies, this drive was concretely realised in two broad dimensions. Firstly, by the growing clustering of economic activities in the central area, especially the exchange and service sectors. Secondly, by improving existing modes of transportation and developing ways of moving people and goods within the city and between London and other national and international regions.'

Thus, in the mid-1850s the Metropolitan Board of Works was set up by Act of Parliament and commenced its extensive programme of street improvements and public works, spending more than £10 million on the creation of new major thoroughfares in central London alone: 'Without . . . this amazing transformation . . . London would have suffocated and stagnated, unable to move the goods and provide the services which were essential to its survival as a great city' (Wohl 1977: 27).[30]

London's massive programme of railway building was approved in Parliament[31] with only token admonitions and unenforceable rehousing orders resultant on the cognisance of the tremendous displacement of working-class people caused by railway building (Wohl 1977: 35–9; Simmons 1973: 297; Gauldie 1974: 247–9; Weinberg 1974: 99–100). The railways

'were often regarded as the panacea to solve the housing question. They were viewed as a double blessing, cutting through slums and whisking the evicted inhabitants off to the fresher air, lower rents and higher mobility of suburban living' (Wohl 1971: 18) although as Weinberg (1974: 99–100) shows, the 'mythical qualities of railway demolitions as slum improvers' were *not* the *cause* of decisions to locate the railways in working-class districts, although they provided this practice with ideological legitimation. Railway building not only improved communications between London and the rest of the country, and Europe and points eastwards, it also provided a profitable outlet for the massive capital surplus being generated in mid-century, and thus kept interest rates from crashing. This resulted in considerable unnecessary duplication of lines, terminals, etc. (Hobsbawm 1969: 110–13; Checkland 1964: 24, 36–8; Church 1975: 33; Dobb 1963: 296). The similar order of magnitude of displacement of the working class caused by street improvement schemes was also accepted in Parliament and by city councils with relative equanimity (Wohl 1977: 30–5). Edinburgh's Lord Provost, for instance, noted the potential hardship caused by evictions resulting from improvement schemes, but commented that the Corporation felt that this was offset by 'the increased health of the community and improvement of the city' (Smith 1977: 39, fn.).

I have argued that government action on housing in London was forced by the realization of threats to public order and threats to the reproduction of the kinds of labour-power necessary to serve London's capital city functions. The reproduction process was threatened by the building of docks, railways, commercial buildings, and new thoroughfares to restructure the spatial form of the city to allow for the growth of commerce, which resulted in the eviction of at least a quarter of a million people from about 1830 to 1880 (Jones 1976: 159–69) and exacerbated overcrowding. But – and this is crucial – even if there *had* been greater concern for the effects of these works on the reproduction of labour-power, this could have made no difference to state policies. For, as Taylor (1974: 51) points out in reference to Liverpool in 1851–71, which was ahead of the rest of the country in its clearance policies,

> 'The main railways and termini, goods yards, dock and warehouse expansion and extension of retail trading were all quietly and continuously nibbling into inner-area housing stocks. These were the very districts where working-class housing was most conveniently situated for access to the chief labour markets.'

The building of docks in particular destroyed large numbers of working-class houses but at the same time increased the size of the labour market that was needed to live in the immediate vicinity of the docks in order that they operated efficiently (Jones 1976: 163–4). Such central locations were essential for these transportation and trans-shipment facilities. Likewise, even if the effects of the Torrens and Cross Acts in increasing overcrowding and class-mixing in districts adjacent to those cleared had been more widely predicted,[32] the policy would not have altered. For although the originators of the sanitary and slum clearance legislation did not see them as tools for the sweeping away of anachronistic land uses and people to make way for commerce, the pressures of land-use 'intensification', impelled by the 'objective' logic of capital articulated through the dynamic of capitalist urbanization, inevitably entailed their use for that very purpose.[33]

The State did nothing – either at central or local levels – to obstruct this process of intensification. Indeed, local state agencies sometimes actively participated in it: but this was not merely due to individual caprice or a desire to indulge in civic glorification. Ostentatious civic monuments, like Birmingham's Corporation Street and some of the more elaborate Victorian town halls, were only superficially the result of delusions of mayoral grandeur about which certain popular social historians tend to wax poetic. They cannot be explained in terms of some abstract 'Progressive Spirit' that descended out of the sky into civic offices. More fundamentally, they enhanced the locational advantages of the central city for capital and were just as essential to the capital accumulation process as the freeing-up ('releasing the potential') of sites formerly occupied by working-class housing for commercial uses.[34] What Wohl (1977: 303) refers to as the 'fundamental laws of supply and demand' – a mystifying phrase, since it obscures the historical specificity of this 'fundamentality' to the *capitalist* mode of production – created *dual* and *contradictory* imperatives for the capitalist State. The growth of London as the imperial centre for the accumulation and circulation of capital in the late nineteenth century depended on the existence of a stratified labour market which included a large and flexible supply of casual labour living near to the centre of the city. Yet the growth of the one continued at the expense of the other. The 'structural' requirement for the reproduction of an 'appropriate' social and technical division of labour, both day-to-day and on an extended basis, required patterns of differential access to housing, transportation, education, public facilities, etc. among the different fractions of the labour force. But this differentiation was threatened by the accumulation dynamic itself, which ultimately constrained decision-making about the pattern and intensity of land uses at different locations in the city, albeit specifically mediated by the dominant leasehold form of land tenure (Weinberg 1974: 64). What appeared as a contradiction[35] between *accumulation* and *reproduction* reached crisis point in the specific conjuncture of London's 'housing crisis' in the 1880s, and the resultant conflicts were played out in the territory of Inner London. The 'losers' were the casual poor themselves and, eventually and to a certain extent, the 'small-fry' owners of working-class housing.

Even after the accelerated rate of suburbanization of the artisans and middle classes in the 1890s had allowed the immediate crisis to subside, the conflict between the London County Council's need to ensure that essential workers were housed in the inner city and its responsibility to guarantee London's continued prosperity as political and financial capital of the British Empire remained, and continued to permeate political debate. The dilemma was crystallized by the Progressive member of the LCC, whose 1898 speech to the Council was quoted in the introduction to this chapter, and is graphically evident in Tables 14.3 and 14.4 and the accom-

panying maps. The escalation of central and inner-city land values, integral
to the dynamic of capital accumulation, had helped to create the situation
whereby the workers whose labour was also essential to that dynamic,
had to live on some of the most expensive land.

In many major Western cities today that dilemma is still unresolved.
It emerges when proposals for commercial redevelopment of London's
Covent Garden, even Brighton's railway station, are publicly debated.
The phrase 'releasing the potential' of inner-city sites remains a popular
one among property developers.[36] Essential service work has tended to
remain very labour-intensive, especially in public transportation, street-
cleaning and other public works.[37] In London, the difficulties associated
with redevelopment are compounded by the age of the inner city and
inner suburban housing stock (half of which was built before 1914), and
its high density, a consequence of the leasehold tenure system which encouraged building speculators to pack as many houses as they could onto
building plots (Weinberg 1974: 82, 154 fn.). Additionally, much of the
inner-suburban housing stock was shoddily built, unplanned and in chaotic
consonance with the expansion of commuter railways, and often inter-
spersed with the remnants of old and declining industries.

Although population decline in inner-cities has accelerated in the mid-
to-late 1970s, bringing an end to acute *shortages* of houses in some 'stress
areas',[38] this fact alone cannot resolve 'inner-city problems'. The logic of
capital accumulation entails industrial restructuring which creates economic prosperity in some areas and decline or stagnation in others – a
historical process of 'uneven development' which operates not only at a
regional scale but also at the heart of seemingly prosperous regions (Community Development Project 1977a; Massey and Meegan 1977; Mellor
1977). It entails the intensification of urban land in central areas and
local 'sub-centres', which usually takes the form of redevelopment for
commercial uses, at the expense of older industrial areas, whose economic
bases, already declining, are often systematically destroyed by asset-
stripping.[39] This process, which is affecting large areas of London, Liverpool and other cities, leaves in its wake in the declining areas high unemployment among a labour-force whose skills are now obsolete owing to
changes in labour processes, thousands of old and deteriorating houses,
derelict land (some awaiting redevelopment, the rest no longer of value
to capital – London's Dockland, for instance), run-down services and a
widening gap between expenditure needs and available resources (Community Development Project 1977a). As a number of major cities are discovering, a 'solution' seems impossible, while palliative measures are extremely
costly and are made more difficult to implement by the exodus of all
but the richest and poorest from many inner-city areas (Shankland *et al.*
1977: 33–5, 127–8).[40] Meanwhile, state policies to clear slums have often
led to the creation of new slums in shoddily-built tower blocks. In addition,
pressure for renewed restructuring of the spatial form of cities by commercial redevelopments and urban motorways, in a climate of financial uncertainty, have created 'planning blight' in many already impoverished areas

on the one hand, and new pockets of 'gentrified' affluence on the other. This continually recreates the dilemma often faced by urban planners, of 'how to destroy blighted areas without aggravating conditions in adjoining districts' (Wohl 1977: 29). Hence the bitterness and intensity of 'community' battles against commercial redevelopments, and the recurrence of conflicts between central and local levels of the State, the former tending to be more concerned with the interests of capital-in-general, and the latter given the job of administering many of the 'socialized' elements of the reproduction of labour-power (Cockburn 1977).[41]

Toward a reformulation of the problem

The foregoing analysis has tried to explore the nature of the 'recurrent crisis of London' by locating its 'urban problems' within the context of the social relations of capitalist commodity production, which set general limits and constraints on the social-spatial form of the city. It has been argued here that in the late nineteenth century the urbanization process in London was shaped by two major imperatives which arose out of the general 'objective' logic of capitalist development but which took historically-specific forms. The first imperative emerged as a need to structure and restructure the physical fabric of the city in such a way as to facilitate directly the process of capital circulation and accumulation in what had become the British Imperial capital. Secondly, a particular form of residential differentiation was necessary for the reproduction of a specific social and technical division of labour to perform the different tasks required to maintain the city's role and the nation's sphere of influence.

However, the analysis also indicated that these 'objective' imperatives could not determine London's development in any automatic or noncontradictory fashion. The physical and social segregation of casual workers from the regularly employed fractions of the working class was threatened by the very process of restructuring the spatial form of the city to enhance its capital accumulation potential, which could only be achieved by the demolition of thousands of working-class dwellings to make way for street improvements, railways, docks and business premises in central and inner-city areas. The specific socio-spatial arrangements of people and land uses in the late nineteenth century resulted from the accumulated actions of various class-fractions, interest groups and individuals, mediated by the intervention of the State. Private bodies made decisions about investment, location and so on according to their own particular and immediate 'subjective' interests, but these interests were in turn constrained by the general logic of the capitalist mode of production – the competitive drive for the appropriation and accumulation of surplus value and the objective class antagonism thus entailed. Similarly, the specific forms taken by state intervention in housing and urban policy in late nineteenth-century London were influenced by particular dominant interests and short-term expedients,[42] but the basic necessity and rationale for such intervention is inscribed in the very nature of capitalism, in that, by definition, capital exists only as individual capitals in antagonistic competition with one another, so that there can be no guarantee that capital will produce and reproduce the conditions of its own existence on an extended scale. (See, e.g., Holloway and Picciotto 1976.)

What appeared in the analysis of London's housing crisis, then, as a contradiction between the imperatives of 'accumulation' and 'reproduction' was *both* an historically-specific conflict between different interest groups and competing land uses *and* an aspect of the more general contradictions of capitalism as a whole. But London's urban problems cannot be simply reduced to those of capitalism as a whole, since urbanization also 'forms part of the "solution" which ongoing capitalist societies create of necessity to further accumulation and to keep the body politic from being irretrievably rent apart' (Walker 1978: 169). The 'solution' or response to the 1880s housing crisis appeared as a new political and ideological commitment, not only to the principle of 'council housing' but also to the practice of trying to co-ordinate the provision of new working-class housing with the provision of cheap public transport and other urban services in an overall concept of 'planning'. Yet it was our initial concern with the apparent *failure* of the 'solution' in contemporary London and elsewhere despite a century of increasingly active and concerted state intervention in the form of housing and planning policies, which led in the first instance to the quest for a deeper understanding through the historical research presented here. As we have seen, state policies have done little for those groups – mainly in the privately rented housing sector – whose living conditions symptomize the 'hard core' of the so-called inner-city problem. The poorest sections of the population tend to remain 'trapped' in the inner cities, either because their labour-power is still regularly or intermittently needed in the essential services sectors, or because the restructuring of capital and changes in the labour process have made their skills obsolete, giving them little chance of moving elsewhere.

In the case of London it appears that state housing and urban policies have not, in the long term, resolved the conflicts they were seemingly addressed to, but on the contrary, have tended to reproduce them after periods of temporary abatement, sometimes between different levels of the State itself. It has been suggested here that we cannot be sure of avoiding the recurrence of such conflicts as long as state policies continue to be guided by the structurally-determined dual imperatives of accumulation and reproduction, since it is these imperatives which periodically produce 'crises' when they are articulated together in urban space.

We are now in a position to replace our initially naïve questions about the nature of state housing and urban policies and the reasons for 'recurrent crises' with some more precise questions.[43] After empirically examining the specific 'urban conflicts' of a city in a particular social formation at a particular conjuncture, we may ask: *how* are these conflicts, which result from the interaction of the private decisions of classes, fractions and interest groups in the city with the public decisions of the capitalist State (through its urban policies) related to and guided by the total 'objective' logic and contradictions of the capitalist mode of production itself? For, although the empirical analysis contained in this paper has pointed to the *necessity* for the survival of capitalism of such a connection between 'agency' and 'structure', the nature and tenacity of the connection remains inadequately theorized. It is to the construction of such theory that we should address ourselves if we wish to evaluate future strategies for 'urban reform' through housing and planning policies.

Notes

1 *The Recurrent Crisis of London: Anti-Report on the Property Developers* (1973; London: Counter Information Services) attracted widespread media coverage and academic attention, being the first serious study of the commercial property boom and its relationship to London's housing crisis.

2 Detailed discussions of sanitary campaigns and early public health legislation can be found in Gauldie (1974) and Wohl (1977). See also Short 1976; Tarn 1971. Lojkine (1976: 144) stresses the impact of the widespread fear generated by the Chartist uprisings on the acceleration of state intervention. I have discussed the reasons for the growth in 'awareness' of the need for the labour force to be 'adequately' housed, and the concept of 'adequate reproduction' of labour power, elsewhere. (See Rose 1978, Chapter 4.)

For a chronology of principal Housing Acts and related legislation from 1851 to 1919, see Table 14.1.

3 These were private organizations with mixed philanthropic/commercial motives who, through careful design, were able to provide estates of housing at basically adequate, though architecturally monotonous and sometimes socially repressive standards, at a cost which only artisans and other better-off sections of the working class could afford (see Gauldie 1974: 221–32).

4 Rubinstein (1974: 156), quoting from Stewart (1900: 137, 142), points out that land had to be sold by the Metropolitan Board of Works to charitable bodies such as the Peabody Trust at great loss, in order that working-class housing might be built. For one scheme alone, involving less than 4,000 people, the Board spent £391,303 on purchase and clearance but received only £76,360 from the building company.

5 Only four towns used the Act to replace the houses demolished by local authority dwellings. These were Nottingham, Devonport, Liverpool and Swansea (Gauldie 1974: 278–9). It is interesting to note that all except Nottingham were ports with a significant number of casual labourers.

6 Dibdin, L. (1884) 'Dwellings of the Poor.' *Quarterly Review* (January): 164.

7 Roweis and Scott (1976) provide a critique of the assumptions behind neo-classical urban land-rent theory.

8 In present-day debates over the proposed use of land in urban redevelopment schemes, the argument that inner-city sites are too valuable to be used for low-cost housing – even if they can be bought at current-use values – is still common. Department of the Environment circulars advising local authorities as to the implementation of the Community Land Act, 1975 have made this quite clear. During a Public Inquiry into the proposed redevelopment of thirty-two acres of land in central Brighton – surrounding the railway station – the

same argument predominated, despite an absolute shortage of low-cost houses and severe physical constraints on peripheral expansion.

9 This argument was used in Hole (1866: 90) and also in evidence to the Royal Commission on the Housing of the Working Classes (Scotland) 1884–5 (quoted in Tarn 1971: 52).

10 Gordon (1971: 357) points out that among liberal and conservative analysts of the American urban housing market the 'filtering' or 'turnover' process is seen as providing 'an essential link in the formulation of many housing policy recommendations'. He quotes a housing economist: 'The turnover process does make sound housing available to income groups which cannot afford new construction. It is this process that is largely responsible for the fact that, in 1960, a majority of families at the lowest income levels were living in standard housing.' A thorough critique of the efficacy of filtering is given by Needleman (1965: 198–9). He points to (i) the difficulty the State encounters in trying to accelerate the rate of private house-building; (ii) the need for all sectors of the population to be highly mobile with respect to where they can live; (iii) the need for a substantial excess of supply over demand for all types of dwelling for all income groups in different areas. According to the theory, moves will be generated if the price of the existing housing stock declines. *But* a decline in prices leads to difficulty in selling new houses, which deters new building. Yet unless the number of households is rapidly declining, a surplus of housing will not occur without new construction. 'Thus the very conditions that are necessary for the rapid working of the filtering process are sufficient to stop the process completely.' Further, new house-building will only take place if there are workers able to pay for it: in North Shields (in N.E. England) a speculative boom in 'by-law' housing – which only a limited number of artisans could afford – crashed in 1906–7 and filtering thus could not continue (Byrne and Damer 1978: 5–6). Another difficulty is that middle and upper-income groups may move into vacated houses and 'gentrify' them. (See also Cullingworth 1966: 35–6.)

   Despite these criticisms of the theory and a paucity of empirical studies on the process in Britain, the argument that the building of more expensive housing will benefit low-income people through a filtering mechanism is still common. It was used, for instance, by a planner in Lewes, East Sussex at a meeting concerning a luxury housing scheme in 1976. It also pops up in the Final Report of the Department of the Environment's Inner Area Study of Lambeth in South London (Shankland *et al.* 1977: 31).

11 The persistence of the 'filtering' argument may be partially explained in ideological terms. Nettlefold, a councillor in Birmingham and close associate of Joseph Chamberlain, argued (1910: 95) that

   'the right policy is to aim upwards, not downwards. The levelling-up process may possibly take longer than the policy . . . of providing houses at low rentals in order to meet low wages, but permanent success in the solution of the housing problem is only to be achieved by those who are patient, and those who are wise enough to hasten slowly. The housing problem is not merely a question of building houses – it is also a question of building up character.'

Such an argument provides a legitimation of the system of 'queuing' on a housing waiting list for council and housing association housing, necessitated by the inability of the State or the private sector to provide housing according to the need under the present system of production relations.

12 The importance of access to credit within immigrant communities in London's East End has been stressed by the writer's great-uncle, whose father ran a grocery and general store in Whitechapel in the 1890s.

13 Weinberg (1974: 88–99) explains in more detail why the railways located along particular routes. (See also Dyos 1955; Kellett 1969.)

14 Weinberg (1974: 102, 158 fn.) argues that one of the main reasons why rail fares stayed relatively high in Britain was that the railway companies themselves could not speculate in freehold land and hence make money out of residential development. In some North American cities, by contrast, landed interests were weak and railway companies often played an active part in suburban development and enhanced their profits on land by keeping fares low and eventually forcing municipalization of the transit lines. (Gord Garland, personal communication; see also Gutstein 1976.)

15 'Pauperism' was commonly thought of as a chronic condition of poverty and unwillingness or inability to help self and family by working for a living. It was seemingly exacerbated by poor relief and other forms of charity. (See 18.)

16 Similar concerns were being voiced in Paris at about the same period (Magri 1977: 8).

17 The provision of cheap consumer goods was essential in order to reduce the value of the labour-power of London's workforce. This kept wages down and encouraged enterprises to remain in London rather than move to the North where land and housing was cheaper, and so helped to maintain London's ability to compete with newer industrial areas. (See Weinberg 1974: 13–14.)

18 Very similar arguments were voiced within the colonial administration of St. John's, Newfoundland, in the earlier part of the nineteenth century (Fingard 1977: 345–6). Charitable organizations in Montréal in the early twentieth century were also criticized on these kinds of grounds (Copp 1974: 116–19).

19 Almost exactly the same approach was taken by the inner-city studies of the late 1960s and early 1970s. It has only recently begun to change, largely as a result of increased structural unemployment and its concentration in areas of industrial decline, and the exposure of this situation by researchers in Government employ. (See Community Development Project 1977a; 1977b.)

20 Some indication of the inability of private enterprise to coordinate the rate of house-building with population growth in particular areas is given by the case of Woolwich, where over the period from 1901–11 the rate of house-building far exceeded population growth, as Table 14.2 suggests. (Woolwich had a strong artisan élite and a number of local-based building societies, as Crossick (1976) points out.) Whereas in a number of other suburban boroughs, speculative building did not keep pace with population growth over this period (it should be noted that comparisons of the percentage change in the housing stock with the percentage change in population provide only a rough illustration of the situation and do not accurately measure changes in levels of overcrowding).

21 The attractiveness of these alternatives is emphasized by Fremantle (1927: 22–3):

> 'Those who had hitherto been the mainstay of the speculative builder by investing their savings in his products had been gradually learning the increasing obligations placed on the houseowner and the value of more profitable industrial, or more secure gilt-edged, securities for investment, and still more had learned to ensure their own future by popular

forms of insurance. And so the War found the main source of supply of houses for the working class rapidly drying up.'

22 Nettlefold (1910: 38) quotes a local authority housing officer of the day: 'Our slums are gilt-edged securities. People who want to get rich quickly and do not care very much what methods they adopt to attain that end, buy slums. The worse the slums, the better the owner's chances of realizing huge profits on his investments.'

23 Child, G. W. (1878) 'How best to overcome the difficulties of overcrowding among the necessitous classes.' *Transactions of the National Association for the Promotion of Social Science:* 502–5.

24 For details of the steps leading directly up to the central state legislation compelling local authorities to build council housing see Bowley 1945: 2–14; Dickens 1977; Gauldie 1974: 295–310; Wilding 1972. Byrne and Damer 1978 attempt a critique of Dickens' and Wilding's arguments.

25 I am grateful to Nathan Edelson (who is researching the history of housing policy in British Columbia) for drawing attention to the importance of these issues. Gray (1977) discusses some of the mechanisms by which different fractions of the Victorian bourgeoisie – particularly landed capital, industrial capital, banking capital and the intellectual/administrative élite – achieved, maintained or lost hegemonic status, in relation to the growth, stabilization and beginnings of decline of competitive industrial capitalism in Britain.

26 The reasons for state encouragement to owner-occupation involve complex economic and ideological factors, which need further unravelling. For existing literature on this topic, see Ball 1976; Boddy 1976; Clarke and Ginsburg 1975. The author of the present paper is now engaged in research on the early development of the British home-ownership movement in the late nineteenth – early twentieth century and its political and ideological significance.

27 Gauldie (1974: 306) suggests that overcrowding was exacerbated by the more vigorous use of clearance powers as encouraged in the 1909 Housing and Town Planning Act at a time when private house-building had slumped drastically. This argument may be corroborated by Tables 14.3 and 14.4 for Inner London, which show the size of the clearance programme and the extent to which working-class housing was replaced by other land uses between 1902 and 1913.

28 An indication of the scale of the so-called 'inner-city problems' facing Liverpool is given by Starkey (1978). The city has lost 200,000 of its population in twenty years; unemployment is very high and one-quarter of all its households are in receipt of social security benefits; 80,000 of its privately owned dwellings date from before 1919 and many of these were shoddily constructed; there is a backlog of 40,000 repairs needed to its stock of council housing, 40 per cent of which was built before the Second World War; there are many acres of derelict land; social facilities are often vandalized; its economic base has all but disappeared and it is faced with a rapidly declining allocation of resources as compared with the money needed to alleviate the situation. (For other examples, on a smaller scale, see Community Development Project, 1977a.)

29 For a critique of 'moralistic liberal' analyses of housing problems, see Rose 1978, Chapter 2.

30 Roweis and Scott (1977: 26–7) discuss a similar situation in mid-nineteenth-

century Paris, although comparisons should not be taken too far owing to important differences in the role of the State in the British and French social formations (see Magri 1977; Mellor 1977: 32–3).

31 Parliament had to grant the railway companies special permission to override the leasehold system of tenure (and the interests of landed capital) so that they could expropriate lands for railway lines and works on a freehold basis (Weinberg 1974: 88–9).

32 In fact, such predictions *were* made, notably by Lord Shaftesbury and by the Charity Organization Society, and even by William Torrens himself – who objected to the enactment of an emasculated version of his Bill for that very reason (Wohl 1977: 90, 93). The 'anti-reformers' had made similar predictions with reference to slum clearance in Liverpool (Taylor 1974: 48).

33 cf. Pritchard (1976: 73) on railway construction and demolition in Leicester. (The usage of slum clearance legislation as a means of facilitating inner-city land-use change from working-class housing to commerce over the period 1902–13 is dramatically shown in Tables 14.3 and 14.4 and the accompanying maps.)

34 Of course, not all the land cleared of slums actually 'intensified' in terms of the density of its use. Between 1902 and 1913 a quarter of the slums demolished were replaced by vacant land (see Table 14.4). But much of this land would have been land awaiting redevelopment for commercial use, or land held by speculators waiting for increases in its value. It might also be argued that street improvement schemes represent a 'disintensification' of uses – but this obscures the essential part these schemes played in facilitating the 'intensification' of other pieces of land and in enhancing their capital accumulation potential. Intensification should rather be viewed as a *dynamic process* integral to capitalist urbanization, rather than as land-use density change measured in abstraction from historical forces.

35 In its Marxist usage, the term 'contradiction' has a very specific meaning. It refers to an antagonism between two tendencies or practices, each of which can only develop at the expense of the other, but both of which are continually produced and reproduced through the historical dynamic of capitalism and both of which are integral to that dynamic. Such contradictions are not, therefore, exogenous, but develop within capitalism itself. They set limits or barriers on the development of the productive forces, limits which demand the development of means to overcome them. But as capitalism develops, the contradictions are reproduced at a 'higher level' and cannot ultimately be transcended without the transcension of capitalist relations themselves (see Godelier 1972: esp. 350–5).

36 For instance, an article written at the height of the early 1970s 'property boom' in Britain reports that 'the developer–entrepreneur–speculator is really concerned with what is described as the development potential of a site or property. . . . The capital value depends very heavily on the income progression, and once that is "released", the value increases commensurately. . . . Releasing the potential usually means changing the use to which some property is being put, and even if it does not, it certainly means bringing the existing residents, tenants and others up to date with the facts of economic life', i.e. steep rent increases (Lester 1973).

37 Early 1970s staff shortages on London Transport which crippled part of the Underground system were largely due to the difficulty of staff in obtaining

affordable housing close enough to the centre for shift work to be feasible (London's public transport system closes at night). The problem was partially resolved by large increases in real wages passed directly on to users and partially by the recruitment of immigrant workers, mainly West Indians, many of whom have to live in overcrowded dwellings in substandard conditions in 'ghetto' areas reminiscent of those occupied by casual workers at the turn of the century. More generally, Mellor (1977: 125) notes that London still needs cheap labour and that this plays an important part in the city's housing crisis.

[38] According to a report in the *Guardian* (14 February 1978), London's housing shortage of 100,000 dwellings in 1971 has been replaced by a surplus of 80,000 dwellings over households, which is likely to increase to 220,000 by 1986, although areas of physical shortage still remain. Many of the 'surplus' dwellings require considerable investment to make them habitable again, however.

[39] The process of asset-stripping has been carefully documented in the case of North Southwark in London (Ambrose and Colenutt 1975).

[40] In response to this situation the Greater London Council is now offering large subsidies to people who are prepared to move into inner-city houses and renovate them (*Guardian* 21 February 1978). This scheme seems to be modelled on the 'urban homesteading' programmes which exist in some North American cities. Central government is also promoting 'partnership' schemes whereby local authorities 'in the several special inner-city partnership areas within the older conurbations' will sell derelict land or land with rundown property on it to the biggest private house-building firms 'at a book loss' (*Guardian* 20 February 1978). In the inner area of Liverpool – where one of these partnership schemes is operating – the subsidies involved are enormous – £312,000 on 5.06 hectares of land which was sold to one of the country's major building firms for one housing development (Fleetwood 1978). The parallels with the implementation of the 1875 Cross Act discussed above are striking indeed.

[41] It can be misleading, however, to generalize about the relationship of the central and local levels of the State (about which very little theoretical work yet exists in Britain, although there is some ongoing work in this area at Sussex University). Many local councils are very 'property oriented', either because of their direct domination by 'property interests' or because they believe that commercial redevelopments will increase their rate base, generate multiplier effects for local industry and commerce and hence create new jobs in all sectors (whether this actually happens is highly questionable). Such an orientation often conflicts with their responsibilities for reproducing labour-power. (See Ambrose and Colenutt (1975) for some local case studies.)

[42] Wohl (1977, Chapter 12) provides a detailed account of this political decision-making process and the interest-group lobbying it entailed.

[43] The analysis attempted here allows us to blow inside out the ideological umbrella-term of 'residual' population, used from the mid-nineteenth century to present-day policy documents to describe these inner-city dwellers – a term which obscures the relationship of their plight to the logic of capitalist development and legitimizes the direction of state policies at *them*, as though they constituted the 'problem', and the administration by bureaucratic agencies of merely allopathic and palliative treatment.

Appendix 1

1851	Common Lodging Houses Act (Shaftesbury Act)	Vestries and boroughs given supervisory public health powers over 'common lodging houses' for the very poor and transients.
1851	Labouring Classes Lodging Houses Act (Shaftesbury Act)	Vestries and boroughs permitted to raise money on local rates or from Public Works Loan Commissioners for building lodging houses for single *working* people.
1853	Common Lodging Houses Act	Police given powers of entry to inspect conditions in common lodging houses.
1855	Dwelling Houses for the Working Classes (Scotland) Act; Labourers' Dwellings Act	Facilitated investment by private capital in philanthropic housing by encouraging formation of companies for the erection of dwellings.
1855	Metropolis Local Management Act	Established Metropolitan Board of Works in London, with wide powers for street-building, etc. Ordered appointments of Medical Officers of Health.
1855	Nuisances Removal Act	Established minimum standards of housing conditions, gave local authorities duty to close houses 'unfit for human habitation'.
1866	Labouring Classes Dwelling Houses Act	Public Works Loan Commissioners empowered to make loans to private companies and local authorities for the erection of labourers' dwellings in populous towns.
1866	Sanitary Act	'Overcrowding' made a statutory nuisance; adoptive powers given to local authorities to make public health regulations and control standards.
1868	Artisans' and Labourers' Dwellings Act (Torrens Act)	Boroughs and vestries given powers for compulsory purchase of insanitary premises for demolition or improvement.

1874	Building Societies Act	Building societies given limited company status, establishment of 'permanent' societies providing investment for savings independent of house purchase.
1875	Artisans' and Labourers' Dwellings Improvement Act (Cross Act)	Metropolitan Board of Works and boroughs with population over 25,000 given powers to purchase and clear large tracts of insanitary property and to lease it to bodies willing to build housing.
1875	Public Health Act (extends Nuisances Removal Act)	Established conditions under which local authorities should take action to purchase premises for clearance purposes.
1879	Artisans' and Labourers' Dwellings Act (1868) Amendment Act	Provisions of compensation to owners of condemned houses added. Rebuilding and rehousing clauses reinstated.
1879	Artisans' and Labourers' Dwellings Improvement Act (1875) Amendment Act	Limits amount of compensation claimable by owners of slum property to value of house after overcrowding abated. Local authorities permitted to rehouse on alternative sites and let cleared land for commercial use.
1882	Artisans' Dwellings Act	Obligation to rehouse under Cross and Torrens Acts reduced to 50 per cent of those displaced.
1883	Cheap Trains Act	Railway Companies in London area made subject to Board of Trade orders to introduce workmen's trains and fares on lines where working-class housing had been built, in return for remission of passenger duty paid to Board of Trade.
1885	Housing of the Working Classes Act	Consolidated and amended Shaftesbury, Torrens and Cross Acts. Lodging houses redefined to include separate dwellings for labouring classes. Interest rates for Public Works Loan Board lowered. Severe limitation of compensation to slum owners.

1888	Local Government Act	Metropolitan Board of Works abolished. London County Council established. Provincial county councils with elected members and salaried officers introduced.
1890	Public Health Act	Extended 1875 Act. Local authorities given powers to strengthen sanitary regulations for private houses.
1890	Housing of the Working Classes Act	Further consolidation of previous housing legislation. Removal of obligation to rehouse displaced tenants in provincial cities. Local authorities allowed to build houses for working class (for disposal within ten years).
1894	Housing of the Working Classes Act	Local authority borrowing powers under 1890 Act extended.
1899	Small Dwellings (Acquisitions) Act	Local authorities empowered to advance money for purchase of small dwellings by their occupiers.
1899	London Government Act	Abolished vestries and district boards. Created twenty-eight London Boroughs with powers to build housing.
1900	Housing of the Working Classes Act	Amended 1890 Act. Local authorities given powers to purchase land outside their own jurisdiction for new house-building in addition to building on cleared sites.
1903	Housing of the Working Classes Act	Amended 1890 Act. Extended period of loan repayment from sixty to eighty years. Limitations on borrowing reduced.
1909	Housing and Town Planning Act	Obligation on local authorities to sell their housing stock removed. Local Government Board given powers to enforce local authorities to build housing. Powers for slum clearance strengthened.
1915	Increased Rent and Mortgage Interest (War Restrictions) Act	Rents of private houses with rateable values not exceeding £35 in London (£30 in Scotland, £26 elsewhere) fixed at prewar level. (Extended to more expensive houses after the war.)

369

1919	Housing and Town Planning Act (Addison Act)	Imposed duties on local authorities to survey housing needs in their districts and to carry out building schemes as needed, with approval of Ministry of Health. All losses in excess of a penny rate to be borne by local authority.

*The information in this Table is summarized and based on Dewsnup (1907), Gauldie (1974) and Stewart (1900), whose works should be consulted for further detail. The table is included merely to indicate the chronology of the principal pieces of legislation discussed in the text, and others. Such a chronology is meaningful only within the wider historical circumstances in which the various pieces of legislation were embedded; it should thus be consulted for purposes of clarification only.

Table 14.2 Change in population and housing stock, County of London, 1901–11

Borough	Inhabited houses 1901	Inhabited houses 1911	Houses being built 1901	Houses being built 1911	Net change 1901-11a No.	Net change 1901-11a %	Population 1901	Population 1911	Net change 1901-11 No.	Net change 1901-11 %
Central London										
City of London	3,865	2,784	166	24	-1,223	-30.3	26,923	19,657	- 7,266	-27.0
Bermondsey	15,817	14,967	17	9	- 858	- 5.4	130,760	125,903	- 4,857	- 3.7
Bethnal Green	14,005	13,649	59	9	- 406	- 2.9	129,680	128,183	- 1,497	- 1.2
Finsbury	9,280	7,830	52	23	-1,479	-15.8	101,463	87,923	-13,540	-13.3
Holborn	4,703	3,754	69	36	- 982	-20.6	59,405	49,357	-10,048	-16.9
St. Marylebone	13,536	11,987	132	105	-1,576	-11.5	133,301	118,160	-15,141	-11.4
St. Pancras[b]	23,715	22,246	21	28	-1,462	- 6.2	235,317	218,387	-16,930	- 7.2
Shoreditch	12,743	11,535	36	22	-1,222	- 9.6	118,637	111,390	- 7,247	- 6.1
Southwark	20,878	18,310	45	45	-2,568	-12.3	206,180	191,907	-14,273	- 6.9
Stepney	31,462	29,172	96	39	-2,347	- 7.4	298,600	279,804	-18,796	- 6.3
Westminster	18,366	17,540	157	105	- 878	- 4.7	183,011	160,261	-22,750	-12.4
Rest of London										
Battersea	23,462	24,321	107	39	791	3.4	168,907	167,743	- 1,164	- 0.7
Camberwell	36,671	36,559	220	68	- 264	0.7	259,339	261,328	1,998	0.8
Chelsea	8,641	7,499	16	40	-1,118	-12.9	73,842	66,385	- 7,457	-10.1
Deptford	15,823	16,102	92	15	202	1.3	110,398	109,496	- 902	- 0.3
Fulham	18,534	19,395	299	6	568	3.0	137,289	153,284	15,995	11.7
Greenwich	14,240	14,772	182	48	398	2.8	95,770	95,968	198	0.2
Hackney	30,634	31,090	79	53	430	1.4	219,110	222,533	3,423	1.6
Hammersmith	15,198	16,121	91	18	850	5.6	112,239	121,521	9,282	8.3
Hampstead	11,294	11,796	116	49	615	5.4	81,942	85,495	4,003	4.3
Islington	38,645	35,778	45	6	-2,291	- 5.9	334,991	327,403	- 7,588	- 2.3
Kensington	22,131	21,399	34	79	- 687	- 3.1	176,628	172,317	- 4,311	- 2.4
Lambeth[c]	41,511	38,634	137	108	-2,906	- 7.0	301,895	298,058	- 3,837	- 1.3
Lewisham	22,750	28,723	598	144	5,519	23.6	127,495	160,834	33,339	26.1
Paddington	17,684	16,586	32	4	-1,126	- 6.4	143,976	142,551	- 1,425	- 1.0
Poplar	22,613	21,107	19	9	-1,516	- 6.7	168,822	162,442	- 6,400	- 3.8
Stoke Newington	7,717	7,471	11	4	- 253	- 3.3	51,247	50,659	588	- 1.1
Wandsworth	37,764	48,432	1,179	377	9,866	25.6	231,922	311,360	79,438	34.3
Woolwich	18,086	21,176	517	38	2,611	14.0	117,178	121,376	4,198	3.6
Total	571,717	573,285	4,621	1,583	-1,470	- 0.26	4,536,267	4,521,685	-14,582	- 0.3

Source: Census of England and Wales, 1911, General Report (1917), Cmnd 8491, table X; Vol. VI, Buildings (1913), Cmd 6577, table 2. London: HMSO. Sanders, W. (comp.) (1903) A Digest of the Results of the Census of England and Wales in 1901. London: Charles and Edwin Layton, 2–3.

a i.e. difference between totals (of inhabited houses plus houses being built) for 1901–11.
b Part of St. Pancras is in 'Rest of London' (see Tables 14.3, 14.4 and maps).
c Part of Lambeth is in 'Central London' (see Tables 14.3, 14.4 and maps).

Table 14.3 Housing demolition related to subsequent use of land, County of London, 1902–13[a] : the number of working-class rooms demolished in London, the purposes for which the sites of the demolished rooms were used and the net loss of dwelling rooms through demolition and land use change.

Borough	Replacement uses, expressed in terms of rooms demolished									Housing losses through clearances		
	Working-class dwellings	Dwellings not working class	LCC educational sites, hospitals, churches, etc.	Government purposes	LCC and Local Authority street improvements, etc.	Railway and dock companies	Business premises	Vacant land	Miscellaneous	Total number of working-class rooms demolished	Net loss of working-class rooms	Net loss of all dwelling rooms
Central London												
City of London						24	64	203		– 291	– 291	– 291
Bermondsey	290	5	77	5	1,236	1,086	289	278	42	–3,308	–3,018	–3,013
Bethnal Green	859	9	187		295		270	22	14	–1,656	–797	–788
Finsbury	4		198	94	761		633	328	14	–2,032	–2,028	–2,028
Holborn	70	12	32	70	965		780	499		–2,428	–2,358	–2,356
Lambeth North	1,632	207	210	442	310	1,817	771	1,777	148	–7,365	–5,733	–5,527
St. Marylebone	179	673	277	36	140		247	575	15	–2,142	–1,963	–1,290
St. Pancras South	558	337	181	40	623	53	834	368	157	–3,151	–2,593	–2,256
Shoreditch	296		54		315	87	897	313	16	–1,986	–1,690	–1,690
Southwark	5,058	14	232	48	1,321	4	1,206	597	128	–8,610	–3,552	–3,338
Stepney	2,045	57	1,198	40	945	112	1,850	1,014	266	–7,619	–5,579	–5,514
Westminster	299	354	152	94	1,994		924	979	76	–4,872	–4,573	–4,219
Total	11,340	1,688	2,801	869	8,904	3,183	8,765	6,953	926	–45,460	–34,120	–32,342

(continued)

372

Table 14.3 (continued)

Borough	Replacement uses, expressed in terms of rooms demolished									Housing losses through clearances		
	Working-class dwellings	Dwellings not working class	LCC educational sites, hospitals, churches, etc.	Government purposes	LCC and Local Authority street improvements, etc.	Railway and dock companies	Business premises	Vacant land	Miscellaneous	Total number of working-class rooms demolished	Net loss of working-class rooms	Net loss of all dwelling rooms
Rest of London												
Battersea	24		113		314	48	300	28	16	– 843	– 819	– 819
Camberwell	383	27	331	15	195	30	303	46	54	–1,354	– 971	– 944
Chelsea	763	522	40	193	369	45	566	952	82	–3,467	–2,704	–2,182
Deptford	14		60	8	8	15	370	92	13	– 618	– 604	– 604
Fulham	76	32	36		186		180	64		– 613	– 537	– 505
Greenwich	412	54	67	30	315		296	473	52	–1,699	–1,287	–1,233
Hackney	391	117	344		348	4	380	105	71	–1,760	–1,369	–1,352
Hammersmith	53	4	45		70	261	357		4	– 796	– 743	– 739
Hampstead	19	9	46		10	35	16		17	– 155	– 136	– 127
Islington	199	14	582	57	110	156	668	13	29	–2,016	–1,817	–1,803
Kensington	120	139	84	41	125	12	162	175	96	– 839	– 719	– 580
Lambeth South	423	194	147		240	31	457	60	62	–2,057	–1,634	–1,440
Lewisham	136	29	61	16	73	15	101	487	10	– 470	– 334	– 305
Paddington	11	14	423	56			119	29	17	– 696	– 685	– 671
Poplar	56		303	43	451	247	158	56	62	–1,491	–1,435	–1,435
St. Pancras North	102		235	36	90	147	340	171	5	–1,131	–1,029	–1,029
Stoke Newington							20			– 20	– 20	
Wandsworth	211	134	18		525	18	208	200	15	–1,338	–1,127	–1,121
Woolwich	340	78	55	298	418	54	280	533	143	–2,189	–1,849	–1,829
Total	3,733	1,367	2,993	743	3,843	1,181	5,301	3,655	748	–23,522	–19,819	–19,071
Grand total	15,073	3,035	5,744	1,612	12,748	4,301	14,066	10,608	1,674	–69,012	–53,939	–52,200

Source: Derived from Wohl (1977:369). *Original source*: London County Council (1914–15) *London Statistics XXV*, table 13:166.

[a]This table shows the number of working-class rooms demolished in London, the purposes for which the sites of the demolished rooms have been used and the net loss of dwelling rooms through demolition and land-use change.

373

Table 14.4 Pressure for land-use change, County of London, 1902–13: the replacement land uses in each borough expressed as a percentage[a] of the total number of working-class rooms demolished in each borough[b].

Borough	Working-class dwellings	Dwellings not working class	LCC educational sites, hospitals, churches, etc.	Government purposes	LCC and Local Authority street improvements, etc.	Railway and dock companies[c]	Business premises	Vacant land[d]	Miscellaneous
Central London									
City of London	8.8	0.2	2.3	0.2	37.4	8.3	22.0	70.0	1.3
Bermondsey	51.9	0.5	11.3		17.8	32.8	1.9	8.4	0.9
Bethnal Green	0.2		9.7	4.6	37.5		16.3	1.3	0.7
Finsbury	2.9	0.5	1.3	2.9	39.7		31.2	16.1	
Holborn	22.2	2.8	2.9	6.0	4.1	24.7	32.1	20.6	2.01
Lambeth North	8.4	31.4	12.9	1.7	6.5		10.5	24.1	0.7
St. Marylebone	17.7	10.7	5.7	1.3	19.8	1.7	11.5	26.8	5.0
St. Pancras South	14.9		2.7		15.9	4.4	26.4	11.7	0.8
Shoreditch	58.8	0.2	2.7	0.6	15.3	0.1	45.2	15.8	1.5
Southwark	26.9	0.8	15.7	5.3	12.4	1.5	14.0	6.9	3.5
Stepney	6.1	7.3	3.1	1.9	40.9		24.3	13.3	1.6
Westminster							19.0	20.1	
Total	25.0	3.7	6.2	1.9	19.6	7.0	19.3	15.3	2.0

(continued)

Table 14.4 (continued)

Borough	Working-class dwellings	Dwellings not working class	LCC educational sites, hospitals, churches, etc.	Government purposes	LCC and Local Authority street improvements, etc.	Railway and dock companies[c]	Business premises	Vacant land[d]	Miscellaneous
Rest of London									
Battersea	2.9		13.4		37.3	5.7	35.6	3.3	1.9
Camberwell	28.3	2.0	24.5	1.1	14.4		22.4	3.4	4.0
Chelsea	22.0	15.1	1.2	5.6	10.6	0.9	16.3	27.4	2.4
Deptford	2.3		11.0	1.3	1.3	7.3	59.9	14.9	2.1
Fulham	12.4	5.2	5.9		30.3	2.5	29.4	10.4	
Greenwich	24.3	3.2	18.5	1.8	27.4		17.4	27.8	3.1
Hackney	22.2	6.7	19.6		19.8	0.2	21.6	6.0	4.0
Hammersmith	6.7	0.5	5.7		8.8	32.8	44.9		0.5
Hampstead									
Islington	9.9	0.7	28.9	2.8	5.5	7.7	33.1	8.7	1.4
Kensington	14.3	16.6	10.0	4.9	14.9	1.4	19.3	7.2	11.4
Lambeth South	20.6	9.4	7.2		11.7	1.5	22.2	23.7	3.0
Lewisham	28.9	6.2	13.0	3.4	15.5	3.2	21.5	6.2	2.1
Paddington	1.6	2.1	60.8	8.1			17.1	8.0	2.4
Poplar	3.8		20.3	2.9	30.3	16.6	10.6	11.5	4.2
St. Pancras North	9.0		2.1	3.2	8.0	13.0	30.1	15.1	0.4
Stoke Newington									
Wandsworth	15.8	10.0	1.4		39.2	1.4	15.6	1.1	15.0
Woolwich	15.5	3.6	2.5	13.6	19.1	2.5	12.8	24.3	6.5
Total	15.9	5.8	12.7		16.3	5.0	22.5	15.5	3.2
Grand total	21.9	4.4	8.3		18.5	6.2	20.4	15.4	2.4

Source: Derived from Wohl (1977:369). *Original source:* London County Council (1914–15) *London Statistics XXV*, table 13:66.

a The percentages may not total 100 owing to rounding errors and the omission of the small amount of condemned property still standing.

b Caution should be exercised in inter-borough comparisons owing to wide variation in the number of houses demolished (see Table 14.3). For this reason Hampstead and Stoke Newington have been omitted.

c Most of the railways and docks were built before 1880 and hence do not appear in this table.

d The high proportion of vacant land in some boroughs could be due to: land undergoing redevelopment; land speculation; the building slump; or creation of public open space.

Map 1: The growth of the County of London, 1800-1955

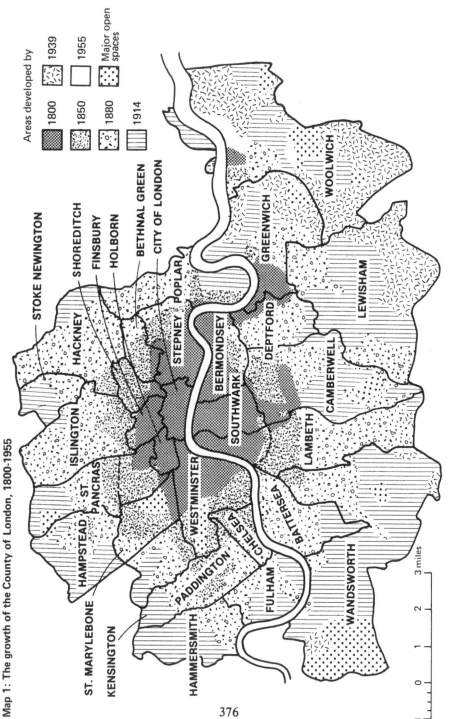

Areas developed by

1800 · 1850 · 1880 · 1914 · 1939 · 1955 · Major open spaces

STOKE NEWINGTON
SHOREDITCH
FINSBURY
HOLBORN
BETHNAL GREEN
CITY OF LONDON
HACKNEY
STEPNEY
POPLAR
ISLINGTON
ST. PANCRAS
HAMPSTEAD
ST. MARYLEBONE
KENSINGTON
WESTMINSTER
BERMONDSEY
SOUTHWARK
PADDINGTON
CHELSEA
BATTERSEA
FULHAM
HAMMERSMITH
WANDSWORTH
LAMBETH
CAMBERWELL
DEPTFORD
GREENWICH
LEWISHAM
WOOLWICH

1 0 1 2 3 miles

Source: Royal Commission on Local Government of Greater London (1960) The Growth of London. MHLG.

Map 2: Percentage change in housing stock and population, London Boroughs/1901-11.

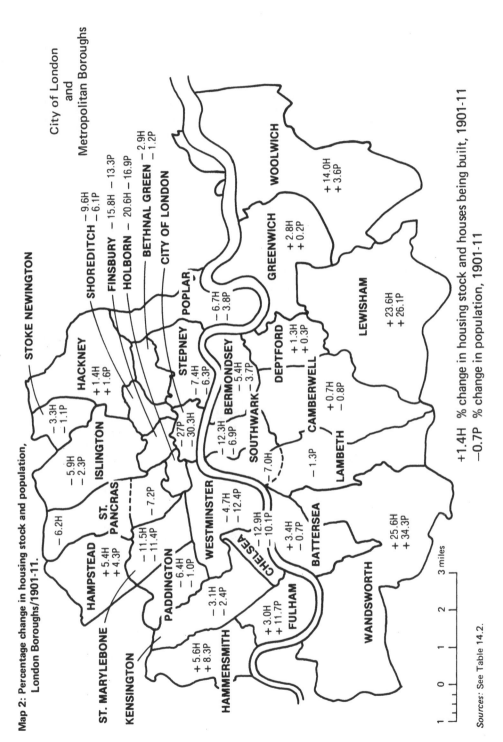

City of London
and
Metropolitan Boroughs

STOKE NEWINGTON

SHOREDITCH — 9.6H — 6.1P
FINSBURY — 15.8H — 13.3P
HOLBORN — 20.6H — 16.9P
BETHNAL GREEN — 2.9H — 1.2P
CITY OF LONDON

HACKNEY
+ 1.4H
+ 1.6P

POPLAR

STEPNEY
— 7.4H
— 6.3P

— 3.3H
— 1.1P

ISLINGTON
— 5.9H
— 2.3P

— 6.7H
— 3.8P

— 6.2H

ST. PANCRAS
— 7.2P

— 27P
— 30.3H

BERMONDSEY
— 5.4H
— 3.7P

SOUTHWARK

— 12.3H
— 6.9P

DEPTFORD
+ 1.3H
+ 0.3P

CAMBERWELL
+ 0.7H
— 0.8P

GREENWICH
+ 2.8H
+ 0.2P

WOOLWICH
+ 14.0H
+ 3.6P

HAMPSTEAD
+ 5.4H
+ 4.3P

PADDINGTON
— 6.4H
— 1.0P

— 11.5H
— 11.4P

WESTMINSTER
— 4.7H
— 12.4P

— 7.0H

LAMBETH
— 1.3P

LEWISHAM
+ 23.6H
+ 26.1P

ST. MARYLEBONE

KENSINGTON
— 3.1H
— 2.4P

CHELSEA
— 12.9H
— 10.1P

BATTERSEA
+ 3.4H
— 0.7P

HAMMERSMITH
+ 5.6H
+ 8.3P

FULHAM
+ 3.0H
+ 11.7P

WANDSWORTH
+ 25.6H
+ 34.3P

+1.4H % change in housing stock and houses being built, 1901-11
—0.7P % change in population, 1901-11

1   0   1   2   3 miles

Sources: See Table 14.2.

377

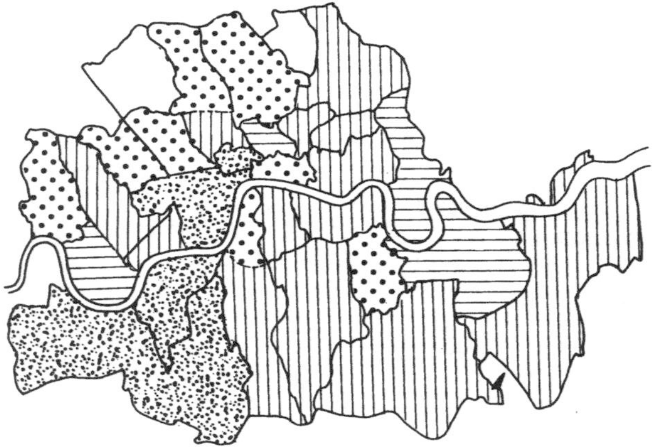

Map 3c
Working-class housing demolition related to subsequent use of land for business premises, London boroughs, 1902-13

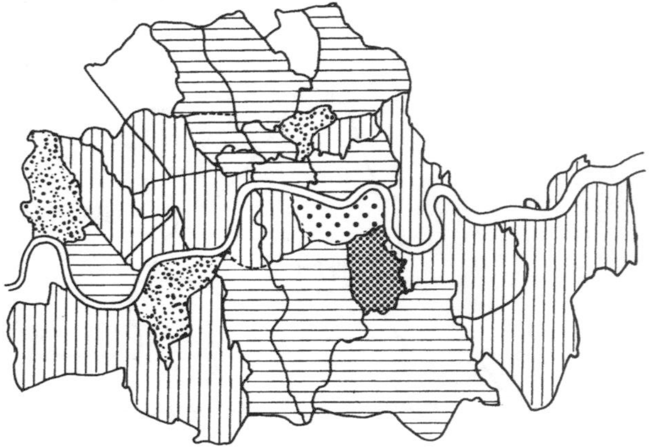

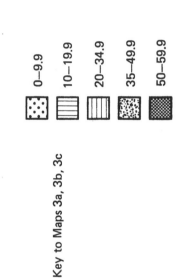

Key to Maps 3a, 3b, 3c

⬚	0–9.9
▤	10–19.9
▥	20–34.9
▦	35–49.9
▩	50–59.9

Replacement land-uses (working-class housing/street improvements/ business premises) in each borough expressed as percentage of total number of working-class rooms demolished in each borough

Source: Derived from Table 14.4.

Map 3a
Working-class housing demolition related to subsequent use of land for working-class housing, London boroughs, 1902-13

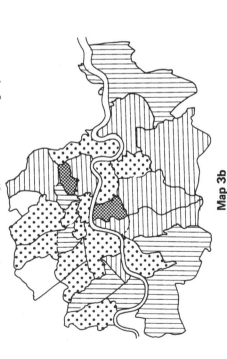

Map 3b
Working-class housing demolition related to subsequent use of land for street improvements, etc., London boroughs, 1902-13

References

Ambrose, P. and Colenutt, B. (1975) *The Property Machine*. Harmondsworth: Penguin Books.
Ball, M. (1976) 'Owner-occupation', in *Housing and Class in Britain*. London: Conference of Socialist Economists Political Economy of Housing Workshop, pp. 24–9.
Barker, G. *et al.* (1973) *Highrise and Superprofits*. Kitchener, Ontario: Dumont Press Graphix.
Boddy, M. (1976) 'Building societies and owner-occupation', in *Housing and Class in Britain*. London: Conference of Socialist Economists Political Economy of Housing Workshop, p. 30–43.
Bowley, M. (1945) *Housing and the State*. London: George Allen and Unwin.
Byrne, D. and Damer, S. (1978) 'The State, the balance of class forces and early working-class housing legislation.' Paper presented to Political Economy of Housing Workshop, London, March 11.
Castells, M. (1977a) 'The class struggle and urban contradictions', in J. Cowley *et al.* (eds) *Community or Class Struggle?* London: Stage 1. (Originally published in *Espaces et Sociétés* 6, 7 (1972).)
—— (1977b) *The Urban Question* (translated by A. Sheridan). London: Edward Arnold.
Chapman, S. D. and Bartlett, J. N. (1971) 'The contribution of building clubs and Freehold Land Society to working-class housing in Birmingham', in S. D. Chapman (ed.) *The History of Working-Class Housing*. Newton Abbot: David and Charles.
Checkland, S. A. and Checkland, E. O. A. (1974) *The Poor Law Report of 1834*. Harmondsworth: Penguin Books.
Church, R. A. (1975) *The Great Victorian Boom, 1850–1873*. London: Macmillan.
Clarke, S. and Ginsburg, N. (1975) 'The political economy of housing', in *Political Economy and the Housing Question*. London: Conference of Socialist Economists Political Economy of Housing Workshop.
Clayre, A. (1974) *Work and Play*. London: Weidenfeld and Nicolson.
Cockburn, C. (1977) *The Local State*. London: Pluto Press.
Community Development Project (1976) *Profits against Houses*. London: CDP Information and Intelligence Unit.
—— (1977a) *The Costs of Industrial Change*. London: CDP Information and Intelligence Unit.
—— (1977b) *Gilding the Ghetto: The State and the Poverty Experiments*. London: CDP Information and Intelligence Unit.
Copp, T. (1974) *The Anatomy of Poverty: The Condition of the Working-Class in Montréal, 1897–1929*. Toronto: McClelland and Stewart.
Counter Information Services (1973) *The Recurrent Crisis of London*. London: CIS.
Crossick, G. (1976) 'The labour aristocracy and its values: a study of mid-Victorian Kentish London.' *Victorian Studies* XIX, 3:301–28.
Cullingworth, J. B. (1966) *Housing and Local Government*. London: George Allen and Unwin.
Davidoff, R. *et al.* (1976) 'Landscape with figures', in J. Mitchell and A. Oakley (eds) *The Rights and Wrongs of Women*. Harmondsworth: Penguin Books.
Dewsnup, E. R. (1907) *The Housing Problem in England*. Manchester: Manchester University Press.
Dickens, P. (1977) 'Social change, housing and the State: some aspects of class fragmentation and incorporation, 1915–46.' Paper presented at Centre for Environmental Studies Urban Sociology Conference, York.
Dobb, M. (1963) *Studies in the Development of Capitalism*. New York: International Publishers.
Dyos, H. J. (1955) 'Railways and housing in Victorian London, I, II.' *Journal of Transport History* 1:11–21; 2:90–110.
—— (1961) *Victorian Suburb*. Leicester: Leicester University Press.
Dyos, H. J. and Reeder, D. A. (1973) 'Slums and suburbs', in H. Dyos and M. Wolff (eds) *The Victorian City: Images and Realities*, Vol. 1. London: Routledge and Kegan Paul.
Fingard, J. (1977) 'The relief of the unemployed: the poor in St. John, Halifax and St. John's', in G. A. Stelter and A. F. J. Artibise (eds) *The Canadian City: Essays in Urban History*. Toronto: McClelland and Stewart, pp. 341–67.

Fleetwood, M. (1978) 'Selling the inner city: the case of Stanfield Road.' *The Architect's Journal* (12 July): 71–9.

Fremantle, F. E. (1927) *The Housing of the Nation*. London: Philip Allan.

Gauldie, E. (1974) *Cruel Habitations*. London: George Allen and Unwin.

Godelier, M. (1972) 'Structure and contradiction in "Capital" ', in R. Blackburn (ed.) *Ideology in Social Science: Readings in Critical Social Theory* (translated by B. Brewster). Glasgow: Fontana. (Originally published in *Les Temps Modernes* 246 (1966).)

Gordon, D. M. (1971) *Problems in Political Economy: an urban perspective*. Lexington: Heath.

Gray, R. Q. (1976) *The Labour Aristocracy in Victorian Edinburgh*. Oxford: Clarendon Press.

—— (1977) 'Bourgeois hegemony in Victorian Britain', in J. Bloomfield (ed.) *Class, Hegemony and Party*. London: Lawrence and Wishart, pp. 72–93.

Gutstein, D. (1976) *Vancouver Ltd*. Toronto: Lorimer.

Harvey, D. (1974) 'Class-monopoly rent, finance capital and the urban revolution.' *Regional Studies* 8:239–55.

—— (1975) 'Class structure and the theory of residential differentiation', in R. Peel *et al.* (eds) *Processes in Physical and Human Geography*. London: Heinemann.

Hobsbawn, E. J. (1969) *Industry and Empire*. Harmondsworth: Penguin Books.

Holloway, J. and Picciotto, S. (1976) 'A note on the theory of the State.' *Bulletin of the Conference of Socialist Economists* V, 2.

Jackson, A. A. (1973) *Semi-Detached London*. London: George Allen and Unwin.

Jones, G. Stedman (1976) *Outcast London: A Study in the Relationship Between Classes in Victorian Society*. Harmondsworth: Penguin Books.

Kellett, J. R. (1969) *The Impact of Railways on Victorian Cities*. London: Routledge and Kegan Paul.

Lester, T. (1973) 'The props beneath property.' *Management Today* (August).

Lojkine, J. (1976) 'Contribution to a Marxist theory of capitalist urbanisation', in C. G. Pickvance (ed.) *Urban Sociology: Critical Essays*. London: Tavistock, pp. 119–46.

Magri, S. (1977) 'State housing policies: demands of capital and class struggles, unpublished mimeo. Abridged translation by E. Lebas. (Originally presented at International Sociological Association Research Committee on Urban and Regional Development 1977 Conference, Regio Calabria.)

Massey, D. B. and Meegan, R. A. (1978) 'Industrial restructuring versus the cities.' *Urban Studies* 15, 3.

Meacham, S. (1977) *A Life Apart: The English Working-Class, 1890–1914*. London: Thames and Hudson.

Mearns, A. (1976) 'The bitter cry of outcast London', in P. Keating (ed.) *Into Unknown England, 1866–1913*. London: Fontana. (Originally published by London Congregational Union, 1883.)

Meller, H. E. (1976) *Leisure and the Changing City, 1870–1914*. London: Routledge and Kegan Paul.

Mellor, J. R. (1977) *Urban Sociology in an Urbanised Society*. London: Routledge and Kegan Paul.

Needleman, L. (1965) *The Economics of Housing*. London: Staples Press.

Nettlefold, J. S. (1910) *Practical Housing*. Letchworth.

Polanyi, K. (1944) *The Great Transformation*. New York: Holt Rinehart and Winston.

Pritchard, R. M. (1976) *Housing and the Spatial Structure of the City*. Cambridge: Cambridge University Press.

Rose, D. C. C. (1978) 'Housing policy, urbanisation and the reproduction of labour power in nineteenth-century Britain: a conceptual appraisal', M.A. research paper, University of Toronto.

Roweis, S. and Scott, A. J. (1976) 'The urban land question.' Paper presented to the CCURR Ontario Forum on Urban Land Management, Toronto.

—— —— (1977) 'Urban Planning in Theory and Practice: A Reappraisal.' Papers on Planning and Design 14. University of Toronto: Department of Urban and Regional Planning.

Rubinstein, D. (1974) *Victorian Homes*. Newton Abbot: David and Charles.

Shankland, G. *et al.* (1977) *Inner London: Policies for Dispersal and Balance: Final Report of the Lambeth Inner Area Study.* London: HMSO.

Short, J. R. (1976) 'Social systems and spatial patterns.' *Antipode* 8, 1:77–82.

Simmons, J. (1973) 'The power of the railway', in H. J. Dyos and M. Wolff (eds) *The Victorian City: Images and Realities.* London: Routledge and Kegan Paul.

Smith, P. J. (1977) 'Planning concepts in the improvement schemes of Victorian Edinburgh.' Paper presented at First Conference on History of Urban Planning, London.

Starkey, B. (1978) 'Liverpool lose.' *Roof* 3, 1:18–9.

Stewart, C. J. (1900) *The Housing Question in London.* London: London County Council.

Tarn, J. N. (1968) 'Housing in Liverpool and Glasgow: the growth of civic responsibility.' *Town Planning Review* 39:319–33.

—— (1971) *Working-Class Housing in Nineteenth-Century Britain.* Architectural Association Paper No. 7. London: Lund Humphries.

Taylor, I. C. (1974) 'The insanitary housing question and tenement dwellings in nineteenth-century Liverpool', in A. Sutcliffe (ed.) *Multi-Storey Living: The British Working-Class Experience.* London: Croom Helm: 41–87.

Tholfsen, T. R. (1976) *Working-Class Radicalism in Victorian England.* London: Croom Helm.

Walker, R. A. (1978) 'The transformation of urban structure in the nineteenth century and the beginnings of suburbanisation', in K. Cox (ed.) *Urbanisation and Conflict in Market Societies.* Chicago: Maaroufa Press.

Weinberg, D. S. (1974) 'The social relations of living, London, 1830s–1880s: The dialectics of urban living and city form.' MCP thesis, Massachusetts Institute of Technology.

Wilding, P. (1972) 'Towards exchequer subsidies for housing, 1906–14.' *Social and Economic Administration* 6, 1:3–17.

Wohl, A. S. (1971) 'The history of the working-class in London, 1815–1914', in S. D. Chapman (ed.) *The History of Working-Class Housing.* Newton Abbot: David and Charles.

—— (1977) *The Eternal Slum.* London: Edward Arnold.

15 A theory of suburbanization: capitalism and the construction of urban space in the United States

Richard A. Walker

Suburbanization is clearly a process of enormous complexity, forming one part of the entire urbanization process, one aspect of the social whole. We approach it as a geographic problem, although it is not 'in essence' this any more than it is essentially an 'economic' or 'sociological' problem. Part of the difficulty in making sense of suburbanization is the power of the myth that it can be defined in a purely geographic fashion, when, in fact, understanding spatial relations necessarily means confronting social relations as a whole.[1]

As a geographic problem, modern suburbanization has three major defining characteristics: spatial differentiation, decentralization, and identification with the waves of urban landscape laid down since the Second World War. Virtually everyone writing on the topic recognizes these three dimensions as empirically valid, with the disputes being over matters of degree and timing. The eminent historical geographer, J. Vance, for example, observes that 'functional decongestion' is 'still the main dynamic force at work in our cities' and that since the Second World War a 'new city' has come into being, which 'differs in both scale and functional structure from the metropolis of pre-war days'.[2] In fact, spatial differentiation, deconcentration and the phenomenon of successive, distinct waves of urbanization have been at work since the capitalist revolution transformed American cities in the nineteenth century and have become, in absolute terms, more pronounced over time. The principle difficulty, of course, is not descriptive specification of the three shaping forces but in coming up with a satisfactory explanation for them.[3]

Putting space at the center of our analysis necessitates a word on the approach to space in historical materialism.[4] Raw land, like raw nature, is of little use to human beings without transformation into an object of utility by means of human labor.[5] Urban space – the built-environment of the city – must be actively created and is obviously a product of considerable amounts of past labor, an enormous humanly-constituted resource and activity system. The construction of urban space is clearly a *constitutive* process, to borrow Raymond Williams's apt term,[6] not a phenomenon to be relegated to an unmaterial, reflective 'superstructure'. To continue Williams's tack, the social relations of the mode of production set limits and create pressures for a certain kind of spatial organization, but they do not determine spatial relations in any unique, non-contradictory or unidirectional way. Indeed, spatial relations are part of the internally structured whole of a mode of production (or social formation)[7] which, to use Ollmans' felicitous metaphor, provides one 'window' through which we may look at the inside of the social process.[8]

Because the construction of space is only one aspect of the labor process, however, we allot spatial relations a subordinate position in the unfolding of human history to the role of the primary social relations of production, and see the primary flow of causality going from the latter to the former. Hence we will try to determine in what ways the logic of capital in the abstract would tend to create a 'suburbanized' form of built-environment. In some instances, this requires only a relatively straightforward deduction from obvious aspects of capitalist behavior, as, for example, cost-minimizing in industrial location; but more often than not we must grapple with more contingent results, specific cultural, political or economic arrangements shaped from the concrete circumstances of history.[9] Such constructions are never random; they are forged amidst the pressure and limits of the structural conditions of society, i.e. the need to reproduce capital and capitalist relations of production through the twin processes of accumulation and class struggle.[10]

Under capitalism the analysis of space begins with the commodity. The built-environment is constructed and circulates like any other commodity – granting, of course, that the nature-given uniqueness of location and immobility of physical structures gives to this commodity certain distinctive characteristics.[11] Following Marx, our analysis begins by dividing the commodity into two sides: use-value and exchange-value.[12] The forces behind the creation of the city operate on both the use and exchange-value sides of the built-environment, and the two should not be collapsed into one.

On the one hand, the built-environment is put into place, and reorganized as part of the process of circulation of capital – movement of value in search of expansion. Capital flows into the built-environment by way of what Harvey calls the 'secondary circuit of capital' – in order to create fixed capital for the production and consumption funds.[13] The part of this circuit which is specifically addressed to real estate and land development (as opposed to the creation of producer and consumer durables), we will call the 'property circuit of capital'. Investment through this circuit is motivated not only by profits from construction, but primarily by the opportunity to appropriate economic rent from new and old property.

On the other hand, the built-environment serves as a physical container for the activities of businesses and people.[14] Its use-values are as diverse as the uses to which urban spaces can be put.[15] The city must therefore be constructed and continually reconstituted to serve those uses, in order to assure the reproduction of capital (accumulation) and capitalist social relations (holding class struggle in check). This cannot always be assured, simply because of the dichotomy between use and exchange value, and the force of capital circulation through the secondary circuit operating somewhat independently of the needs of spatial organization. Both sides of the 'spatial commodity' are constitutive but we ordinarily assign priority to the use-value side, as before.

This approach allows us to surmount the inadequacies of two countervailing modes of analysis in the conventional literature. The main body of 'location theory' conceives of the organization of space purely in terms of use-values, or the demand for space, with the supply side seen as totally passive.[16] Opposed to this prevailing academic approach is a literature of sound criticism – largely non-academic in origin – which seeks to expose the power of land developers and real-estate interests to shape the urbanization process.[17] While less convincing theoretically than formal location theory, this refreshing literature blows holes in the strictly passive view of land development. The point, of course, is that both demand and supply play a role, and, as Marx observed in another connection, capital works at both ends.[18]

With this introduction to spatial analysis, we can return to the question of the generation of the three major characteristic aspects of suburban spatial form in US cities. The first part deals with the process of spatial differentiation, the second with decentralization, and the third with stages and waves of urbanization.[19] The first two parts concentrate on what might be called the 'secular' logic of the location, while the third reinjects the real historical process of constructing cities in waves, and how this contributes to locational dynamics. In each case, the discussion is meant only to indicate the crucial dimensions of the problem in a logical fashion, not to provide a full treatment.[20]

Spatial differentiation

Spatial differentiation is a universal characteristic of capitalist urbanization, although its degree and form depend on the specific conditions of different societies. Conventional treatments of American urbanization take 'functional' differentiation of space for granted, but never explain why these 'functions' arise in the first place.

At the broadest level, spatial differentiation is the product of the capitalist development of the division of labor.[21] Pre-capitalist modes of production had division of labor, of course, and consequently functional differentiation in their cities, but the extent of this was sharply limited.[22] The antecedent mode of production in the northern US was petty commodity production (combined with mercantile exchange) and American cities prior to the middle of the nineteenth century reflected the kind of undifferentiated repetition of household units of production, trade, and consumption on which such a mode of production is based.[23] When the capitalist revolution took hold in the big mercantile cities in the mid-nineteenth century, spatial differentiation became a dominant force in the urban landscape.[24] By revolutionizing the mode of production, capital revolutionized the spatial relations of production and social life.[25]

Briefly, the historical development of the large grain functional differentiation in US cities is as follows.[26] The big change began with the transfer of production out of the household and into the factory, owing to competitive destruction of handicrafts and concentration of the means of production into the hands of a small class of capitalists.[27] The first and most fundamental result of this 'formal subsumption of labor by capital' was to separate production from consumption as functions and as places – the workplace versus the home. The division between production and selling which was already partly developed under mercantile relations became rapidly augmented by the same transfer of activity from household to capitalist, as well as the sheer increase in goods to be circulated. Warehousing districts, joining wholesaling and small-scale manufacturing assumed the dominant role in the creation of a distinctive 'central business district' in the second half of the century.

Large-scale factories were rare until late in the nineteenth century and only became the norm in the twentieth. But the few that did exist, such as ship and locomotive, already formed distinct nodes of employment, usually somewhat apart from the central business district. The huge industrial complexes which were built in number by the turn of the century had a dramatic impact on the spatial organization of cities. Successive clusters of industry have emerged at the urban fringe and been absorbed into an expanding, fragmented metropolis.

Except for a few specialty shops catering to the rich, an important 'retailing' function only came into its own after the Civil War, with the appearance of the first mass-retailing outlets organized on strictly capitalist lines. These took up a position in distinctive 'shopping' districts downtown. If the second half of the nineteenth century was 'the age of the warehouse' the first half of the twentieth century saw the peak of downtown retail shopping.

Distinctive financial districts within the Central Business District first appeared in New York and Boston in the 1830s, as appendages to a well-developed merchant capital, and sprouted rapidly thereafter in all major cities by the turn of the century. They grew rapidly (but unevenly between cities) owing to financial concentration at the turn of the century. Separate 'headquarters' buildings and districts specializing in management and allied professional functions did not appear until the very end of the nineteenth century coincident with the rise of the modern corporation. Clusters of corporate and professional offices have since nestled close to those of banks and insurance companies in the CBD.

Today the city center is given over almost entirely to office functions, encompassing finance, management and professions. Some secondary office centers have been springing up in the suburbs. Retail shopping centers are a prominent part of the urban, especially suburban, landscape. The division of the metropolis into industrial subregions is also readily apparent. Large spaces are given over to transport systems, including highways, parking, railyards and ports, and the 'public facilities' of the State have become increasingly significant units in the urban landscape.

The rise of these basic divisions of social labor, let alone of the finer grain of intra-sectoral specialization, cannot be detailed here. Understanding the development of the capitalist division of labor involves nothing less than an analysis of capital itself. What can be said incontrovertibly is this: the division of labor has, and will continue to develop with the advance of capitalism, because it rests on such fundamental features of this mode of production as the development of labor productivity, extension of the market, and centralization of capital, i.e. on the restless drive to accumulate.[28]

The social division of labor only gives us the *possibility* of 'functional' differentiation of urban space, however. What translates this possibility into a locational imperative? The same capitalist logic which led to the division of labor in the first place; first of all the drive to rationalize production, consumption and circulation in cost and revenue terms. This rationalization is the subject of conventional location theory, which deals with tendencies to minimize costs of inputs (markets given), to maximize sales revenues (production costs given) or to minimize transportation/communication costs. Separate branches of location theory have grown up around these partial views of the problem.[29] Never, however, do they grasp the full locational import of Marx's concept of the circulation of capital in general, the unity of production and circulation, or the non-economic aspects of social life and class struggle involved in even these 'economic' functions.[30]

The drive of rationalization[31] means that capitalists will try to locate their particular 'functional unit' (or units within a larger corporation) in some sort of optimal way with respect to transportation, markets, suppliers, labor, etc., i.e. all the traditional locational factors. As new functions have arisen they have been relocated in accordance with their own needs, instead of being merely appended to something else, as e.g. leaving management offices in the factory. Since these needs are all somewhat different, the various functions will be arranged differentially in space. The same will be true *within* functional categories, as, for example, the distinct locational imperatives of industries with strong relations to markets versus those with high dependence on bulky raw materials or a need for special labor skills.[32] The specifics of industrial location, retail location and so forth cannot, however, be taken up here.

As the individual units of economic activity move to their respective corners of the city, they frequently group together to form functional 'districts' which generate the large grain spatial differentiation of the city.

This may occur simply because their independent locational requirements are similar[33] but it is usually owing to more than the sum of individual logics. That is, it depends on the mutual relations among users, their collective logic, or what is conventionally lumped under the heading of 'economies of agglomeration'.[34] Indeed, mutual interactions are the essence of spatial relations. However, over time, as capitalism develops, economies of agglomeration have diminished; they are a historically contingent force.[35] But they are in part replaced by economies of (organizational) scale with the concentration of capital, so that gigantic nodes of activity still structure the urban landscape; where yesterday stood warehousing districts today stand petrochemical complexes.[36]

No less important are the forces of repulsion among uses. To some extent this process is mediated by the land market, which erects barriers of rent around nodes of activity to keep away those who cannot or need not pay to be close by. But repulsion is communicated by extra-market means as well, as, for instance, in the case of the fear of 5th Avenue merchants in New York before the First World War that workers from the encroaching garment district would destroy their fashionable trade.[37] This example points up the fact that the land market may have to be circumvented altogether, as the merchants succeeded in doing by generating political pressure – and ultimately the seminal New York Zoning Ordinance – to maintain spatial separation. The force of repulsion among uses has been unduly slighted by urban geographers, despite the grand twentieth-century practice of using the powers of the State – local boundaries, zoning, etc. – to enforce it.

One reason repulsive forces have been overlooked is that they fall less neatly into models which focus on competitive efficiency – the narrow sense of economic rationalization. One effort to break out of this straitjacket has been to focus on the effect of monopolistic behavior in space.[38] But this, too, is a limited conception which ignores the larger issues of class monopoly, class struggle and social control as they affect locational choices. We need to consider, for example, the whole question of worker control as a factor in industrial location: spatial separation can be useful for capital to avoid concentrations of militant workers, to avoid 'inappropriate' mixing of different worker strata, or to control one's own workers better, as in a company town.[39] This matter of control actually extends to the whole field of reproduction of certain kinds of labor-power and of political hegemony in space.[40]

The latter point brings us necessarily to the role of the State, an essential factor in any theory of locational differentiation (and itself part of the division of labor). As a tool of political independence and control over space, the territorial form of local government in the United States has proved enormously useful to capital; indeed, it has been actively promoted by the latter in the twentieth century. The possibility of escaping political control, on everything from smoke-ordinances to zoning decisions, by seeking out 'friendly' jurisdictions or even drawing up new ones, has been fundamental to spatial segregation and the suburbanization process in

US cities.[41] Local government has also offered advantages to industry in terms of cost-revenue rationalization, but principally with respect to land development (discussed below).

Transportation and communication networks play a central role in spatial separation, but not the simple causal role ordinarily ascribed to them. By breaking down spatial barriers and increasing locational freedom these networks obviously give the *possibility* of separation, but they do not create spatial differentiation – least of all by some technological *deus ex machina*. This can be shown to be historically inaccurate in several cases where the means were at hand but unused for want of demand.[42] Transportation and communication have, in fact, developed to a large degree directly in *response* to emergent demands for putting space between activities, for sorting space into distinct groupings, and, most generally, for reducing the time of circulation of capital.[43] Capital operates on the supply side, too. It not only generates a demand for improvements, it virtually guarantees their emergence through the progress of industrial innovation in general – but especially among firms competing in the field – and by channeling surplus capital into the secondary circuit of fixed capital formation and the tertiary circuit of research and development.[44]

Furthermore, the elimination of spatial barriers cannot be seen solely in terms of transportation and communication; it includes everything from improved financial networks to more flexible energy sources.[45] I have elsewhere dubbed this process as a whole 'spatial generalization of capital.'[46]

In sum, non-residential differentiation of urban space is made possible by development of the social division of labor and by the spatial generalization of capital. It is made necessary by the logic of capitalist rationalization and control and is enhanced by the opposing forces of attraction (economies of agglomeration and (organizational) scale) and repulsion among uses. The role of the State embraces all these dimensions of the problems. On both theoretical and empirical grounds we conclude that these tendencies of capital either increase or stay constant over time and hence spatial differentiation will continue to increase.

Residential differentiation poses somewhat different problems of analysis, but our approach is still to search for the necessary forces of *social* differentiation and then ask under what circumstances space has become inextricably bound up in them. We start once more from the social division of labor, but from a different angle. To begin with, we posit the fundamental split in capitalist life between production and consumption – between working and 'living', between reproduction of capital and reproduction of labor – which creates a distinct residential sphere in the first place. This must be explained in terms of the logic of capital, but I will not repeat arguments made at length elsewhere.[47] What of subsequent divisions within the residential sphere?

The basis of social differentiation under capitalism is the class structure, itself a particular hierarchical dimension of the social division of labor.[48]

Class separates people in a variety of ways: income, leisure time, work tasks, status, wealth, education, modes of behavior, and so forth,[49] and these are translated into – or, rather, characteristically take shape as – spatial forms. There are three main aspects of this translation, corresponding to three dimensions of the division of labor in space between production and consumption. They are, in roughly reverse order of importance: (i) the employment linkage, i.e. the necessity of reasserting the spatial linkage between homeplace and workplace; (ii) the pursuit of consumption, i.e. maximizing the rewards (use-values that can be purchased) of wages and revenues; and (iii) the reproduction of labor-power, i.e. the necessity of returning each day and each generation a workforce appropriate to the needs of production and circulation.[50]

The employment linkage subsumes the whole set of immediate relations to the workplace in space, such as location of employment, hours of work and their regularity, need to seek periodic employment and ability to pay for commuting.[51] It sets certain limits on spatial distribution of the classes. Historically, the working class has had to remain closer to the place of work and to depend on different modes of transportation than capitalists, although certain small businessmen and managers may remain even closer because of the need to be 'on call'; women workers usually work closer to home than men, owing to responsibilities in the home, lack of mobility, etc.[52] In general workers remain more spatially tied, and spatially sorted, by their place of work than the dominant classes. Despite its historic importance, however, the employment linkage is generally overlooked in the contemporary literature on residence and suburbia, except for general references to the need to commute. Obviously the development of auto transport, multiple workplaces within the family, higher incomes and more restricted work-hours than in the past, and so forth, have rendered the employment linkage more tenuous, producing the well-known complexity of criss-cross commuting patterns and making the consumption factor more important in housing than in the past.[53] Nonetheless, clustering of workers' residences around employment centers remains significant in shaping the face of metropolitan areas.[54] The detailed division of labor, in which the linkages are more subtle, also appears to be statistically significant.[55]

The perquisites of class (i.e. uneven distribution) also translate into differences in levels and modes of consumption, which consumers hope to realize through appropriate strategies. This becomes a matter of intense personal concern, moreover, given the degree of alienation in production and from production and the consequent self-realization through commodities in capitalist society, not to mention the sheer promotion of a consumerist way of life by capital out of its own realization problem.[56] Social differentiation in consumption is sought therefore because it enriches and even defines the individual.[57] But class antagonism, individual competition, deviant behavior and even production itself (e.g. pollution) put the perquisites of consumption in potential jeopardy; despite the effort to 'banish the facts of production' socially and spatially through consumption, they have a way of reasserting themselves.

What remains problematic is that a differentiated mode of consumption takes on a spatial form as it does in American suburbanization. Historically the wealthy with assets to protect have frequently tended to withdraw in space to enjoy the fruits of their position, but they have not always done so. Certainly space provides a buffer of social control and defense when other social mechanisms are lacking, i.e. if the lower classes do not 'know their place' socially at least they may be confined to it literally.[58] A spatially oriented mode of consumption also becomes increasingly likely for the mass of people with the focus of activity on the home and with the spread of consumer durables, especially the house, which attaches one to a spot. The emphasis in American culture on the single-family house, in particular, has put a premium on space and chiefly in the horizontal dimension, given a well-honed taste for the Arcadian look. In Europe, on the other hand, vertical space has typically been a better indicator of class position; but this appears to be giving way to an 'American' pattern with the growth of consumer durables, automobile culture and so forth, there.

As the objects and mode of consumption become increasingly socialized, moreover, it becomes harder to withdraw into a strictly private space, so collective consumption spaces of the kind which shape city and suburb today become a matter of moment. Increasing dependence on collective consumption goods provided by the State and state regulation of private activities in the interest of the collectivity have figured large in residential clustering in US cities and, in turn, have contributed to the shaping of 'balkanized' political units at the local level. The successful emergence of localized government within the metropolis has played an irreducible role in the development of a spatially differentiated mode of consumption.[59]

Collective regulation of private consumption and the search for mutually reinforcing homogeneity within communities takes on greater importance as the mode of consumption assumes a more 'public' display mode and as residents come to imbue residential landscapes with great significance.[60] The wealthy of Europe were long content to build high walls and use interior courtyards as highly private places, whereas the culture of American homes calls for a certain show of public display, with open facades and yards presenting the household to the world. Under such circumstances one becomes highly concerned with neighborhood effects which reflect badly on oneself and hence with keeping out those who might lower the appearance of the neighborhood.[61] Spatial homogeneity is further advanced in such a situation when conformity is hard to enforce by traditional class values.

Consumption is also the means of reproduction of labor-power (and of other classes), preparing the current adult generation and its children for re-entry into the workplace. Spatial differentiation aids in the reproduction of the widely divergent types of people needed to fit the various slots in the division of labor by creating differential access to social resources, from schools to health care, and reinforcing very different experiences, behaviors and ideologies among sectors of the population.[62] It has come about through a complex historical practice of millions of people pursuing their own self-interest and of interventions by capital in-general to reassert the requirements of capitalist reproduction, all acting under the pressures and limits given by the class structure and accumulation process.[63]

The spatially fragmented form of residential life we identify most with suburbanization has been constructed around a definite *mode* of reproduction carved out by the early bourgeoisie and carried forward since by the upper middle class (new petty bourgeoisie or managerial-professional class) since the later nineteenth century.[64] This mode has as its basic building blocks the so-called nuclear family, the single-family home, homeownership, the neighborhood school, and a certain limited type of 'community', conjoined with a localized political jurisdiction.[65] All are constitutive social constructs about which volumes could be written.[66] But in writing such volumes it must not be forgotten that they did not emerge separately, but evolved in a related way in the midst of a fragmenting urbanism. The dominant classes have, in fact, constructed a veritable set of 'cults' around the above pillars of their mode of life: a 'cult of domesticity', to borrow Ryan's felicitous phrase,[67] and complementary cults of the home-as-castle, of neighborhood schools and of community. The term 'cult' is useful because it connotes the element of ideology and experience embodied in these institutions of socialization and social reproduction. The purpose of these institutions has been to send the working father back into the fray rejuvenated, to optimize the life-chances of children, to avoid falling back into the working class, and to reproduce a mode of life – for what is involved here is more than a refreshed worker and a well-tutored child, but a whole constellation of mutually supporting values, experiences and social resources which go into the construction of human beings.

The nuclear family, homeownership, the homogenous, inward-looking community, balkanized government and the competitive struggle to place children in a class-structured workforce are all socially divisive modes of organizing life. Given these specific elements of social fragmentation above and beyond a general requirement to reproduce the division of labor, it is a short step to spatial fragmentation.

The spatial emphasis in American cities is, in one sense, culturally contingent, as is shown by the different avenues by which differential reproduction is carried out in, say, England. Lack of the neighborhood school in the latter country has, as Cox has shown, been especially important in reducing the degree of spatial fragmentation compared with the US.[68] Ours is a culturally unique, if not surprising, solution to the structural problem of class reproduction. On the other hand, spatial differentiation is in no way incidental to the American middle-class mode of life. Spatial and social division have evolved hand in hand, as the particular cultural solution to the problem of class reproduction. Such things as imbuing neighborhood landscapes with class values and putting boundaries around the experience of children and adults have been essential in the creation of socially separated experiences.[69] The process of suburbanization has, in fact, been one and the same with the process of creating a certain form of middle-class lifestyle.[70]

This form of living and of spatial differentiation, which dominates suburbanization, does not apply fully either to the truly wealthy, including the present big bourgeoisie, or to much of the working class. The working class has developed its own characteristic urban life and communities[71] – aided, in fact, by the withdrawal of the dominant classes into their own (usually suburban) enclaves.[72] This, too, has not been unified, for a variety of reasons which cannot be pursued here, since it leads us away from the suburbs. But certain strata of the working class have been substantially absorbed into the fragmented middle-class lifestyle, with profound implications for class consciousness, as part and parcel of the social construction of suburbia. We take this up in the second part, below.

Let us shift now to the exchange-value side of space, to the property circuit of capital and formation of the built-environment, to see how it reinforces and even actively creates spatial differentiation. We start where it appears to have the most influence, in residential space, since it is there that demand and supply are most clearly mediated by property capital. Four types of actors are involved in this: dealers in property (brokers, speculator-developers), financial institutions, the State (and to a lesser extent other suppliers of infrastructure) and users who also have an interest in exchange values (homeowners, industrialists, etc.).

To begin with, simple risk avoidance generates a reinforcing conservatism; real-estate agents, financiers, etc. find it wiser to go with trends than to buck them, and channel people and capital into segregated residential districts. This is typically carried to the point where the institutions themselves becomes promoters of discrimination. But there is more than lower risk to be had from residential differentiation. For one, it pays to promote it as a kind of product differentiation to increase sales. It also pays to promote mutually supportive land values in order to secure the maximum level of (differential) rents; property investors would be foolish to dilute such values by randomly mixing people. Moreover, if people are sorting themselves out from the top down, suppliers have a parallel interest in helping the cream rise to the top in their areas so it can be skimmed off. Finally, building large tracts of homogenous housing units helps to lower costs.[73]

But property capital wants more. As Harvey has pointed out, the search for monopolistic rents and rapid turnover leads to an explosive process of neighborhood change; property capital freely exploits artificial scarcity in the housing market, and has at its command a variety of devices for creating and reinforcing the market rigidities which generate such scarcity. The result is 'to devastate existing communities and to create new ones in their place'. Moreover, to divide and conquer the residential market in this context is useful for fragmenting political opposition and allowing the profit-making to continue in a hit-and-run fashion.[74]

Homeownership has had an impact on spatial differentiation by its conversion of two-thirds of the city's householders into mini-property speculators. The home functions not only as a use-value but as a source of exchange value, usually the most important single asset of a family. As a result, people are torn between conflicting views of their houses.[75] In order to preserve and enhance their investment, homeowners must take a conservative view of land values and change in their neighborhood, and exclude all who might diminish it. Using local government to secure the property values of the community is also called for, given the influence of taxes, public service provision and zoning.

With respect to both the individual homeowner and to property capital, the existence of a generalized housing market itself enhances residential discrimination, as Cox has argued.[76] The way in which the market rapidly translates use-value changes into property value changes and shifting capital flows forces everyone to be on their toes if they are not to lose out. Indeed, the way in which future expectations are incorporated in investment decisions and the mutually reinforcing nature of expectations means that changes in the market are normally amplified. In the American context, moreover, the pursuit of gain from property is so much a part of the culture that one cannot reasonably expect anyone to resist the flux of the market on their own – much less expect property capitalists to do so. Out of all this comes a logic of conservatism toward land-use change and good reason for the mutual defense of community space.[77]

The State has played several roles as promoter of residential segregation through the support of property developers. Local jurisdictions relate not only to use values but to the stabilization and promotion of land values.[78] The federal government's FHA program for years operated by 'bankers' rules' of conservation and under an explicit racial segregation clause.[79] Less well known is the maze of financial regulations which support market segmentation.[80]

The point is that consumers are not sovereign and the individual resident chooses a home within sharply circumscribed bounds which are determined to a considerable degree by property suppliers and the State. While the latter are by no means free from the dictates of consumer demand or political struggle, they respond in light of what is profitable or what is considered 'best' for home-seekers and have enormous power to steer demand in such directions.[81]

Much the same logic operates with respect to non-residential land uses, i.e. conservatism towards existing demand for spatial differentiation and active promotion of it to enhance land values (rents), although here the users are often their own developers. Landowners, property developers and large commercial and industrial capitalists have a particular interest in promoting aggregations of high-value functions in their vicinity in order both to internalize external economies (to secure differential rents) and to exploit an overblown desirability of the site (monopoly rent).[82] Local governments reinforce advantages of separation and clustering not only in terms of political access, but in their role as land promoters. They offer favorable cost and service conditions and enforced zonation through land use and jurisdictional arrangements: e.g. by providing special industrial parks, by serving as pure industrial suburbs, by competing among themselves to attract industry, or by simple land-use zoning.[83]

All of the above forces generating spatial differentiation from the exchange-value side have tended to increase over time, with the concentration of property capital, improvement of financial markets, generalization of home ownership and increased role of the State.[84]

Decentralization[85]

The preceding discussion of spatial differentiation is rather a formalistic shell until we combine it with decentralization in a model which more realistically recreates the historical spiral of 'functional decongestion'.[86]

Why have the different users of land sought out the urban fringe so eagerly in US cities? The conventional answer consists of a trilogy: cheap land (and more space), better transportation and more space-extensive land uses.[87] In so far as such models are not tautological,[88] a considerable amount remains unexplained. Why do users want to occupy more space or put more space between themselves and the city center? What else besides transportation development makes decentralization possible (and why did transportation develop)? Did the land market contribute anything besides cheap land? Our model, on the contrary, rests on three pillars: (i) diminishing restraints on location of all kinds (generalization of capital); (ii) push-pull forces between uses at the center and periphery, with capital working at both ends; and (iii) the way the property circuit propels the whole process.[89]

Decentralization cannot be conceptualized independently of the opposing and historically prior trend of concentration of activities in the city, especially in what came to be known as the Central Business District. Decentralization of uses arises partly out of the dissolution of forces of aggregation (i.e. diminishing restraints on location) and partly out of repulsion from the center (and attraction to its opposite, the fringe). Particularly because of the latter, decentralization has an indissoluble element of historical geography and cannot be reduced only to a reflection of the spatial requirements of current uses.[90]

The morphologic history to be accounted for is briefly this.[91] During the second half of the nineteenth century most of the economic activities and people congregating in the cities sought proximity to the central business district.[92] This 'implosion' of the space-economy was the primary fact of industrial urbanism. Decentralization was still a secondary movement, which was led by upper-income residents. From the 1830s, the big bourgeoisie abandoned their previous downtown residences in number for suburban life principally by means of trains and ferries. Horse-drawn streetcars and omnibuses provided a way for the small middle class of clericals, professionals and petty entrepreneurs to begin sorting itself out to the peripheral zone of the built-up area as well. The latter movement had developed sufficiently by the end of the Civil War, that the middle class in several cities began adopting the single-family detached house

style imitative of the bourgeois suburban cottage. The few large-scale factories that existed did, additionally, show a tendency to seek fringe locations,
but their numbers were not sufficient in the nineteenth century to have
a decisive impact and they were rapidly engulfed in the expanding city.[93]

The explosive growth of residential suburbs came after 1890, with the
beginning of the era of corporate capitalism. The movement was led by
the rising 'new middle class' of managers and professionals, who took
the mantle of suburban pioneer away from the bourgeoisie in the twentieth
century; they were followed by a new army of white-collar office workers.
These suburbanites rode on the rails of the trolley car and had their homes
financed by a new breed of mortgage banks. Lower middle-class and working-class decentralization began before the First World War in the modest
form of what was called 'the Zone of Emergence' by bourgeois reformers,
as well as around certain industrial satellites. Large-scale working-class
suburbanization, especially that of the white-collar workers, is chiefly a
post-Second World War phenomenon, and has remained rather closely
allied to paths of industrial dispersal. The automobile, along with other
consumer durables, did not become generalized until this era. Although
the suburban norm has extended a considerable way down the social ladder,
it nonetheless remains dominated even in our day by the middle class
and by a lifestyle largely established in the nineteenth century.

Decentralization took additional impetus from the wave of dispersal
of industry to satellite towns around the turn of the century.[94] Each subsequent wave of factory-building has renewed the outward flight, especially
around the Second World War and in the last decade. The impact of
this has been offset by the steady shift of employment from manufacturing
to so-called 'tertiary' and higher activities, i.e. large-scale retailing, management, finance, related professions and government, and continuing tendencies to aggregation among these functions.[95] Only since the Second World
War have exchange and office-type functions become significant forces
in suburbanization. In the process they have further helped suburbia's
transformation from an élite residential exodus to a mass movement –
indeed, a veritable reconstruction of the city as a 'suburban city'.[96] By
the 1960s density gradients of retailing, wholesaling and manufacturing
finally 'caught up' in degree of decentralization with residences.[97]

Why have residences decentralized? This began with the rise of an industrial bourgeoisie,[98] whose pioneering example subsequently became an important cause in itself.[99] Without teasing out a full model, suffice it to
say that the demand for the fringe was generated chiefly by repulsion
from the growing concentration of productive activities and the associated
districts of the working class, in an effort to 'banish the facts of production'
from the landscape, i.e. pollution, noise, crime, the threat and the misery
of the working class. This repulsion worked at both ends, taking on a
mystified form as an attraction to the Arcadian ideal of ruralized living
at the edge of the city.[100] Thus was established the basic pattern of escapism
from capitalist reality which still colors suburbia deeply. Its realization
was made feasible by a loose employment linkage, due to growing affluence,

less direct need to supervise business (at all hours) and improved transit.

The fringe of the city was sought out, moreover, because it was the easiest place to carve out an appropriate space within which to fashion a whole mode of consumption and reproduction – i.e. lifestyle, institutional complex, class-support system. The suburbs are not middle class simply because the middle class lives there; the middle class lives there because the suburbs could be made middle class.[101] They could construct the socially differentiated and defensible landscape discussed in the first part (pp. 389–93). This is largely a question of control: at the fringe the dominant classes have found it easier to control their conditions of life free from the impact of industry, the working class and even of each other. This is clearly not simply a matter of the dangers of proximity, but also of class weakness, of inability or unwillingness to forge an alternative solution to the problems of controlling industry, the working class or the land market.[102] Furthermore, the obsolescence of the past, frozen into stone in the older urban core and solidified by social practice in the permanence of older neighborhoods, makes it more practicable to create a new city, resting on new division of labor, new wants and needs, etc., outside the old.[103] As a single example of what this means, Young and Wilmott have pointed out that spatial separation from parents was virtually essential to the successful establishment of the middle-class nuclear family.[104]

In terms of control, the role of the State is especially important. Freedom from political control through city government and ability to mould government in the desired form and for the desired ends could best be achieved outside city boundaries. This has to be balanced with the benefits of urban services provided by the city government before the turn of the century, but it seems that even then the bourgeoisie could frequently afford to move out ahead of annexation.[105] But during the progressive era the necessary changes in financing, zoning powers, city-manager government and so forth were put into place by and for the dominant classes to make suburban governmental autonomy possible. Annexation rapidly gave way thereafter to the politically fragmented metropolis, which supported the construction of the suburban way of life.[106]

Not only have the dominant classes sought the urban fringe, they have installed a particularly space-extensive form of city-building there.[107] The Arcadian ideal took the form of a detached house surrounded by a garden, and as this has been passed down to the mass market of home-buyers, vast acreage has been given over to residential use.[108] Add to this the parklands (and parkways) coming out of the same naturalistic tradition, the space devoted to the automobile as means of transportation and as cultural artifact, and space-using methods of land use control, such as large-lot zoning, and it compounds the uniquely space-extensive format of the American suburbs.

The situation of the trailing lower middle class and working class is somewhat different, given their different wants and needs and lesser ability to initiate and manipulate social institutions. But the social construction of suburbia intertwines with a profound transformation from the working-class life of the nineteenth century to that of the twentieth, including a substantial absorption of workers into the 'middle class' mainstream of American life.[109]

The working class has developed its own characteristic life and communities, which are underrepresented in the suburbs, owing partly to exclusion by the dominant classes.[110] When the working classes have suburbanized they have typically done so more to be near employment than in pursuit of the suburban dream, and have remained more closely tied to suburban industrial areas.[111] Politically too, they have remained much more dominated by nearby employers and the local petty bourgeoisie.[112] And where workers have made the move to suburbia, it has not automatically caused them to become middle class (in lifestyle or otherwise) as Berger, Gans and others have shown.[113]

Nonetheless, in criticizing the myth of suburban culture, Berger and his followers have led us to believe, falsely, that suburbanization has been wholly without effect on working-class culture. As Berger himself admits, workers following their employer to Fremont felt they had 'moved up' in life, due especially to the acquisition of a house.[114] Dobriner notes adopted middle-class values and customs by working-class people in 'lower middle class' suburbs, e.g. homeownership, keeping up house appearances, little class consciousness, mixed in with continuing threads of blue-collar culture.[115] There is no question that, however different it is from that of the middle class, working-class life in the suburbs is quite different from that of the traditional ethnic districts of the inner city. This difference can be seen clearly in the interfamilial tensions between parents in the inner city and children moving to the suburbs, as described by Sennett and Cobb.[116]

Conventional sociologists have futher underplayed the impact of suburbanization of the working class by defining the latter narrowly as blue-collar workers. In fact, the largest impacted group has been the 'rising' strata of white-collar workers, tied to a changing social division of labor between production and circulation, mental and manual labor and private/public sectors, whose personal life-experience and, indeed, class-experience has been intensely colored by developing within a 'middle class' mode of living.[117] This, in addition to differing situations within the workplace, has served to divide the working class and rupture its class traditions, and class struggle, in this country.

This effect has not been lost on the bourgeoisie who have carried out a systematic campaign of urban 'reform' to promote 'decongestion', homeownership, the cult of domesticity, 'Americanization' and so forth, which I have documented at length.[118] At the same time, workers have actively participated in the making of suburbia and the 'great American middle class' by pursuing their own rationale. The rationality of the choice is, in certain respects, clear enough: better housing at good prices, search for better schools and so forth. But such 'choices' and 'rationality' are always conditional, taking place within the prevailing hegemony of capitalist society. This bourgeoisie hegemony[119] consists of at least three cooperative elements: (i) sheer imitation, because the bourgeois suburban ideal of where and how to live is the ruling ideal;[120] (ii) inability to forge a class-alternative, making individual pursuit of entry into the great American middle class an apparently rational choice; i.e. class struggle has been 'displaced' in the suburban direction;[121] (iii) because conscious interventions by the dominant classes have steered workers in this direction, as, for example, by tax-breaks for homeownership.

Thus the suburbanization of the working class has also been a constitutive process,[122] a part of the forging of new modes of living containing a large dose of 'bourgeoisification'.[123]

Despite the historical leading-role of residential decentralization, only with the dispersal of employment – of production and circulation activities as a whole – was mass suburbanization possible and the reformulation of the city into the twentieth century metropolis realized. The urban implosion of the industrial era could not have been reversed and the 'explosive' form of today begun without the part taken by industrial decentralization, beginning around the turn of the century, at the beginning of the corporate stage of capitalist development and coincident with the suburban tide of the new middle class.[124] Since then major new industrial loci have spiraled outward, at the expense of investment in the old manufacturing districts. A look at the causes (from the use-values side) of the initial leap to the suburbs in the first decade of this century can be instructive as to the logic of industrial decentralization.[125]

What is involved is, once again, the historical process by which capital becomes more 'generalized', or universal, in space.[126] As it does so, it frees itself progressively from the constraints of locating in any one spot, but particularly from the traditional locus of production in the city center. Strong forces of polarity are at work. These processes involve both the dynamics of accumulation and those of class struggle, or, to put it more narrowly, they have aspects both of cost/revenue calculation and concern with social control. Specifically, what had been achieved by the beginning of the century was as follows:[127]

The merger movement created the giant corporation which made a new generation of large factory complexes possible, because it could provide the capital, the market share, oligopoly prices and raw material to make large-scale production possible, and because it had the management scope, research divisions and scientifically-managed labor to make technical breakthroughs feasible. The large corporation and the large factory-complex together internalized many of the economies of agglomeration which made central-city location attractive to small manufacturers. Both were more self-contained, incorporating more aspects of production, special labor and professional skills, more infrastructure. The industrial giants were also better able to attract to their chosen sites complementary factories and a supply of workers, instead of depending on central location for access to these things. At the same time, those corporate functions demanding central location could be separated into downtown offices, leaving factories more freedom to disperse.

Development of telephone communication made the detachment of factory and headquarters feasible. The railroad network had developed nationally – and was suffering from sufficiently low prices – that it opened up a truly national market, freeing industry from its local, urban base. More immediately, a large number of spur lines had been built which allowed factories to locate at isolated fringe sites.[128] The trolley system had also been expanding rapidly, providing better intra-urban mobility, but particularly making reverse-commuting by inner-city workers possible.

Greater labor productivity in large factories, besides ultimately supporting the whole edifice, reduced the number of workers needed (sometimes absolutely) and hence reduced reliance on the central-city labor pool. Crucial here was the breaking down of corporate dependence on skilled labor, a matter of considerable class interest.[129] It also allowed large producers a competitive edge in expanding markets.

The provision of infrastructure and housing in general also played a role in decentralization. Concentration had occurred in part because use of the accumulated stock of housing, warehouses, etc. was financially attractive to industry at the dawn of capitalist industrialization – that is, any one company could act as a 'free-rider' on the total social capital in the built-environment. Corporations could rely on the property developers to supply housing for workers, or even do the job themselves, while institutional changes in local government were freeing business from dependence on big-city governments for infrastructure. Companies could even build their own rail spurs or ports, if necessary.

Developments such as the above allowed factories greater freedom to migrate but clearly barriers remained; the suburban fringe was a kind of compromise location. Given the possibility of dispersal, what polar forces actually produced the compulsion to decentralization? The enlarged scale of factories certainly made sites near the center harder to find and the open spaces at the periphery attractive. Conversely, the old warehouse districts were inappropriate to this scale and type of production, indeed, much of the urban infrastructure was obsolete in terms of changing needs and could be more easily begun from scratch at the fringe.[130] Traffic congestion was increasingly intolerable, and the national market to which large producers had to relate could be reached more easily from the fringe.

More important than this relation to the built-environment was level of class tension in big cities by the end of the nineteenth century. The logic of industrial decentralization was intimately tied to elements of escape from a geographically concentrated working class in the industrial centers and to an effort to improving capitalist class control over the conditions of production. As the urbanized working class became more organized, more independent of capital in its daily life, it became a threat to profits, factory discipline and social control generally. Capitalist consciousness of the threat was sharpened by the labor struggles of the 1880s and 90s. Suburbanization was thus seen as a way to attract a better brand of labor, removed from the 'bad moral atmosphere' of the inner city, and promising the stability of homeownership for the 'better class' of workers.[131]

To bring about this 'suburban solution' to class conflict and to cultivate a less combative, more stable workforce, the industrialists not only relocated, but frequently took their trained workers along with them and provided for land, houses and credit so that the latter could buy homes.[132]

Similarly, capital's relation to the State became more favorable at the urban fringe, where working-class political access was less, social regulation such as smoke-pollution ordinances absent, and older forms of government structure not entrenched. Wholly new units of local government could be carved out or old ones transformed, and while the strict company town, like Pullman, proved too oppressively obvious, the capitalists had great success with more selective intervention in governments whose day-to-day politics were buffered against working-class influence by a petty-bourgeois political alliance (merchants, professionals, real-estate interests) and a managerial form of government.

A similar logic of decentralization appears to have operated up to the present, with successive waves of industrial concentration, improvements in transportation-communication networks, and rapid extension of the general urban infrastructure, and intensified class conflict, followed by waves of factory dispersal. Local government forms and political practices forged in the Progressive era remain in place today, with the additional features of intensive competition among balkanized localities for industry and the interventions of the federal government.[133] That nothing has changed with respect to a desertion of the inner city is shown by the following:[134]

> In a 1967 survey of 700 major corporations conducted by the Dun and Bradstreet subsidiary, the Fantus Company, only 112 of the 700 firms 'indicated a willingness to build new plants in or near slum areas'. Moreover, not one of them would make the move unless certain conditions were met:
> - A large and non-aggressive labor pool.
> - Programs to train local workers with new and useful skills.
> - A moratorium with local unions that would agree to stabilize new construction costs and open their ranks to minority workers.
> - Ample land at reasonable prices, adequate roads and utilities, and new zoning codes.
> - Pleasant environmental conditions, including quality housing in the immediate area and assurances that 'ghetto blight' would be removed.
> - Lower real-estate taxes.
> - Adequate security and fire protection.
> - Relations only with 'responsible community groups', not with militants.

The Fantus report adds: 'These companies are interested in good publicity and profits. None of them indicated any interest in philanthropy.'

Only recently – since the Second World War but especially during the 1960s and 1970s – have other functions besides industry become important in decentralization. Retailing played a follow-the-leader role behind residences until the rise of shopping centers – now giant regional shopping complexes – transformed it into a leading factor.[135] Office decentralization, more in its infancy, appears to turn on one factor above all: access to cheaper, more docile female labor in the suburbs.[136] It would seem, there-

fore, still to be a lagging element. Federal offices may well be more autonomous and have been aggressively decentralized in Washington, DC and
elsewhere. Nonetheless, an analysis of leading and lagging factors misses
an essential point, which is that the kind of mass suburbanization of today
is strongly mutually reinforcing, and would not be possible without a
whole ensemble of uses moving outward together. This leads us to the
macrologic of waves, considered in the third part (pp. 405–9) below.

How has the exchange value aspect of urban space, the property circuit
of capital, contributed to the decentralization of urban areas? As passive
'sorter and arranger' of uses, the land market spreads users out according
to their willingness to pay, as in all conventional (differential) rent-location
models. This subsumes both the pull of cheap land (and, concommitantly,
lower taxes)[137] at the fringe and the push of high-priced land at the center,
owing to general concentration or to competition from higher-valued uses,
e.g. offices out-bidding manufacturing. Similarly, the generalization of capital, the removal of spatial barriers, depends on a flow of capital through
the property circuit into housing, infrastructure, transportation and so
forth which is at least consistent with demand from the use-value side,
technical capabilities and so on. For our purposes we assume that these
mechanisms function in at least a neutral way,[138] i.e. that decentralization
is at least profitable, capital is moving into the secondary circuit, the
State is playing its role – in short, that certain institutions, attitudes and
behaviors are operative. This constellation of mutually reinforcing linkages
is typically called a 'growth coalition'.[139]

What is more significant for this inquiry is the active way in which
the property capital and the land development process have propelled
suburban decentralization. That is, the activities of the property circuit
have actually created and exaggerated demand by three main methods:
by 'land speculation', by channeling capital flows and by state support
for both.

Land speculation has been a hallmark of American urbanization,
equalled nowhere else in the advanced capitalist countries until the recent
past.[140] The role of speculation has been recognized by urban geographers,
but without systematic analysis.[141] What is loosely known as land speculation – where it is not merely a perjorative for land investment – consists
of two active forces: the manipulation of land and land uses to create
rents (the main source of profits in property investment) and self-sustaining
property bubbles (which will be discussed below, pp. 408–9). The former
falls into three categories: the directing of users, infrastructure and buildings towards one's own property (for differential rents), the creation of
artificial scarcity (for monopolistic/absolute rents), and the redistribution
of state revenues and related phenomena (redistributional rents).[142]

The property investor – appearing in various guises as housing developer,
landowner, shopping center builder, transit line owner or industrialist –
operating at the urban fringe, where the city is being most actively constructed, has as his/her aim the maximization of return from capitalized
rent, i.e. land-value appreciation. This may be achieved, of course, simply

by correctly 'anticipating' the market, buying cheap land beyond the fringe and waiting for the city to follow or fill in. By the apparently 'passive' act of waiting, of idling land until it is 'ripe' for development, the class of property investors can, as a class, have the active effect of either forcing up land values by creating artificial scarcity or spreading out the city by forcing other developers, users and infrastructure to leapfrog over their idle lands.[143] Instead of waiting quietly, however, most property speculator-developers are actively working to pull the city along behind them. They can do this in several ways, which cannot be discussed in detail here: by directly drumming up demand, or selling suburbia (differential rent); by creating a focus with mutually reinforcing advantages, as is done by big industry or big shopping mall developers (differential and monopolistic rent); by steering transit lines, sewer lines, etc., i.e. infrastructure, in their direction (differential and monopolistic rents); by underpricing transit or state services through flat-rate charges to every location, as e.g. the notorious 5¢ fare which helped bankrupt many trolley lines (redistributional rents); and by zone-breaking (a redistribution of rents from one landowner to another).[144]

Property investment also has its effect at the center, the other main locus of speculative activity;[145] traditionally this has helped propel uses outward. Land values are easily inflated near the center, owing to monopolistic behavior, idling land, mutually reinforcing expectations and periodic investment booms,[146] which drive out users who might otherwise be able to pay for central space. The effect of property speculation is clearest in the 'grey areas' around the expanding CBD, which are being held for conversion to higher uses but will not be committed for interim uses other than parking lots. A similar result obtains where inner city residential districts are considered marginal and risky investments – even where no conversion to higher uses is imminent – and investment funds are consciously withheld, e.g. through red-lining. Of course, the generalization of the speculative ethic and the pursuit of exchange value by all property owners has long made it almost impossible to hold the line in defending residential areas against encroachment of business uses, forcing people to seek out the suburbs to establish new enclaves of exclusion and stability.[147] But the actions of large property investors, particularly of the financial institutions which back most property transactions, compound the general effect either by withholding funds from profit-poor areas or by throwing funds into neighborhood conversion. Harvey has shown how in Baltimore the machinations of financiers and real-estate brokers, in search of profit-opportunity from neighborhood change (turnover, shift to higher-paying users, elements of monopoly rent), contribute to the process of driving whites to the suburbs by pushing blacks in behind them – what he terms 'blow out'.[148]

The role of financial institutions is further magnified in that they are the key link in bridging investment activities across space. This means that the inner city disinvestment and suburban investment processes are intimately related. For reasons of relative profitability, finance capital has systematically backed mortgages in the suburbs while redlining inner city districts and funneling their savings into suburbanization.[149] Such a process is, moreover, self-reinforcing, as poor neighborhoods deteriorate and suburbs prosper. Historically, the lack of new low-income housing in the inner city, which began to be noticed in the 1920s (at the same time that the biggest suburban homebuying binge up to that time was underway) appears to be intimately linked to the rise of large-scale mortgage capital which could, for the first time, successfully bridge the whole housing market of the city.[150] But whatever the adverse impact on the inner city of suburbia, it is nonetheless true that American financial institutions have been uniquely successful in mobilizing savings and funneling them into low-cost suburban housing.[151] [152]

The role of the State in facilitating the work of developers has been fundamental. Indeed, the principal mode of intervention by the State specifically addressed to the urbanization process has been through support of the property circuit. Local governments in the twentieth century have been converted into specialists in land development. This includes both general service municipalities and the large class of special service and school districts, which now make up over half of all local government units in the US. They provide clear boundaries, infrastructure, police powers to organize space, low taxes, favorable debt and professional administration.[153] And they are readily available to political manipulation by property developers as well as users; partly by direct infiltration of the latter, partly out of the political aspirations of individuals in government and partly out of competition among localities to attract development.[154] But manipulation is perhaps less important than that growth boosterism (growth coalition) has been the spirit of the time, which looked upon suburbanization as a universal good.

The federal and state governments have also eagerly greased the wheels of suburbanization, chiefly by financial aid in construction/redevelopment of the built-environment. The major federal programs involved are certainly well known by this time: highways; mortgage-insurance and housing subsidies (FHA, VA, GNMA, FNMA, 135–136, etc.); tax benefits (subsidy to homeownership, low capital gains tax limits until 1968, rapid depreciation since 1964, tax-free local bonds, etc.); grants-in-aid to local governments to support capital investment (sewer-building programs, school support, etc.); and urban renewal.[155] The state played a complementary role in most of these federal programs, as well as being immediately responsible for the system of local governments. These activities of the government were by no means accidental, nor, for all their importance, can we ascribe to the State a procreative role in suburbanization, since we can discover in the origins of every program a conscious effort to sustain a budding suburbanization process.[156]

In sum, the role of land speculation, mortgage finances and government has been to control the dynamics of urban property development in the United States. It is not coincidental that the US exceeds any other advanced capitalist nation in successful *private* building (and abandonment) and the degree of suburbanization (and central city decay).[157]

Stages of accumulation and waves of urbanization

We are now in a position to take account of the temporal dynamics of capital accumulation in our model of the suburbanization process. The basic propositions concerning accumulation cycles, developed by David Harvey and myself at length elsewhere, are these:[158]

1 Accumulation takes a cyclic form of growth – pause (or stagnation) growth.
2 Each period of growth requires 'balanced growth ensemble', not only in the economic sense of a 'balanced growth path' but including political, cultural and economic arrangements as a whole: specific institutions, mode of behavior, types of industry, etc.[159]
3 The progress of accumulation necessarily raises contradictions within the growth ensemble which ultimately generates crisis and a cyclic downturn (or other, lesser forms of adjustment).
4 Crisis produces a response, or 'forced rationalization,' of elements of the ensemble, from the conditions of the labor process to industrial organization and beyond, to political institutions and culture. Sooner or later these again achieve a workable balance – a new growth ensemble – on which basis accumulation can proceed again.
5 This pattern of accumulation – growth, crisis, rationalization, growth – divides up the history of capitalist development into distinct 'stages', whose growth ensembles look rather different.
6 Successive stages are, of course, linked by the fact that the features of one stage are forged through a selective process of building on the past, continuing its strengths and transcending its contradictions.[160]

The specification of the various growth ensembles and the periodization of capitalist history is an enormously difficult task, which is less clear cut than our simple model makes it sound. Still, the evidence for such a periodization of capitalist history into stages of accumulation is compelling.[161]

How, then, does urbanization relate to this pattern of accumulation? The basic propositions, again, are these:

1 The city is a 'container' for the expanded reproduction of capital and capitalist social relations, i.e. the built-environment and spatial organization of the city embodies the growth ensemble of any stage of accumulation.[162]
2 There is, then, a spatial dimension to the general development of contradictions and crises – in other words, the crises of social reproduction have one dimension which may be called 'urban crisis'.

3 Part of the rationalization process for overcoming crisis consists in reorganizing spatial relations; but this presents particular problems because the built-environment literally freezes the past – and its contradictions – in stone. Hence, the built-environment is both 'crowning glory' and 'prison' of capital.[163] It, too, must be rationalized by a selective process of continuity and rejection.[164]

4 The built-environment comes into being through the movement of capital into the secondary circuit of fixed capital formation (including the property circuit) as overaccumulation of capital begins to afflict a period of growth.

5 Thus, the use-value side of space does not instantly call into being an appropriate built-environment without the mediating role of the property circuit (exchange-value side). In fact, the latter tends to lag behind growth of output, employment, etc., and then erupt in a period of furious building, topped off by speculative excesses.[165]

6 The laying down of the built-environment is thus characterized by waves of growth which reflect the use-value ensemble of their period, but also contribute their own forces by virtue of the mediation of the property circuit.[166]

The periodization of American urban history is again hardly unequivocal. Empirically, waves of urbanization appear to be an uneven overlay of three to five-year short cycles, fifteen to twenty-year Kuznets cycles, and fifty-year Kondratieff waves.[167] As a broad first cut, the division of urban history into roughly fifty-year stages is extremely compelling and has been arrived at by several different observers working independently.[168] The basic periods are mercantile (1780–1840), national industrial (1840–90), early corporate (1890–1940) and advanced corporate (1940–90?).[169] A finer cut would get down to the critical fifteen to twenty-year Kuznets stages, which also differ considerably one from another in their growth ensembles, and which would tighten up on the broad generalizations one has to make with respect to fifty-year periods.[170]

Periodization is not merely a scholastic exercise. Different types of cities were constructed during each period. Our cities are literally museums of the past and of past social relations.[171]

The above framework adds some new dimensions to the analysis of suburbanization, including shedding new light on certain aspects of what was said in the first and second parts of this article.

The idea of stages of development forces us to confront the problem of history – and the irreducible element of historical contingency – head on, much in the same fashion as we approached the questions of spatial and cultural contingency before.[172] What I have elsewhere dubbed 'the suburban solution' – meaning that suburbanization has necessarily been consistent with the needs of capitalist reproduction in the US – actually has three distinct aspects: its relation to certain fundamental 'structural' relations of capitalism of long standing;[173] its relation to the ever-changing character of capitalist society, i.e. to specific growth ensembles; and the relation of the present to the failure of earlier 'urban solutions', particularly to the kind of concentrated industrial city built in the national industrialization period of the late nineteenth century and carried over well into the early corporate period. In the first and second parts of this paper we emphasized the first aspect, which is of primary importance; but some ambiguity crept in because the other two elements could not be ignored.

With respect to the second element, growth ensembles, it is correct to see post-war expansion as more than a culmination of 'emerging' suburban tendencies, but as unique in its own right and as a process of building a fundamentally new form of city.[174] A teleological view of history which sees the present as the necessary endpoint of history must be avoided. Indeed, we recognize that the very definition of suburbanization changes over time and that ours is located in post-war experience.[175] As for the third element, suburbanization has been a 'solution' forged with an eye on the past, as when the Progressive-era industrialists sought to escape a spatially-concentrated working class, or when New Deal financial reformers looked back longingly on the housebuying boom of the Twenties, which they sought to restore.[176] Thus, in the second part of this essay, we posed the congested industrial city as a counterpoint to decentralization. A further complication introduced by both elements is that American cities of varying ages embrace different histories: there are newer cities, especially in the South-west, which are almost wholly 'suburban cities' and there are older cities, chiefly in the North-east, which contain a contradictory past and present.[177]

This raises the issue of urban 'obsolescence'. The built-environment must change with the changing nature of industry, housing, etc. As conditions alter, however, what was the crowning glory of a prior stage of accumulation becomes obsolete, literally freezing the past into stone, where it can become a trap for capital and labor.[178] It no longer serves current uses well and also traps invested capital where it can no longer return a competitive profit, given use-value deterioration. Labor and capital will therefore find it cheaper, easier and more profitable to start anew in open territory, at the suburban fringe, than to try and remake the social and physical environment of the old city.[179][180]

What was presented in the second part, in a piecemeal way, as to the logic of escape and of constructing new relations in the suburbs should, then, be seen as a general process, which inheres in the necessity of building a city around certain historically transient social relations. What users are running away from is, besides present problems, the built-environment itself and the past made manifest. Urban renewal, in its various guises, is costly, conflict producing, and requires collective effort – all things which capital shuns. It may fairly be said, then, that capital readily adopts a 'slash and burn' approach to cities.

The wave/stage model has further implications for the way we comprehend suburbanization. An urban growth ensemble must have internal coherance, that is, the 'pieces of the puzzle must fit'. While the pieces are not random by any means, still the logic of the whole is to some degree

wholly internal to itself. Because of this, the phenomenon of suburbanization cannot be reduced simply to the sum of its parts or even to this or
that relation between the parts, which is how we have approached it up
to now. We must step back and see it as a whole – a mutually structured
and self-reinforcing whole, which shapes the parts at the same time as
the parts shape it.[181] Literally and figuratively it pays to climb aboard
the investment bandwaggon; or, as Joan Robinson says, 'the possibility
of profits only depends on what the other fellow is doing'.[182] Thus, quite
apart from individual motives, agglomeration economies, obsolescence
problems, etc., once a critical mass of location shift to new territory is
underway it tends to become self-supporting and self-exaggerating. Without
overstating the importance of this, it must nonetheless be admitted that
to some degree there is no logic whatsoever to suburbanization: it was
just the road down which society and economy headed.[183]

   Waves of urbanization are waves of capital investment in the built-
environment. In the first and second parts we repeatedly emphasized the
role of the property circuit of capital in sustaining and propelling suburbanization, but without linking that circuit to the dynamics of accumulation
cycles. Yet it is precisely the process by which 'overaccumulated' capital
pushes into the property circuit which provides the capital necessary to
property investment. In order for this to happen there must exist, at
the minimum, the requisite financial and governmental institutions
to funnel off surplus capital as part of the growth ensemble. But
what we observe with every major wave of urbanization is a surge of
investment towards the end of the cycle, ending in a peak of speculative
excess before the collapse of building.[184] That is, overaccumulation comes
to afflict the property circuit as well.

   Indeed, it is impossible to explain the periodic waves of overextension
(and underpricing) of transit lines,[185] public utilities and highways, housing,
mortgage credit, etc. without reference to the general dynamics of the
accumulation process and the forcing of too much surplus capital into
the secondary circuit in hopes of staving off crisis. This 'push' of capital
is helped along, of course, by the 'speculative bubble' atmosphere which
eventually overtakes the property circuit, and lures capital into its web
of erroneous but mutually reinforcing expectations. Even though expectations do not pan out, the overextended city is already built and a commitment already made to the use-values put into place. Capital will then
have to be written off and funds transferred in the subsequent crisis and
rationalization period – e.g. through municipal bond defaults, bankruptcy,
consolidation or municipalization of trolley lines, or mortgage foreclosures.
But the consequences for the built-environment have already been felt.
There is no question that the successive property booms which have marked
the history of American urbanization have contributed materially to the
hyperextension of the city.[186]

   We should, however, now widen our focus from the property circuit
to the secondary circuit of capital as a whole – the overall process of
fixed capital formation and the creation of the built-environment – which

includes everything from factory construction to purchases of consumer durables. A mass of capital will necessarily flow into the activities in order to enhance labor productivity, consumer buying power, the speed of circulation and so forth – in short, to sustain accumulation. It has been pointed out many times that post-war suburbanization has served as a vast outlet for capital in all its forms: as direct investment in factories, infrastructure and housing productions; as consumer buying; as credit-creation.[187] To put the matter the other way around, it would be very hard indeed to imagine how American capitalism would have fared if it had had to make do with cities as they were in 1949, or even with a mere extension of the kind of city which existed then, rather than the much more expansive, capital-absorbing form that it took.

The suburbs are thus a vast conduit of concrete use-values through which value flowed and expanded. Progressive accumulation not only provides the *means* of supporting suburbanization, it practically *demands* that the city expand to keep pace, else a crisis of accumulation develops.[188] Of course, it is not required that the city alone take on this role, and, in fact, defense expenditures and other things have contributed their share; but it is hard not to see it as playing a big part. The city is certainly more conducive to sustained accumulation than, say, bombs – unless it becomes socially explosive. It is, of course, problematic exactly how this role of the city contributes to decentralization or spatial differentiation. However, in an interesting study, Ullman and his colleagues have judged that a more spatially concentrated form of urbanization could not very easily have accommodated the mass of consumer durables which the suburbs grew up around.[189]

A final influence of the wave-form of city-building is that different users have different specific relations to the timing of accumulation and urbanization cycles, in addition to their various atemporal location rationales, and this temporal element contributes to spatial differentiation. As Whitehand has shown for a section of Glasgow, the fact that housing and public institutions are not synchronized over the course of the building cycle results in alternating rings of development.[190]

Conclusion

This paper develops a framework for the analysis of suburbanization as a geographic process. It by no means fills all the theoretical blanks concerning spatial processes under capitalism, nor provides the necessary historical evidence to sustain all that has been said. Moreover, we have closed our eyes to the question of contradictions in the 'suburban solution' and to changes taking place in the suburban process over the last decade. What we have tried to do here is to see the problem from several angles in all its interlocking complexity. It is time that we advanced beyond merely listing the factors contributing to suburbanization – whether couched in Marxist theory or not – and approach the problem systematically, probing the various causal elements and establishing basic principles of analysis to link suburbanization to the structure of capitalism. Needless to say, constructing a satisfactory theory of urbanization and suburbanization is a remarkably long and arduous process toward which we move at an uncertain gait. The work continues.

Notes

[1] See Castells' (1977) critique of the 'urban myth'.

[2] Cf. Vance 1964: 54, 68, Muller 1975: 8, 12, Castells 1977: 386.

[3] We eschew the question of contradictions in suburbanization altogether in this paper. Because of this our presentation will appear artificially static, missing the continuing flux in suburbanization today. On the contradictions of suburbanization, see Ashton 1978, Sobin 1971, Kaplan 1976, Masotti and Hadden 1973, 1974, Castells 1977. On recent trends in suburbia, see also Muller 1975, Johnson 1974, Schwartz 1976, Sternlieb and Hughes 1976.

[4] This is a response to the criticisms of Marxist treatments of space by Soja 1978 and Evans 1978.

[5] Marx 1967 ed., Vol. I, Chapter 7, Schmidt 1971.

[6] See Williams 1977 but replace the term 'culture' with 'space' in his formulations.

[7] 'Social formation' is the Althusserian term for a particular concrete society, combining elements of more than one mode of production. On space and social formation, see Santos 1977.

[8] Ollman 1971.

[9] It is common among Marxists today to refer to the triad of economic, political and cultural (or ideological) instances of society, rather than base and super-structure. Cf. Castells 1977, Williams 1977.

[10] See Harvey in this volume; Walker 1977. Because we are dealing with structuring forces (what Williams (1977) refers to as 'limits and pressures') instead of strict determinations, it is necessary to employ concepts such as 'displaced class struggle', 'mystified relations to nature', etc. to unlock the relation between capitalist structure and concrete forms of urbanization. (See Harvey 1978 and in this volume.) Such formulations are not merely twists and turns to create a deterministic analysis, for they do not eliminate what Williams (1977) calls the 'constitutive' element of human practice in concrete historical conditions.

[11] Actually land only takes the commodity *form* (it has a price but not a value) and even human-made structures ordinarily circulate only as titles on paper, not as physical commodities. The real subsumption of urban construction and landed property by capital does not happen automatically, of course; its history is a long and interesting one.

[12] See Harvey's (1972, 1973) treatments of the housing market and rent theory, especially his critique of one-sided approaches. (However, as Harvey has since shown in an unpublished chapter, the commodity divides into three: use-value, exchange-value (which becomes price) and value. This further division is beyond our purposes here.)

[13] Harvey in this volume. I would modify his analysis to include a category of 'circulation' or 'exchange' fund of fixed capital.

[14] As Vance puts it, 'Urban buildings, and districts, are artifacts whose shape is derived from use and is ever subject to modification to facilitate that activity' (1967: 95).

[15] Do not confuse the use-value–exchange-value dichotomy as it applies to the space-commodity with the parallel division with respect to the use of space. That is, the principal use-value of space for the capitalist is to accumulate (exchange) value, while the principal use for labor is for the direct consumption of commodities as use-values. Cf. Harvey 1978.

[16] See e.g. Mills 1972, Alonso 1964, Muth 1969.

[17] See e.g. Downie 1974, Paulson 1973, INFORM 1976–8, Marriott 1967. For more scholarly treatments of the supply side, see, besides the works of Harvey and myself, those of Gaffney 1967, 1973 and Checkoway 1978.

[18] Marx 1967 ed., Vol. I, Chapter 25. Unfortunately, Marxists frequently repeat the error of one-sidedness. E.g. Castells (1977) treats only the use-value side. See also Gale and Moore 1976.

[19] Spatial differentiation and decentralization are, of course, integrally related – indeed, they have been part of a unified process of 'functional decongestion' (Vance) since at least 1900 (decentralization was not felt strongly until this time). But the former is historically and analytically prior. As long as urban space is relatively homogeneous and repetitive it is impossible to speak of concentration or decentralization; space must first (or simultaneously) be broken into different units which may cluster and/or spread apart differentially.

[20] The inevitable criticism of essays which take on as big a field as this one is that they should be turned into books, since the necessary documentation and supporting theory is lacking. I will anticipate by saying that a book is indeed in the future. But for now ample documentation and argumentation can be found in the works of Harvey and myself, supplemented by the other references cited in this paper. The idea here is to pull together and summarize much of this work.

[21] Cf. Castells 1977: 15. Even Castells takes the functional divisions (in Marxist terms, however) as given rather than as evolving.

[22] And the form was different, as in the gild districts of medieval towns. See Vance 1977.

[23] Walker 1977, 1978, Pred 1966.

[24] Cf. Vance 1964.

[25] The historical record simply does not accord with Soja's (1978) view that spatial relations of production and social relations of production should be assigned equal weight.

[26] This summary is drawn from Ward 1971, Hoyt 1933, Vance various works, Pred 1966, Armstrong 1972 and other sources cited in Walker 1977, 1978, 1978 ms. Self-evident as these functional divisions of urban space may seem, they are in continuous flux with the development of capitalism and must not be frozen into universal divisions.

[27] The destruction of the household economy proceeded rapidly in the nineteenth century (see Tryon 1917) but is not by any means complete today.

[28] See generally Marx 1967 ed. Capital accumulation is made possible by, first, the class ownership of the means of production and second, by competition

among private capitals (Cf. Harvey in this volume). The division of labor is thus intimately connected to the development of control over the means of production.

[29] The school following Weber 1909 focuses around minimizing production costs, while the school following Christaller 1933 and Losch 1954 concentrates on markets in space. The neo-classical/von Thunen models are, despite their pretensions, chiefly transport-cost models.

[30] This is not to say that Marxists have always done so either; we are still groping in this direction. An important beginning to the analysis of the geography of accumulation is Harvey 1975a. The dynamics of circulation must be seen in light of the drives and limits set by accumulation crisis and class struggle (see above pp. 384–5).

[31] Seen here in a micro-locational sense. Crisis and rationalization in the accumulation process and the built-environment as a whole are treated on pages 405–9.

[32] These distinctions are common to both conventional and Marxist industrial location theory. Cf. Castells 1977; Massey 1977.

[33] 'In retail districts the tie between stores was largely one of similar location requirements rather than functional ties one with another, so that changed conditions of access could bring quite different location requirements' (Vance 1960a: 486).

[34] 'Collective logic' is used here in a narrow sense, since it ultimately subsumes the logic of exchange value/use value, of accumulation cycles, etc.

[35] See Walker 1978.

[36] At the same time, other supposed 'economies' of agglomeration, being based more on class and monopoly power and access than on competitive cost-revenue rationalization, are particularly resistant to dissolution under the impact of communication improvements – particularly what is euphemistically known as 'the need for face-to-face contacts' among the highest ranks of capital in the CBD. Cf. Harvey 1973. Marriott's (1967) discussion of the way certain pubs function as nerve centers for property developers is revealing here.

[37] Toll 1969.

[38] See e.g. Chamberlin 1933. Cf. Harvey 1973 and Walker 1974.

[39] See Gordon 1977, 1978b, Walker 1977, 1978 ms, and above pages 392–3.

[40] As an example, it is quite in the interests of San Francisco financial district capitalists to promote the conversion of that city to a purely middle-class, white-collar workplace and residential area, since the residual blue-collar working class and lumpenproletariat present a physical and political threat to the former's hegemony.

[41] Markusen 1976, 1978, Gordon 1977, 1978b, Greer 1976–7, Walker 1977, 1978 ms, Walker, Storper and Gersh 1979. This includes the role of special districts. See Bollens 1961, Downie 1974, Eichler and Kaplan 1967.

   While the fragmented form of local government rests on the unique basis of American federalist tradition, it cannot simply be ascribed to this 'cause' of a mythical continuity with past New England town meetings. The specific form of metropolitan fragmentation was a specific construction of the twentieth century, which was nurtured in connection with the logic of spatial differentiation as a whole, by a 'selective use of traditions' in the sense used by Williams 1977.

42 I have shown this with respect to horse-drawn streetcars in the mid-nineteenth century and cars/trucks at the turns of the century. See Walker 1977, 1978a, 1978 ms, and Gordon 1977, 1978b. (Cf. Note 128 below.)

43 See Harvey 1975a on reducing time of circulation.

44 Harvey in this volume. Investment is just as crucial as technique, and often lags behind the latter. See Walker 1978a.

45 Harvey 1975a.

46 Walker 1978b. Actually it depends on the developing social forces of production and circulation which capital has at its command, in general and by individual capitals (owing to concentration). These forces consist of various concrete elements: fixed capital (including past accretions); modes of organization (corporations, financial networks, the State); knowledge (science, technique); an industrial proletariat (with certain capabilities); social wants and needs (consumer power of society), etc. These 'powers of capital' are, of course, the alienated powers of social labor.

47 Harvey 1978, Walker 1977, 1978.

48 The category 'class' is meant in the classic Marxist sense of relations of production. The various elements of everyday life and consumption around which people build their lives, such as differences in income levels, status, education and so forth, obscure but do not transform the underlying class structure. Nonetheless, the ambiguities in production relations of class (see Wright 1978, on 'contradictory class locations'), combined with the complexities of everyday life in the sphere of consumption, residence and personal reproduction, are such that, in the American historical experience, perceived class boundaries do not correspond to Marxist theory. The working poor, so many of whom are non-whites, have been dropped off the bottom of the working class to form a strata of 'poor and minorities', while the top of the working class, consisting of professionals, supervisors and better-paid craft workers (and sometimes even wage-earning white-collar workers), have been elevated to the status of 'middle class'. The 'working class' has ended up as a residual: the traditional blue-collar factory workers. These categories of consciousness have then been distilled into the categories of conventional social science.

For the purposes of this essay, I adopt conventional usage of 'middle class' versus 'working class', because I am trying to capture the way suburbanization fits into the formation of an effort by certain ranks of workers to rise to 'middle-class' status, with the eager support of the capitalist class. In retrospect, my use of terms is too casual and the matter is deserving of greater thought than I have given it here. But it must stand, awaiting future revision.

49 This simplifies the actual intersection of class with race/ethnicity and sex, which are themselves the product of a whole set of antecedent conditions of biology and history. They are integrated into the capitalist structure out of specific collisions with pre-capitalist and colonial societies, from the household economy to slavery, and subsequently reproduced for reason of their usefulness to a capitalist dominated society (or the inability of capital to break down such distinctions).

Racial differences partly parallel and partly compound class differentiation in space, as has been shown repeatedly. (See e.g. Taeuber and Taeuber 1965,

Schnore 1972, Harrison 1973, Kerner Commission 1968). That is, most blacks are bottom rung workers and lumpenproletarians, but they are not well integrated with white workers; at the same time, working-class and middle-class blacks are also well segregated.

Sex differences have a close alliance with the division of labor between production and consumption/reproduction of labor-power, with women's unpaid labor providing a pillar for the maintenance of a viable residential sphere. But women, like blacks, have a complex history of being pulled in and out of the wage-labor workforce as it suits capital. Household labor is both a residual form of the household economy, which appears to be breaking down over time, and a necessity to capitalist reproduction.

I will not try to deal with the elements of tension and transformation in race and sex divisions in this simplified model of residential life.

[50] Social reproduction through consumption cannot be reduced to the reproduction of labor-power only, although this is the bulk of it. Capitalists must be reproduced, too. Castells (1977) makes this reduction.

[51] Feldman 1977, Vance 1960b, 1966, 1967.

[52] Vance 1967 shows this to be a practice of long standing.

[53] Vance (1966, 1967, 1977) calls this process the 'generalization of the housing market' or the move from a 'determinant' to a 'contingent' employment linkage. While historically irrefutable as a trend, this process cannot be assumed to have gone all the way even at this late date.

[54] In the San Francisco area, for example, we find a remarkable division between the East and West bay areas, which have distinct employment centers and relatively little cross commuting (Vance 1964). Within the East Bay we find successive clusters of workers' residences around Richmond, in north Contra Costa County, in Berkeley, South Oakland, Fremont. I do not agree with Vance's (somewhat contradictory) statement that workplace-residence ties have been broken in our time (1964: 65). Schnore (1965: 142) notes the general association of suburban industry and typically working-class housing (though some of this is left over from the past).

[55] Feldman 1977. Thus we have the influence of both the social/territorial division of labor among places of employment and the detailed division of labor on the employment linkage.

The journey-to-work as a whole, however, embraces more than the employment linkage as defined here. It must overcome the spatial divisions caused by the forces of consumption/social reproduction as well.

[56] Gintis 1972, Ewen 1976, Ashton 1978, Harvey 1978, etc.

[57] Seeley, Sim and Loosely 1956 thus refer to the 'commodity self-image' of suburbanites (cited in Sobin 1971: 31).

[58] In pre-capitalist America, spatial exclusion in the form of banishment was practiced regularly on those people who could not be controlled internally by social conventions and punishment. As the problem of social deviance and control increased in the early nineteenth century, the dominant classes turned to institutionalization in asylums, penitentiaries, etc. These were typically located in a suburban setting (Rothman 1971). As these institutions proved unable to cope with the social revolution taking place, the bourgeoisie chose to suburbanize themselves instead of the deviants (Walker 1977, Chapter 3).

59 See Note 41 above.

60 These things actually evolved hand-in-hand. On the importance of landscape significance in suburbia, see Duncan 1976 and Evans 1978, who rightly point out the way landscape attachment is wrapped up with personal identity, emotional responses and individual psychology. Schwartz (1976a) remarks that suburbs are a 'meaningful, that is, moral order' (325).

61 It is well known that working-class people in the US generally attach less importance to exterior appearances than the middle class.

62 See Harvey 1974a. Also Peet 1977. Harvey only deals with the class reproduction issue, which is insufficient.

63 It has not been a process without contradiction, owing to the constantly changing needs of capital, the struggle by labor to define its own life, and the division between production and consumption in life and in space.

64 But actively promoted and defined by capital acting in various roles as producer, seller, financier, state manager, etc. (Cf. Castells 1977: 388).

65 Some useful literature on the sociology of middle-class suburbs includes: Schwartz 1976, Sobin 1971, Seeley, Sim and Loosely 1956, Thorns 1972, Marshall 1973, Gans 1967, Whyte 1956, Muller 1975, Kaplan 1976.
One might add a 'cult of mobility' in which 'one can move geographically from one suburb to another but stay tied to the same culture' (Schwartz 1976a: 328). This is certainly less true of working-class communities: 'what is important about the life-style of this (middle-class) culture is its universality and freedom from particular relationships and localities. Suburbia is in this sense the negation of the traditional community' (Schwartz 1976a: 328).

66 See, besides the literature cited in Note 65, such works on the family, home and local community as Sennett 1970, 1973, 1977, Sennett and Cobb 1972, Mitchell and Oakley 1977, McKenzie 1977, Ryan 1972.

67 Ryan 1975 and Note 49 above.

68 Cox 1978.

69 The influence of spatial arrangements on human consciousness and practice in suburbia is a much-discussed topic; see especially Schwartz 1976a, for a review of the literature. The dominant view since Berger (1960) has been that locating in suburbia has no impact, but this is based on a hasty rejection of earlier facile generalizations that suburbia made one middle class. Schwartz concludes with observations such as:
'More than any other type of arrangement, suburbanism regulates distance and closeness among groups by its inherent ability to minimize social variation within communities and maximize differences between them.' (331)
'Under the influence of this effect, middle-class people may be expected to behave differently than middle-class people in communities less isolated . . . . "The suburb," says Riesman, "is like a fraternity house at a small college . . . in which like-mindedness reverberates upon itself as the potentially various selves in each of us do not get evoked or recognized." ' (331)
'Part of the anti-heroic character of suburbia becomes intelligible in that the fervent intensity of its affairs . . . is muted by the triviality of their content. His new-found "republic in miniature" allows the subur-

banite to take part in the debate on whether to put a fence around the community tennis court. Society's central issues, on the other hand, seem distant. . . . In a dominantly urban society the previous condemnation of the "idiocy of rural life" is reserved for the suburban provinces.' (336) (Cf. Sennett 1976)

One observes a kind of 'social amnesia' among students who have grown up in such a world. (Cf. Donaldson's (1968: 316) remarks on suburbanites disliking to hear bad news.)

Castells (1977: 385) says:

'Difference in cultural style, rooted mostly in social class and family practices, will be symbolically reinforced by the social-spatial distance and by the environmental imagery. The two worlds of the suburbs and the inner city increasingly ignore each other except insofar as they develop reciprocal fears, myths and prejudices, often articulated as racial and class barriers.'

[70] See above, pp. 397–8, where it is argued that not only is space incorporated in the mode of life, but spatial relocation was essential to its construction.

[71] The literature on traditional urban working-class communities is voluminous. Especially good is Fried 1973. For a comparison of this with middle-class suburban life, see the two studies by Young and Wilmott 1957a and 1960.

[72] I have argued the importance of this middle-class withdrawal in Walker 1977, 1978 ms. See also Warner 1967.

[73] See Checkoway's (1978) discussion of Levitt.

[74] Harvey 1977: 137–8, 1974b, Harvey and Chatterjee 1974.

[75] See Harvey 1972.

[76] Cox 1978 ms.

[77] Some areas isolated from the general finance-housing market do exist, typically in traditional working-class areas (Harvey 1977); this provides a different kind of community differentiation based on a 'conservative' resistance to the property market itself (zoning is not the only way).

[78] This leads Gaffney to refer to them as 'cartels of landowners', a provocative hyperbole (Gaffney 1973).

[79] Gelfand 1975.

[80] Harvey 1977, 1974b, 1975b.

[81] See discussion by Checkoway 1978. Cf. Walker and Large 1975.

[82] See Walker 1974 for explanation of rent terms.

[83] On industrial parks, see Muller 1975; on industrial suburbs, see Sobin 1971, Schnore 1965; on competition for industry, see Walker, Storper and Gersh 1979; on planning/zoning, Vance notes that, for example, land-use planning has helped 'to do away with the isolated shop and has reinforced the economic trends toward integration of shopping facilities' (Vance 1960a: 487).

[84] There are contradictions in the tendency to spatial differentiation, owing to the opposing tendencies of spatial generalization of capital and homogenization of the built-environment.

[85] The term decentralization is somewhat confusing since it refers both to accretion at the fringe and to deconcentration, i.e. the decline in rent/density gra-

dients since the nineteenth century. (Though exactly when this decline began, and when concentration peaked, remains problematic. See Walker 1978a. It is also confusing because both of these aspects are consistent with absolute increase in density at the center. We are mainly interested in deconcentration here, since the passive operation of the land market is sufficient to explain why some activities move to the fringe.

86 Our model will remain incomplete until the temporal dynamics are added in the third part (pp. 405–9). Obviously the discussion in the second part relies on points somewhat artificially segregated into the first part of the paper. See above, Note 19.

87 The terminology is confusing because in economics a process which used more space would be called 'land intensive.'

88 That is, land is always cheaper at the fringe but this does not mean declining density over time; also more space-extensive uses will tend to move to the fringe, to be absorbed and intensified later, without the whole of the city becoming more space-extensive; and, finally, both cheaper land and space-extensive uses are usually explained in such simple models by transportation improvements and the operation of the land market alone. Transport innovation is generally introduced as *deus ex machina*.(See above, p. 389.)

89 These are the same forces introduced in the first part – complemented by spatial differentiation itself.

90 See above p. 383–4. Two difficulties arise in a polar conception of decentralization. Most decentralization does not consist in facilities or people picking up and moving to the suburbs, but on new investments and new families locating at the fringe (Wood 1974: 135, 150, Keeble 1976). This does not negate the connection between center and periphery, which are specifically linked by corporate structures, capital markets, and family ties.

Second, cities on the national periphery, growing up in the twentieth century, have taken on a predominantly suburban form without the bother of a nineteenth century urban core. This, too, does not sever the links of unity and contrast between a national urban core and a national urban/suburban periphery.

This forces us not to buy too simple or literal a 'push' hypothesis and to incorporate a theory of urban waves/stages. See below, pp. 405–9. (For opposing views which reject 'push' theories altogether, see Guterbok 1976 and Wood 1974.)

91 For details and references, see Walker 1977, 1978a, 1978 ms.

92 The causes of concentration were, principally: (i) destruction of household economy; (ii) economies of agglomeration among small manufacturers and traders; (iii) powerful employment linkages which kept workers (and managers) close; (iv) poor intraurban transport and ability to pay (but good enough *inter* urban transport to reach large markets); (v) creation of a 'housing market' which forced workers to minimize their occupation of space; (vi) steam power – especially the availability of coal brought by canal and rail (Walker 1977, 1978a).

93 Wood (1974: 133) is correct but overstates the case for continuity, when he says, 'In summary, the urban fringe has provided the milieu for manufacturing growth since the mid-19th century, so that current developments may be

regarded simply as a continuation of long-established trends.' Cf. Pred 1966.

[94] This occurred just after the climax of industrial centralization from small towns to big cities at the end of the nineteenth century. See Pred 1966.

[95] See above, Note 36.

[96] Cf. above pp. 405–9. This has spawned the 'urbanization of the suburbs' literature of the last few years, e.g. Massotti and Hadden 1973, Muller 1975. While there is an essential truth here, it can be overstated such that the urban core is virtually written off as unnecessary ('the city as sandbox' view) (see e.g. Muller); it glosses over the fact that suburbs have managed to retain an élite cast and to assume more and more urban functions without becoming the city as a whole. NB Schwartz's provocative exaggeration: 'the suburbs are becoming industrialized without the bother of an industrial population' (1976a: 329).

[97] Mills 1972; cf. Muller 1975: 7.

[98] Previously the mercantile bourgeoisie lived near the city center.

[99] As Vance notes, 'American cities have always expanded in directions that the wealthy and powerful have pioneered' (1964: 24).

[100] See Walker 1977, 1978, Harvey 1978, Schmitt 1968 and Williams 1973 on landscape ideology and images of country and city which went into suburbanization. On banishing the working class from the scene, see Engels 1844 and Harvey 1973.

[101] Summarizing the sociological debate on whether suburbs create a lifestyle or only reflect it, Schwartz remarks that, if nothing else, residential differentiation is important, i.e. 'we all know that "recruitment mechanisms", that is, a community's method of attracting certain types of people and repelling others, is one of its most important *constitutive* features' (1976a: 329–30).

[102] Control of industrial excesses, the land market and the working class have all been tried (pollution laws, zoning, urban reforms) but have never been wholly successful. Moreover, such 'solutions' take a great deal of effort and class-conscious, class-integrative activity (i.e. major 'reform' movements). So it has always looked more reasonable to the individual to choose an individualist solution, or to seek out homogeneous political enclaves where some unanimity can be achieved. The suburbs, then, are not exactly a tribute to a love for political activity among the middle class; quite the contrary.

[103] See above pp. 505–9. Cf. Harvey 1974a.

[104] Young and Wilmott 1973. Cf. Sennett and Cobb 1972.

[105] Jackson 1975, Walker 1978a.

[106] References, Note 41. The reform mode of government was disproportionately adopted by the suburbs. Greer and Greer 1976: 209.

[107] Made possible by a supply of cheap land but not explained by supply conditions.

[108] Though the appearance of the mass version of the suburban ideal has perennially stood as 'a cruel joke', as Warner (1962) puts it.

[109] The degree of 'bourgeoisification' of the working class – indeed, the definition of the working class – remains a lively topic in Marxist theory today. See Poulantzas 1975, Wright 1978, Braverman 1974.

[110] The references here are numerous. See e.g. Schwartz 1976a, Kasarda 1976.

However, the general overrepresentation of the poor and the rich in the central city is less marked in the newer cities of the sunbelt, an indication of the lesser importance of residual nineteenth-century cores there. Castells 1977: 177 and references there. See also above, pp. 405–9.

[111] Blue-collar suburbanization away from rail corridors is almost wholly a postwar phenomenon. The effect of this was that 'prior to the Second World War American suburbs tended to fall into two classes rather distinct from each other. There were strictly residential areas of generally high income housing and there were industrial satellites of the central city with workers' housing' (Vance 1960a: 496). Muller 1975: 16. See also references in Note 51, above.

[112] This accounts for the paradoxical finding (for some) that it is not the middle class but the working class who have the higher rates of (measurable) 'political participation' and dissent in local politics (Greer and Greer 1976: 211–12). The middle class has their class in power and class interests served, so 'participation' is less essential.

[113] Gans 1966, Berger 1960, Fischer and Jackson 1976.

[114] Berger 1960: 82. Cf. Sennett and Cobb 1972. Castells 1977, like many others, simply repeats Berger's mistake of missing the constitutive element in suburbanization of workers.

[115] Dobriner 1963: 58–9.

[116] Sennett and Cobb 1972.

[117] Braverman 1974. Braverman perhaps overstates the case, but not much, when he says,

> 'The apparent acclimatization of the worker to the new modes of production in the 20th century grows out of the destruction of all other modes, the striking of wage bargains that permit a certain enlargement of the customary bounds of subsistence for the working class, the weaving of the net of modern capitalist life that finally makes all other modes of living impossible.' (151)

[118] See Walker 1977, 1978 ms.

[119] See Gramsci 1971 ed., who coined the term. Also discussion in Williams 1977.

[120] The whole question of the myth of suburbia and its imperfect realization by all but the élite is an interesting aspect of the contradictions of suburbanization, not taken up here. Cf. Schwartz 1976a: 326–7, Sobin 1971, Kaplan 1976.

[121] See above, Note 10.

[122] See above, p. 397.

[123] Hence the possibility of a bourgeois ideology of 'modern' versus 'traditional' modes of living, with the former identified with the suburbs and the latter with the older inner city. See Castells 1977: 78.

[124] Cf. Note 111, however.

[125] The following is only suggestive of the key issues. A full treatment would have to distinguish among types of industries with different locational requirements. See above, p. 387.

In fact, the mainstay of industrial suburbanization has been large, capital-intensive manufacturing, usually within large corporations. Labor-intensive and small manufacturing, with special needs for support or highly exploitable (sweatshop) labor have remained in the city, as have certain highly innovative companies with special information needs. Many of the labor-intensive and risky types of production have been moved to the South or abroad in recent years, but capital-intensive plants continue to remain at the urban fringe (conversation with Allan Pred).

126 See above, p. 389.

127 This section is based on Walker 1977, 1978 ms, Gordon 1977, 1978b, Markusen 1978, Taylor 1915, Pratt 1911, Pred 1966 and Braverman 1974.

128 The movement of goods by truck did not become important until after the First World War, despite popular belief that cars/trucks caused suburbanization. Industry used the rails and could have continued doing so had not the automobile era transcended the railroad era.

129 Regarding deskilling in industry at the turn of the century, see Braverman 1974 and Stone 1975. What we see here is a shift in which the advantages of increased labor productivity began to overhaul the advantages of urban agglomeration in the cost calculations of industrialization.

130 Cf. pp. 405–9.

131 This strategy for damping class struggle was paralleled by a great many other reform movements in which the industrialists regularly joined, aimed at 'domesticating' working-class culture and neutralizing working-class power (Walker 1977, 1978 ms). Even changes in production methods to reduce the number of workers were motivated in part by the need for class control (Braverman 1974).

132 Indeed, being uprooted, finding frequently that no housing was immediately available, and even having to reverse commute from inner-city districts caused the workers considerable consternation (Pratt 1911, Taylor 1915). These conditions, which were common in the pre-war period, argue against the view that the decentralization of manufacturing is actually caused by prior suburbanization of labor. It was not a labor pool which industry sought in the suburbs but a 'better sort of labor', even if it had to be forged on the spot. The social construction of new relations is an irreducible aspect of suburbanization here, as in the case of residential decentralization.

133 The role of the federal government in industrial siting during and immediately after the Second World War is especially interesting. Also, an additional locational consideration in dispersing factories – national defense – came into play at this time. (See Wood 1974: 136, note 13.)

134 Frappier 1977. Cf. Sobin 1971: 53. However, it should be noted that 'the forces pushing industry out of old areas are of diminishing importance in relation to the attractions and advantages of new locations' (Wood 1974: 136).

135 Muller 1975. Cf. Dawson (1974) who takes less account of large shopping malls (perhaps because of an English bias) and gives retailing a more passive role. He, too, emphasizes the role of a part-time female labor force in the suburbs.

136 See Daniels 1974, Armstrong 1972.

137 So long as property taxes are based on land assessments they are a reflection of the land market, though clearly there are distortions here (which appear to aid the fringe).

138 Although this is a big assumption, the 'obvious' correctness of which only

obtains in the US, where the land market has been in full effect for the longest of any capitalist country and where property capital has so readily greased the wheels of expansion that it is hard to imagine it failing to do so. In Europe, this has by no means been the case, where walls and landlords have blocked urban expansion (Vance 1978: 94; Castells 1977; see also Note 143, below).

139 Mollenkopf 1977, and references in Note 144, below. Cf. Warner 1962 on the nineteenth century. The formation of an operative growth coalition is an important topic of study in itself.

140 This has been discussed at length by Harvey in an unpublished chapter.

141 See e.g. Vance 1964: 15 and 54, Pred 1966 and Warner 1962. The literature on land speculation in the US is vast. See especially Hoyt 1933 and Sakolski 1932. The Greers comment, for example, that 'the United States became one of the biggest uncontrolled real estate developments in history' (1976: 206).

142 For further discussion, see Walker 1974.

143 The best discussion of this is by Gaffney 1967, 1973. In Europe the tendency has been to generate monopolistic/absolute rents by this means (Hall *et al.* 1973, Castells 1977, Vance 1978). In the US the availability of land from rural owners, the well-developed financial structure and the eagerness of government to promote land development has channeled most of the demand into rapid and sprawled development (Vance 1978: 105). This has kept down prices and allowed land value appreciation to substitute, in part, for high rents to landlords and to help finance purchase.

Nonetheless, this process has its costs, as Gaffney insists; a major one is the hyperextension (and subsidy) of infrastructure.

144 See generally Walker 1974, Gaffney 1966, 1973, Clawson 1971, Neutze 1968, Downie 1974, Paulson 1973 on land-developer interventions in local government.

145 'The two frontiers for land speculation lie at the center and at the edge of the city' (Vance 1964: 69).

146 The center is most prone to the kind of artificial scarcity European cities have been plagued by. Cf. Notes 138 and 143, above.

147 Cf. remarks on the role of the land market pp. 393–4.

148 Harvey 1974b, 1977, Harvey and Chatterjee 1974.

149 Bradford and Rubinowitz 1975, Harvey 1972, 1975b, Walker and Large 1975.

150 See Walker 1977, 1978 ms.

151 Vance 1978: 96–7. See various works by Harvey and also Stone 1978. The US has thus not only been the most suburbanized of nations but the only one where substantial public housing has not been deemed necessary.

152 Of course, the activities of property investors need not run entirely in the direction of the suburbs, as the recent gentrification boom shows. This boom is hardly a historical reversal of suburbanization, but a specific response to opportunities created by the central-city development of offices, prior disinvestment in housing, collapse of new housing construction, rapidly inflating property values, demographic and cultural changes among occupants, etc. It shows, however, that suburbia is not the only place where property investment can move mountains to promote development. Actually gentrification helps suburbanization by continuing to drive out the working class. See Kucherenko 1978.

[153] The debt issued is much overlooked. In the slash-and-burn approach of capital, localities which have been overencumbered with debt in the past can be avoided by moving to suburban jurisdictions with clean slates. Special districts are, in fact, partly an invention to circumvent debt-limits on general governments (Bollens 1962, Hillhouse 1936).

[154] On the behavior and manipulation of local governments and special districts see references in Notes 41, 139, 143 and 144 above.

[155] Overviews are Gelfand 1975, Walker 1977, US Dept. of HUD 1975, and National Commission on Urban Problems 1968.

[156] I have done this in large part in Walker 1977.

[157] Cf. Vance 1978, Castells 1977 and various works by Harvey.

[158] Harvey in this volume, and 1975a, 1978, Walker 1977, 1978 ms.

[159] On Keynesian growth theory, see Kregel 1972. In a similar formulation to mine, Gordon 1978a refers to 'the social structure of accumulation'. In the more conventional Kuznets-cycle literature, Abramowitz 1976 has a similar concept of basic economic conditions for growth.

[160] There is, therefore, a constant dialectic between creation of new urban space at the fringe and the rehabilitation/conversion of old buildings and districts. This relation has been little studied, but appears to have a cyclic character. (Conversation with Paul Groth, Department of Geography, University of California, Berkeley.)

[161] See Walker 1977, Gordon 1978a; cf. Schumpeter 1939 and Mandel 1975.

[162] The relation between form and function is not confined to capitalist cities, of course. See Vance 1977: 23.

[163] Harvey 1975a: 13.

[164] Again, on adaptation of urban structure over time, see Vance 1977. Vance stresses the role of physical structures, which must be adapted to new uses over time (Vance 1977: 81).

[165] Thus urbanization waves are produced from the exchange value side of the built-environment and tend to be more exaggerated than growth waves in general.

[166] This temporal distinction parallels and embraces our earlier static formulation of the distinction between the use and exchange-values of the built-environment. See Introduction, above.

[167] The lack of synchronization between the primary and secondary circuits or between the needs of use and pressures of the property circuit naturally complicates the correlated periodization of waves of urbanization and stages of accumulation.

[168] See Walker 1977, 1978, Gordon 1977, 1978b, Hill 1975. The periodization is a little tricky here, because each stage of accumulation and coincident urban form takes time to come to maturity. Hence early in the stage the legacy of the past is still strong – possibly overwhelming in visceral terms. Thus we find it common for urban historians and geographers to date the major stages of urbanization as 1830–70, 1870–1920, 1920–70, i.e. from mid-point to mid-point of the stages we adopt. See e.g. Borchert 1967, Adams 1970, Muller 1975. This is by no means unreasonable, but it invariably relies on simplistic models of causality (chiefly transportation changes) and misses the fact that

the basic economic transformations which created these city forms took place in the 1840s, 1890s and 1940s. See Schumpeter 1946, Gilchrist and Lewis 1965, in addition to works cited above and in Note 161.

[169] The types of cities associated with each stage may be called 'mercantile', 'classic industrial', 'metropolitan' and 'suburban' or perhaps 'megapolitan'.

[170] The Kuznets wave is the most apparent empirical phenomenon in urbanization, subject to little dispute. See Gottlieb 1976, Kuznets 1960, Abramowitz 1961, Thomas 1973 and Hoyt 1933.

[171] See Vance 1978.

[172] See the Introduction, above. I do not mean by 'historical contingency' that everything is historically unique or accidental. I mean to take account of time as a factor in human practice, just as we account for culture, space, politics, etc.

[173] This does not mean that capitalism is forever 'the same thing' and only the form changes, in some kind of one-way view of determination (a common Marxist mistake). This is undialectical. Capitalism as a whole changes, develops. But the basic structuring relations are reconstituted in the process of changing everyday practice, in an active interaction. (See Harvey 1973, Conclusion.)

[174] See references in Notes 2 and 3.

[175] See above, Note 3.

[176] See Walker 1977, Chapter 6.

[177] See Note 90, above.

[178] Under certain circumstances, of course, the past can be a benefit; this was most true in the early stages of accumulation.

[179] The difficulty of remaking the city versus the ease of building anew at the fringe, discussed previously in connection with residences and industry, is, in fact, a general condition affecting society, including whatever social planning is undertaken. Indeed, it forces capital and labor to look to collective solutions if cities are to be maintained. Cf. Vance 1978: 95.

[180] The obsolescence problem is partly overcome by remaking the city, of course, but this tends to be a selective process which reinforces spatial differentiation: parts of the city built at different times become ready receptacles for different functions and different people.

[181] Cf. Harvey 1973, Conclusion.

[182] Robinson 1973: 101.

[183] Walker and Large 1975.

[184] Generally coincident with Kuznets cycles, but stronger toward the end of Kondratieffs, e.g. 1830s, 1880s, 1920s and 1970s. See especially Hoyt 1933, Chapter 7.

[185] Discussed in detail in Walker 1977, 1978a, 1978 ms.

[186] They have also contributed to spatial differentiation in some strange ways, as, for instance, the case of Harlem – which was pre-First World War white suburban development that turned sour. Seeking to recover their speculative investment, the developers sought black buyers – an unheard-of break with discrimination. Thus began a new ghetto (Osofsky 1966).

[187] This idea has been suggested, in various forms, by Harvey 1973, Baran and Sweezy 1966, Ashton 1977 and Castells 1977. Unfortunately I have yet to see anyone work out the figures. Ashton estimates 50 per cent of GNP, but it is only a guess. Harloe (1977) mistakenly lumps Harvey in with Baran and Sweezy as a 'surplus absorption' theorist, however. On this point, see the critique of O'Connor by Mosley 1978.

[188] What was said earlier about overaccumulation and overextension of the city through the property circuit holds for the secondary circuit and suburbanization as a whole. For instance, the whole structure of suburbanization today is sustained artificially by what *Business Week* (1974, 1978) calls a 'mountain of debt'. Cf. Harvey 1975b, Stone 1978, Ashton 1978.

[189] Ullman 1977. Cf. Castells 1977, who says,

'the shopping centers and the supermarkets were made possible by suburban sprawl, as were the new leisure activities (from the drive-in restaurant or cinema to the private swimming pool). But even more important was the role of the single-family house in the suburbs as the perfect design for maximizing capitalist consumption. Every household has to be self-sufficient, from the refrigerator to the TV. . . .' (388)

[190] Whitehand 1974.

References

Abramowitz, M. (1961) 'On the nature and significance of Kuznets' cycles.' *Economic Development and Cultural Change* 9: 225–48.
—— (1976) 'Likeness and contrasts between investment booms of the postwar period and earlier periods in relation to long swings in economic growth', in Hamish Richards (ed.) *Population, Factor Movements and Economic Development.* Cardiff: University of Wales, pp. 22–49.
Adams, J. (1970) 'Residential structure of midwestern cities.' *Annals of the Association of American Geographers* 60:37–62.
Alonso, W. (1964) *Location and Land Use.* Cambridge, Mass.: Harvard University Press.
Armstrong, R. (1972) *The Office Industry.* New York: Regional Plan Association.
Ashton, P. (1978) 'The political economy of suburban development', in W. Tabb and L. Sawers (eds) *Marxism and the Metropolis.* New York: Oxford University Press, pp. 64–89.
Baran, P. and Sweezy, P. (1966) *Monopoly Capital.* New York: Monthly Review Press.
Berger, B. (1960) *Working Class Suburb.* Berkeley: University of California Press.
Bollins, J. (1961) *Special District Governments in the United States.* Berkeley: University of California Press.
Borchert, J. (1967) 'American metropolitan evolution.' *Geographical Review* 57: 301–32.
Bradford, C. and Rubinowitz, L. (1975) 'The urban-suburban investment-disinvestment process: consequences for older neighborhoods.' *Annals of the American Academy of Political and Social Science* 422: 77–86.
Braverman, H. (1974) *Labor and Monopoly Capital.* New York: Monthly Review Press.
Business Week (1974) *The Debt Economy* Special Issue, 12 October.
—— (1978) *The New Debt Economy* Special Issue, 16 October.
Castells, M. (1977) *The Urban Question.* Cambridge, Mass.: MIT Press.
Chamberlin, E. (1933) *Monopolistic Competition.* Cambridge, Mass.: Harvard University Press.
Checkoway, B. (1978 ms) 'Large builders, federal housing programs and postwar suburbanization.' Unpublished manuscript, Department of Urban and Regional Planning, University of Illinois (forthcoming in the *International Journal of Urban and Regional Research*).
Christaller, W. (1933) *Die Zentralenen Orte in Suddeutschland* (Central Places in Southern Germany). New Jersey: Prentice-Hall, 1966.

Clawson, M. (1971) *Suburban Land Conversion in the United States*. Baltimore: Johns Hopkins Press for Resources for the Future.

Cox, K. (1978) 'Local interests and urban political processes in market societies', in K. Cox (ed.) *Urbanization and Conflict in Market Societies*. Chicago: Maaroufa Press and London: Methuen, pp. 94–109.

—— (1978 ms) 'The politics of exclusion in the United States.' Paper presented at the Conference of the Institute of British Geographers (5 January).

Daniels, P. (1974) 'New offices in the suburbs', in J. Johnson (ed.) (1974: 177–200).

Dawson, J. (1974) 'The suburbanization of retail activity', in J. Johnson (ed.) (1974: 155–76).

Dobriner, W. (1963) *Class in Suburbia*. New Jersey: Prentice-Hall.

Donaldson, S. (1976) 'The machines in Cheever's garden', in B. Schwartz (ed.) (1976b: 309–24).

Downie, L. (1974) *Mortgage on America*. New York: Praeger.

Duncan, J. (1976) 'Landscape and the communication of social identity', in A. Rapoport (ed.) *The Mutual Interaction of People and the Built-Environment*. The Hague: Mouton.

Eichler, E. and Kaplan, M. (1967) *The Community Builders*. Berkeley: University of California Press.

Engels, F. (1844) *The Condition of the Working Class in England*. Moscow: Progress Publishers Edition, 1973.

Evans, D. (1978) 'Demystifying Suburban Landscape'. Paper presented to Pacific Coast Geographers Association, Portland, Oregon, 13–16 June.

Ewen, S. (1976) *Captains of Consciousness: Advertising and the Social Roots of Consumer Culture*. New York: McGraw-Hill.

Feldman, M. (1977) 'A contribution to the critique of urban political economy: the journey to work.' *Antipode* 9, 2: 30–50.

Fischer, C. and Jackson, R. (1976) 'Suburbs, networks, and attitudes', in B. Schwartz (ed.) (1976b: 279–308).

Frappier, J. (1977) 'Chase goes to Harlem: financing black capitalism.' *Monthly Review* 28, 11: 20–36.

Fried, M. (1973) *The World of the Urban Working Class*. Cambridge, Mass.: Harvard University Press.

Gaffney, M. (1967) 'Land rent, taxation and public policy.' *Papers of the Regional Science Association* 23: 141–53.

—— (1973) 'Tax reform to release land', in M. Clawson (ed.) *Modernizing Urban Land Policy*. Baltimore: Johns Hopkins Press for Resources for the Future, pp. 115–53.

Gale, S. and Moore, E. (eds) (1976) *The Manipulated City*. Chicago: Maaroufa and London: Methuen.

Gans, H. (1967) *The Levittowners*. New York: Random House.

Gelfand, M. (1975) *A Nation of Cities: Federal Urban Policy from 1933–1965*. New York: Oxford University Press.

Gilchrist, D. and Lewis, W. (eds) (1965) *Economic Change in the Civil War Era*. Greeneville, Delaware: Eleutherian Mills-Hagley Foundation.

Gintis, H. (1972) 'Consumer behavior and the conception of sovereignty: explanations of social decay.' *American Economic Review* 62, 2: 267–78.

Gordon, D. (1977) 'Class struggle and the stages of urban development', in D. Perry and A. Watkins (eds) *The Rise of the Sunbelt Cities*. Beverly Hills: Sage, pp. 55–82.

—— (1978a) 'Up and down the long roller coaster', in Editors of URPE *US Capitalism in Crisis*. New York: Union of Radical Political Economists, pp. 22–35.

—— (1978b) 'Capitalist development and the history of American cities', in W. Tabb and L. Sawers (eds) *Marxism and the Metropolis*. New York: Oxford University Press, pp. 25–63.

Gottlieb, M. (1976) *Long Swings in Urban Development*. New York: National Bureau of Economic Research.

Gramsci, A. (1971) *Selections from the Prison Notebooks*. New York: International Publications Edition.

Greer, A. and Greer, S. (1976) 'Suburban political behavior: a matter of trust', in B. Schwartz (ed.) (1976b: 203–20).

Greer, E. (1976–7). 'Monopoly and competitive capital in the making of Gary, Indiana.' *Science and Society* 40, 4: 465–79.

Guterbok, T. (1976) 'The push hypothesis: minority presence, crime and urban deconcentration', in B. Schwartz (ed.) (1976b: 137–64).

Hall, P. *et al.* (1973) *The Containment of Urban England.* Beverly Hills: Sage.

Harloe, M. (1977) 'Introduction' to M. Harloe (ed.) *Captive Cities.* London: John Wiley, pp. 1–48.

Harrison, B. (1973) *Metropolitan Suburbanization and Minority Economic Opportunity.* Washington, DC: The Urban Institute.

Harvey, D. (1972) *Society, the City and the Space-Economy of Urbanism.* Washington, DC: Association of American Geographers, Resources Paper #18.

—— (1973) *Social Justice and the City.* London: Edward Arnold.

—— (1974a) 'Class structure in a capitalist society and the theory of residential differentiation', in R. Peel, M. Chisholm and P. Haggett (eds) *Processes in Physical and Human Geography.* London: Heinemann, pp. 354–89.

—— (1974b) 'Class monopoly rent, financial capital and the urban revolution.' *Regional Studies* 8, 3: 239–55.

—— (1975a) 'The geography of accumulation: a reconstruction of the Marxian theory.' *Antipode* 7, 2: 9–21.

—— (1975b) 'The political economy of urbanization in advanced capitalist societies – the case of the United States', in G. Gappert and H. Rose (eds) *The Social Economy of Cities.* Beverly Hills: Sage.

—— (1977) 'Government policies, financial institutions and neighborhood change in US cities', in M. Harloe (ed.) (1977: 123–41).

—— (in this volume) 'The urban process under capitalism.'

—— (1978) 'Labor, capital and class struggle around the built-environment in advanced capitalist societies', in K. Cox (ed.) (1978: 9–38).

Harvey, D. and Chatterjee, L. (1974) 'Absolute rent and the structuring of space by financial and government institutions.' *Antipode* 6, 1: 22–36.

Hill, R. (1975) 'Capital accumulation and urbanization in the United States.' *Comparative Urban Research* 4, 3: 39–61.

Hillhouse, A. (1936) *Municipal Bonds: A Century of Experience.* New Jersey: Prentice-Hall.

Hoyt, H. (1933) *One Hundred Years of Land Values in Chicago.* Chicago: University of Chicago.

INFORM (1976–8) *Promised Lands* (3 volumes), Jean Halloran (ed.). New York: INFORM, Inc.

Jackson, K. (1975) 'Urban deconcentration in the nineteenth century: a statistical inquiry', in L. Schore (ed.) *The New Urban History.* Princeton: Princeton University Press, pp. 110–42.

Johnson, J. (ed.) (1974) *Suburban Growth: Geographical Processes at the Edge of the City.* London: John Wiley.

Kaplan, S. (1976) *The Dream Deferred: Politics and Planning in Suburbia.* New York: Seabury Press.

Kasarda, J. (1976) 'The changing occupational structure of the American metropolis: apropos the urban problem', in B. Schwartz (ed.) (1976b: 113–36).

Keeble, D. (1976) *Industrial Location and Planning in the United Kingdom.* London: Methuen.

Kerner Commission (1968) *Report of the National Advisory Commision on Civil Disorders.* Washington, DC: US Government Printing Office.

Kregel, J. (1972) *The Theory of Economic Growth.* London: Macmillan.

Kucherenko, A. (1978) 'A view of neighborhood transition.' Unpublished master's thesis, Department of City and Regional Planning, University of California, Berkeley.

Kuznets, S. (1960) *Capital in the American Economy.* Princeton: National Bureau of Economic Research.

Lösch, A. (1954) *The Economics of Location.* New Haven: Yale University Press.

McKenzie, S. (1977) 'The changing role of women in the nineteenth-century Canadian city.' Unpublished manuscript, May.

Mandel, E. (1975) *Late Capitalism*. London: New Left Books.
Markusen, A. (1976, 1978) 'Class and urban social expenditure: a Marxist theory of metropolitan government', in W. Tabb and L. Sawers (eds) *Marxism and the Metropolis*. New York: Oxford University Press, pp. 90–112.
Marriott, O. (1967) *The Property Boom*. London: Hamish Hamilton.
Marshall, H. (1973) 'Suburban life styles: a contribution to the debate', in L. Masotti and J. Hadden (eds) (1973: 123–48).
Marx, K. (1967) *Capital*. New York: International Publications Edition.
Masotti, L. and Hadden, J. (eds) (1973) *The Urbanization of the Suburbs*. Beverly Hills: Sage.
Masotti, L. and Hadden, J. (eds) (1974) *Suburbia in Transition*. New York: New Viewpoints.
Massey, D. (1977) 'Towards a critique of industrial location theory', in R. Peet (ed.) (1977: 181–98).
Mills, E. (1972) *Studies in the Structure of the Urban Economy*. Baltimore: Johns Hopkins Press for Resources for the Future.
Mitchell, J. and Oakley, A. (1977) *The Rights and Wrongs of Women*. Harmondsworth: Penguin Books.
Mollenkopf, J. (1978) 'The postwar politics of urban development', in W. Tabb and L. Sawers (eds) *Marxism and the Metropolis*. New York: Oxford, pp. 117–52.
Mosley, H. (1978) 'Is there a fiscal crisis of the state?' *Monthly Review* 30, 1: 34–45.
Muller, P. (1975) *The Outer City: Geographical Consequences of the Urbanization of the Suburb*. Washington, DC: Association of American Geographers, Resource Paper #75–2.
Muth, R. (1969) *Cities and Housing*. Chicago: University of Chicago.
National Commission on Urban Problems (Douglas Commission) (1968) *Building the American City*. Washington, DC: USGPO.
Neutze, M. (1968) *The Suburban Apartment Boom*. Baltimore: Johns Hopkins Press for Resources for the Future.
Ollman, B. (1971) *Alienation*. Cambridge: Cambridge University Press.
Osofsky, G. (1966) *Harlem: The Making of a Ghetto*. New York: Harper and Row.
Paulson, M. (1972) *The Great Land Hustle*. Chicago: Henry Regnery.
Peet, R. (1977) 'Inequality and poverty: A Marxist-geographic theory', in R. Peet (ed.) *Radical Geography*. Chicago: Maaroufa Press and London: Methuen, pp. 112–24.
Poulantzas, N. (1975) *Classes in Contemporary Capitalism*. London: New Left Books.
Pratt, E. (1911) *The Industrial Causes of Congestion of Population in New York City*. New York: Columbia University Press.
Pred, A. (1966) *The Spatial Dynamics of U.S. Urban-Industrial Growth, 1800–1914*. Cambridge, Mass.: MIT Press.
Robinson, T. (1973) *Economic Heresies*. New York: Basic Books.
Rothman, D. (1971) *The Discovery of the Asylum*. Boston: Little, Brown and Co.
Ryan, M. (1972) 'American society and the cult of domesticity: 1830–1860.' Unpublished Ph.D. dissertation, University of California (Ann Arbor: University Microfilms).
Sakolski, A. (1932) *The Great American Land Bubble*. New York: Harper and Row.
Santos, M. (1977) 'Society and space: Social formation as theory and method.' *Antipode* 9, 1: 3–14.
Schmidt, A. (1971) *The Concept of Nature in Marx*. London: New Left Books.
Schmitt, P. (1968) *Back to Nature: The Arcadian Myth in Urban America*. New York: Oxford University Press.
Schnore, L. (1965) *The Urban Scene: Human Ecology and Demography*. New York: Free Press.
—— (1972) *Class and Race in Cities and Suburbs*. Chicago: Markham.
Schumpeter, J. (1939) *Business Cycles*. New York: Harper and Row.
—— (1946) 'The decade of the twenties.' *American Economic Review* 36, 2:1–10.
Schwartz, B. (1976a) 'Images of suburbia: some revisionist commentary and conclusions', in B. Schwartz (ed.) *The Changing Face of Suburbia*. Chicago: University of Chicago Press, pp. 325–42.
—— (ed.) (1976b) *The Changing Face of the Suburbs*. Chicago: University of Chicago Press.
Seeley, J., Sim, R. and Loosely, E. (1956) *Crestwood Heights*. New York: Basic Books.
Sennett, R. (1970a) *Families against the City*. Cambridge, Mass.: Harvard University Press.

—— (1970b) *The Uses of Disorder.* New York: Knopf.
—— (1977) *The Fall of Public Man.* New York: Knopf.
Sennett, R., and Cobb, J. (1972) *The Hidden Injuries of Class.* New York: Vintage.
Sobin, D. (1971) *The Future of the American Suburbs.* Port Washington: Kennikat Press.
Soja, E. (1978 ms) 'Topian Marxism and spatial praxis: a reconsideration of the political economy of space.' Paper presented to the Annual Meetings of the Association of American Geographers, New Orleans, 12 April.
Sternlieb, G. and Hughes, J. (eds) (1976) *Post-Industrial America: Metropolitan Decline and Interregional Job Shifts.* New Brunswick: Center of Urban Policy Research, Rutgers University.
Stone, K. (1975) 'The origin of job structures in the steel industry', in R. Edwards, M. Reich and A. Gordon (eds) *Labor Market Segmentation.* Lexington, Mass.: D. C. Heath.
Taylor, G. (1915) *Satellite Cities: A Study of Industrial Suburbs.* New York: D. Appleton and Co.
Thomas, B. (1973) *Migration and Economic Growth.* Cambridge: Cambridge University Press (2nd Ed.).
Thorns, D. (1972) *Suburbia.* London: MacGibbon and Kee.
Toll, S. (1969): *Zoned American.* New York: Grossman.
Tryon, R. (1917) *Household Manufactures in the United States, 1640–1860.* Chicago: University of Chicago Press.
Ullman, J. (ed.) (1977) *The Suburban Economic Network.* New York: Praeger.
US Department of Housing and Urban Development (1975) *Housing in the Seventies.* Washington, DC: USGPO.
Vance, J. (1960a) 'Emerging patterns of commercial structure in American cities.' Proceedings of the IGU Symposium in Urban Geography, *Lund Studies in Geography* B-24: 485–518.
—— (1960b) *'Labor-shed, employment field, and dynamic analysis in urban geography.'* Economic Geography 36, 3: 189–220.
—— (1964) *Geography and Urban Evolution in the San Francisco Bay Area.* Berkeley: Institute of Governmental Studies.
—— (1966) 'Housing the worker: the employment linkage as a force in urban structure.' *Economic Geography* 42, 4: 294–325.
—— (1967) 'Housing the worker: determinative and contingent ties in nineteenth-century Birmingham.' *Economic Geography* 43, 2: 94–127.
—— (1977) *This Scene of Man.* New York: Harper's College Press.
—— (1978) 'Institutional forces that shape the city', in D. T. Herbert and R. J. Johnston (ed.) *Social Areas in Cities: Processes, Patterns and Problems* (vol. 1). London and New York: John Wiley, pp. 91–108.
Walker, R. (1974) 'Urban ground rent: building a new conceptual framework.' *Antipode* 6, 1: 51–8.
—— (1977) *'The Suburban Solution: Urban Geography and Urban Reform in the Capitalist Development of the United States.* Unpublished Ph.D. dissertation, Johns Hopkins University, Baltimore.
—— (1978a) 'The transformation of urban structure in the nineteenth century and the beginnings of suburbanization', in K. Cox (ed.) (1978:165–211).
—— (1978b) 'Two sources of uneven development under advanced capitalism: spatial differentiation and capital mobility.' *Review of Radical Political Economics* 10, 3: 28–37.
—— (1978 ms) 'Corporate capitalism, urban morphology and suburbanization in the United States at the turn of the century.' Unpublished manuscript, Department of Geography, University of California, Berkeley.
Walker, R. and Large, D. (1975) 'The economics of energy extravagance.' *Ecology Law Quarterly* 4: 963–85.
Walker, R., Storper, M. and Gersh, E. (1979) 'The limits of environmental control: The saga of Dow in the Delta.' *Antipode* 11, 2: 48–60.
Ward, D. (1971) *Cities and Immigrants.* New York: Oxford University Press.
Warner, S. (1962) *Street-car Suburbs.* Cambridge, Mass.: Harvard University Press.
—— (1967) *The Private City: Philadelphia in Three Periods of Growth.* Philadelphia: University of Pennsylvania Press.

Weber, A. (1909) *Über die Standorten der Industrien.* Tubingen. (Transl. C. J. Friedrick, University of Chicago Press, 1929.)

Whitehand, J. (1974) 'The changing nature of the urban fringe: a time perspective', in J. Johnson (ed.) (1974: 31–52).

Whyte, W. (1956) *The Organization Man.* Garden City, N.J.: Doubleday.

Williams, R. (1973) *The Country and the City.* Cambridge: Cambridge University Press.

—— (1977) *Marxism and Literature.* New York: Oxford University Press.

Wood, P. (1974) 'Urban manufacturing: a view from the fringe', in J. Johnson (1974: 129–55).

Wright, E. (1978) *Class, Crisis and the State.* London: New Left Books.

Young, M. and Wilmott, P. (1957) *Family and Kinship in East London.* Harmondsworth:Penguin Books.

—— —— (1960) *Family and Clan in a London Suburb.* Harmondsworth: Penguin Books.

—— —— (1973) *The Symmetrical Family.* Harmondsworth: Penguin Books.

16 Capitalism and conflict around the communal living space[1]

Kevin R. Cox

Introduction

In this paper I attempt to define and situate with respect to the capitalist mode of production a category of conflicts that are not in the classical Marxist sense, class conflicts: conflicts around the communal living space. Briefly, one can define a set of conflicts between those capitals involved in urban development and labor over resources providing utility to labor and consumed in common in the living place. These conflicts are evident at both neighborhood and municipal levels. Development impacts on residential amenity, for example, and results in controversy. It also has impacts on municipal tax rates and consequently one can envisage a set of fights around fiscal issues.

These are some of the conflicts which Harvey has described elsewhere as displaced class conflicts.[2] They express the reverberations of class conflict in the workplace throughout the totality of social relations in those social formations dominated by the capitalist mode of production. Further, the same mystifications of class relations recur in the context of the urban living place as in the workplace. Just as struggle in the factory so often tends to be conceived in distributional or cultural terms, so too in the communal living space.

The paper is divided into three major sections. In the first and briefest section a point of departure is provided by a discussion of the meaning of the communal living space for labor. Two meanings apparent in the historical experience of capitalist societies are sketched out: communal living space as community and communal living space as commodity.

In the second section the relations between labor's communal living space, and those capitals which see the development of urban land as a means of accumulation are outlined. Communal living space as community, it is argued, is an obstacle to accumulation through urban development. Urban land development as a means of accumulation presupposes a definition of the living space as a consumption artifact, as a way-station between less desirable, socially (rather than physically) obsolescent housing packages and the yet-to-be-fashioned. As the meaning of life becomes redefined around consumption, occupational achievement and acquisition, so the meaning of the communal living space is transformed in a manner stimulating development both directly and through numerous multiplier effects. But the problem with defining consumption as the meaning of life is that sooner or later consumers will take their consumption seriously[3] and object to those new urban developments detracting from the value of the communal living space as a commodity. It is in this context that one can begin to understand conflicts between labor and capital, in the neighborhoods over residential amenity, and at the municipal level over issues of public provision.

The capitals through which development proceeds have to reproduce themselves in a context where ability to do so is constantly at risk due to the competition of others. In striving to reproduce themselves they are guided, through this structural imperative, into channels which make conflict with labor totally forseeable. Thus, they are driven to transform the spatial form of the city, substantially at labor's expense, and in a manner so spatially and temporally concentrated as to leave localized fractions of labor without a glimmer of doubt that development is impacting adversely upon them.

Conflict, however, and the attempts of labor to repel development are clearly a threat to the ability of capitals to continue reproducing themselves. Labor presents barriers which must be suspended. The third and final section of the paper outlines the ways in which capitals work both through the market and through the State to achieve these goals.

The meaning of the communal living space for labor

The concept of a communal living space summarizes those resources important to labor and consumed in common, and access to which is contingent on living in a particular neighborhood or municipality. They include schools, local aesthetics and domestic architecture, quiet, fiscal resources, traffic safety and public safety, and public parks; they also include those others who share the same meaning system, both normative and cognitive, and with whom are shared bonds of trust, mutual obligation and sociability. Some of these resources, such as public safety and quiet, are relevant more at the neighborhood level. Others, however, occur at a municipal level; fiscal resources would be the most salient but municipal school policies are also significant.

Historically in capitalist societies it is possible to identify, among labor, two perspectives on the communal living space: the communal living space as community; and the communal living space as commodity consisting, nevertheless, of resources which are only imperfectly commodified.[4]

Communal living space as community

Here the meaning of the communal living space and its values derive in the first place from the people living there and from the mutually supporting ties of trust, friendship, sociability and predictability that they have created for each other. People are valued for their own sake and for the contribution they make to a place-bound community. Status is not so much measured in terms of the occupational achievement standards of the market; rather one acquires distinction and worth as a neighborhood handyman, an adept domino player, or as a repository of information on pregnancy and child-birth for neighborhood women.[5]

A local neighborhood economy reflects these ties of reciprocity. Buying and selling are indelibly stained by considerations of community. The shopkeeper is not only regarded as a tradesman; he is also a member of the community and expected to give credit to accepted members of that community.[6] Selling to someone in the community or who has community ties is more important than getting the highest price; homes are bought by kin or children of existing community members at a lower price than that which could be obtained in a metro-wide real-estate market.[7] Local credit institutions aggregating community savings feel first obligation to the community rather than to maximizing profits.[8]

Interactions occur within the context of a specific place. Local buildings, parks, even views of a smoking chimney, elicit responses which reflect that they are indissolubly bound to a localized community commanding an emotionally profound attachment. Neighborhood as community is not only people: it is people relating to one another in the context of a specific place.[9]

In brief, the concept of communal living space is a concept of home spilling out of the individual household and its dwelling and projecting itself on to neighbors, streets, local businesses, schools and other institutions. Consequently, the meaning of the communal living space as community is one which is so utterly personalized and particularized as to put it beyond the money system. For those experiencing it there can be no adequate substitutes anywhere else. It is a concept of the communal living space as noncommodified because it *cannot be* commodified.

On the other hand there is no way in which communal living space as community can be understood outside of the context of capitalism. While cultural traditions and ties provide the necessary matrix of resources it is the exigencies of a society in which people relate to each other more and more as buyers and sellers which compel use of those resources in the construction of community. Germane are the ties of mutuality providing some insurance against job market vicissitudes. This function has been widely noted in the literature.[10] In addition, local community status systems independent of occupational achievement and income provide some insulation against the cruel self-images imposed by capitalism. And finally the communal living space as community acts as an intermediary between rural areas and city workplaces. It is a source, for example, of job-market information, housing assistance and that home-away-from-home that eases adaptation into the harsh environment of industrial capitalism and makes it bearable.

Communal living space as commodity

To the extent that neighborhood functions as a 'haven in a heartless world'[11] alternative responses to communal living space as community are not feasible. But if it is the individual home which functions as haven the way is open for an alternative perspective. There is, in short, a view of the communal living space as a commodity, access to which has been purchased by rent or home purchase. Value for money becomes the critical standard: if the commodity should deteriorate in terms of its capacity to provide consumer satisfaction, then rights of access to it can be sold and rights of access to a more adequate substitute purchased elsewhere.

A variety of property-value studies take this as their fundamental behavioral tenet.[12] They find that such resources as school quality, public safety, quiet, and even views of Pacific sunsets, are indeed reflected in home values.[13] This is a perspective, therefore, which sees the communal living space as, *in principle,* commodifiable. It is a conception, moreover, within which the ideas of externality, compensation and formal cost-benefit analysis rest most comfortably; while for those adhering to a community perspective these ideas have absolutely no validity at all.

The communal living space, capitalist accumulation and conflict

Conflict between capital and labor over the communal living space presupposes the existence of capitals for which a relation to that living space is of critical significance as they attempt to reproduce themselves. These are the capitals which accumulate through the development, leasing and sale of urban real estate. They include: developers, both residential and commercial; construction companies; land companies; small speculators, and landowners; and financial institutions such as savings and loans, and insurance companies for whom urban real estate is a major investment. Their goal, in brief, is to reproduce themselves by the diversion of surplus value in the form of rent. They are defined in this paper as *property capitals.* We consider now the significance for property capitals of the meaning of the communal living space for labor.

Capitalist accumulation and the communal living place

For property capitals communal living space as community presents itself as a barrier. The colonization of urban land as a means of accumulation depends on capital imposing upon labor its own definitions of the meaning of the communal living space. Accumulation presupposes a relatively footloose consumer willing to cut community ties and take advantage of, e.g., new housing packages in the suburbs. It presupposes a concern over property values and a willingness to protect investments by liquidating at one point and investing elsewhere. In brief, accumulation presupposes neighborhoods as relatively open systems: people move in and people move out uncluttered by the baggage of community consideration.

To some degree this is accomplished by a transformation of the forces of production which, embedded in capitalist development,[14] makes community progressively less and less place-bound. Automobile and telephone facilitate the maintenance of ties of community with friends and relatives over much longer distances than hitherto feasible. Voluntary organizations, clubs, churches, school associated activities, all experience substantially enhanced membership areas. The extent to which this has occurred has been revealed in a recent paper on a neighborhood in South Chicago.[15]

Further, mass education and the mass media tend to homogenize meaning systems, so that culture has become less geographically restricted. A blue-collar household comes to feel almost as much as home in Buffalo as in San Jose. More specifically the cultural ties that reinforce a sense of community have tended to be reworked so that they occur more within status groups than between highly localized groupings.[16]

At the same time there is a careful and sustained cultivation of the values of bourgeois individualism, acquisitiveness and occupational achievement. Careers come before friends.[17] The neighborhood, through the mechanism of appreciating home values, becomes a stepping stone to higher levels of consumption elsewhere.[18] Or alternatively it serves, through the local school, as a mechanism by which children can leave the community rather than return, give it sustenance and reproduce it.[19]

In a context of general convergence of meaning systems and the drive towards the elimination of distance barriers there is an inevitable result: levels of residential mobility and an integration of the real-estate market that put any place-bound community at risk. Instead of arranging sales informally with friends-of-friends or co-ethnics, sellers seek to maximize prices through the services of a professional real-estate broker or, at the very least, by advertising in the local newspaper. The local savings and loan institution reorients its activities away from the immediate neighborhood and towards the metropolitan area as a whole.[20] As community becomes more and more divorced from place, therefore, so the metropolitan housing market becomes more and more integrated and people in the neighborhood are increasingly exposed to its vicissitudes and insecurities.

By sweeping away the barriers posed by place-bound community and by simultaneously cultivating labor as a commodity-consumer, capital, through advertising, policies of homeownership and the like, opens up a vast field for accumulation for itself. Suburbanization can occur because the concept of 'suburb' no longer has to compete with that of community. And neighborhoods can be made obsolescent: a process that is directly contrary to the logic of place as community.

Further, the emergence of the communal living space as a commodity rather than as a community permits an intensification of demand through a plethora of multiplier effects. Changing residential compositions and/ or increasing densities threaten use values and exchange values alike for long-term residents, particularly given the spatially concentrated form in which they tend to occur. These factors lend an additional thrust both to suburbanization on the metropolitan periphery; and to declining residential densities, neighborhood obsolescence, and abandonment in the inner city.[21] The decisions of financial agencies regarding loans for new construction and home purchase both reflect and further intensify these multiplier effects.

It would be a mistake, however, to regard commodity man as a mere puppet responding to market forces in a mechanical fashion: selling a home in an area which is no longer fashionable in order to buy in areas where addresses convey status; or selling where home values are declining so as to buy into areas regarded as good investments. As he comes to regard the living place as a commodity affecting market capacities and individual consumption so there is increasing resentment at the threat posed by property capitals and the urban development process to the value of that commodity. And where the losses threatened are sufficiently large the free-rider problem fades into the background and coalitions emerge[22] to oppose development. This is particularly true for the homeowner whose stake is of an altogether different kind than that of the renter. Property values are a major source of concern and provide a rallying-point around which households concerned about different aspects of the communal living space as commodity can gather.[23]

Controversy is most apparent at the neighborhood level. Particular issues include the impacts of new residential development and business expansion on landscape aesthetics, traffic hazard, noise and, consequently, residential property values. But social as well as physical impacts – or their threat – provide the necessary datum. New low-income apartment developments raise the associated specters of increased crime and school problems.[24] Conversion of single-family homes to apartments in inner-city areas elicits an image of impending and deleterious change in public safety, standards of home maintenance and neighborhood property values.[25]

It is in this light that one can begin to understand the policy repertoire of the suburban 'no-growth' movement: resistance to annexation; resistance in the voting booth to the sale of bonds for expansion of sewer or water capacity; defensive incorporation; resistance to public housing; 'gold plated' subdivision regulations; and, of course, the notorious exclusionary zoning.[26]

Within the already built-up area of the city there is the usual opposition to rezonings seen as facilitating neighborhood change. In addition neighborhood organizations lobby for school improvements enhancing the image of the neighborhood for the middle and upper-income homebuyer. Elsewhere the issue may be the construction of new highways or traffic management schemes impinging on neighborhood amenity. Or alternatively, organization may be stimulated by the threat of displacement generated by plans for urban renewal[27] or large scale middle-class rehabilitation.

Beyond the neighborhood, and at the municipal level, labor has a basis for concern over issues of public provision, taxation and the funding of public services and improvements: a basis, moreover, which was highlighted with great clarity during the recent New York City crisis. There is an on-going debate in most cities, lively at times, quiescent at others, about public priorities. Property capitals tend to push for growth-enhancing investments: a new convention center, downtown urban renewal, a mass transit system focusing on the downtown, or a municipally subsidized underground parking facility. On the other hand labor tends to lobby for a diversion of public resources to labor's consumption fund: increased school spending, mass transit subsidies and expansion of the city's public housing stock, either by new construction or purchase.

Related are conflicts of a fiscal character. Conflicts over tax holidays for new buildings, preferential tax treatment of agricultural land[28] and subdivision exactions provide instances.[29] Outcomes of these policy issues have a palpable effect on the costs and conditions of existence of labor in the communal living space. For their result is to shift the burden of public provision to labor, a problem frequently articulated by the opposition of public service agencies, such as school boards, who find their revenues proportionately diminished.

Hence at both neighborhood and municipal levels labor forms coalitions to protect, from the depredations of capital, a commodity yielding a stream of utility and to which it has purchased rights of access. Labor joins with others to protect an individual interest rather than the interest of some community with which it might once have identified. Eloquent in this regard is the manner in which labor relates to capital in ensuing conflicts. For while labor enters conflicts as guardian of a common resource, it is willing to bargain with capital for compensation, environmental improvements in the development scheme at issue, land for public housing, etc. The posture is rational, utility-maximizing and in its assumption of arbitration by the money system, remarkably analogous to market place relations. And once labor concedes the idea of the relation as a commodity relation it becomes vulnerable to fragmentation through tradeoffs capital is willing, indeed anxious, to make between the communal living space and other putative benefits flowing from the project to diverse fractions of labor.

But just as we would be in error to assume commodity man as a non-resistor, ever adapting, through buying and selling and relocating, to the impacts of the urban development process, so we would be wrong to predict the total dissolution of a relationship to the communal living space that is felt, in some albeit residual manner, as a community relationship. There is always an ambivalence, weak in some, and strong in others, about defining the communal living space purely in commodity terms. This is particularly true of those not totally assimilated to the meaning system dominant in the social formation: recent immigrants and religious groups for whom prejudice has been a real and recent experience, for example.[30] Likewise, partly for reasons of socialization and partly for reasons of financial constraint, it is the old who are more likely to resist the world of the commodity and seek comfort in the old and familiar.[31] Yet more generally the longer a household stays in a particular neighborhood the more it develops an attachment to co-residents and to the place itself.[32] While for some, response to the threats of the development industry is more in commodity terms, for others there is a concern for the endurance of ties, more or less deeply felt, with others in the area, with a particular turf and the like. And as households come together to oppose rezonings, new traffic plans, redlining, etc., so these ties are cemented further by a common plight.

Nevertheless, the fight has to be fought on capital's own terrain: that of the commodity. In a context of general mobility and integration of metropolitan real-estate markets the community can only be saved by treating the communal living space as a commodity. Banks redlining the area, for example, have to be convinced that the neighborhood is a good investment; only if the area can continue to be one of middle-class home owners – dependent, in turn, on the availability of mortgage credit – can existing residents negotiate the obstacles created by their own ambivalence in a manner sufficiently satisfactory to keep them there. Community organizations form their own realty organizations to steer in new residents, residents who will be socially and culturally compatible with existing ones and at the same time, assure those existing households that they *can* have both community and commodity.[33] And those whose greatest concern is community have to appeal for the support of those whose primary interest is neighborhood as commodity through a scenario terminating in massive property value impacts.

Accumulation and the inevitability of conflict around the communal living space

Labor, then, is pushed progressively in the direction of a relation to the communal living space as a commodity. To some degree this is apparent in residential choice behavior. It is also evident, and necessarily so, in the defensive battles labor wages against developers, e.g. in the suburbs over residential amenity, and in the inner city over urban renewal.

Essential context for comprehending the inevitability of these conflicts over the communal living space is the competition and conflict between property capitals. The problem of property capitals is to reproduce themselves in a context where ability to do so is constantly threatened by the activities of competitors. Each has as its goal the maximization of net revenue from land assembly, subdivision, leasing, construction, speculation, etc. In pursuing this objective each pushes simultaneously in diverse directions. Included are: attempts to stimulate the demand for urban land in general; and to deflect demand in the direction of one's own investments in particular; attempts to minimize competition; and risk-reducing policies, all within the general context of cost-minimization.

To some extent capitals pursue these strategies independently of one another; and to some degree they work collectively, expressing commonalities of interest that stem from, for instance, locational considerations. The end product, however, is the same. In attempting to reproduce themselves property capitals revolutionize the spatial form of the city in a way which cannot help but impact on the communal living space; for it is a revolutionizing which is carried out to a large degree at labor's expense and it impacts, necessarily, in a highly concentrated and uneven manner. We discuss these in turn.

REVOLUTIONIZING THE SPATIAL FORM OF THE CITY   First, then, there is the revolutionizing of the spatial form of the city. This includes urban development at the periphery and abandonment at the center; the implantation of suburban shopping centers; the construction of a freeway network; and downtown urban renewal. As developers on the city's edge succeed in their goals of obtaining annexation, sewer line and water line extension, and as the metropolitan population increases and the consumer yields to the ad-man's definition of an acceptable way of life, so the city expands at its edge; and as central-city population densities decline, so the apprehension of downtown booster lobbies increases and campaigns to float urban renewal and highway construction projects emerge.

This is not to assert, however, that the geographic changes are purely at the level of the built environment. Development also imposes a dynamic upon the social geography of the city, the suburban wave leaving behind not merely an area of abandonment in the central city, but areas of rapid and disorienting social change further out and even into the older, inner suburbs.

Partly these changes reflect the independent actions of individual property capitals. The imperative for self-reproduction leads to the fashioning, refinement, and marketing of new concepts in housing: developments for the retired or for the mature adult, for the recreation-oriented, or simply the concept of suburban living. At the same time demand is cultivated by, e.g., the development of new mortgage instruments with longer terms of repayment; or by integration with local transportation. In the construction of the Victorian suburb, for example, there was a close interlocking of the business interests of suburban developers, streetcar-line companies and railroad companies.[34]

Yet the revolutionizing of urban form cannot be purely a matter of individual initiative. For urban land cannot be produced by the individual firm; it depends on the decisions of others, and in particular on those of local government: the extension of sewer and water lines, and new highways in particular.[35] Moreover, land as a commodity has the serious disadvantage of immobility. By its very nature it cannot be moved so as to adapt to those decisions of local governments or of others – businesses, residents – helping to spark off urban development.

As a consequence conflict between property capitals owning real estate in different parts of the metropolitan area, and indeed in different metropolitan areas, can be anticipated. This suggests that urban development is necessarily highly politicized as property capitals attempt to manage and stimulate the urbanization process to the advantage of their localized and immobile investments.

Consider, in this light, the booster lobby, a loose coalition of downtown property capitals.[36] A general context of suburbanization has made the protection and enhancement of investments in central city land one of its major goals. In North America effort has been directed to the restructuring of the urban environment: urban renewal[37]; parking garages; and the creation of freeway networks and mass transit systems[38] focusing on the downtown.

But while the booster lobby attempts to give the momentum of development a centripetal thrust there are other coalitions of property capitals just as anxious to impose on it a more centrifugal emphasis. Property capitals with investment in particular subsectors of the metropolitan periphery come together to lobby for waterline and sewerline extensions or to promote annexation. And where the necessary commitments from city engineers and the like are not forthcoming alternative avenues to the establishment of the necessary infrastructure may be pursued: municipal incorporation, for example, or the creation of water and sewer districts. Often established for purposes of promoting development, such local authorities are of critical concern for developer interests[39] and are frequently joined by local retailing and merchant interests standing to gain from population growth.

Yet scale considerations alter political calculation. While capitals may enter coalitions against one another at one level in an attempt to stimulate and deflect growth onto their own particular turfs, there is a broader commonality of interest at the level of the metropolitan area as a whole. After all, having growth to deflect depends to a considerable degree upon the ability of the metropolitan area to grow in population. All property capitals, therefore, unite behind proposals to secure, e.g., large federal and state office buildings or military installations; to guarantee future water supplies; or to attract new major private employers. And of course, to the extent that they pursue, by infrastructural investment, and succeed in, their growth programs, the spatial form of the city will, of necessity, be transformed.

THE EXTERNALIZATION OF COSTS   This is of substantial concern to labor since many of the costs of development are externalized, either by default or design, on to labor and experienced in the communal living space. Suburban development congests public schools and highways, a congestion which must be relieved at public expense. On the other hand, development leaves behind it in the central city underused public facilities – schools, firehouses – which continue to be an item in city debt payments long after the last house in the neighborhood has been seized by city government on grounds of property tax delinquency. New transportation systems focusing on the central business district become another charge on the public; and urban renewal itself generates a plethora of negative externalities.[40] More specifically development leaves behind it a trail of homes the declining values of which testify in precise terms to the magnitude of the uncompensated costs heaped by capital on to labor in the communal living space.[41]

But this is necessarily how urban development under capitalism must proceed. To the extent that capitals can foist on to labor the costs of development so their chances of reproduction are enhanced. There is therefore, in urban development, as we saw in the preceding section, a history of struggle over subdivision exactions; over the variables entering into the cost-benefit analyses on which urban renewal decisions are based; over tax breaks for developers; and, more generally, over the degree to which labor should be asked to underwrite the costs of urban land development.

THE CONCENTRATION OF DEVELOPMENT   The problem of externalization is intensified by the concentrated form assumed by urban development. The development process, and, consequently, the external costs which it inevitably entails, is highly concentrated, and necessarily so, in both space and time. Urban development is not only concentrated in some cities rather than in others; it is also concentrated *within* cities. At any one time suburban development will occur at only a very few locations around the urban periphery, development in one particular suburb tending to spark off further rounds of development in the same suburb. Middle-class rehabilitation exhibits the same spatially focused character, speculators searching out properties for renovation either inside, or immediately adjacent to, gentrifying neighborhoods.

To some degree concentration of development has to do with internal economies of scale in development.[42] In housing construction a larger development facilitates more efficient deployment of labor and more advantageous terms for materials purchase. Advertising costs can be reduced; and the costs of special facilities enhancing the marketability of the project – a golf course or private park for a housing development, party houses and swimming pools for apartment developments – spread over a greater number of units.

Development in a given location increases the likelihood of development in the immediate vicinity. Consequent is a demand for complementary land uses: shopping centers focus traffic creating demands for strip development – gas stations, fast food outlets, discount stores – on adjacent highways; residential development creates a demand for shopping, medical and recreational facilities, and shopping facilities increase the attractiveness of the area for apartment construction.[43,44]

Consequently sharp gradients in profitability and risk encourage sharp geographical discontinuities in urban development and, hence, in impaction of existing residents. Temporal considerations reinforce the resultant sense of loss. Steep profit gradients over space are necessarily accompanied by steep profit gradients over time – an assertion to the validity of which any self-respecting blockbuster or speculator will readily testify. Capitals converge on the honey pot in a race against competitors so that consequent development is highly compressed in time. The use of housing as a

Keynesian regulator does nothing to mitigate this, of course; rather it intensifies it. For instead of, e.g., resultant congestion increasing by almost imperceptible amounts, it increases in quantum leaps.

In sum, conflict appears endemic to the urban development process as it unfolds in capitalist societies. To reproduce themselves capitals have to continually transform the built form of the city; and in a manner which both imposes costs on labor in the communal living space and concentrates those costs geographically, and in time. Slow accommodation by existing residents to the urban development process under capitalism is something precluded by the very nature of that process. Labor is left in absolutely no doubt of its impaction in the communal living space and vigorously resists those acts – annexation, rezonings, sewage works expansion – which enhance the possibility of development. The capitals involved in urban development, therefore, inevitably excite barriers to their own reproduction, barriers which have to be, and which are, suspended.

Suspending the barriers of labor resistance in the communal living space

That capital will attempt to suspend these barriers to accumulation goes without saying. On the one hand capital accepts community resistance as a new market fact. It channels itself into areas and into forms where resistance will be least likely; and subverts resistance through exploitation of the market incentives so frequently created. Resistance is thereby suspended as a problem through strategies of market adaptation.

On the other hand capital has investments which may be costly to liquidate and where adaptive tactics are not feasible. The strategy in this context is to harness to its goals the power of the State. By careful definition of community and community interest capital attempts to create coalitions with fractions of labor and with government agencies sufficiently powerful to overcome, or even co-opt, local resistance. The approach in this instance, therefore, is of a political character. These strategies are considered in turn.

Market adaptation

Capital can adjust to the fact of resistance by incorporating it into its investment calculus. Capital, therefore, avoids areas which have developed a successful 'no-growth' track record. Alternatively investment may shift into those forms eliciting least resistance. In this way housing development in the suburbs may become more emphatically geared to the upper end of the housing market; housing provision for those of lower income is then confined to the filtering mechanism. Exclusion of low-income housing developments from the suburbs can also stimulate investment in high rise apartment buildings in the central city; this appears to have been a market response in the United Kingdom[45] and a major stimulus to the industrialized building industry. In short, resistance can enter market calculations through the costs of delay, of hiring attorneys for court fights and the like. In addition, however, it creates market incentives which stimulate its own subversion.

Resistance to public works programs, for example, can negate itself, courtesy of planning blight. The uncertainty arising from resistance to highway development or urban renewal creates reluctance on the part of the homebuyer to purchase into the area. In addition financial agencies will come to regard the area as a poor investment until uncertainty is resolved. If there should be a market for home purchase in the neighborhood withdrawal of credit will make consummation of transactions or even home repairs difficult. The result is, on the one hand, disruption of normal processes of residential turnover and a growing pool of residents who 'want out'. And on the other hand, the locality assumes a more and more dilapidated air which merely serves to intensify the desire to sell. If local government should, as it often does, purchase some of the properties and clear their sites in anticipation of the project eventually succeeding, the air of dereliction will be further enhanced. Not surprisingly, those who were initially most resistant become increasingly willing to accept any compensation the local authority offers. And, of course, what will be offered will reflect the decline in market values registered in the interim.

Likewise, exclusionary zoning notwithstanding, suburbs get development. It could not be otherwise. For suburban exclusion has two effects. On the one hand it increases the desirability of a suburb as a place in which to live, an effect of scarcity of which developers will be well aware; on the other hand, it produces a surplus of land for low-density uses and a serious shortage of land for high-density uses. Consequently, rezonings from low-density to high-density uses provide the prospect of considerable capital gain. They acquire market values yet are themselves not marketable. The effect is to create a quasi-market, the existence of which is readily expressed in the widespread corruption prevailing on suburban zoning boards.[46] Rezonings are bought and sold, though in specific instances it may be difficult to discern the nature of the payment. Perversely, therefore, exclusion creates market pressures resulting in the very same developments which are anathema.

Ideology, the State and suspending the barriers

Alternatively capitals may, through ideological formulation and coalition building, succeed in imposing their own goals on those of the State and, through resultant policy, again suspend the barriers presented by local community resistance.

Necessary context for an approach of this nature are cultural changes which make the ideological formulations of capital meaningful and which are, themselves, embedded in the trajectory of capitalist development. We have related how, with the accumulation process and the socially-solvent effects of capitalist development men relate less and less to each other as community members sharing bonds of solidarity, trust, obligation and local, insulating, meaning systems; and more and more as commodity buyers and sellers, benefit-cost calculators and maximizers. This is not to assert that community relations are totally voided. Rather they are dominated by, and redefined in terms of, the commodity relation. We illustrated this earlier for example with the attempts of neighborhood groups to save what remains of a place-bound community by harnessing the property market to local purpose. The commodity link becomes the ultimate arbiter of social relations. Community becomes a mere shadow both of its pre-capitalist self and of the forms it assumes under competitive capitalism.

This change, cultural in its manifestation but a mere surface expression of a deeper structural logic, is tied, by mutual implication and conditioning, to another: that of the separation of public and private spheres. The dissolution of pre-capitalist formations is accompanied by a shift of functions away from community to State. What had previously been taken care of by mechanisms of social control, by a sense of community obligation or in an informal family context is transferred to government in general and to its specialized bureaucracies in particular. Katz has, in these terms, explained the emergence of what he calls the institutional State[47]: a State which assumes responsibility through specialized institutions and a technically qualified staff for health care, care of the indigent and the unemployed, education and care of the aged and of the mentally incapable. The functions of public safety, education, housing, etc. are assumed to have a public content which can no longer be left to citizens who define their lives more and more in private terms. The State must therefore intervene and assume responsibility for functions hitherto satisfied by informal community mechanisms.

Consequent to this separation community interest is defined more and more as public interest. The social fragmentations induced by the logic of the commodity, however, ensure that there can be no single definition of the community qua public interest. Labor, for example, dissolves into distributional groupings according to race, income, location or any other social cleavage created by the necessarily uneven development of capital. The State can, and does, become a battleground for contending consumer interests. Fractions of labor are willing to identify with those sharing a similar consumption status and develop consciousnesses opposing their claims to those of other groupings. Political parties come to represent broad coalitions of consumer groupings. The community qua public interest is then transitory and represents the ability of one distributional grouping or coalition to impose its own definition on those of groups holding quite different, antithetical conceptions.

Having nurtured the social forces that created this context, capital re-enters it in order to suspend the barriers labor erects to its reproduction. Capital's goal is defined as nothing more vile than that of public service[48]: building houses for tomorrow's families, constructing highways to relieve congestion, building heated shopping malls for the comfort of the shopper and generally indulging in an orgy of altruism and community spirit.[49]

But community service presupposes both a community and a community interest. In this regard capital has only one bias: definitions must be such as to facilitate coalitions which will, by smashing the opposition – or by co-opting it – further the cause of capital accumulation.

Consider, for example, the politics of opening up the suburbs. Conflict came to the fore in the late sixties and early seventies. A policy ostensibly aimed at expanding housing opportunity for the low-to-moderate income groups took the specific form of interest and rent payment subsidies (the so-called '235' and '236' programs). The resultant increase in demand, however, could only be satisfied by construction of new housing in suburban locations. This was vigorously opposed in a host of rezoning incidents receiving widespread publicity. In the resultant attempts – judicial and carrot-and-stick – to break down the barriers, capital developed and exploited definitions of both community and community interest.[50]

Community was defined as the whole metropolitan area rather than the immediate suburbs projected to receive new housing development. Community interest was defined in both equity and efficiency terms. Equity would be satisfied by increasing the housing opportunities to the poor and the black of the inner city; by increasing their access to job opportunities in the suburbs; and by so rearranging the geography of demand for public expenditures as to substantially alleviate the central city-suburb fiscal disparities problem.[51] Efficiency considerations, on the other hand, would receive attention through a reduction of labor shortage in the suburbs as blue-collar workers moved close to places of employment. Opening up the suburbs, therefore, was portrayed as a policy from which most would gain, simply through the increased productivity of the community; and from which some would gain to a very large degree what was legitimately theirs with respect to other community members.

That this argument was meaningful to those fractions of labor for whom it was intended is apparent from subsequent political postures. Black pressure groups – the National Association for the Advancement of Colored People and the National Committee Against Discrimination in Housing – were prominent in support of attempts to break down the barriers.[52] Likewise the interest of central-city populations in general in the fiscal, and indeed social, consequences of opening up the suburbs was communicated by the support of central-city mayors and legislators, both state and federal.

Controversies over urban renewal provide a second example of the attempts of capital to define and exploit a definition of community and of community interest congenial to its reproduction. Typically the conflict is between, on the one hand, the urban renewal authority, the downtown booster lobby, and large developers and, on the other hand, residents who will be displaced. The strategy of capital here is to define the community of concern as the central city locked in combat with other local governments for tax base, and having serious implications for the central-city taxpayer. Alternatively the development package may be so designed as to appeal to other constituencies among labor: few, after all, can find fault with the side-payment represented by an expansion of higher education or medical facilities.

Both these examples might suggest that capital works to suspend the barriers by allying with one fraction of labor against another. This, however, would be to misread the ultimate goal. The ultimate goal, after all, remains that of reproduction. A coalition with one fraction of labor against the resistant elements may be sufficiently powerful to do this. Alternatively it may not be. Co-optation through the State provides an additional tactic.

Thus, the definition of a community of beneficiaries from an urban development may be used not to crush the opposition politically but as a basis for compensating and so defusing it. As beneficiaries, it might be argued, it is only equitable that they should compensate in some way those other members of the 'community' sustaining the costs of urban development. There is, therefore, an attempt to use the theory of externalities as a rationalization for public compensation of those who would otherwise resist development projects. Instead of the immediately impacted bearing the uninternalized costs of development, there is an attempt to shift them on to some putatively beneficiary public. And, of course, to the extent that the resultant costs will be diffused and spread out over many rather than a few, electoral and/or legislative resistance will be negligible.

There is no doubt, for example, that consequent to citizen pressure, levels of public compensation to those displaced by urban renewal have substantially increased.[53] Likewise, one of the tactics pursued in the 'opening up the suburbs' controversy was to call for federal compensation, in the form of school subsidies and property value insurance for impacted suburban jurisdictions.[54] More generally, there is substantial externalization of the costs of the urban development. A variety of state and federal monies result in heavy subsidization of local government investments in water works, sewage capacity and highways. In addition tax-free municipal bonds give local governments access to credit for capital improvements at below-market rates of interest. Without these types of 'compensation' for the locally experienced costs of urban development, antagonism to growth would be considerably more intense than it already is, and to the greater disadvantage of developer interests.

On the other hand, there is no desire here to depict the State as a passive creature, meekly acceding to the definitions and meanings put forth by capital and ignoring those alternatives proposed by labor. The State is active rather than passive in the creation of those policies facilitating the suspension of the barriers; and necessarily so.

The separation of public from private spheres is expressed in a diversity of forms. Not the least is the technical division of labor represented by both legislature and executive. Legislators are interested in getting reelected because their livelihood depends on it. Likewise, there is the emergence of specialized bureaucracies and public agencies: the educational bureaucracy, the local housing authority, welfare agencies, transportation departments, planners and the like. Each is dependent upon the public fisc and must compete with other agencies for a share of it. The stakes are not only jobs. Enhanced career lines and the prestige of an expanding bureaucracy also figure prominently in the public agency calculus.

With the goals of suspending the barriers the interests of government intersect in numerous ways. On the other hand for legislators and city councillors, there are the side payments offered by the development industry itself. Campaign finance is common in this regard.[55] But so is the diversion of business to the councillor's law practice or insurance agency. And many councillors themselves have property interests[56] in the constituency to which they are electorally responsible.

On the other hand there is the electorate to attend to. Campaign finance from developers helps but does not make re-election inevitable. Clearly 'undesirable' development must be kept out, not necessarily in a negative manner but by indicating more appropriate locations. In this way local governments help developers suspend the barriers elsewhere. The support given by central city mayors to the attempts of large developers to 'open up the suburbs' is an interesting case in point.

However, the concept of 'undesirable' development includes the concept of 'desirable'. Legislators have substantial latitude in articulating concepts of what is desirable and considerable power, due to their electoral legitimacy and their access to the public arena, in defining these concepts. And that they will define these concepts in a manner consistent with 'progress', 'growth – but with due regard to the quality of life', derives from their relations with developers. The old slogans are therefore taken down from the shelf and tarted up in another bid to defuse opposition: 'development is inevitable', 'it will help the tax base', 'our citizens deserve improved shopping facilities', and so on. And if that doesn't work a little log rolling may do the trick. Public health may be the Trojan horse through which development is introduced in to the area. Why is it for example that a *modernized* sanitary system so often seems to come in a package that also includes an *expanded* sanitary system?

Similar remarks could be made with regard to public agencies. Conflict of interest is just as common among sewer, water and transportation engineers and urban renewal authorities as it is among city councillors. And unlike legislators they have access to an arcane body of technical expertise which can be bent in the direction of justifying before legislative bodies their own particular design for suspending the barriers.[57]

A concluding comment

The perspective on the communal living space as commodity is, of course, one further expression of what Marx defined as the fetishism of commodities: the tendency to view social relations as relations between commodities. It is this which has been termed 'both ingredient in, and constitutive of, economic exploitation' under capitalism;[58] and it is at the basis of that ideological unity between capital and labor which in the final analysis permits capital to repeatedly suspend the barriers to its reproduction.

More particularly the perspective yields to what James Anderson has dubbed 'the geographer's particular conceit'[59]: the fetishism of space. Social relations – between developer and labor in the communal living space, for example – come to be viewed as relations between areas; between city and suburb, between one suburb and another, and between redlined and non-redlined areas. The fetishism of space, as Marris has indicated,[60] becomes a self-fulfilling prophecy: the local defensive reactions of labor deflect capital into other areas verifying that the problem is indeed one of space relations. This in turn stimulates an institutional response – defensive incorporation, fiscal mercantilism, cost-revenue analysis – which again reinforces the spatially uneven character of urban development and the apprehension of location as the problem.

Unsurprisingly spatial fetishism also stamps contemporary social science outside of the field of geography. On the one hand it is possible to identify liberal views which focus on inequality in the communal living space and see it as the outcome of competition between spatially based communities: competition to attract in utility-enhancing developments and residents and to keep out the detracting.[61] On the other hand public choice theories, deriving from the socially conservative postures of neo-classical economics, see spatial variation in the communal living space as reflective of differences in preference and *ipso facto*, efficient: in other words, different communal living spaces represent qualitatively different commodities rather than the homogeneous, but quantitatively varying commodity assumed in the liberal view.

Yet in the fundamental assumption underlying these viewpoints on the communal living space, both liberal and conservative views are remarkably convergent. For both it is seen as a commodity. The – albeit different – policy implications which they derive, therefore, merely serve to facilitate capitalist urban development, generate the data which verify their theoretical perspectives and at the same time reproduce those conflicts around the communal living space which it has been the objective of this paper to both define and elucidate.

Notes

[1] I would like to gratefully acknowledge helpful comments on an earlier draft of this paper from Jeff McCarthy and Frank Nartowicz of The Ohio State University. In addition I would like to thank Alan Hooper of Reading University, and Dick Walker of the University of California at Berkeley. Although neither of them have seen drafts of this paper they have given stimulating comments on previous papers of mine. I have found this most useful in developing the conceptual framework within which this present paper falls.

[2] D. W. Harvey (1978) 'The urban process under capitalism: a framework for analysis.' *International Journal of Urban and Regional Research* 2, 1: 125–6.

[3] D. W. Harvey (1976) *The Political Economy of Urbanization in Advanced*

Capitalist Societies. Baltimore, Maryland: Johns Hopkins University Center for Metropolitan Planning and Research, pp. 34–5.

4 This classification was originally suggested to me by an attempt of Alan Hooper to categorize neighborhood-based political action in the city. See A. Hooper (1973) 'Class, coalition and community: dimensions of power in the city.' Unpublished ms., Faculty of Urban and Regional Studies, Reading University (November).

5 Standard references in this context include: H. J. Gans (1962) *The Urban Villagers: Group and Class in the Life of Italian-Americans*, New York: Free Press; M. Young and P. Willmott (1962) *Family and Kinship in East London.* Harmondsworth: Penguin Books. A more recent study of such a place-bound community in Philadelphia is: R. A. Cybriwsky (1978) 'Social aspects of neighborhood change.' *Annals, Association of American Geographers* 68, 1: 17–33.

6 R. Roberts (1973) *The Classic Slum.* Harmondsworth: Penguin Books, pp. 81–3.

7 Cybriwsky 1978: 22–3, 28.

8 Harvey 1976: 34–5.

9 Remarkably eloquent in this regard is the opposition generated by a plan to rationalize the settlement pattern in County Durham, England. County planning officers envisaged a program of public disinvestment from small, visually ugly, erstwhile colliery villages and relocation to larger urban centers where economies of public provision could be achieved. Houses in public ownership, therefore, have been left unrepaired, and, for want of new water, sewage and school facilities, moratoria on new private development declared. This has brought tremendous opposition from residents of the villages affected, despite a (in my experience, altogether realistic) characterization of the local environment by one commentator in highly negative terms: 'Patches of rubble and weeds are dumping grounds where houses once stood. Boarded up corner shops and cafes and unmade roads of ankle-deep mud complete a landscape of shut-down pits and leftover spoil tips.' See J. Barr (1969) 'Durham's Murdered Villages.' *New Society* (3 April), pp. 523–5.
  Webber has also commented perceptively on this type of identification with place, particularly as it occurs among income groups: 'The physical place becomes an extension of one's ego. The outer worlds of neighborhood-based peer groups, neighborhood-based family, and the *physical* neighborhood place itself, seem to become internalized as inseparable aspects of one's inner perception of self. In the highly personalized life of the working-class neighborhood, where one's experiences are largely limited to social contacts with others who are but minutes away, the physical space and the physical buildings become reified as aspects of the social group. One's conception of himself and of his place in society is thus subtly merged with his conceptions of the spatially limited territory of limited social interaction.' (M. M. Webber (1964) 'Culture, territoriality and the elastic mile.' *Papers and Proceedings of the Regional Science Association* 13: 63.)

10 Gans 1962, Young and Willmott 1962. E. Wolf and C. N. Lebeaux (eds) (1969) *Change and Renewal in an Urban Community.* New York: Praeger, pp. 196–8.

11 C. Lasch (1977) *Haven in a Heartless World.* New York: Basic Books.

[12] A review of some of these studies, illustrating the type of insights they yield, has been provided by M. J. Ball (1973) 'Recent empirical work on the determinants of relative house prices.' *Urban Studies* 10 (June): 213–33.

[13] One suspects, however, that a study which included market prices of homes sold privately rather than through realtors would provide less impressive results. See Cybriwsky 1975: 22–3.

[14] 'The more production comes to rest on exchange value, hence on exchange, the more important do the physical conditions of exchange – the means of communication and transport – become. Capital by its nature drives beyond every spatial barrier. Thus the creation of the physical conditions of exchange – of the means of communication and transport – the annihilation of space by time – becomes an extraordinary necessity for it.' (K. Marx (1973) *Grundrisse.* Harmondsworth: Penguin Books, p. 524.)

[15] R. Taub et al. (1977) 'Urban voluntary associations, locality based and externally induced.' *American Journal of Sociology* 83, 2 (September): 425–42. Also, regarding the low level of intra-neighborhood contact in some suburban developments see F. Rabinovitz and J. Lamare (1971) 'After suburbia, what?: The New Communities Movement in Los Angeles,' in W. Z. Hirsch (ed.) *Los Angeles: Viability and Prospects for Metropolitan Leadership.* New York: Praeger, p. 198.

[16] For documentation of a national 'working class culture' in Britain see R. Hoggart (1971) *The Uses of Literacy.* London: Oxford University Press.

[17] M. Seeman, J. M. Bishop and J. E. Grigsby (1971) 'Community and control in a metropolitan setting,' Chapter 6 in W. Z. Hirsch (ed.) (1971: 146, 161).

[18] The use of the home as an investment to facilitate mobility aspirations is graphically documented in C. Werthman et al. (1965) *Planning and the Purchase Decision: Why People Buy in Planned Communities.* University of California, Berkeley: Institute of Urban and Regional Development, Center for Planning and Development Research.

[19] B. Jackson and D. Marsden (1966) *Education and the Working Class.* Harmondsworth: Penguin Books.

[20] Harvey 1976: 34–5.

[21] A major conclusion from a recent study of the St. Louis housing market is highly germane: 'Observing the apparent pattern of neighborhood change, householders began to devalue their property in an effort to "get out" even *before* the (racial) boundary had shifted. That is, in neighborhoods that lay across the apparent path of transition *expectation alone brought a drop in values.* The effect of this, of course, was to accelerate the entire process.' (J. T. Little et al. (1975) *The Contemporary Neighborhood Succession Process.* St. Louis: Institute for Urban and Regional Studies, Washington University, p. 40.)

[22] R. Hardin (1971) 'Collective action as an agreeable n-prisoners' dilemma.' *Behavioral Science* 16, 5 (September): 472–81.

[23] For a particularly revealing study of the significance of property value concerns for neighborhood action see: J. A. Agnew (1978) 'Market relations and locational conflict in cross-national perspective', in K. R. Cox (ed.) *Urbanization and Conflict in Market Societies.* Chicago: Maaroufa Press and London: Methuen, Chapter 6.

[24] An interesting case study of this type of conflict is provided by M. Cuomo

(1975) *Forest Hills Diary: The Crisis of Low Income Housing*. New York: Vintage Books.

25 Case studies of these apprehensions and resultant neighborhood action include a study of the Bagley neighborhood in Detroit: E. P. Wolf and C. N. Lebeaux (eds) (1979) *Change and Renewal in an Urban Community*. New York: Praeger, Part 1; and a study of the Mattapan neighborhood in Boston: Y. N. Ginsberg (1975) *Jews in a Changing Neighborhood*. New York: Free Press.

26 A review of exclusionary mechanisms can be found in M. Gaffney (1973) 'Tax reform to release land', in M. Clawson (ed.) *Modernizing Urban Land Policy*. Washington, DC: Resources for the Future, pp. 115–29; on exclusionary zoning see M. N. Danielson (1976) *The Politics of Exclusion*. New York: Columbia University Press. For exclusion in the British context see K. Young and J. Kramer (1978) 'Local exclusionary policies in Britain: the case of suburban defense in a metropolitan system', in K. Cox (ed.) 1978, Chapter 10.

27 J. H. Mollenkopf (1976) 'The post-war politics of urban development.' *Politics and Society* 5 (Winter).

28 J. Kolesar and J. Scholl (1972) *Misplaced Hopes, Misspent Millions*. Princeton: Center for Analysis of Public Issues.

29 R. H. Platt and J. Moloney-Merkle (1973) 'Municipal improvisation: open space exaction in the land of "Pioneer Trust".' *Urban Lawyer* 5, 4 (Fall): 706–28.

30 Cybriwsky (1978).

31 Jeremy Seabrook in a study of Blackburn, a declining English textile town, has beautifully evoked the conflicts to which the old are subjected, as the world of the commodity slowly encroaches and destroys old localized, face-to-face subcultures; he is worth quoting at length:

> 'Most old people I met expressed resentment of the forces in society which have robbed them of the consoling certainty that all their neighbours shared the same poverty and the same philosophy, and were as uniformly helpless and resourceless as themselves. They are always sadly evoking long summer nights, when the afterglow in the western sky darkly illuminated the faces of neighbours on the doorsteps, serene and certain of the permanency of their predicament. But now they feel they were deceived. The long warm evenings return to the streets, but the neighbours do not. The values and habits that grew out of their poverty have been abolished with the poverty itself. While they were still striving for social justice and economic improvement, they took no account of any accompanying change that would take place in their value-structure: they simply transposed themselves in imagination into the houses of the rich, and it was assumed that they would take with them their neighbourliness and lack of ceremony, their pride in their work, their dialect and common sense. What they did not foresee was the estrangement and departure of the generation which should have risen to replenish the community, but who abandoned them like thankless step-children; sold their houses and moved out of town, immured themselves in a diminished, though impregnable, family structure, and devoted themselves to an endless accumulation of goods and possessions – a sterile therapy in dutiful response to altered circumstances. Instead of imposing their own will upon changing conditions, they allowed themselves to be manipulated

by them, not preserving anything of their past, but surrendering it, like the victims of a great natural disaster, who flee before the elements and abandon all that they have painstakingly accumulated. Perhaps, if they had understood what was happening, they could have preserved something of the old culture; but instead, they raise their voices in wild threatening querulousness against the young, or the immigrants or any other fragment of a phenomenon that is only partially and fitfully visible to them.' (J. Seabrook (1971) *City Close-Up.* Harmondsworth: Allen Lane and Indianapolis: Bobbs-Merrill, pp. 58–9.)

[32] 'Mobility creates city-wide ties while residential stability fosters local ties; most city dwellers have both.' (P. Meadows (1973) 'The idea of community in the city', in M. I. Urofsky (ed.) *Perspectives on Urban America.* Garden City, NY: Doubleday, p. 8.) On the impact of length of residence on community attachment see J. D. Kasarda and M. Janowitz (1976) 'Community attachments in mass society.' *American Sociological Review* 39, 3: 328–39.

[33] A remarkable feature of residents' organizations is, in fact, their direct involvement in the real-estate market. A common conception of neighborhood problems assigns major significance to the activities of the conventional real-estate market. Realtors, it is alleged (and often with a good deal of truth), are trying to blockbust the neighborhood by steering white, middle-income households away from it. An obvious response to this analysis is for residents to form their own realty company, the goal of which will be to steer *in* the desirable rather than steer them *away*. See H. Molotch (1972) *Managed Integration: Dilemmas of 'Doing Good' in the City.* Berkeley and Los Angeles: University of California Press; and Wolf and Lebeaux 1969. In other cases residents' organizations have worked to secure a relaxation of mortgage restrictions imposed on their area. See A. J. Naparstek and G. Cincotta (1976) *Urban Disinvestment: New Implications for Community Organization, Research and Public Policy.* Washington, DC: National Center for Urban Ethnic Affairs.

[34] Consider, for example, the real-estate speculation activities of the early street car pioneers in the USA: K. H. Schaeffer and Elliott Sclar (1975) *Access for All.* Harmondsworth: Penguin Books, pp. 30–1, 38–9 and 77–8. In Britain railroad companies were forced to sell land surplus to construction requirements on completion of a scheme; land development activities were not permissible and this limited greatly the speculative profits the railroad companies could claim. Nevertheless, 'There was, on the other hand, nothing to stop the directors of railway companies, or their contractors and solicitors, from using their business knowledge to make private investments in land; and a great deal of such speculation went on in the course of railway building.' (J. R. Kellett (1969) *The Impact of Railways on Victorian Cities.* London: Routledge and Kegan Paul, p. 397.) In addition landowners attempted to structure the location of railroad lines by actually donating land for the railroad companies or putting up much of the capital necessary for line extension (Kellett 1969: 400).

[35] S. T. Roweis and A. J. Scott (1978) 'The urban land question', Chapter 2 in K. R. Cox (ed.) *Urbanization and Conflict in Market Societies.* Chicago: Maaroufa Press and London: Methuen, pp. 57–61.

[36] Useful discussions of the downtown booster lobby include: H. Molotch (1976) 'The city as a growth machine: toward a political economy of place.' *American Journal of Sociology* 82, 2: 309–32; and T. K. Barnekov and D. Rich (1977)

'Privatism and urban development.' *Urban Affairs Quarterly* 12, 4 (June): 431–60.

[37] Mollenkopf 1976. See also C. Stone (1976) *Economic Growth and Neighborhood Discontent.* Chapel Hill: University of North Carolina Press; and C. Hartman *et al.* (1974) *Yerba Buena: Land Grab and Community Resistance.* San Francisco: Glide Publications.

[38] D. Beagle, A. Haber and D. Wellman (1971) 'Rapid transit: the case of BART', in David M. Gordon (ed.) *Problems in Political Economy: An Urban Perspective.* Lexington, Mass.: D.C. Heath, pp. 437–9.

[39] Consider, for example, Gaffney, who has described local government (LG) as
' . . . a group of landowners in league to preside over the collective capital that they use jointly. The LG is a halfway house between the individual landowner and the state. Landowner control is modified by democracy, which gives the whole system some of its characteristic tensions and compromises. But landowners, as the permanent party of every LG, take a strong and steady interest in local government out of proportion to their numbers. It is reasonably accurate for many purposes to think of the LG as a collective landowner, maximizing land income.' (Mason Gaffney (1973) 'Tax reform to release land', in M. Clawson (ed.) *Modernizing Urban Land Policy.* Washington, DC: Resources for the Future, p. 117.)
Substantial documentation of the close involvement of property capitals in Canadian local politics is provided in J. Lorimer (1972) *A Citizen's Guide to City Politics.* Toronto: James Lewis and Samuel.

[40] On some of the uninternalized costs of urban renewal see A. Downs (1970) 'Uncompensated nonconstruction costs which urban highways and urban renewal impose upon residential households', in J. Margolis (ed.) *The Analysis of Public Output.* New York: Columbia University Press, pp. 69–106. For a documentation of struggle around the uninternalized costs of urban renewal see C. Hartman and R. Kessler (1978) 'The illusion and reality of urban renewal: San Francisco's Yerba Buena Center', in W. K. Tabb and L. Sawers (eds) *Marxism and the Metropolis.* New York: Oxford University Press, pp. 153–78.

[41] Paul Ylvisaker has recognized the problem of compensation in this context but in terms which make it an issue between fractions of labor rather than between capital and labor. In commenting on suburbanization he has written:
'What is worse by my reckoning is that they (those moving to the suburbs) don't have to pay the costs of their choice. They can leave for greener pastures and (for a while) lower costs – whether it is the suburbs or the far reaches of the metropolitan area, or the Sunbelt or whichever – they can go there without paying the costs of what they leave behind.
We have not set up a public accounting system that forces them to pay the full measure of the choices they make.' (P. Ylvisaker (1977) 'Some political difficulties for city savers', in *Toward a National Urban Policy:* Subcommittee on the City of the Committee on Banking, Finance and Urban Affairs, House of Representatives, 95th Congress, 2nd Session, Washington, DC: USGPO, p. 83.)

[42] For some sense of the physical scope of some of the larger developments see Rabinovitz and Lamare 1971.

[43] E. J. Kaiser and S. F. Weiss (1975) 'Public Policy and the Residential Development Process', in S. Gale and E. G. Moore (eds) *The Manipulated City*. Chicago: Maaroufa Press and London: Methuen, p. 195.

[44] Public infrastructural investments reinforce this agglomerative tendency. Under pressure from developers and land companies with investments in particular areas, water lines and sewer lines are extended in certain localities sparking off development. Thus a large speculative shopping center may gain a large sewer and water line extension which stimulates development between it and the built up edge of the city. As development proceeds and congestion emerges as a significant problem so the same localities will become priority areas for highway widening, construction of new freeway ramps, expansion of sewage pumping stations, etc. As Roweis and Scott have shown, such increases in infrastructural capacity tend to stimulate further rounds of new development and, ultimately, additional investments in public works. The common assumption of planners that existing trends will continue into the future adds further thrust to a congestion-relieving bias in the spatial allocation of public expenditures (Roweis and Scott 1978: 64).

[45] M. Clawson and P. Hall (1973) *Planning and Urban Growth*. Washington, DC: Resources for the Future, p. 49.

[46] The prevalence of corruption is well documented. See, for instance: A. Balk (1966) 'Invitation to bribery.' *Harper's Magazine* (October); and L. Downie, Jr. (1974) *Mortgage on America*. New York: Praeger, pp. 90–2. On the market incentives to corruption see D. G. Hagman (1974) 'Windfalls for wipeouts', in C. L. Harriss (ed.) *The Good Earth of America: Planning Our Land Use*. Englewood Cliffs, NJ: Prentice-Hall, p. 114; also see W. J. Stull (1974) 'Land use and zoning in an urban economy.' *American Economic Review* 64, 3 (June): 347.

[47] M. B. Katz (1978) 'Origins of the institutional state.' *Marxist Perspectives* 1, 4 (Winter): 6–22.

[48] If one remains sceptical a brief perusal of the arguments presented by capital in Senate Committee hearings should dispel illusion.

[49] An instance recently identified in the literature is the use of the idea of filtering and 'trickle down' to legitimate the tendency of the private housing industry to focus on provision for the more affluent. See M. Boddy and F. Gray (1979) 'Filtering theory, housing policy and the legitimation of inequality.' *Policy and Politics* 7, 1 (January): 39–54.

[50] My treatment of the 'opening up the suburbs' controversy in this section is based heavily upon Danielson (1976), Chapter 6.

[51] As an example of the numerous ideological tracts written and adopting this viewpoint see A. Downs (1973) *Opening Up the Suburbs*. New Haven and London: Yale University Press.

[52] Danielson 1976, Chapter 6.

[53] The improved provisions for relocation of the 1970 Uniform Relocation and Real Property Assistance Act are significant in this regard. See, Congressional Research Service (1973) *The Central City Problem*. Washington, DC: USGPO, pp. 58–9.

[54] Downs 1973: 163.

[55] See J. Margolis (1974) 'Public policies for private profits: urban government',

in H. Hochman and G. Peterson (eds) *Redistribution Through Public Choice.*
Washington, DC: Urban Institute, Chapter 11. Also see G. J. Stigler (1971)
'The theory of economic regulation.' *The Bell Journal of Economics and Management Science* 12, 1 (Spring): 3–21.

56 For an intriguing case study of the implications of this see J. Barr (1969) 'Divided
city.' *New Society* (8 May): 701–4.

57 On the role of technical expertise in the formulation and legitimation of city
policy see W. H. Cox (1976) *Cities: The Public Dimension.* Harmondsworth:
Penguin Books, Part 4; and J. G. Davies (1972) *The Evangelistic Bureaucrat.*
London: Tavistock Publications.

58 R. Lichtman (1975) 'Marx's theory of ideology.' *Socialist Revolution* 5, 1 (April):
56.

59 J. Anderson (1973) 'Ideology in geography: an introduction.' *Antipode* 5, 3
(December): 3.

60 P. Marris (n.d.) 'The Ideology of Human Settlements.' Unpublished ms., Department of Architecture and Urban Planning, University of California at Los Angeles.

61 An excellent instance of this viewpoint is contained in a recent paper by the
sociologist J. R. Logan:
    'In conflicts over boundaries, constitutional powers, allocations of public
    resources, taxation policies, land use controls, etc., places compete for
    development outcomes which would maintain or improve their relative
    position in the hierarchy of places. More precisely coalitions of local
    interests – recruited and organized along territorial lines determined by
    political boundaries – compete for outcomes in which coalition partners
    have a mutual interest, even when at another level their interests may
    diverge . . . My point here is that, by providing a communality of interests among internal groups, place accounts for political behavior which
    cannot be understood in terms of class conflict (e.g. the cooperation of
    banks, municipal unions, and city government to forestall bankruptcy
    of New York City).' (J. R. Logan (1978) 'Growth, politics and the stratification of places.' *American Journal of Sociology* 84, 2: 409).
But Logan is by no means alone in this conception; an example in geography
is provided by a recent textbook of my own in which I apply the idea of spatially-
based competing communities not only to the urban realm but also to regional
and international scales: K. R. Cox (1979) *Location and Public Problems: A
Political Geography of the Contemporary World.* Chicago: Maaroufa Press and
Oxford: Blackwell.

17 Homeownership and the capitalist social order

J. A. Agnew

The intellectual and political problem of how different 'types' of society (e.g. feudal, capitalist) are maintained over time is of long-standing interest to social scientists. Prevalent theories in American social science appear to require acceptance of one of two mutually exclusive positions: widespread societal consensus concerning the nature of a social order ('social order' refers to the social context for individual action) (e.g. Parsons 1951) or overt coercive dominance by an élite group (e.g. Mills 1956a; Domhoff 1967).[1] A viewpoint focusing on the historical and practical basis for the *dominance* of a given social order, however, may be more accurate for some societies. For example, contemporary societies in which market exchange relationships predominate in everyday life ('capitalist' societies) can be characterized as having both coercive institutions and fairly widespread consensus concerning their social order (Polanyi 1944; Baldus 1977). Following Marx several writers have argued for the importance in these societies of a process of 'practical incorporation': the expansion of commitment to the prevalent social order by the development of personal stakes in its survival (e.g. Israel 1972; Macpherson 1962). Recent research suggests, however, that the extent of incorporation differs between capitalist societies, that it is far from complete in any of them, and different social groups are incorporated to different degrees (e.g. Parkin 1971; Moorhouse and Chamberlain 1974; Sallach 1974; Abercrombie and Turner 1978).

The object of this paper is not to document the net extent of incorporation in various capitalist societies. Rather, I want to examine the specific proposition that widespread homeownership is a means of practical incorporation in such societies. The aim is to offer an evaluation of the process of incorporation in a concrete context from everyday life.

The paper commences with a consideration of the link between everyday life, individual consciousness and the capitalist social order. I then move to an analysis of the extent and importance of homeownership in capitalist societies. The third section explores the role of homeownership as a means of incorporation in capitalist societies. A final section offers some tentative conclusions.

Everyday life and the capitalist social order

In a capitalist society relations between people are based largely upon 'mutual exploitation' (Ollman 1971: 209) where men do not view themselves and others as anything other than private and competing units for scarce resources. In this connection Marx and Engels (1942: 58) wrote of the growth of what they called the 'personal life.' This is the development of a view of life in which the individual believes that he is affected by and affects no one except insofar as he competes with others, is in fact a socially isolated and apparently autonomous self, and conceives of himself apart from his socio-economic ties to other people (Brittan 1977). The practice of the personal life by masses of people serves to anchor them to the capitalist social order by legitimizing in their everyday lives the privatistic and competitive basis of capitalist social relations, provides a ready population of workers and consumers, and effectively retards any restructuring of the social order on some alternative basis. But how does such a practice come about?

In the *Grundrisse* (1953) Marx argued that an essential characteristic of capitalist society is that all activities and products are of no value for an individual if they cannot be used in exchange for other activities or products. This arises because under the division of labor people become independent of and indifferent to one another in their personal relations, yet are impersonally dependent upon one another as a consequence of the complementary nature of their economic activities. Thus the

> 'mutual and comprehensive dependence of individuals, who are indifferent to one another, forms their social context. This social context is expressed in the *exchange-value*, through which for each individual his own activity or product is first transformed into an activity or product; the individual must produce a general product – the *exchange-value*.' (Marx 1953: 74)

The inter-personal separation and indifference engendered by the division of labor, then, necessitates the creation of an objective and quantifiable measure of value – the exchange-value.

Marx maintains that the transformation of all activities and products into exchange-values leads to the neglect of human, social relations as man becomes a means in the economic process. 'In the exchange-value the social relations of persons are transformed into the social conduct of objects' (Marx 1953: 75). This process has significant effects upon man's self-image and self-evaluation. Marx argued that one of man's basic needs is the need for self-evaluation and self-estimation and that this need is mediated by the evaluations of other people known to the individual (Israel 1971). Yet in capitalist society, both at work and outside the labor process, social relations have become reified as objective, impersonal conditions of everyday life. Little real basis for a self-evaluation mediated by the evaluation of others exists. As a consequence, people turn to the things they own or use as a means of self-evaluation (Marx 1953). When these things are relatively scarce and appreciated by others, and one can tell by looking around, status is endowed and the basic need of self-evaluation is satisfied. However, the objects which endow status have a certain exchange-value.[2] They acquire use-value as status-objects through their exchange-value. In other words, one needs to command exchange-value to obtain status-value. In everyday life, therefore, individuals evaluate themselves and are evaluated by others through objects which have a certain exchange-value and status-value.[3] The quest for self-sufficiency and the expression of identity through objects are, then, the marks of a capitalist social order. For Marx the logic of private property has been extended to all human relations and, to quote Wikse (1977: 33), 'ownership [of self and objects] is the authentic metaphor for identity.'[4]

The process of reification that Marx emphasized has probably become deeply rooted. People often come to accept what exists as necessarily legitimate – not consciously as a result of moral reflection but unthinkingly as a result of continuous involvement in everyday life. Thompson (1967: 60) offers an example from nineteenth century England:

'The first generation of factory workers were taught by their masters the importance of time; the second generation formed their short time committees in the ten-hour movement; the third generation struck for overtime or time-and-a-half. They had accepted the categories of the employers and learned their lesson, that time is money, only too well.'

Over several generations, then, there was an increasing acceptance of certain practices and an active framing of expectations in terms of an increasingly dominant set of meanings and values. When this final 'sense of reality' (Williams 1973: 9) comes to guide conduct it is difficult to see the original process of reification which lies masked behind it. In everyday life, therefore, our personal behavior is based upon acceptance of the world as we immediately experience it. There is a relatively unconscious incorporation into practice of the personal life which characterizes capitalist social relations.

Some recent writing, however, suggests that commitment to dominant values may be far from complete in any capitalist society and that although there may be a 'dominant ideology' the value systems of many people are underdeveloped, fragmented and inconsistent (Sennett and Cobb 1972; Livingstone 1976; Parkin 1971; Williams 1973). In the first place, at the level of the individual the determinacy between material existence and personal consciousness is problematic. C. Wright Mills (1962: 114) has maintained that:

'In a satisfactory model of social structure we must allow a considerable degree of autonomy to the formation and role of ideas. We must trace the ways in which ideas are related to individuals and to institutions with more sophistication than Marx was able to achieve in his general model.'

Certainly, we should note the endurance of 'pre-capitalist' patterns of thought which coexist with and contribute to dominant ideologies and 'emergent' ideologies which challenge the existing social order in all capitalist societies (Gramsci 1971; McNeill 1963; Williams 1973). Also the role of certain institutions, such as the family, schools, churches and mass media, in propagating values favorable to the dominant ideology of the personal life in a sense independently of the process of individual reification cannot be ignored (Gramsci 1971; Williams 1973).

In the second place, people do not experience life in the abstract context of 'mass society'. Their knowledge is acquired and they live their lives in the context of 'social worlds', in which meaning is attributed to acts and events through communication and interaction with rather limited numbers of people. In everyday life such social worlds provide the setting for definition of material needs and the identification of status-objects. The spread of the practice of the personal life, therefore, can be either enhanced or impeded depending on the relative insulation of social worlds from outside influences.

Marx was aware of the essentially social process whereby human needs are defined (Marx 1962: 93–4; Friedman 1974: 326–7). For national societies he also argued that needs can be characterized as being more or less based upon capitalist social relations (Avineri 1970: 157–60). Hence, we can expect less of a commitment to the personal life in those societies and social worlds which have maintained pre-capitalist or developed what Marx would have regarded as post-capitalist values and beliefs.

In the third place, and finally, survey research indicates that across a range of social groups in both North American and European contexts there is considerable muddle and inconsistency about values and beliefs (Sennett and Cobb 1971; Lane 1962; Blackburn and Mann 1975; Lipsitz 1974). Among American and British manual workers Lipsitz (1974) and Blackburn and Mann (1975) found their respondents generally unable to articulate clear interpretations of various social issues in *any* ideological context. The limited amount of research into the values and beliefs of such groups as national political leaders and corporation executives also points to the existence of an 'ambiguous' rather than clearly ordered personal consciousness (e.g. Newman 1975; Davies 1966; Luttbeg 1968; Brown 1970). Conclusions about 'élite' groups, however, should be tempered by the knowledge that some research has found them to have less fragmented, more consistent belief systems (Converse 1965). Mann (1970: 435) argues that available evidence suggests that 'only those actually sharing in societal power need develop consistent values' and that fragmentation and inconsistency in belief systems on the part of 'mass publics' help ensure the continuance of the social order and prevent the emergence of 'oppositional' ideologies (but see Moorhouse and Chamberlain 1974).

In summary for this section, we can expect contemporary capitalist societies to be characterized by fairly widespread incorporation through the practice of the personal life. However, this practice and its necessary underpinnings, indifference to others and reification, may be affected by the existence and practice of antithetical ideas, ideological fragmentation which may or may not inhibit attachment to the capitalist social order, and the social definition of needs which may in different social settings retard or enhance the whole-hearted practice of the personal life.

The extent and importance of homeownership

The need for shelter is a basic element in everyday life. Yet there is a tremendous difference between societies in the ways in which housing is provided and the types of houses that are built (Abrams 1964). Even within a group of industrialized capitalist societies there is considerable heterogeneity in forms of tenure and methods of provision (Nevitt 1967; Abrams 1964; Kemeny 1978). However, in each of these societies, to either a greater or lesser extent, a sizeable proportion of the total housing stock is owner-occupied (Kemeny 1978: 43). This owner-occupancy differs from that found in largely peasant societies in being typically based upon a complex financial transaction which usually involves the purchaser in long-term indebtedness and *chosen* as an alternative to renting. Until the 1920s renting characterized the major urban areas of most capitalist societies. A major focus of the next section of this paper is on the implications of widespread homeownership for the capitalist social order using a range of information collected by researchers in Britain, the United States, France, Sweden and Australia. The aim is to understand the extent to which homeownership incorporates people into capitalist societies in the context of the argument presented in the previous section. In the present section the extent and importance of homeownership in capitalist societies is discussed with particular reference to Britain and the United States for which more extensive data are available than many other countries.

With respect to the relative extent of homeownership, Table 17.1 shows the considerable range in the proportion of owner-occupied dwellings within a small group of capitalist societies. Of significance in Table 17.1 is the lack of any strong correlation between the incidence of homeownership and per capita GNP, an indicator of the material standard of living. The wealthier countries are not inclined to have more homeownership than the poorer ones. Also of some importance, there is no general tendency for the proportion of homeowners to increase over time. This is not true for Britain and the United States, however, where homeownership has grown steadily during the period indicated.

Though Britain and the United States are similar, then, in having increasing rates of homeownership they do differ somewhat in the total extent of homeownership and the relationship between income and owner-occupancy. In the United States in 1974 65 per cent of all households were living in houses they owned or were buying. Moreover, although low-

Table 17.1 Selected capitalist societies by percentage of homeownership and rank order of per capita national income

	Percentage of homeownership		Average annual inter-censal change (%)		Rank order per capita GNP (1970)
	1970–71	*Previous census*	*Increase*	*Decrease*	
Iceland	70.3[a]	not available			12
Australia	68.7	70.8 (1966)		−0.42	8
New Zealand	68.1	68.9 (1966)		−0.18	14
USA	62.9	61.9 (1960)	+0.10		1
Canada	60.2	65.4 (1967)		−1.35	3
Belgium	55.9	49.7 (1966)	+1.55		11
UK	50.1	47.8 (1966)	+0.46		16
France	43.3[b]	42.7 (1962)	+0.10		10
Sweden	35.0	35.5 (1965)		−0.10	2
W. Germany	34.3[b]	29.4 (1961)	+0.70		5
Switzerland	27.9	33.7 (1960)		−0.58	4

Source: J. Kemeny 1978:43. National census sources were used when UN data were unavailable.

[a] 1960

[b] 1968

income households are less likely to be homeowners, there are nevertheless relatively large numbers of low-income homeowners (see Table 17.2). In Britain in 1972 51 per cent of all households were living in houses they owned or were buying. However, evidence from a number of surveys suggests that in Britain homeownership has not spread to all strata of society as it has in the United States (Murie 1974). Data from the General House-

Table 17.2 Homeownership by income of households in the United States, 1974

Income group	*Total households (in thousands)*	*Per cent owners*
$		%
0– 4,999	15,653	48
5,000– 9,999	16,875	55
10,000–14,999	15,878	68
15,000–24,999	15,519	79
25,000 or more	6,907	87
Total	70,832	65
Median income	$10,831	$12,800

Source: Congressional Budget Office (1977), *Homeownership: The Changing Relationship of Costs and Incomes, and Possible Federal Roles.* Washington, DC: US GPO, Table 2, p. 9.

Table 17.3 Homeownership and renting from public authorities by socio-economic group of head of household in Great Britain, 1971

Group	Per cent total households (sample = 11,664)	Per cent owners (5,749)	Per cent renters from public authorities (3,611)
	%	%	%
Unskilled manual	7	21	56
Semi-skilled manual	19	31	44
Skilled manual	33	45	38
General non-manual	20	59	20
Employers and managers	15	74	10
Professional	4	85	3
Others	3	45	17

Source: A Murie *et al.* 1976:22–23. Based on HMSO (1973) *The General Household Survey: Introductory Report*, Table 5.18. London: HMSO, p. 109.

hold Survey of 1971 (Table 17.3) indicate that manual workers have not shared as greatly in the expansion of ownership as have other groups and that the availability of rental housing from public authorities may be partly responsible for the lack of 'penetration' (also see Clawson and Hall 1973; Crawford 1975: 12; Clarke 1976; Cox and Agnew 1974).

Other data suggest that within Britain there is also considerable regional variation in the extent of homeownership with Scotland having a low of 30 per cent and south-west England a high of 58 per cent (Murie *et al.* 1976: 9). This pattern is partly related to the availability of public housing. In the United States there was, as of 1960, much less regional variation (Beyer 1965: 139). The suburbs of American metropolitan areas, however, are marked by a considerable preponderance of owner-occupation that stands in contrast to the lower, if recently increasing, proportion of home-owners in central cities (Table 17.4). Clawson and Hall (1973) intimate that suburban housing in England, particularly that built since the 1950s, is also preponderantly owner-occupied.

Table 17.4 Homeownership in US central cities and suburbs, 1970–75

	Per cent of households owning their own home			
	1970	1973	1974	1975
	%	%	%	%
Central cities	48.1	49.3	49.6	49.6
Suburbs	70.3	70.8	70.9	71.0

Source: Senate Committee on Banking, Housing, and Urban Affairs (1977), *Hearings on Problems of Dislocation and Diversity of Communities Undergoing Neighborhood Revitalization Activity*. Washington DC:US GPO, p. 144.

Table 17.5 Type of housing owner-occupied in the United States, 1970

Type of housing	Total number of dwellings (in millions)	Per cent owned
		%
Detached single-family houses	44.8 (66%)	83
Attached single-family houses (including townhouses)	2.0 (3%)	56
Mobile homes	2.1 (3%)	85
Multi-family dwellings	18.8 (28%)	17

Source: *Census of Housing: 1970 Detailed Housing Characteristics.* Washington, DC: US GPO, 1972.

The term 'homeownership' stands in need of further clarification. In particular, what is it that is owned? In the United States the bulk of owner-occupied dwellings are detached single-family homes (Table 17.5). Nevertheless the category homeownership in the United States does cover a wide range of dwellings – from plush mansions to austere tract and mobile homes. In Britain the detached house is much less characteristic and owner-occupancy tends to be associated with semi-detached (duplex) and terraced (row) housing (Table 17.6). In sum, owner-occupied houses in Britain tend to be at higher density and are more likely to be joined to other houses than are owner-occupied houses in the United States.

Access to a specific type of house, then, varies as a function of what is available. Clearly, in different countries land availability, construction methods and styles, the rate of new building, building codes, zoning regula-

Table 17.6 Type of housing owner-occupied in Great Britain, 1971

Type of housing	Per cent total dwellings (sample = 11,823)	Per cent owned	
		Outright (sample = 2,633)	Mortgaged (sample = 3,185)
	%	%	%
Detached house	16	30	25
Semi-detached house	33	29	45
Terraced house	30	30	24
Flat/maisonette (purpose built)	13	4	3
Other flat/rooms	6	4	2
Other	2	3	1

Source: A. Murie *et al.* 1976:13. Based on HMSO (1973) *The General Household Survey: Introductory Report*, Table 5.12. London: HMSO, p. 99.

tions and, over time, the 'preferences' of people for types of housing differ and determine what is available for purchase. However, the purchase, as opposed to the renting of houses, is itself also contingent.

In 1890 48 per cent of occupied housing units in the United States were occupied by owners. As a result of the Depression during the 1930s the proportion declined to a low of 44 per cent in 1940 (Beyer 1965: 119). By 1974 64 per cent of all housing units were occupied by their owners. There is then a fairly lengthy history to homeownership in the United States. Bodfish (1933) has argued, however, that fairly widespread homeownership only became possible with the development of building and loan organizations in the period immediately before the Civil War. Indeed the rise of urban homeownership is closely associated with the increased use of the mortgage as a credit instrument for home purchase. Between 1890 and 1940 the proportion of owner-occupied nonfarm homes which carried mortgages increased from 27.7 per cent to 45.3 per cent (Dean 1945: 69). The Depression experience of financial collapse and fore-closed mortgages brought drastic change in the mortgage system through government intervention. This intervention has had the net effect of vastly expanding the size of the population that can afford to buy homes (Winnick 1958: 45). Concomitantly, a huge residential real-estate industry has developed with a vested interest in expanding ownership (see Stone 1975; Harvey 1977). More specifically, construction is the largest industry in the United States and building single-family homes is the largest category in construction, real-estate brokering is the nation's largest licensed occupation, and more than half the assets of all banking institutions are mortgage or construction loans for 'homebuilding' (Mayer 1978: 6).

In Britain the growth in homeownership came later and more dramatically. In 1919 less than one-tenth of households were owner-occupiers. By 1947 the proportion had risen to over 25 per cent and in 1971 stood at just over 50 per cent (Murie *et al.* 1976: 139). This dramatic increase in the rate of homeownership is, as in the United States, related to the growth of building societies as depositories of savings and lenders on mortgages. In the years since the Second World War government policies have encouraged homeownership as an end in itself and in order to stimulate the construction industry and through it the national economy. In this way a residential real-estate industry much smaller but in some ways similar to that in the United States has grown up in Britain (Murie *et al.* 1976; Clawson and Hall 1973).

It would be a mistake, however, to credit the growth in homeownership in either the United States or Britain *solely* to government policies or the development of a real-estate industry-government axis. Even in Britain where the increase in homeownership is more recent and closely tied to government subsidy policies such as homeowner mortgage interest payment claims against income tax and local authority mortgage lending, economic historians note the growth of owner-occupancy prior to the First World War (Hobson 1953; Jackson 1973). For instance, Jackson (1973: 35) writing of London's fringe areas in the 1900–14 period maintains that:

'most new property was still built for letting [renting], but there was a growing tendency to erect speculatively for sale. In some measure this was a response to relaxations by certain building societies, but it also reflected the increasing prosperity and size of the middle classes.'

Whatever the outcome of debates over the role of government policies in stimulating homeownership, however, it is clear that owner-occupancy has become a major element in the everyday lives of millions of people in capitalist societies such as the United States and Britain. Moreover, survey evidence points out that many renters in both the United States and Britain would prefer to own if they could, or intend to own when they feel they are able (Cullingworth 1969; Ineichen and Brown 1977; Mayer 1978; Lansing *et al.* 1969). Rather than simply a desire for ownership, however, this may also reflect a preference for the types of housing, such as detached single-family dwellings, which are difficult to obtain other than by purchase (Rosow 1948; Michelson 1968; Michelson *et al.* 1973; *Professional Builder* 1975; Handlin 1972; Harris 1973).

Homeownership and the capitalist social order

A basic premise of the practical incorporation thesis outlined earlier is, to quote Macpherson (1973: 191), that

'As an institution, property – and any particular system of property – is a man-made device which establishes certain relations between people. Like all such devices, its maintenance requires at least the acquiescence of the bulk of the people, and the positive support of any leading classes.'

In the previous section we have seen how there has been a dramatic growth in the phenomenon of homeownership in some capitalist societies, particularly Britain and the United States. The purpose of the present section is to examine the incorporation thesis with respect to homeownership by exploring (i) the links between homeownership and practice of the personal life and (ii) the relationship between homeownership, the personal life and the growth of local or community consciousness. As a working hypothesis it seems reasonable to postulate that homeownership creates a circumstance in which the capitalist ethic of 'possessive individualism' can develop and flourish and, further, that in combination with the uncertain conditions produced by urban land markets, homeownership encourages a consciousness of local events that effectively precludes much in the way of a 'larger' social consciousness. Evidence from several capitalist societies, but particularly the United States and Britain, is employed in order to offer a wider base for generalization and investigate the caveats to the incorporation thesis outlined earlier.

Homeownership and the personal life

There are two major ways in which homeownership can contribute to practice of the personal life: first, the possession of a house offers a major physical object for use as an indicator of status and source of personal autonomy and, second, the house is an exchange-value insofar as it is a commodity that can be bought and sold. To the extent that it is 'commodified' – regarded as an investment – the past labor which went into building the house is masked as personal capital. This capital serves as a 'guarantee' of personal independence and future security. Each of these contentions is examined separately and in turn.

STATUS AND PERSONAL AUTONOMY   There is a considerable and growing literature on the house as a status-object and the key role of tenure in the allocation of status (e.g. Cooper 1972; Handlin 1972; Duncan 1973; Perin 1977; Rakoff 1977; Seeley *et al.* 1963; Kemeny 1978). In her study of social order and land use in the US, Perin (1977) argues for the importance of a financial basis to social obligation as the key element in American definitions of status. She sees homeownership as vital to this process and writes that (1977: 66):

'. . . although the single-family house is itself a possession and an asset, often enough a concrete symbol of higher income, it is less that fact which confers the high status culturally and more the achievement of a social relationship with the banker. That is, the banker "qualifies" the homebuyer with a credit rating that is a major threshold of American social personhood crucial in the correct traversal of the ladder of life . . . Being "able to own" is a threshold criterion of social personhood that renters, by definition, do not meet; they partake of less citizenship and on that account have lower status.'

One's status then, is determined in large part by the long-term debt represented in homeownership. Renters having an impermanent legal tie have lower status. Why should this be so? Perin maintains that long-term debt enhances certainty in the housing and banking industries whereas renting involves short-term leases and limited financial obligations which create unpredictability. Concomitantly, when there is nothing owing there is no social tie. Sahlins (1965: 177–8) expresses this argument in general terms as follows:

'In so far as the things transferred are of different quality, it may be difficult ever to calculate that the sides are "even-steven". This is a social good. The exchange that is symmetrical or unequivocally equal carries some disadvantage . . . it cancels debts and thus opens the possibility of contracting out. If neither side is "owing" then the bond between them is comparatively fragile. But if accounts are not squared, then the relationship is maintained by virtue of "the shadow of indebtedness", and there will have to be further occasions of association, perhaps as occasions of future payment.'

Tenants as members of a transient and debtless social group, therefore, become 'pariahs.' They are not integrated into society through the threat of mortgage foreclosure, debt-commitment and acquisition of social esteem associated with homeownership. They are an out-group who are unsettled and unsettling. Owners to the contrary have invested themselves in their houses through debt and obtained the social esteem and the 'freedom' from landlords that renters cannot acquire.

Other research has focused more on the nature of the house itself as a possession and its consequent role as a source of personal autonomy. Rakoff (1977), for instance, provides considerable empirical evidence for the 'meanings' that are attached to owner-occupied houses by their owners. Although he finds conflicting meanings and a variety of experiences there is a shared emphasis (p. 101) ' . . . on the importance of the individual's unceasing search for a realm of personal control in a world where he or she generally feels impotent.' The single-family detached house with its greater isolation and insulation from others was a particularly appreciated symbol of self-sufficiency and personal autonomy. In general, however, the *owned* house symbolized personal control in two ways (p. 102):

> 'First, having control of one's own private space gave people a feeling of freedom from the control and intrusion of others. . . . Second, and more importantly, people felt that by being in control of their own private space, they had the power and opportunity to make something of themselves, to be "more of an individual", to achieve a kind of self-fulfillment.'

In owning a house, therefore, people both provide a means for communicating their identity as autonomous individuals and offer a 'meaning-contribution' which represents the practice of the personal life.

It is apparent from the writing of Perin (1977) and Rakoff (1977), and those they reference, that there is something of a consensus among researchers in the United States concerning the strong association between owning and designations of the house as a status-object and source of personal autonomy. However, neither Perin and Rakoff nor many other writers on the topic, would deny that 'incorporation' through homeownership as a symbol of status and source of personal autonomy is rarely complete, has required considerable stimulation and even need not prevail. Indeed, Rakoff notes a tension in the responses of his informants between their apparent egoism and a desire for what he terms ' . . . the lost fraternity of old family networks and neighborhood ties.' In this he sees a parallel with the individualism-sociality contradiction in everyday life noted by Sennett and Cobb (1972) in a study of American blue-collar workers. Other writers have noted the importance of government programs and housing industry propaganda in creating the context for the growth of a homeownership ideology. Dean (1945) and Perin (1977) both note the pervasiveness of public sentiment in favor of homeownership and the constant efforts of homebuilding interests to promote homeownership. Finally, some research has indicated that homeownership motives are fairly complex and may not always have a strong basis in status considerations (e.g. Berger 1960: 84; Rosow 1948). In particular, scarcity of rental opportunities can 'force' people into owning and they may not regard ownership as being superior to renting (Dean 1951).

Outside of the United States homeownership as a means of allocating status and source of personal autonomy has not been examined to any great extent. However, there is some literature which is suggestive of the approximate extent if not the precise nature of these relationships (e.g. Kemeny 1978; Mellor 1978; Hamilton 1967; Zweig 1961; Chapman 1955; Runciman 1966). Runciman (1966) and Zweig (1961) suggest that in Britain there has been a progressive 'nationalization' of status allocation criteria over time such that by the 1960s there was across all classes a fairly uniform status structure based on consumption. In particular, they note the development of egoistic *aspirations* of status on the part of working-class families. Homeownership provides a means of channeling these aspirations and both Zweig (1961) and Runciman (1966) observe a shift towards a 'house-centered' existence. Runciman (1966: 136) writes:

' . . . it has been noticed that the sense of community, and the frequent public, although not intimate, social contacts which characterize it, have given way to a greater reserve and homecentredness and an increasing awareness of distinctions of status within the working class itself.'

Studies by Mogey (1956), Young and Willmott (1957), and Ineichen (1972) point towards the same conclusion and recent research by Roberts *et al.* (1977: 56–8) suggests that homeownership is important in the adoption of a 'prestige imagery' based upon consumption.

The extent to which a uniform status structure focused upon homeownership has developed in Britain, however, is open to question. For example, research reported by Couper and Brindley (1975) discovered a range of 'housing values.' Many of their survey respondents rejected homeownership as an ideal tenure and instead preferred renting. Zweig (1961) and Young and Willmott (1957) also report a variety of housing values from their research. In part this range of values seems to be related to the security of tenure associated with renting in Britain (compare Perin 1977, on the USA). Kemeny (1978) has argued that the type of landlordism one finds in different countries, particularly the extent of government provision and scale of private landlordism, determine the relative security and attractiveness of renting. In Britain, and certain other societies such as France and Sweden, there is apparently much less social stigma attached to renting than is the case in Australia and the United States where homeownership is the *only* really secure form of tenure (see Kemeny 1978; Hamilton 1967; Popenoe 1977; Rosow 1948; Perin 1977; Winnick 1958: 45; Dean 1945). However, in both France and Sweden there is evidence that the idea of owning appeals despite the difficulty of financing homeownership (Kemeny 1978: 53; Todd 1978). In Britain the nationally strong Labour movement, in the form of the Labour Party and its trade-union allies, has been committed to public ownership and renting of housing. Although in recent years the Labour Party has been an active supporter of homeownership (Nevitt 1978) its historical emphasis on public housing has undoubtedly contributed an ideological justification and argument for renting as opposed to owning.

Among homeowners in Britain research has shown a wide spread of meanings attached to ownership (e.g. Couper and Brindley 1975). Only a minority of respondents to the Couper and Brindley survey (30 per cent) valued ownership mainly for autonomy and independence. Moreover, there was no consensus concerning the ranking of types of housing owned (detached, terraced, etc.). This variety of meanings would seem to indicate the existence of housing reference groups or social worlds in which different definitions of the meaning of homeownership are arrived at. Research into the housing values of Asian and West Indian immigrants to Britain reinforces this conclusion (Burney 1967; Deakin 1969). Whereas West Indians tended to see owning a house in terms of its value for keeping possessions in a 'contained' space, Asians saw owning in terms of its status-value and as a source of profit (Deakin 1969). Finally, it is apparent that in Britain attachment to homeownership can co-exist with a variety of antithetical commitments. For example, there is evidence that homeownership does not correlate highly with commitment to established political and economic institutions, *ceteris paribus* (Jessup 1974: 197–9). This is also the case in France (Hamilton 1967: 160–85).

EXCHANGE-VALUE   Buying a house is an expensive proposition for most people who undertake it. Yet in capitalist societies in order to obtain use-values (including status-values) one must command exchange-values. So a house, any house, requires considerable financial investment and takes on a value as a commodity. As such it can become a potential source of profit and in the form of built-up equity a source of financial security.[5] Sternleib (1972: 39) has made this point as follows:

> 'For all but the most affluent in our society, a house is not only a home, it is typically a major repository of capital investment and stored equity. As any imaginative architect will testify, houses are purchased to be sold not to be lived in. Their ultimate sale represents the edge which makes Social Security and Old Age Pensions endurable.'

A consequence of homeownership other than its status-value and contribution to personal autonomy, therefore, is its role as an investment. Once involved in owning and as it consumes a large proportion of household income, the home takes on a distinctive financial aspect.

Empirical research into the 'house as an investment' is sparse and mainly American in focus. Available evidence, however, suggests that homeownership does not invariably entail treatment of the house as an investment (Wertham *et al.* 1965; Coons and Glaze 1963; Couper and Brindley 1975; Perin 1977; Agnew 1978).

In an extensive study of house purchase decisions in several Californian communities Wertham *et al.* (1965) found that their informants held two concepts of investment. The first of these was 'making a profit' and the second was 'protecting the equity.' All respondents accepted the notion of their houses as investments in the second sense and those buyers who were 'lower middle' and 'working class' also tended to express interest in their houses as investments in the first sense. Rather than a simple reflection of 'poorer' homeowners having a greater commitment to the profit potential of their houses, however, this also reflected the greater possibilities for profit in the local settings in which the 'poorer' respondents were concentrated (Wertham *et al.* 1965: 137–7).

Perin (1977) in her study of 'real estate professionals' also emphasizes the pervasiveness of an investment posture on the part of American homeowners. They cannot avoid becoming caught up in the basic goal of the real-estate industry of which they must, sooner or later, become a part: to realize a profit. She argues that housing consumers are also major housing producers when the used-property market represents 70 per cent of total housing activity as it did in 1976. Her informants made clear to her that (1977: 129):

' ... whenever home buyers move and whenever they think of selling for any reason, they expect to recoup their equity in order to use it in buying the next house, they expect the sale to cover the costs of the transaction, and always, they hope to realize some profit or a fair return on their investment.'

Recent newspaper and business magazine articles also report on the particular attraction of the house as an investment in an era of inflation and provide an image of many homeowners as small-scale speculators. For example, the *New York Times* (11 June 1978) reported the case of a man who at forty-three years of age had already owned eight homes. His seventh house cost him $176,000 last summer (1977) and he sold it this summer (1978) for $230,000. Other newspapers, particularly those in Southern California and the north-east United States, frequently contain similar stories. Readers may know of such cases personally.

Table 17.7 The importance of profit. Responses tabulated by city.

City	Very important	Important	Unimportant	Very unimportant
	%	%	%	%
Leicester	0.0	14.4	42.2	43.3
	(0)	(13)	(38)	(39)
Dayton	12.9	60.0	24.3	2.9
	(9)	(42)	(17)	(2)

Source: Agnew 1978:131.

Some cross-national research carried out in 1975 suggests, however, that British homeowners do not share their American counterparts' investment orientation (Agnew 1978). For a sample of homeowners in the British city of Leicester and the American city of Dayton, Table 17.7 shows quite different perceptions of the house as an investment.[6] Analysis reveals that these perceptions are not the result of compositional differences between the two national components of the sample (Agnew 1978: 132). In their research in Bath (UK) Couper and Brindley (1975) also found a relatively small proportion of the homeowners in their sample (25 per cent) primarily committed to the home as a financial asset and investment. It should also be noted, however, that not all American research has found a commitment to the home as an investment. A study by Coons and Glaze (1963) found that a sample of homeowners in Westerville, Ohio were relatively unconcerned with their houses as investments. The authors used a rigorous definition of investment similar to that used by bankers and required something of a 'portfolio' mentality from their respondents for 'investment' to register as a response. They may, therefore, have under-estimated a more 'casual' attachment to the home as an investment.

Apparently, the different historical contexts in which homeownership has developed have affected the extent to which the home has come to be treated in investment terms. In particular, the fact that the British population is relatively immobile when compared to the population of the United States would indicate that since they are less likely to move and, hence, to sell their houses, British homeowners are less likely to be concerned with capital gain from sale than American homeowners (Agnew 1978). Moreover, the *idea* of the house as an investment does not appear to have been reinforced as aggressively in Britain as in the United States by government and real-estate industry propaganda. For example, few British or other European leaders have made the public declarations on behalf of the financial and other benefits of homeownership that one finds in the speeches of leading American public figures (Dean 1945; Perin 1977: 78). American 'family' magazines also regularly extol the financial rewards of homeownership whereas their British and European counter-parts seem to concentrate more on the owner-occupied house as a status-symbol and setting for realizing one's interior design fantasies.[7]

Such considerations, however, may be secondary to the relative impor-tance of residential real estate as a locus of investment and speculation as a source of profit in different capitalist societies. Perin (1977) and Mayer (1978) both note the tremendous financial significance of the residential real-estate industry in the United States. It has been maintained as a major locus of investment perhaps because other sectors of the economy and overseas investment have been less attractive and less secure. The absence of regional land use planning may have also served to encourage speculative endeavors. Perin (1977: 133) maintains that:

> 'Those "unplanned" and "chaotic" land-use patterns of metropolitan areas, where hunks and chunks of land are put into development without regard for coordinated facilities or services – these represent as well the gamble inherent in homeownership as a readily accessible speculative investment. *The small-time investor is the market for the big-time speculator who subdivides and moves on.'* [my emphasis added]

In Britain, and since the 1930s, there have been successive measures taken by governments in the form of tax and planning legislation which have reduced the attractiveness of residential real estate as an investment and restricted the potential for speculative building (Mellor 1977: 115–

17). Historically, there has also been resistance on the part of some rural landowners to adoption of a 'commercial' attitude towards their land (Clawson and Hall, 1973: 49–50). This and English law on land inheritance and subdivision have at times restricted the land available for private residential development (e.g. Kellett 1969: 414–17). Investment in commercial real estate has become increasingly attractive, however (Broadbent 1977: 124). Indeed, commercial property has become synonymous with 'property' in Britain to the extent that when *The Economist* publishes a survey on 'Property and Financial Institutions' no mention is made of residential real estate (*The Economist*, 10–18 June 1978). Competition for prime, mainly central-city commercial sites has obscured the relative decline of residential real estate as a focus of investment. Although the individual house remains a hedge against inflation it is not set in a very active market context where use-values (and status-values) are being constantly created and destroyed in the search for exchange-value. Consequently, homeownership is less likely to involve a commitment to the house as an investment in the sense of 'making a profit'.

Homeownership, the personal life and community consciousness

People not only live in houses, they also live in local communities or neighborhoods. Practice of the personal life through homeownership requires that such areas facilitate personal autonomy, help realize social esteem and maintain or enhance exchange-value. It is necessary, therefore, to be concerned with events which threaten any or all of these symbolic and material interests. One such event could be a proposed high-density housing development which would intrude into a homeowner's personal space, and lower the status and exchange-value of his house in the eyes of significant others such as relatives and potential purchasers. 'Community consciousness' is then, in terms of the practical incorporation thesis, a concomitant of practice of the personal life.

Evidence from the United States is, as might be expected from earlier arguments, largely consistent with this portrayal of the grounds for 'community consciousness' particularly when applied to middle-class neighborhoods (Perin 1977; Wertham 1968; Wertham *et al.* 1965; Danielson 1976; Agnew 1978; Harvey 1978). This community consciousness is also reinforced by the essentially *local* nature of property-tax administration and much public service provision in the United States (Cox 1978; Cox 1973; Gaffney 1973). Protecting 'tax base' from erosion by the arrival of 'low assessment per capita' in-migrants and 'school quality' from the admittance of students from lower-income families are typical concerns in many American urban neighborhoods and suburban jurisdictions (Gaffney 1973; Cox 1973).

Various writers, however, would dispute the pervasiveness throughout the United States of community consciousness in the sense that the term has been defined above (Gans 1962; Suttles 1972; Berger 1968). But evidence for this viewpoint is available largely from renter communities. Gans

(1962: 21) in his study of Italian-Americans in the West End of Boston observed that:

> 'Housing is not the same kind of status symbol for the West Enders that it is for middle class people. They are as concerned about making a good impression on others as anyone else, but the people to be impressed and the ways of impressing them do differ. The people who are entertained in the apartment are intimates. Moreover, they all live in similar circumstances. As a result they evaluate the host not on the basis of his housing, but on his friendliness, his moral qualities, and his ability as a host.'

In such working-class neighborhoods, therefore, performance in primary group relations has retained its significance for the allocation of status. There are none of the concerns which are central to a definition of community consciousness based upon defense of the personal life. It is easy to infer that the absence of homeownership is associated with the lack of such community consciousness.

In Britain, evidence for community consciousness as a defense of the personal life is present but rather meager. Young and Kramer (1978) in a review of extant literature and as a result of research in suburban London do argue for the existence of a community consciousness similar to that noted for American homeowner communities (also see Swann 1975). Other research, however, suggests that among homeowners and renters in Britain there is much less of this type of community consciousness (Agnew 1978; Agnew 1976). The relative decline in private residential real-estate activity and the advent of comprehensive and enforced land use planning have combined to reduce the importance of exchange-values and limit the uncertainty facing homeowners (Broadbent 1977: 164; Mellor 1977: 117). Consequently they have less need to be concerned about their local environments. At one time community consciousness may have been more widespread in Britain. For example, there probably were more controversies over the location of public housing in the 1930s, 1940s and 1950s than in the 1960s (Agnew 1976: 112–14; Huttman 1969). This may indicate a decline in community consciousness in rhythm with the increase of governmental intervention in the housing market. At any event, community consciousness based upon defense of the personal life does not appear to be as widespread in contemporary Britain as it is in the United States.

Conclusion

Much more empirical research is necessary before any *definitive* conclusions concerning the subject-matter of this paper are in order. However, some preliminary judgments can be made on both the theme of homeownership as a means of practical incorporation into a capitalist social order and, more generally, on the incorporation thesis itself.

The first point to note about homeownership is that it appears to serve more of an incorporative role in the United States than in other capitalist societies. This is in part a reflection of the greater extent and wide distribution of homeownership through all social strata in the United States. More importantly, however, American homeownership involves a much greater commitment to the personal life through which incorporation takes place than homeownership in Britain or some other countries. This is the indirect result of the greater significance of the residential real-estate industry in the United States, the absence of vigorous institutional controls over land conversions, and the relative weakness of ideas antithetical to the personal life. Moreover, the greater and more widespread commitment to the personal life generates a community consciousness that further retards the growth of any wider social, such as class, consciousness.

Second, the situation described in the preceding paragraph is only 'more or less' the case in the United States. There is some evidence to suggest that among American homeowners there is an ambivalence about the rewards to be gained from their practice of the personal life. This ambivalence or 'contradictory consciousness' persists in the face of the everyday realities of homeownership and the tremendous propaganda on behalf of the personal benefits of homeowning.

Both American and other evidence suggests, therefore, that the practical incorporation thesis is more applicable in some contexts than others and not entirely applicable anywhere. This is not unexpected nor does it invalidate the incorporation thesis as a *model* of how capitalist social relations come to dominate a social order. We should remind ourselves, following Avineri (1970: 160), that:

> 'Since all historical reality is always in a process of becoming, the model [of capitalism] is either a criterion for a reality developing towards it – or, if adequacy between model and reality is maximized, internal circumstances have given rise to a reality that has overtaken the model and moved farther and farther away from it.'

In the contemporary United States historical circumstances have kept the population more closely wedded to the classic capitalist social order. In other societies, which we often conventionally designate as 'capitalist', there is less correspondence between the capitalist model and the reality which people face in their everyday lives.[8] That much our investigation of homeownership and maintenance of the capitalist social order would indicate. Homeownership helps maintain and reflects the social order *most* when and where that social order is not constrained by institutions and practices which are antithetical to the practice of the personal life. The meaning and significance of homeownership for the capitalist social order, therefore, is predicated upon the extent to which that social order has been *changed* by the advent of powerful new ideas and institutions and the maintenance of past practices and ideas antithetical to those of capitalism.

Notes

[1] Included under the latter position are those who propose coercive dominance through manipulation, propaganda, and mass deception as well as violence and

military repression. For a critical review of some writing in this tradition see Swingewood 1977.

[2] Goldmann (1959) has argued for the central importance of exchange-value and the use of quantitative judgments based upon exchange-value in individuals' evaluations of objects and people. This does not account satisfactorily for the 'conspicuous consumption of objects', however.

[3] This process of evaluation can be equated with the notion of 'the struggle for recognition or respect' associated by Kojève (1947) with the writings of Hegel and Marx. On this see Harrington 1976: 380–1 and Poster 1975: 13.

[4] C. Wright Mills (1956b: 9) sees this perspective as underpinning the individualism of American political thought. He quotes one A. Whitney Griswold in an interpretation of Jefferson's doctrine: 'Who would govern himself must own his own soul: To own his own soul he must own property, the means of economic security.'

[5] There is no space here to enter into the debates over whether the home *really* is a good investment or whether a considerable redistribution of wealth has taken place as a consequence of the extension of homeownership. Relevant studies on these issues include (i) on the home as an investment, Ball 1976, Saunders 1977, Luria 1976, Shelton 1968, Williams 1961, Rosefsky 1977, Topalov 1973, and Angell 1976; and (ii) on the redistribution of wealth, Luria 1976 and Williams 1961.

[6] See Agnew (1978) for a discussion of the survey research upon which this finding is based.

[7] This observation is the result of perusing a range of 'house and family' type magazines for the years 1973–8. These magazines included *Better Homes and Gardens* (circulation of eight million copies per issue), *House Beautiful, House and Garden, Maison et Jardin, Votre Maison,* and *La Maison Individuelle.* Clearly, a more detailed analysis of cross-national homeownership 'images' is required.

[8] E. P. Thompson (1965: 350) has argued eloquently for the need to see the capitalist model in dialectical terms:

'Nothing is more easy than to take a model *to* the proliferating growth of actuality, and to select from it only such evidence as is in conformity with the principles of selection . . . Even in the moment of employing it the historian must be able to regard his model with radical scepticism, and to maintain an openness of response to evidence for which it has no categories. At the best – which we can see at times in the letters of Darwin or Marx – we must expect a delicate equilibrium between the synthesizing and the empiric modes, a quarrel between the model and actuality. This is the creative quarrel at the heart of cognition. Without this dialectic, intellectual growth cannot take place.'

References

Abercrombie, N. and Turner, B. S. (1978) 'The dominant ideology thesis.' *British Journal of Sociology* 29, 2: 149–70.

Abrams, C. (1964) *Man's Struggle for Shelter in an Urbanizing World.* Cambridge, Mass.: MIT Press.

Agnew, J. A. (1976) 'Public policy and the spatial form of the city: the case of public housing location.' Ph.D. dissertation, Ohio State University.

—— (1978) 'Market relations and locational conflict in cross-national perspective', in K. R. Cox (ed.) (1978).

Angell, W. J. (1976) 'Housing alternatives', in C. S. Wedin and L. G. Nygren (eds) *Housing Perspectives: Individuals and Families.* Minneapolis: Burgess.

Avineri, S. (1970) *The Social and Political Thought of Karl Marx.* Cambridge: Cambridge University Press.

Baldus, B. (1977) 'Social control in capitalist societies: an examination of the "Problem of Order".' *Canadian Journal of Sociology* 2, 3: 247–62.

Ball, M. (1976) 'Owner occupation.' *Housing and Class in Britain* 2. London: Political Economy of Housing Workshop.

Berger, B. M. (1968) *Working Class Suburb: A Study of Auto Workers in Suburbia.* Berkeley: University of California Press.

Beyer, G. H. (1965) *Housing and Society.* New York: Macmillan.

Blackburn, R. and Mann, M. (1975) 'Ideology in the non-skilled working class', in M. Bulmer (ed.) *Working-Class Images of Society.* London: Routledge and Kegan Paul.

Bodfish, H. M. (1931) *History of Building and Loans in the United States.* Chicago: United States Building and Loan League.

Brittan, A. (1977) *The Privatised World.* London: Routledge and Kegan Paul.

Broadbent, T. A. (1977) *Planning and Profit in the Urban Economy.* London: Methuen.

Brown, S. R. (1970) 'Consistency and the persistence of ideology: some experimental results.' *Public Opinion Quarterly* 34: 60–8.

Burney, E. (1967) *Housing on Trial: A Study of Immigrants and Local Government.* London: Oxford University Press.

Chapman, D. (1955) *The Home and Social Status.* London: Routledge and Kegan Paul.

Clarke, M. (1976) 'Who Owns a Home?' *New Society* (9 December): 513.

Clawson, M. and Hall, P. (1973) *Planning and Urban Growth: An Anglo-American Comparison.* Baltimore: Johns Hopkins University Press.

Converse, P. (1964) 'The nature of belief systems in mass publics', in D. Apter (ed.) *Ideology and Discontent.* New York: Free Press.

Coons, A. E. and Glaze, B. T. (1963) *Housing Market Analysis and the Growth of Home Ownership.* Columbus: Ohio State University Bureau of Business Research.

Cooper, C. (1972) 'The House as Symbol.' *Design and Environment* 3, 3. Reprinted in C. S. Wedin and L. G. Nygren (eds) (1976).

Couper, M. and Brindley, T. (1975) 'Housing Classes and Housing Values.' *Sociological Review* 23: 563–76.

Cox, K. R. (1973) *Conflict, Power and Politics in the City: A Geographic View.* New York: McGraw-Hill.

—— (1978) 'Local interests and urban political processes in market societies', in K. R. Cox (ed.) *Urbanization and Conflict in Market Societies.* Chicago: Maaroufa Press and London: Methuen.

Cox, K. R. and Agnew, J. A. (1974) *The Location of Public Housing: Towards a Comparative Analysis.* Columbus: Ohio State University, Department of Geography, Discussion Paper No. 45.

Crawford, D. (ed.) (1975) *A Decade of British Housing: 1963–1972.* London: Architectural Press.

Cullingworth, J. B. (1969) *Owner-Occupation in Scotland and in England and Wales: A Comparative Study.* Edinburgh: NHBRC (Scotland).

Danielson, M. N. (1976) *The Politics of Exclusion.* New York: Columbia University Press.

Davies, A. F. (1966) *Private Politics.* Melbourne: Melbourne University Press.

Deakin, N. (1969) 'Race and human rights in the city.' *Urban Studies* 6, 3: 385–407.

Dean, J. P. (1945) *Home Ownership: Is It Sound?* New York: Harper and Bros.

—— (1951) 'The ghosts of home ownership.' *Journal of Social Issues* 7: 59–68.

Domhoff, G. W. (1967) *Who Rules America?* New Jersey: Prentice-Hall.

Duncan, J. S. (1973) 'Landscape taste as a symbol of group identity: a Westchester county village.' *The Geographical Review* 63: 334–55.

The Economist (1978) 'Property and financial institutions: a survey' (10–18 June).

Friedman, J. (1974) 'Marx's perspective on the objective class structure.' *Polity* 6, 3: 318–44.

Gaffney, M. (1973) 'Tax reform to release land', in M. Clawson (ed.) *Modernizing Urban Land Policy.* Washington, DC: Resources for the Future.

Gans, H. J. (1962) *The Urban Villagers: Group and Class in the Life of Italian-Americans.* New York: Free Press.

Goldmann, L. (1959) *Recherches Dialectiques.* Paris: Gallimard.

Gramsci, A. (1971) *Selections from the Prison Notebooks.* Edited and translated by Q. Hoare and G. Nowell Smith. New York: International Publishers.

Hamilton, R. (1967) *Affluence and the French Worker in the Fourth Republic.* Princeton: Princeton University Press.

Handlin, D. P. (1972) 'The detached house in the age of the object and beyond.' *Environmental Design: Research and Practice,* Proceedings of the EDRA 3/AR8 Conference.

Harrington, M. (1976) *The Twilight of Capitalism.* New York: Simon and Schuster.

Harris, L. (1973) *A Study of Public Attitudes Toward Federal Government Assistance for Housing for Low Income and Moderate Income Families.* Springfield, Va.: National Technical Information Service, PB-228 903.

Harvey, D. (1977) 'Government policies, financial institutions and neighborhood change in United States cities', in M. Harloe (ed.) *Captive Cities: Studies in the Political Economy of Cities and Regions.* London: John Wiley.

—— (1978) 'Labor, capital, and class struggle around the built environment in advanced capitalist societies', in K. R. Cox (ed.) (1978).

Hobson, O. R. (1953) *A Hundred Years of the Halifax: The History of the Halifax Building Society, 1853–1953.* London: Batsford.

Huttman, E. (1969) 'Stigma and public housing: A comparison of British and American policies and experience.' Ph.D. Dissertation, University of California, Berkeley.

Ineichen, B. (1972) 'Home ownership and manual workers' life-styles.' *Sociological Review* 29: 391–412.

Ineichen, B. and Brown J. (1977) 'Men of property.' *Town and Country Planning* (July–August): 352–4.

Israel, J. (1971) *Alienation: From Marx to Modern Sociology.* Boston: Allyn and Bacon.

Jackson, A. A. (1973) *Semi-Detached London: Suburban Development, Life and Transport, 1900–39.* London: George Allen and Unwin.

Jessup, B. (1974) *Traditionalism, Conservatism and British Political Culture.* London: George Allen and Unwin.

Kellett, J. R. (1969) *The Impact of Railways on Victorian Cities.* London: Routledge and Kegan Paul.

Kemeny, J. (1978) 'Forms of tenure and social structure: a comparison of owning and renting in Australia and Sweden.' *British Journal of Sociology* 29, 1: 41–56.

Kojève, A. (1947) *Introduction à la Lecture de Hegel.* Paris: Gallimard.

Lane, R. (1962) *Political Ideology.* New York: Free Press.

Lansing, J. B., Clifton, C. W., and Morgan, J. N. (1969) *New Homes and Poor People: A Study of Chains of Moves.* Ann Arbor: Institute of Social Research, University of Michigan.

Lipsitz, L. (1974) 'On political belief: the grievances of the poor', in A. Wilcox (ed.) *Public Opinion and Political Attitudes.* New York: Wiley.

Livingstone, D. W. (1976) 'On hegemony in corporate capitalist states: material structures, ideological forms, class consciousness and hegemonic acts.' *Sociological Inquiry* 46, 3–4: 235–50.

Luria, D. D. (1976) 'Wealth, capital and power: the social meaning of home ownership.' *Journal of Interdisciplinary History* 7: 261–82.

Luttbeg, N. (1968) 'The structure of beliefs among leaders and the public.' *Public Opinion Quarterly* 32: 398–409.

Macpherson, C. B. (1962) *The Political Theory of Possessive Individualism.* London: Oxford University Press.

—— (1973) *Democratic Theory: Essays in Retrieval.* London: Oxford University Press.

Mann, M. (1970) 'The social cohesion of liberal democracy.' *American Sociological Review* 35: 423–39.

Marx, K. (1953) *Grundrisse der Kritik der Politschen Ökonomie (Rohentwurf).* Berlin: Dietz Verlag.

Marx, K. and Engels, F. (1942) *The German Ideology.* London: Lawrence and Wishart.

Mayer, M. (1978) *The Builders: Houses, People, Neighborhoods, Governments, Money.* New York: Norton.

McNeill, W. (1963) *The Rise of the West: A History of the Human Community.* Chicago: University of Chicago Press.

Mellor, J. R. (1977) *Urban Sociology in an Urbanized Society.* London: Routledge and Kegan Paul.

Michelson, W. (1968) 'Most people don't want what architects want.' *Transaction* (July/August).

Michelson, W., Belgue, D. and Stewart, J. (1973) 'Intentions and expectations in differential residential selection.' *Journal of Marriage and the Family* (May): 189–96.

Mills, C. Wright (1956a) *The Power Elite.* New York: Oxford University Press.

—— (1956b) *White Collar.* New York: Oxford University Press.

—— (1962) *The Marxists.* New York: Dell.

Mogey, J. M. (1956) *Family and Neighbourhood.* Oxford: Oxford University Press.

Moorhouse, H. F. and Chamberlain, C. W. (1974) 'Lower class attitudes to property: aspects of the counter-ideology.' *Sociology* 8, 3: 387–405.

Murie, A. (1974) *Household Movement and Housing Choice.* Birmingham: University of Birmingham, Centre for Urban and Regional Studies, Occasional Paper No. 28.

Murie, A., Niner, P. and Watson, C. (1976) *Housing Policy and the Housing System.* London: George Allen and Unwin.

Nevitt, A. (ed.) (1967) *The Economic Problems of Housing.* London: Macmillan.

—— (1978) 'British housing policy.' *Journal of Social Policy* 7, 3: 329–34.

Newman, P. (1975) *The Canadian Establishment.* Toronto: McClelland and Stewart.

The New York Times (1978) 'Home Buyers Persevere' (11 June).

Ollman, B. (1971) *Alienation: Marx's Conception of Man in Capitalist Society.* New York: Cambridge University Press.

Parkin, F. (1971) *Class Inequality and Political Order: Social Stratification in Capitalist and Communist Societies.* London: Paladin.

Parsons, T. (1951) *The Social System.* Glencoe, Ill.: Free Press.

Perin, C. (1977) *Everything in Its Place: Social Order and Land Use in America.* Princeton: Princeton University Press.

Polanyi, K. (1944) *The Great Transformation.* New York: Rinehart.

Popenoe, D. (1977) *The Suburban Environment: Sweden and the United States.* Chicago: University of Chicago Press.

Poster, M. (1975) *Existential Marxism in Postwar France: From Sartre to Althusser.* Princeton: Princeton University Press.

Professional Builder (1975) 'Professional builder's national consumer builder survey on housing' (January), pp. 81–97.

Rakoff, R. M. (1977) 'Ideology in everyday life: the meaning of the house.' *Politics and Society* 7, 1: 85–104.

Roberts, K., Cook, F. G., Clark, S. C. and Semeonoff, E. (1977) *The Fragmentary Class Structure.* London: Heinemann.

Rosefsky, B. (1977), 'Blowing away the great equity myth.' *Apartment Life* (May): 24–32.

Rosow, I. (1948) 'Home ownership motives.' *American Sociological Review* 13: 751–6.

Runciman, W. G. (1966) *Relative Deprivation and Social Justice: A Study of Attitudes to Social Inequality in Twentieth-Century England.* London: Routledge and Kegan Paul.

Sahlins, M. (1965) 'On the sociology of primitive exchange', in M. Banton (ed.) *The Relevance of Models for Social Anthropology.* London: Tavistock Publications.

Sallach, D. L. (1974) 'Class domination and ideological hegemony.' *The Sociological Quarterly* 15: 38–50.

Saunders, P. R. (1977) *Housing Tenure and Class Interests.* Brighton: Urban and Regional Studies Working Paper No. 6.

Seeley, J. R., Sim, R. A. and Loosley, E. W. (1963) *Crestwood Heights: A Study of the Culture of Suburban Life.* New York: Wiley.

Sennett, R. and Cobb, J. (1972) *The Hidden Injuries of Class.* New York: Random House.

Shelton, J. P. (1968) 'The cost of renting versus owning a home.' *Land Economics* 44: 59–72.

Sternleib, G. (1972) 'Death of the American dream house.' *Society* 9, 4: 39–45.

Stone, M. E. (1975) 'The housing crisis, mortgage lending, and class struggle.' *Antipode* 7, 2: 22–37.

Suttles, G. (1972) *The Social Construction of Communities.* Chicago: University of Chicago Press.

Swingewood, A. (1977) *The Myth of Mass Culture.* London: Macmillan.

Swann, J. (1975) 'The political economy of residential development in London.' *Political Economy and the Housing Question.* London: Political Economy of Housing Workshop.

Thompson, E. P. (1967) 'Time, work-discipline and industrial capitalism.' *Past and Present* 38: 58–75.

—— (1965) 'The peculiarities of the English', in R. Miliband and J. Saville (eds.) *The Socialist Register, 1965.* London: Merlin.

Todd, O. (1978) 'Proudhon is dead.' *Encounter* (August): 71–2.

Topalov, C. (1973) *Capital et Propriété Foncière: Introduction a L'Etude des Politiques Foncières Urbaines.* Paris: Centre de Sociologie Urbaine.

Wertham, C. (1968) 'The social meaning of the physical environment.' Ph.D. Dissertation, University of California, Berkeley.

Wertham, C., Mandel, J. S. and Dienstfrey, T. (1965) *Planning and the Purchase Decision: Why People Buy in Planned Communities.* Berkeley: Institute of Urban and Regional Development.

Wikse, J. R. (1977) *About Possession: The Self as Private Property.* University Park, Pa.: Pennsylvania State University Press.

Williams, R. (1973) 'Base and superstructure in Marxist cultural theory.' *New Left Review* 82: 3–16.

—— (1961) *The Long Revolution.* London: Chatto and Windus.

Winnick, L. (1958) *Rental Housing.* New York: McGraw-Hill.

Young, K. and Kramer, J. (1978) 'Local exclusionary policies in Britain: the case of suburban defense in a metropolitan system', in K. R. Cox (ed.) (1978).

Young, M. and Willmott, P. (1957) *Family and Kinship in East London.* London: Routledge and Kegan Paul.

Zweig, F. (1961) *The Worker in an Affluent Society: Family, Life and Industry.* New York: Free Press.

18 Social and spatial reproduction of the mentally ill*

Michael Dear

The tendency for recently-discharged mental hospital patients to cluster in geographically limited parts of our inner cities has attracted increasing attention in the geographic and psychiatric literature. In a recent review, I suggested that the tendency toward the 'ghettoization' of the mentally ill was not unlike the development of ghettos of other disadvantaged or minority groups. Specifically, the growth of the 'asylum without walls' was linked to several forces: (i) the formal assignment of patients to institutional aftercare facilities, which tend to congregate in downtown locations because of planners' actions in siting and because of the limited availability of large convertible properties; (ii) the extent of citizen opposition to community mental health in other residential neighbourhoods; and (iii) an informal process of spatial filtering by which patients tend to gravitate toward transient areas of rental accommodation in the inner urban core (Dear 1977). The policy response to the mental patient ghetto has been diverse, but always pragmatic. Some have viewed the ghetto positively, as creating the kind of supportive environment necessary for the ex-patient to survive without the formal support of the hospital. Others have viewed the saturation of certain neighbourhoods with alarm, and have agitated for tighter zoning control over the location of aftercare facilities.

In the absence of a precise definition of the structural limits of the

* This is a revised version of a paper first presented at the 1978 annual meeting of the Association of American Geographers; the original paper was written while the author was on leave at the School of Architecture and Urban Planning, Princeton. Thanks to Gordon Clark, Richard Peet, Allen Scott and Jennifer Wolch for valuable critical comments.

ghettoization process, the policy-maker's response is always likely to be partial and incomplete. The purpose of this essay is to situate the ghettoization problem in a more adequate theoretical framework. The point of departure for the paper is that the mentally ill are one of many social groups in competition for space in the city. Like the elderly or disabled, they are restricted environmentally in their choice of residence, workplace, and so on. Part of the reason for environmental restrictiveness may be found in the particular social characteristics of the group in question (e.g. limited work skills). However, there is also a significant spatial factor, which tends to facilitate and to reinforce the environmental isolation of the disadvantaged group. It is precisely this link between social process and geographic process which is at issue in the ghettoization phenomenon. Why do patients released from hospital replicate their asylum ward in a downtown location? What is the interplay between social and spatial forces permitting the development of the ghetto? The answers to these questions lie in the theory of social reproduction and in the way space acts to mediate the reproduction of social relations. In order to demonstrate this argument, the paper first examines the notion of reproduction in the social formation, and the role of space in reproduction. Secondly, the actions of mental health professionals and of the community are shown to have significant impacts on mental health care. These impacts are motivated by the necessity for the reproduction of social relations and are, in both cases, mediated through the use of space as a means of isolation and exclusion.

Reproduction in the social formation

The concept of reproduction

The notion that society devotes large energies toward reproducing itself is due to Marx, who noted that 'every social process of production is, at the same time, a process of reproduction' (Marx 1971, Vol. 1: 531). Therefore, any production process must not only produce material objects, but also reproduce continually the production relations and the corresponding distribution relations (Marx 1971, Vol. 3: 857). Note that reproduction implies much more than the mere repetition of existing production processes. A theory of reproduction implies a triple continuity: (i) a link between individual capitals, or economic subjects; (ii) a link among the different levels of the social structure, including the non-economic conditions of the production process; and (iii) a link between successive historical production processes (Althusser and Balibar 1970: 258–9). Hence, reproduction is the method by which the total social 'ensemble', including modes of circulation, distribution and consumption, is protected and repeated through time.

A theory of reproduction is an indispensable element of the theory of production. Reproduction is a dynamic concept, emphasizing historical continuity in the transition from one mode of production to another. In such a transition, reproduction allows for the replacement and transformation of things, but *retains the fundamental relationships indefinitely.* Hence, the perpetuation of political, legal and other institutions in support of the economic order may be anticipated, as well as the key relations in the economy itself.

Given the theoretical significance of the concept, it is important to determine exactly *how* the reproduction of social relations is secured. Traditionally, the answer has been sought in the functioning of the 'legal-political and ideological superstructure' (Althusser 1971: 148). Historically, important elements of this superstructure have been the Church, which has now been superseded by the educational establishment. The role of the family in the reproduction of social relations has also been stressed (Engels 1975: 71–2). More than any other relationship, capitalism requires that class *structuration* be maintained over time.

The existence of differential market capacities (based on ownership of property in the means of production, on educational or technical skills, or on manual labour-power) is the source of class structuration. Two factors are important in the structuration of class relations: the mediate and the proximate (Giddens 1973, Chapter 6). The *mediate* factors are governed by market capacities and the distribution of mobility chances in society, since the greater the limits on mobility, the more likely are identifiable classes to form. The lack of intergenerational movement reproduces common life experiences, and such homogenizing of experience is reinforced by limitations on an individual's mobility within the labour market. The effect of 'closure' generated by the mediate structures is accentuated by the *proximate* factors of class structuration, according to which the basic within- and between-class structures are intensified and further differentiated. These more 'derivative' characteristics are generated by the need to preserve the processes of capital generation (Harvey 1975). There are three groups of proximate factors.

1 The *division of labour* within capitalism which is both a force for consolidation and for fragmentation of class relationships. It favours the formation of classes according to the extent to which it creates homogeneous groupings. On the other hand, the profit-motivated drive for modernization and efficiency often implies a specialization of labour functions and hence, a fragmentation within an otherwise homogeneous group.

2 *Authority relations* are a second force for class structuration. These may occur as a hierarchy of command within the productive enterprise, although as Harvey (1975: 359) emphasizes, it is equally important that the non-market elements in society be so ordered that they sustain the system of production, circulation and distribution.

3 The third source of the proximate structuration of classes, *distributive grouping,* is an aspect of consumption rather than production. Distributive groupings are those relationships (and their concomitant status implications) which involve common patterns of consumption of economic goods. They act to reinforce the separations initiated by differential market capac-

ity, but 'The most significant distributive groupings . . . are those formed through the tendency towards community or neighborhood segregation' (Giddens 1973: 109). This tendency is based on many factors, including income and access to the mortgage market, and ultimately gives rise to distinct 'working-class' or 'middle-class' neighbourhoods. It is at this level that the spatial dimension of the social class structuration process is manifest.

The role of space in social processes

Transformations in the social structure inevitably imply transformations in spatial structure. For instance, the simple precepts of class structuration tend to lead to the creation of relatively homogeneous environments in terms of social status. Such symbolic differentiation of urban space reflects choice in associates and opportunities for interaction in a class-differentiated society. The fundamental theoretical issue is the extent to which spatial form determines, or is determined by, social forces. Castells (1976: 77) has argued that technical progress reduces the role of space as a determinant; it is not that space is external to the social structure and unaffected by it, but that its specific importance may be diminishing. Elsewhere, Castells has criticized the tendency to make a fetish of space in urban analysis (Castells 1977). However, in this essay, the relationship between space and society is viewed as truly dialectical. As Soja (1978: 10) argues, once the organization of space is regarded as a purposeful social product, then it can no longer be regarded as a separate structure, with its own rules of construction and transformation independent of social practice. The production of space is both an ideological and a political process which is a vital mediator in capitalist production and reproduction (Lefebvre 1976).

In his analysis of inequality and poverty, Peet (1975) has developed the formal links between social and spatial theory. Having shown that inequality and poverty are endemic to capitalism, Peet proceeds to demonstrate how space encourages their reproduction through generations. Central to his thesis is Hägerstrand's notion of a 'daily-life environment', composed of residence and/or workplace, and defined by the physical friction of distance and the social distance of class. Each social group operates within a typical daily 'prism', which, for the disadvantaged, closes into a 'prison of space and resources' (Peet 1975: 568). Deficiencies in the environment – limitations on mobility, and the density and quality of social resources – must clearly limit an individual's potential, or market capacity; similarly, low income limits access to more favourable environments. A self-reinforcing process thus sets in, and it is easy to understand how an individual can carry an 'imprint' of a given environment, and how the daily-life environment can act to 'transmit' inequality.

In reproducing the ensemble of socio-spatial inequality and poverty capitalism normally produces a class-differentiated society each stratum of which is allowed to reproduce itself, using varying proportions of its income to raise the next generation (cf. Engels 1972). Since the amount of money spent by each stratum varies, unequal resource environments, which perpetuate the class system, are produced. The city is thus composed of a differentiated hierarchy of resource environments which reflect the different hierarchical labour demands of the capitalist economy (Peet 1975: 569). The social and the spatial dimensions of reproduction are, therefore, inextricably fused.

A similar approach to spatial residential differentiation has been taken by Harvey (1975). He emphasizes that differentiated space is an outcome of the capitalist production and reproduction processes, and not (as more typically construed) a product of the aggregate of individual consumer preferences. Harvey (1975: 362–3) advances four hypotheses linking residential differentiation and social structure: (i) residential differentiation is part of the reproduction of social relations within capitalism; (ii) residential areas provide the locus for social interaction through which market capacities are transmitted; (iii) the fragmentation of population into distinct communities also fragments class consciousness; and (iv) patterns of residential differentiation incorporate many contradictions in capitalism, and are consequently a source of instability and contradiction themselves.

Reproduction and the social formation

So far, the analysis of reproduction has been developed using fairly traditional Marxist categories. For instance, the key concept of class structuration has implicitly accepted class status *vis-à-vis* the means of production as the determining cause of individual market capacities. However, in this essay, I am seeking an explanation of the status and reproduction of decidedly *non-economic* agents in society. These include the mentally ill and mental health professionals neither of whose status can be explained simply by reference to their control over the means of production. In fact, the service sector and its clients have received little direct attention in Marxist analysis. Welfare services are typically dismissed as elements of the superstructure charged with maintenance and social control of the labour-force, especially during times of unemployment. However, as we shall see, the situation is at once more subtle and more complex. In order to progress, it will be useful to introduce certain extensions of the historical materialist method, focusing in particular on the concept of the 'social formation'.

In Marxist analysis, the traditional point of departure has been the *mode of production,* which has been defined as a '. . . combination of relations and forces of production structured by the dominance of the relations of production' (Hindess and Hirst 1975: 9). Out of this economic base or foundation a whole superstructure of legal, administrative and political machinery develops in order to facilitate the productive and reproductive functions of society. The totality of this social order, consisting of the economic, ideological and political, is commonly termed the *social formation* (Hindess and Hirst 1975: 13). The precise character of the social formation has historically been viewed as being determined and dominated by the economy. Hence, the existence of a particular mode of production and social formation is dependent upon a particular form of material production.

More recently, a fundamental critique of these traditional concepts has been prepared by Hindess, Hirst and their associates. They have suggested that over-concentration on modes of production has restricted analysis to a limited range of economically-determined class relations, with the consequent neglect of the more complex social relations. For instance, the simple necessity for economic existence is clearly insufficient to explain the non-economic forces of class structuration. These (and similar) observations have caused a fundamental reorientation of analytical attention. There is henceforth no basis for analytical priority of the economic over other levels of the social formation. Moreover, as the concept of mode of production diminishes in significance, so the concepts of relations of production and social formation attain new theoretical significance (Hindess and Hirst 1977). The social formation should be conceived as a totality and it is not structured by the primacy of the economy. Henceforth, connections among the economic, political and cultural should be sought not in terms of causality and determination, but in terms of '. . . conditions of existence and the forms in which they may be satisfied' (Cutler *et al.* 1977, Vol. 1: 172).

In summary, Cutler *et al.* (1977, Vol. 1: 314) argue for the displacement of mode of production as a primary object of Marxist conceptualisation in favour of social formation conceived as a definite set of relations of production together with the economic, political, legal, and cultural forms in which their conditions of existence are secured. It is obvious from this that the structure of social relations cannot solely be conceived as a relation between 'direct producers' and their 'exploiters' (those appropriating the surplus product from the direct producer). The analysis of class in the social formation requires a fuller clarification of the division of social labour into distinct branches producing specialized product categories (Hindess and Hirst 1977: 67). Such a requirement is the theoretical premise of the remainder of this essay. Therefore in what follows, I am guided by three assumptions:

1 that the form of the reproduction process is determined by the structure of the social formation, which is itself not fully preconditioned by the primacy of the economic;
2 that non-economic forces in the social formation condition the division of social relations, including those among mental health professional, mental health client, and the community;
3 that the social processes defining the relationships among professional, client, and community are necessarily mediated by, and constituted through, space.

Professional and client in mental health care

In this section, two basic themes are explored: first, the nature of the class relationship between professional and client, and the dominant role of reproduction in this relationship; and secondly, the role of space in mediating the professional/client relationship.

Reproduction of social relations

The class relationship between the mental health professional and the mental health care client is determined by two dimensions. First, there is a clearly defined *authority* relation separating the two. Secondly, there is some form of 'exploitative dependence', where exploitation is defined as a socially conditioned form of asymmetrical production of market capacities (Giddens 1973: 92 and 130). Notice that there is a *mutual* dependency in this relationship: the client needs professional care, but the professional also needs a client to serve. However, the asymmetry is clear, in that the professional has a much greater power over the individual client's life-chances than vice versa. The structural requirements for the reproduction of the professional and client classes are defined by this authority/exploitative relationship. Hence, it is worth outlining in more detail.

Consider first the client. Either by choice or by persuasion (including referral), the client agrees to seek psychiatric help. This is the first source of asymmetry in the relationship between professional and client. In accepting the 'sick role', the client has been labelled as mentally ill, and has implicitly or explicitly given up certain civil rights or freedoms (even if it is only the right to privacy which is surrendered). There seems little consensus on what physical, social, or emotional problems may be appropriately labelled 'mental illness'. For the present, however, it is more important to emphasize that the labelling process identifies the client as a social deviant, with some form of 'illness' which can presumably be 'cured'. This is notwithstanding the fact that deviance is not a property inherent in certain forms of behaviour, but is a property conferred upon these forms by those who witness them (Erikson 1967: 296).

In short, the client needs help, and has to accept the label of the sick role and dependency upon a professional to obtain it. The exploitative/ authoritative social relationship is thus established.

The professionals, for their part, are in the business of providing care or a cure. Their objective, simply stated, is 'normalization', or resocialization of the client. This involves five tasks (cf. Foucault 1977: 182–3):

1 referring individuals to a normative social model which acts as a basis of comparison;
2 labelling and thereby differentiating amongst individuals in the system with respect to the normative model;
3 measuring and categorizing the specific defects and potential of the individual;
4 establishing the level of conformity which has to be achieved by each individual; and
5 defining by these tasks the operational limits of the social model of the abnormal.

Hence, the professional service offered compares, labels, categorizes, homogenizes and excludes; that is, it normalizes. The normalization principle operates on two levels. For *clients,* normalization entails a degree of mastery over internal drives and the immediate environment, as well as a realization of social and vocational potential (however limited this may be). In addition, client normalization implies integration within a normal social network and community setting. For *service delivery systems,* the normalization principle guides development of the delivery system toward strategies aimed at normalizing client behaviour and life situation. Theories on normalization indicate that the most effective way to promote resocialization is to alter (to the minimum extent possible) the client's everyday living situation, concomitant with treatment needs. The treatment setting itself must also be normalized so as to assume the characteristics of normal community settings, and to facilitate social activity and interaction (Wolfensberger 1972). In spite of these admirable objectives, it is often conceded that much client normalization occurs at the lowest of functional levels, consigning many individuals to undemanding custodial treatment settings.

The power of the psychiatric profession over clients derives from its function as a major institution of social control, not necessarily through political power, but through the 'medicalizing' of much of daily living. This has been caused by expansion of areas deemed relevant to the practice of medicine, retention of control over technical procedures (e.g. drugs), the exclusive access to 'taboo' areas (the body's organs, the mind) among other things (Zola 1977: 51–61). In fact, with a judicious use of the term ILLNESS, there seems to be an infinite expansion potential for the caregiving business. It is easy to be persuaded by McKnight (1977: 73) who argues that behind the mask of the medical services lies '. . . a business in need of markets, an economy seeking new growth potential, professionals in need of an income'. However, it is important to point out that the mask of service is not a false or conspiratorial mask. The caregivers' motivation in the delivery of care can only rarely be questioned. However, the service relationship within which the professional and the client interact has an independent effect in structuring their interrelationship. This relationship may be defined via the notion of 'need'. Clients have needs; professionals meet needs; psychiatric services are designed to bring clients and services together. In a society where service is a major business, McKnight (1977: 74–5) concludes that

> '. . . the political reality is that the central "need" is an adequate income for professional servicers and the economic growth they portend . . . the client is less a person in need than a person who is needed. . . . The central political issue becomes the servicers' capacity to manufacture needs in order to expand the economy of the servicing system.'

If this is the case, then professional power to manufacture needs is unbounded. To see this, consider the professional judgements which are commonly made regarding needs and the meeting of needs (cf. McKnight 1977: 78–89). Needs are first translated into a deficiency, *which is located in the individual,* and which requires specialized service to meet. In meeting such needs, the professional defines the problem, the prescription, and (usually) the satisfactory outcome by which intervention will be judged. This process involves complex tools and procedures, both shrouded in a mystical language. The circle of dependency is completed.

Once established, the class relation in professional-client dependency is maintained in many ways. For example, the tendency toward increasing inmate dependency upon an institution, as length of stay in the institution is extended, has frequently been noted. The longer the client remains in care, the less likely he/she is to leave. As Erikson (1967) points out, police often seem to regard ex-convicts as a permanent population of offenders; in the same way, psychiatric institutions could not operate without viewing former patients as a group unusually susceptible to mental illness. In addition, patients often respond appropriately to professional expectations that they will remain sick (Illich *et al.* 1977). These and other factors tend to contribute toward the development of 'patient careers', as various expectations set off a chain of dependency-inducing events. At each stage of such a career (e.g. outpatient, inpatient, long-term hospitalization) individual freedoms are increasingly surrendered to the professional care-giver. In some instances, all semblance of 'freedom' is lost. The role of mental patient may be created and perpetuated by the very groups and institutions whose responsibility is its eradication (Magaro *et al.* 1978, Chapter 6).

In an intriguing study of the social history of the French penal system, Foucault (1977) has illustrated a parallel 'professional-client' interdependency in the prison. He points out that in the space of only eighty years (1757–1837), punishment changed from the 'art of unbearable sensations' (i.e. torture) to an 'economy of suspended rights' (i.e. imprisonment for varying time periods). As a result, a whole army of wardens, doctors, chaplains, psychiatrists, psychologists and educationalists took over from the executioner. By the beginning of the nineteenth century, public executions preceded by torture had almost entirely disappeared. In their place, a whole panoply of assessing, diagnostic, prognostic and normative judgements concerning the criminal was lodged in the penal framework (Foucault 1977: 11–19). Prison had failed to eliminate crime; indeed 'So successful has the prison been that, after a century and a half of "failures", the prison still exists, producing the same results, and there is the greatest reluctance to dispense with it' (Foucault 1977: 277). Foucault concludes that through the centuries the supposed failure of the prison is part of its true function; that, in effect, prisons participate in fabricating (or, at the very least, merely manage) the delinquency they are supposed to combat. In penal systems, as in other institutions, loyalty to the profession and its institutions often seems to transcend the needs of the client. Little attention is paid toward attacking the genesis of crime, somewhere in the domain of socio-economic relations.

Role of space in mental health care

The history of treatment of the mentally ill is a study in isolation. Every culture appears to have had its 'madness', although this was not always easily distinguishable from other behaviours. Significantly, most cultures also appear to have devised some principles for the spatial isolation of the mentally ill. In classical Greece, Plato advocated that atheists, whose lack of faith derived from ignorance and not from malice, should be confined for five years in a 'house of sanity' (Simon 1978: 32). In medieval Europe, the mad were driven out of the city enclosures, and forced to roam in distant fields. In addition, two modes of ritual exclusion were developed: the 'ship of fools', where the insane were entrusted to sailors of chartered ships, and dropped off in uninhabited places; and pilgrimages to holy places, in the hope of recovery (Kittrie 1973, Chapter 2). In the Renaissance period, the previous isolation through exclusion was replaced by a philosophy of isolation through confinement or separation. The 'great confinement' of indigent, old, and physically and mentally disabled began in mid-seventeenth century Paris (Foucault 1965). The purposes of the great 'hospitals' of Salpêtrière and Bicêtre were economic, social and moral. They were intended to increase manufactures, to provide productive work, and to end unemployment; to punish idleness, restore public order, and remove beggars; and to relieve the needy, ill and suffering while providing Christian instruction (Rosen 1968: 162–3).

The true birth of the asylum occurred toward the end of the eighteenth century when the distinctive qualities of madness led to a call for separate institutions for the insane. At Bicêtre, for example, the reformer Philippe Pinel began the classification of patients to calculate needs, observe symptoms and establish treatment. In England, the principles of 'moral treatment' led to further classification and isolation of patients, with concomitant change in hospital/asylum architecture (Thompson and Goldin 1975). During the nineteenth century, there was a large-scale expansion of asylums throughout Europe and North America. This expansion took the form of massive hospital structures situated on extensive rural campus. Once again, a spatial exclusion was being practised, albeit with an entirely defensible rationale for facilitating patient cure.

By the mid-twentieth century, asylums were overcrowded, and were reduced, in the majority of instances, to purely custodial care. As Thompson and Boldin (1975: 68) observe: 'Hospitals have a way of being conceived in glory, executed with ingenuity and humanity, then subjected in use to misuse and abuse, finally to be overcrowded and understaffed and forever plagued by insufficient funds.' Then, in the 1950s, a revolution in mental health care occurred. This was the time when a strong thrust toward a community-based mental health care was being experienced. The pressure for a community-based care derived from several sources, more especially the burgeoning evidence of the ill-effects of extended hospital confinement on the patient, and the counter-belief that a community-based care would aid in the normalization of the mentally ill. At the same time, large advances in chemotherapy enabled the effective treatment and symptomatic management of chronic patients without the need for confinement. These changes in treatment philosophy and capacity were sanctioned at the federal level by government intervention, on a cost-sharing basis, to promote a non-asylum based community mental health care. In both Canada and the

United States, the infusion of federal funds enabled local officials to shift the fiscal burden while simultaneously satisfying contemporary psychiatric and civil rights philosophies. The effect of the shift away from asylums to the community has been radical. In the US, for instance, there were 559,000 patients in state and county hospitals in 1955; this had dropped to 193,000 by 1975. Although the resident hospital census had dropped markedly, the rate of admissions and discharges to hospitals has skyrocketed. Additionally, a majority of patient care episodes now occur in community mental health facilities.

In short, the care of the mentally ill has been shifted into the community. Or has it? The development of the mental patient 'ghetto' prompts a fuller analysis of the isolationist philosophy in mental health care. Foucault (1977: 141–7) again provides insights. He points out that prison discipline – and by analogy, mental health treatment – proceeds from the isolation and separation of individuals in space. In penal systems, the most intense architectural manifestation of this principle was Bentham's 'Panopticon', which arranged individuals in isolated cells on tiered circles about a central observation area. More generally, the spatial separation of individuals for treatment requires four principles: (i) *enclosure,* the definition of a protected place of treatment; (ii) *partitioning,* an elementary principle of internal spatial organization in which each unit has its specific place; (iii) *functional sites,* in which internal architectural space is coded for several different uses, reflecting for instance, the need for therapeutic, administrative and work areas; and (iv) *rank,* the definition of the place one occupies in a classification hierarchy, in which status is not so much defined by place as by position in a network of relations. These principles enable professionals to describe a functional analytic space, and to allocate patients for treatment within that space. In short, the *isolation of clients is still practised, but a new spatial partitioning has been devised, based in the 'community'.* The partitioning uses spatial separation to control the elements in the mental health system, so that the roles and activities within the larger social environment become manageable. Thus, the social need for client treatment and differentiation is translated into a policy of community-based spatial isolationism.

The community and mental health care

The 'community' is placed in a highly ambiguous role by the requirements of the mental health care system. It is required, under the rubric of normalization, to act as an 'accepting host' to the mentally ill, and to provide some sort of support network to aid in their resocialization. However, large numbers of communities are simply 'rejecting' the mentally ill. Community attitudes towards the mentally ill are mediated through a complex set of predominantly non-economic relations in the social formation. These relations are symbolic in character, being organized as a system of signs with its own internal logic. The relationship between community and mentally ill is mediated by this set of symbolic representations, and these interdependent mediations define the situation of class structuration and conflict (cf. Habermas 1971, Chapter 12).

Community as normalizer

Little is known about the qualities which make for a good 'host' community for a community-based mental health service. It is generally acknowledged, however, that successful resocialization of the mentally ill will require a certain input from the community members. This does not mean that the community is involved in 'curing' the patient, but that its members are likely to provide the necessary supportive 'infrastructure' to help the patient back to independent living. Such support services may include assistance in shopping, or even home visits. Segal and Aviram (1978) have argued that the degree of client external and internal integration is significant in determining social integration in the community. External integration refers to those factors which facilitate the client's outreach into the community, and vice versa. These include the extent to which clients interact with neighbours, storekeepers and other service personnel outside the mental health profession. Internal integration refers to the extent to which an individual focuses upon life within the facility or lives a life mediated by the facility. Segal and Aviram concede that, for the chronic patient, the social support system offered by an institution may best suit the rehabilitative needs of the client. For the mentally ill who are mildly disturbed and asymptomatic, however, there is a great potential for integrating the client totally into the community. Segal and Aviram (1978, Chapter 12) have determined three basic components which act to produce a positive relationship between external and internal integration. In order of their importance, these are: (i) *Community Characteristics* including positive response of neighbours; (ii) *Resident Characteristics* including client satisfaction with living arrangement and therapy, and control over financial arrangements; and (iii) *Facility Characteristics* including the facility as an ideal psychiatric environment and the integration of clients with residents from the external community.

Smith (1976) also explored the qualities of residential neighbourhoods as 'humane' environments. He successfully used three descriptors of neighbourhood as predictors of the rate of recidivism in a cohort of discharged mental patients: commercial/industrial character; housing density; and population transience. Similarly, Trute and Segal (1976) tried to identify supportive communities for the severely mentally incapacitated. Such communities appeared to be those in which there exists neither strong social cohesion nor severe social dislocation. The former neighbourhood tends to 'close ranks' against the client; the latter tends to be too chaotic and threatening.

Community attitudes

The success or failure of a community-based mental health care will depend upon the community's attitudes toward the mentally ill. Research on attitudes suggests a contradictory mixture of sympathy and rejection. On one hand, we sympathize with the 'sick' person in need of care; on the other, we seek to maintain our social distance from the social outcast who manifests deviant behaviour. This confusion of motives was evident in a 1962 study of hospital personnel's attitudes toward mental illness. The OMI (Opinions about Mental Illness) was a multidimensional scale which resolved into five attitudinal factors: (i) *authoritarianism,* which implied a view of the mentally ill as an inferior class requiring coercive handling; (ii) *benevolence,* a paternalistic, kindly view of patients, derived from humanistic and religious principles; (iii) *mental hygiene ideology,* a medical model view of mental illness as an illness; (iv) *social restrictiveness,* viewing the mentally ill as a threat to society; and (v) *interpersonal aetiology* reflecting a belief that mental illness arises from stresses in interpersonal experience (Rabkin 1972).

In the wider (non-professional) community, it has proven extremely difficult to predict attitudes to mental illness and mental patients. Many characteristics of the patients themselves are important factors in community response; especially significant are anticipations that the mentally ill may be unpredictable and dangerous. Needless to say, however, the characteristics of the perceiver are also important parameters in attitude formation. For instance, the socioeconomic status of a community is important in predicting attitudes, as well as less traditional considerations such as a community's awareness and familiarity with mental illness (Jones 1978).

Since care for the mentally ill is usually associated with some kind of facility-based service, the opinions of the host community regarding the mental health facility itself are also important in securing client normalization. While attitudes towards patients may be the major determinant of community behaviour, it has frequently been observed that community opposition has been withdrawn following concessions in the form of design or structural alterations to a facility. These concessions have included the building of screening walls or enlarging the internal waiting area of a facility.

The dimensions of community attitudes toward mental health facilities have not been extensively researched. A range of tangible and intangible external effects seem to act as a major catalyst in behaviour. The former refers to a range of clearly definable (usually quantifiable) impacts of the facility in question, including anticipated property decline and increased traffic in the facility's vicinity. The latter group includes a wide variety of nonquantitative externalities, typically involving fear of personal or property security, and dislike of loitering clients (Dear *et al.* 1977).

Community exclusion of the mentally ill

Community attitudes and behaviour toward the mentally ill appear to depend upon resolution of an interdependent perceptual 'trade-off'. On one hand, there is a positive psychological benefit of helping a group in need; on the other, there is a negative, protectionist attitude toward one's daily-life environment.

The mentally ill, like other minority social groups such as the poor, are restricted in their selection of residence, workplace and recreational outlets. Their continued isolation, in asylums historically, and now in ghettos, can be interpreted as part of a wider system of socio-spatial organization which causes the separation of antagonistic groups. Thus, just as the processes of residential differentiation cause the appearance of class and ethnically-separated neighbourhoods, so similar processes tend to isolate and exclude the mentally ill. The community in opposition to mental health facilities employs two indirect sources of power to exclude the mentally ill: the power of socio-spatial exclusion; and the power of state authority as manifest through planning policy.

The power of socio-spatial exclusion operates at two separate levels: individual and group. First, the mentally disabled person is subject to a series of informal and formal exclusionary forces which operate at the individual level. Informally, a mental disability often tends to make the individual distinguishable in a social setting. Moreover, individuals have been observed to make personal behavioural adjustments to exclude the offending individual. More formally, organizational exclusion can occur, as when an individual is disciplined for aberrant behaviour in the workplace, for example. Secondly, and more important for present purposes, is the set of mechanisms of group exclusion. This refers to the generic ability of communities to exclude undesirable or noxious objects and people from their neighbourhoods. In an early study of exclusion of the mentally ill, Aviram and Segal (1973) recognized several strategies used by communities to place 'social distance' between them and the mentally ill. These included formal strategies, e.g. the use of legal (especially zoning) ordinances, and informal strategies, e.g. physical abuse of facility or client.

The community not only uses its own power, but also evokes the power of the State to exclude the mentally ill. Mental health planners have responded to increasing community opposition by developing locational strategies which minimize conflict over facility-siting decisions. As a consequence, the formal mechanism of state planning policy is brought to bear on the exclusion process. While some neighbourhoods are excluding the mentally disabled, other neighbourhoods (with less political clout) are being saturated by mental health facilities (Dear 1977).

If it is true that a limited environment of social resources has a significant impact on one's life chances, then it is evident that the household has an enormous stake in the local environment. Hence the need to protect one's environment from any undesirable negative impact becomes paramount. It seems likely that the entrance of the mentally disabled into a community is perceived as a threat to the environmental resource base of the neighbourhood, and hence the market capacities contained within it. Accordingly, the community's power for spatial exclusion is often marshalled to prevent the incursion of the mentally ill. State power, in the form of planning policy, is also implicated. When these forces are matched by the mental health professional's use of spatial isolation in care of patients, an inevitable force for ghettoization is constituted.

Conclusion

The problems associated with the ghettoization of the mentally ill are a manifestation of the requirements of reproduction in the social formation. The fundamental class relations are non-economic in character, in the sense that they are only indirectly related to the economic mode of production. The relationship between professional and client is both authoritarian and exploitative; that between community and client is exclusionary. In both cases, space is an integral force through which class relations are constituted and mediated.

The fundamental relationships among client, professional and community have undergone several mutations since the birth of the asylum. For instance, the assessments of client needs have changed drastically over the centuries. The 'manufacture of madness' has not occurred as a result of some capricious whim, but as a consequence of specific contemporary societal pressures. As societal norms alter, it becomes necessary to redefine needs constantly as the operational capacity of the sick individual is reassessed. This reassessment is the responsibility of the professional, and we have already discussed the proliferation of programs, the burgeoning bureaucracy, and the other manifestations of the expanding domain of the medical/psychiatric professions. The objective of the service professions frequently seems less directed toward client 'cure' and more concerned with regulating and managing the flow of clients through the case network. In spite of such developments, however, the fundamental relations (based in exploitative dependency, isolationism and exclusion) have remained unaltered through time.

This clarification of the structural limits of the ghettoization 'problem' poses three political problems.

1 *The growing tendency to 'psychiatrize' a wide variety of social, emotional and mental problems.* Service providers are a different class from service clients, and they have a vested interest in the continuing existence of mental health problems. The mental health system has tended to produce disabling effects in a population as a prerequisite for receiving care. It can be argued that the system does not 'cure' mental illness; that it 'produces' illness in clients and their social networks (especially the family); and that it encourages long-term dependency in those who enter the system. On the other hand, it must be emphasized that mental illness is not a chimera. For many, psychiatric care is literally a matter of life or death. But the temptation to label everyday 'problems in living' as 'mental illness' is surely to be resisted?

2 *The implication of the State in the reproduction process,* both in support of mental health professionals, and through the urban planning machinery. Illich (1977: 16) has argued that professionals have turned the '. . . state into a holding corporation of enterprises which facilitates the operation of their self-certified competencies'. The State, however defined, sanctions and protects professionals' power. Some

suggest that, in return, the psychiatrist acts as a conservative agent of social control and repression. This may be true. (It is certainly evident that psychiatry tends to encourage the individual to adjust *to* society, rather than addressing the social context itself.) However, what is of greater concern here is the precise nature of the links between the state élite and the professional élite, and how state policy (as in urban planning) is directed toward the reproduction of social relations.

3 *The creation and reproduction of 'client class'.* How is the client class induced to accept the label and the sick role, and to tolerate being 'treated'? How is client dependency created and reproduced? How do some deviants respond differently to signs which induce conformity in others? It may or may not be true that psychic and social repression is involved. What is certain, however, is that the power to provide treatment has been increasingly deeply inserted into the social system (cf. Foucault 1977: 80–2). At issue is whether or not the limits of this system have been appropriately defined. How do we design a caring social system which is not simultaneously coercive?

References

Althusser, L. (1971) *Lenin and Philosophy and Other Essays.* New York: Monthly Review Press.

Althusser, L. and Balibar, E. (1970) *Reading Capital.* London: New Left Books.

Aviram, V. and Segal, S. P. (1973) 'Exclusion of the mentally ill.' *Archives of General Psychiatry* 29:126–31.

Castells, M. (1976) 'Theory and ideology in urban sociology', in C. G. Pickvance (ed.) *Urban Sociology: Critical Essays.* London: Tavistock Publications.

—— (1977) *The Urban Question: A Marxist Approach.* London: Edward Arnold.

Cutler, A., Hindess, B., Hirst, P. and Hussain, A. (1977) *Marx's CAPITAL and Capitalism Today* 2 vols. London: Routledge and Kegan Paul.

Dear, M. (1977) 'Psychiatric patients and the inner city.' *Annals, Association of American Geographers* 67:588–94.

Dear, M., Fincher, R. and Currie, L. (1977) 'Measuring the external effects of public programs.' *Environment and Planning A* 9:137–47.

Engels, F. (1972) *The Origin of the Family, Private Property and the State.* New York: International Publishers.

Erikson, K. T. (1967) 'Notes on the sociology of deviance', in T. J. Scheff (ed.) *Mental Illness and Social Process.* New York: Harper and Row.

Foucault, M. (1965) *Madness and Civilization: A History of Insanity in The Age of Reason.* New York: Vintage Books and London: Tavistock Publications.

—— (1977) *Discipline and Punish: The Birth of the Prison.* New York: Pantheon Books.

Giddens, A. (1973) *The Class Structure of the Advanced Societies.* London: Hutchinson University Library.

Habermas, J. (1971) *Knowledge and Human Interests.* Boston: Beacon Press.

Harvey, D. (1975) 'Class structure in a capitalist society and the theory of residential differentiation', in R. Peel, M. F. Chisholm, and P. Haggett (eds) *Processes in Physical and Human Geography.* London: Heinemann.

Hindess, B. and Hirst, P. (1975) *Pre-Capitalist Modes of Production.* London: Routledge and Kegan Paul.

—— —— (1977) *Modes of Production and Social Formation.* London: Macmillian Press.

Illich, I. *et al.* (1977) *Disabling Professions.* London: Marion Boyars Publishers Ltd.

Jones, K. (1978) 'Public attitudes toward mental patients and discharged mental patients.' Unpublished paper, Department of Geography, McMaster University, Hamilton.

Kittrie, N. N. (1973) *The Right to be Different: Deviance and Enforced Therapy.* New York: Penguin Books.

Lefebvre, H. (1976) 'Reflections on the politics of space.' *Antipode* 8:30–7.

Magaro, P. A., Gripp, R. and McDowell, D. J. (1978) *The Mental Health Industry: A Cultural Phenomenon.* New York: Wiley Inter-Science.

Marx, K. (1971 edition) *Capital* 3 vols. Moscow: Progress Publishers.

Peet, R. (1975) 'Inequality and poverty: a Marxist-geographic inquiry.' *Annals, Association of American Geographers* 65:564–71.

Rabkin, J. (1972) 'Opinions about mental illness.' *Psychological Bulletin* 77:153–71.

Rosen, G. (1968) *Madness in Society.* Chicago: University of Chicago Press.

Segal, S. P. and Aviram U. (1978) *The Mentally Ill in Community-Based Sheltered Care.* New York: Wiley.

Simon, B. (1978) *Mind and Madness in Ancient Greece.* Ithaca: Cornell University Press.

Smith, C. J. (1976) 'Residential neighbourhoods as humane environments.' *Environment and Planning A* 8:311–26.

Soja, E. W. (1978) 'Topian Marxism and spatial praxis: a reconsideration of the political economy of space.' Paper presented at the annual meeting of the Association of American Geographers, New Orleans.

Thompson, J. D. and Goldin, G. (1975) *The Hospital: a Social and Architectural History.* New Haven: Yale University Press.

Trute, B. and Segal, S. P. (1976) 'Census tract predictors and the social integration of sheltered care residents.' *Social Psychiatry* 11:153–61.

Wolfensberger, W. (1972) *The Principle of Normalization in Human Services.* Toronto: National Institute on Mental Retardation.

Zola, I. K. (1977) 'Healthism and Disabling Medicalization', in I. Illich *et al.* (1977).

VI Urbanization and the political sphere

19 The analysis of state intervention in nineteenth-century cities: the case of municipal labour policy in east London, 1886–1914*

N. H. Buck

Introduction

This paper is concerned with the application of new approaches to the study of class relations and state intervention in nineteenth-century cities.

In the first part we shall consider the work of two groups of writers, the recent French Marxist urban sociologists, notably Castells and Lojkine, and the British Marxist social historians studying the nineteenth century. The major theoretical problem we shall consider concerns the specificity of the French analyses to the contemporary social formation, state monopoly capitalism, and hence the issue of whether they have anything useful to say about earlier periods. Our conclusion will be that their general orientation, the application of Marxist analysis to the study of the urban, and hence their conception of the urban as an arena for class conflict, and urban politics and urban policy as expressions of this conflict is extremely valuable. However it will also be argued that the concepts they make use of in analysing contemporary cities cannot be simply transferred to the urban in earlier periods. In particular, we shall suggest that the structuralism which dominates Castells' work leads him into a conceptualization of the urban based on a static view of the social formation of advanced capitalism. One of the benefits to be derived from a historical study of the urban is to show that the overall framework of urban policy

* This paper is based on research for a Ph.D. thesis at the University of Kent at Canterbury on class structure and local government activity in West Ham. I should like to acknowledge the assistance of my supervisor, Chris Pickvance, in the preparation of this paper.

and the definition of the urban which Castells adopts is itself the product of the previous development of the class struggle.

This point will be amplified in the second part of the paper in which we shall present a case study of state intervention in the borough of West Ham, in east London during the period 1886 to 1914. We shall focus on the local council's labour policy, which includes its approach to unemployment and municipal employment, an area which falls outside some French writers definition of the urban. We will use this example – which shows the failure of labour policy to become a permanently established local policy – to show that the scope of intervention by the local State is not a given, but results from the successes and failures of efforts by the working-class movement to extend it.

The analysis of class formation and state intervention in nineteenth-century cities

We shall begin the first part of this paper by discussing the growth of cities in nineteenth-century Britain, and its impact on contemporary observers. We shall then briefly consider the ways in which non-Marxist historians have studied aspects of this growth, particularly state intervention in the urban, and use this to establish a distinction between different forms of relation between theory and empirical data in history. Next, we shall discuss the two approaches to the analysis of urban politics – British Marxist social history and French Marxist urban sociology. We shall conclude the first part by contrasting these.

The growth of cities in nineteenth-century Britain

The impression which the new cities of the British industrial revolution made on contemporaries was immense. This was a result not only of their size, for they continued to be dwarfed by London, which was in many senses not an industrial city, but also because they appeared to be qualitatively new. These were not cities of trade or administration, nor cities of luxury expenditure or conspicuous consumption, but rather appeared to be purely centres of economic production. This reversed the traditional dichotomy (which was only partly valid) between the country, in which wealth was produced, and the city in which it was distributed and consumed.

The facts of the situation were startling enough. The number of cities in England and Wales with populations over 100,000 increased from one (London) in 1801 to forty-four in 1911. Cities of over 20,000 population contained 17 per cent of the total population in 1801 and 61 per cent in 1911. Individual cities experienced huge rates of growth, which census data if anything tended to underestimate, based as they were on static boundaries. Manchester doubled between 1801 and 1831, and doubled again in the next thirty years. Birmingham did the same, accelerating in the later period. Other towns, particularly suburban communities around conurbations experienced similar, or even higher rates of growth in the later part of the century.

The cities were large and growing, and their dynamism provoked an ambivalent response. On the one hand there was shock and hostility at the living conditions they offered. On the other, in addition to the eulogies of the bourgeois apologists, radicals recognized that the city could also be a source of change. Thus Engels wrote in *The Housing Question:*

> 'That the situation of the workers has on the whole become materially worse since the introduction of capitalist production on a large scale is doubted only by the bourgeois. But should we therefore look back longingly to the (likewise very meagre) fleshpots of Egypt, to rural small-scale industry, which produced only servile souls, or to the "savages"? On the contrary. Only the proletariat created by modern large-scale industry, liberated from all inherited fetters including those which chained it to the land, and herded together in the big cities, is in a position to accomplish the great social transformation which will put an end to class exploitation and class rule. The old rural hand weavers with hearth and home would never have been able to do it; they would not have been able to conceive such an idea, not to speak of desiring to carry it out.' (1975: 25)

This opportunity, or as others saw it, this dynamism and economic growth was bought at a substantial cost, which many were not slow to see. The city was giving rise to new forms of human association and to new social problems – though there is some question whether they were really new, or simply more visible: rural housing was quite as bad in some areas as urban. These newly-perceived urban problems were highly diverse – the new city was variously seen as being responsible for dreariness or new levels of squalor, an ugly uniformity of physical surroundings, a uniformity of people and thus the creation of a mass society, and as a consequence, the breakdown of local community, and hence individual isolation. Also following from this massing of people was a perceived threat to the social order – the mass could become the mob. Beyond this was the isolation, not of individuals, but of classes, the social segregation which progressively split the city into rich and poor residential districts, and split place of residence from place of work.

All these reactions came from a diversity of ideological positions, and they were instrumental in creating a new urban ideology which saw the source of problems in the city, rather than in the capitalist process as a whole. Engels, in *The Condition of the English Working Class,* admitted that the concentration of problems did make it appear that the city itself was responsible:

> 'The brutal indifference, the unfeeling isolation of each in his private interest becomes the more repellent and offensive, the more these individuals are crowded together, within a limited space. And, however much one may be aware that this isolation of the individual, this narrow self-seeking is the fundamental principle of our society, it is nowhere so shamelessly barefaced, so self-conscious as just here in the crowding of the great city.' (1973: 64)

The impression must not be given that once established the industrial city did not change in character, nor that new processes did not start. Indeed in this paper we shall discuss just such change, particularly the developing relation between intervention by the State in the urban and the urban working-class movement. The last part of the nineteenth century saw the beginnings of the former, and the resumption of the latter following a forty-year lapse after 1848. Apart from these changes over time, there was also a very great diversity of types of cities. In particular, London was very different from many northern English cities, in its size, in the diversity of activities taking place within it, and in the fact that, unlike many other cities of the period, it was not dominated by factory production. Unlike the northern industrial cities it was thus not the major focus of attention during the early part of the century. During the latter part of the century however it drew more and more attention, because of its size, and because so many of the 'urban' phenomena seemed to be shown up there in starkest relief. From the 1860s it is possible to write the history of government intervention in the city in terms of London – though of course the problems of other areas did not necessarily diminish.

Three types of history

Within the context of these changes in the urban, and the responses to them, we will be focusing on political responses, either in the form of central or local state action, or political conflict around this action. It is possible to write purely descriptive analyses of these processes, with relatively little attempt to explain them or link them to the development of society as a whole. Indeed there has been much work along these lines, some of which describes the processes very fully and effectively. One could quote here Hennock's work (1973) on the ideology and practice of recruitment of councillors in nineteenth-century Birmingham and Leeds, or Wohl's study (1977) of the development of housing policy in London, or Ashworth's book (1954) on the origins of town planning in the nineteenth century. It would be overstating the case to say that these writers provide no explanations at all, but they are all products of an Anglo-Saxon school of historiography, which has, in general, abstained from theory, and concentrated on the technicalities of historical research. Stedman Jones (1976) has argued that, even in the case where in recent years the historian has begun to import sociological theory to redress the balance, it has still been assumed that there was a division of labour, the sociologists providing the theory, the historians 'the facts', and the historians have been largely uncritical of the sociology that they do borrow.

There seem, in fact, to be three types of history, depending on the priority that is placed on theory, and the sort of relation established between theory and empirical data. These will be labelled descriptive, explanatory and theoretical history. To illustrate the difference between the first two one may compare Hennock's *Fit and Proper Persons* with Hays's paper (1964) on the politics of reform in American municipal government. These two works concern very similar themes in different national contexts, the changing character of municipal government and particularly the changing ideologies of representation and efficiency and the practices that lay behind them. The crucial difference is not the empirical data they have at their disposal, but rather the interpretations they draw from it.

Hennock clearly charts the growth of an ideology of municipal service, with its demand that more substantial businessmen could be represented on the councils, together with the changing composition of these councils. Unlike Hays, however, he fails to link this clearly to the struggle for control of local councils by large business, due to their growing need for local government services in the latter part of the century. The change in business attitudes to local government followed from their needs being thwarted, in the English case by 'economist' parties who were resisting high expenditure, and in the American case by the high cost of 'machine politics', with its focus on very localized needs. This class dimension is something which Hennock misses, although it is a tribute to the strength of his empirical work that his material can be readily reinterpreted in the light of this theme. He also fails to ask some questions which would be relevant, in looking further at the class roots of opposition to the municipal movement, and the relation between this movement and the working-class movement. Much of this Hays is able to do, and he is thus able to make a contribution to a theoretical understanding of intervention in the urban, which is only very indirectly possible for Hennock.

However, beyond this distinction between descriptive history and explanatory history there is a history which reverses the priority between theory and empirical material, and is concerned first and foremost with the establishment of a theoretical history. Marx's *Capital* is arguably the primary instance of this, constructing a theory of capitalist development out of a study of capitalist production in Britain in the nineteenth century and a debate with other theorists of the period, particularly the British political economists. The importance of this theoretical enterprise is twofold. On the one hand it transcends the atheoretical descriptive history, with its division of labour referred to above, by which historians do not produce theory, but import it from other disciplines. On the other hand it enables a dynamic theory of modes of production and social formations to be produced, as opposed to a static and formalistic structuralism which risks missing the processes which change social formations.

These three types of history are not hard-and-fast categories. In particular, leaving aside descriptive history, there is no firm dividing line between explanatory history and theoretical history in Marxist analysis, and these may be more appropriately seen as lying on a continuum. This continuum may be projected further, beyond theoretical history to cases where theory is used in an ahistorical manner, examples of which would be some forms of structuralism. Theoretical history represents an ideal position at the centre of this continuum.

We will now move on to discuss the two groups of writers, British Marxist social historians and French Marxist urban sociologists, and evaluate their relevance for the study of nineteenth-century cities. We will see that within the continuum proposed above, the British historians tend towards atheoretical history, while the French tend towards ahistorical theory.

British Marxist social history

Stedman Jones, at the beginning of his review of Foster's book *Class Struggle and the Industrial Revolution* says:

> 'If the best work of English social historians has largely grown within a Marxist tradition, that Marxism has been lightly worn. Mainly as a reaction against the positivism dominant within social science, English social historians have tended to disguise sharp analytical distinctions and eschew sophisticated quantification or explicit theorization. If their guiding lines have been Marxist, they have also drawn much from a native socialist tradition, a tradition which remembered *Capital* as much for its moral passion as for its theoretical achievement.' (1975: 35)

There have been more recent exceptions to this low priority on theory, particularly Foster's book, and Stedman Jones's own study, *Outcast London* (1971), but the point still stands that there has not been much sign of a theoretical debate among practising historians about the relation between their work and Marxist theory as a whole.

The theme which this group of historians, including Stedman Jones and Foster, has been specifically concerned with is the process of class formation of the British working class. They have focused much less on social processes in general or on the capitalist class. There has perhaps been an unhealthy split, in which the behaviour of the working class is social history, and may be susceptible to Marxist analysis, whereas the behaviour of the capitalist class is economic history, and has not been very closely studied by British Marxists in recent years, except in so far as this behaviour provides a stimulus to working-class action. It may be that this is because the latter type of study requires a more developed theoretical apparatus than British Marxists have in general been prepared to use.

One of the strongest developments in this field in recent years has been a movement away from studying working-class organizations, particularly trade unions, and working-class leaders – an approach pioneered by the Webbs around the turn of the century – towards a study of the local foundations of collective action and to the development (or not) of class consciousness. The central concern has varied between those studying different periods. Thompson, and others looking at the early part of the nineteenth century have been concerned with, in the words of Thompson's title, *The Making of the English Working Class,* its first formation of a political consciousness. On the other hand, those studying the period after the 1840s have to come to grips with the failure in further development in this consciousness, or even its collapse. From the end of the century there is again a political movement to be considered, but one flawed by reformism, and in Lenin's terms, a trade-union consciousness. Thus for the period with which we are concerned here (i.e. the latter part of the nineteenth century) the central focus is on the absence of class consciousness. Marxists are thus in a rather difficult position in arguing with non-Marxists, who have no expectation of the development of such a consciousness, and need to distance themselves from some theories of working-class incorporation, such as those of Marshall (1950) and Bendix (1964), which stress the idea of citizenship.

The main concept developed to meet this situation is that of the 'labour aristocracy', which emphasizes the division of the working class into two or more strata, one of which was in some sense bought off by higher wages or a level of control over the labour process, and developed its own ideology and characteristic set of institutions. This stratum, it is argued, while not incorporated, developed in isolation from the rest of the working class, and to the extent that it did provide leadership to the rest of the working class, provided a reformist leadership. This is, in brief, the argument that Foster uses to explain the collapse of the previously very strong working-class movement in Oldham in the 1850s. It is a theory which has a very long pedigree, stretching back to Engels and Lenin, and revived by Hobsbawm (1964, especially Chapter 15, 1970).

Some critics of the labour aristocracy theses, for instance Moorhouse (1978), have pointed out the high level of confusion which surround the concept, its periodization, the group of workers it applies to, and the extent to which it resulted from a conscious policy on the part of the bourgeoisie. Moorhouse suggests that it provides only a very partial explanation of the failure of the British working class, and he suggests that more attention needs to be devoted to mechanisms such as state education which reproduce the social formation at the ideological level. Nevertheless the feature of this debate which needs to be emphasized is that it is focusing on the nature and behaviour of the working class – a perspective which we will find is missing in the case of the French work.

Two other individual studies will be considered here. Foster's book, referred to above, is primarily devoted to a study of one town, Oldham, in Lancashire, during the industrial revolution, though comparative material is drawn from two others, Northampton and South Shields. Foster argues that by looking at the class consciousness of a town's inhabitants one is able to look at the town as a whole, 'because the degree to which labour was politically and socially united very largely determined a community's mass social structure – housing and marriage, language and politics' (1968: 281). In developing this emphasis on consciousness, Foster seeks to apply Lenin's distinction between trade-union consciousness and revolutionary class consciousness, and argues that the Oldham working class developed the latter in the 1830s and 1840s, being able to thwart efforts of social control by the State and the bourgeoisie, at least partly, by its control of the local political apparatus including the police. Foster has been criticized by Stedman Jones for forcing the concepts of class consciousness, revolutionary class consciousness and labour aristocracy onto his evidence, a criticism which does seem well founded. He does, however, point the way towards a form of community power study in which class conflict could be the organizing concept, as distinct from the American literature which treats power in an individualistic manner.

Stedman Jones's own study is very different from Foster's, and indeed from much other British Marxist social history, since he is concerned primarily with middle-class responses to the working class, moreover to a working class which was largely lacking in class consciousness – the casual labourers of inner London. He seeks to chart the growth of social policy in terms of these responses, particularly the fear of social disorder which casual labourers provoked. He is then able to show with some success the political and ideological processes behind this aspect of state intervention, and gives considerable substance to a view which would see the growth of such intervention as a response to a social threat, rather than as directly in the economic interest of a fraction of the bourgeoisie. The limitation of his analysis is that he is considering a very special case, since inner London lacked a substantial industrial bourgeoisie, and one may not be able to extend his argument to areas where this fraction had a direct interest.

It will be seen from the previous discussion that the British social historians do not form a unified group, and have many different theoretical and empirical interests. They do however have some common themes. Their great strength is their focus on working-class behaviour and working-class consciousness. As a corollary, however, they do not focus very clearly on the workings of the capitalist system as a whole, and there is a notable absence of theoretical debate about how their work relates to Marxist analysis in general. These points will be developed later, in a contrast with the French urban sociologists.

French Marxist urban sociology

In discussing the French Marxist urban sociology it is necessary to note that it has two distinct but related aims. On the one hand it attempts to found a new 'science' of urban sociology in place of the previous work in the area, which is seen as ideological; and on the other, it seeks to delimit the urban as having some intrinsic significance for a Marxist analysis of the capitalist process. Clearly certain problems follow from this, in that there is only a fine distinction between delimiting an area for study, and giving that area priority in the analysis of the development of the capitalist system as a whole, as for instance Lefebvre (1970) appears to come very close to doing. The approach may involve an acceptance of at least a modified version of what Castells calls the 'urban ideology' (i.e. giving causal priority to the urban in explaining social phenomena, rather than seeing cities as themselves the product of economic and social development). It will be argued that Castells does do this in some of his hypotheses relating to contemporary capitalism, but that this does not invalidate his whole approach.

Needless to say members of this school do not form a homogeneous group, but display significant differences in their approaches. We shall contrast the work of two writers here, Castells and Lojkine, beginning with the former, since he has been more explicitly concerned with the epistemological status of the discipline.

Castells (1976) argues that previous urban sociology has been ideological, being based on a false formulation of the nature of the urban, which gives it neither a specific real object, nor a theoretical object. He argues for a reformulation on the basis of a sociology of space and a sociology of collective consumption. The urban system is to be used to mean a spatially defined unit, delimited in terms of the economic, rather than the political or the ideological. Within the economic, however, it is to concern only one element of the capitalist process, that relating to the reproduction of labour power, whilst the process of production is to be studied at the regional level, in view of the lack of self-containment of the urban system in respect of this element. This restriction has been severely criticised by other French writers, particularly Lojkine, and it will be argued later that the definition is an instance of the effects of Castells's structuralism on his work.

Castells, however, then goes on to argue that this urban system cannot be understood as a static structure but only in the light of its transformations. Thus he says in *The Urban Question:*

> 'The heart of the sociological analysis of the urban question is the study of urban politics, that is to say, of the specific articulation of the processes designated as "urban" with the field of the class struggle and, consequently, with the intervention of the political instance (state apparatuses) – object and centre of the political struggle and what is at issue in it.' (1977a: 244)

Castells then distinguishes between two perspectives,

> 'depending on whether one places the stress on the structures or on the practices or, to put it more clearly, whether the analysis bears on a modification of the relations between the instances in the logic of a social formation or on the processes of its transformation. . . .' (1977a: 260–1)

He re-labels these two possible objects of research in urban politics 'urban planning' and 'urban social movements'.

There are two points about Castells's work that need to be brought out at this stage, which apply more particularly to his empirical work, and that of writers closely associated with him. In the first place his view of the State and of power in capitalist societies follows closely from the work of Poulantzas (1973), stressing the relative autonomy of the State from the capitalist class, and the dual function of the State, that is to say, the enforcement of the dominance of the bourgeoisie and the maintenance of the social cohesion of the social formation as a whole. The latter of these includes the regulation of contradictions between fractions of the bourgeoisie, and, to the extent that it is possible, between the bourgeoisie and the working class. Studies within this approach, such as *Monopolville* (Castells and Godard 1974) or *Institution Communale et Pouvoir Politique* (Biarez *et al.* 1974) thus focus extensively on struggles between fractions of capital, and particularly on the political expression of these conflicts.

Perhaps a fault that may be criticized is that they internalize most such conflict within the state apparatus. This may be even more serious when extended to the working class, where they tend to concentrate on politically organized expressions of working-class activity (except in the more specialised case of urban social movements) and perhaps ignore the material base for this activity, particularly where it is a working-class failure to act that is being considered.

There seems to be a division in the approach to working-class political power between the two areas of study 'urban planning' and 'urban social movements', which rests on the assumption that formally organized working class movements are in some sense incorporated when they are involved in conflicts over 'urban planning', whilst real 'urban effects' (i.e. gains from working-class struggle) are, by implication, only achieved where organizations operate outside the formal political system in 'urban social movements'. This raises a theoretical problem when a part of the state apparatus is controlled by a working-class party.

There would seem to be three alternative hypotheses about what happens in this situation. In the first place the relation between party and class interest may be weak, and the party may be in some sense co-opted by the capitalist State. This is the argument that Miliband uses in *Parliamentary Socialism* (1972) to explain the failure of the British Labour Party to achieve a fundamental transformation of British society. Secondly, there is Poulantzas's argument (1973: 115–17) that the power of an institution is not intrinsic to it, but is a function of the class interests controlling it. It would follow from this that working-class control of a part of the state apparatus (even where one could ignore the distance between party and class interest) would not imply a fundamental change unless it reflected a favourable balance of forces in the class struggle. The third hypothesis would be that working-class control could achieve significant effects. These hypotheses do not exclude one another, but they do need to be the subject of empirical investigation, and it is perhaps a legitimate criticism of Castells that he does not squarely face the issue of the possibility of working-class gains within the area 'urban planning'.

The second point is that, in addition to the methodological propositions which Castells makes about the formulation of a new urban sociology, he is also making a number of substantive statements about tendencies in the social formations of monopoly capitalism. In addition to the thesis that the reproduction of labour power is a valid independent object of study, he is also saying that it is becoming increasingly important in advanced capitalism, and that there is a transition from simple reproduction of labour power (e.g. food, shelter, etc.) to extended reproduction (e.g. education, health, environmental and cultural facilities) and moreover that the State is becoming increasingly involved in this process.

There is a diverse group of arguments for this process. In the first place the labour process is increasing in complexity, and whereas in early capitalism undifferentiated labour may have been sufficient, in later phases higher skills are necessary. There are serious problems with this argument.

On the one hand it does make sense to argue that a higher level of the organic composition of capital, and the necessity of higher levels of productivity do indeed demand a higher level of work discipline and regularity. Thompson (1968), for instance, has convincingly argued that there is a problem in the transition to a factory economy of imposing a time discipline on a workforce used to agricultural routines. When one comes to consider skills, however, certain problems arise, since most evidence about the production process itself suggests a process of *deskilling* rather than increasing skills. Many of the most serious struggles waged by capitalists in the nineteenth century, and since, were aimed at destroying skilled artisan élites who had strategic control over points of the production process, and routinizing these processes to reduce the differentiation of labour. Braverman (1974) shows that this strategy is being used in the twentieth-century American economy, and applies to the office as well as the factory. It must be noted that Freyssenet (1977), a French writer associated with the urban sociologists, has also been concerned with this issue, and has developed the idea of *deskilling* by showing how it relates to a process of *overskilling* of a small part of the labour force. There has of course been a substantial application of scientific knowledge to the production process and an enormous increase in the size of a strata of professional and technical non-manual workers, but this seems an insufficient basis for explaining the scale of the expansion of the extended reproduction of labour power.

There remain two subsidiary arguments. On the one hand, the changes in the reproduction of labour power are to be seen as gains of working-class action; and/or (the two are not exclusive) extended reproduction is doing more than simply improve the quality of labour applied in the production process, it is also reproducing a set of social relations, in other words, ideological reproduction is the key. Whichever is the case, it does seem to be a valid criticism that neither Castells nor indeed any other writer in this group devotes enough attention to the reproduction of labour power in this social relations sense.

Castells does however consider an alternative explanation for the changes in this area, through the argument that: 'The economy of advanced capitalist societies rests more and more on the process of consumption, i.e. the key problems are located at the level of the realization of surplus value or, if one prefers, on the extension of the market' (1977b: 63). As a consequence the State is involved in supporting new social forms which promote consumption – e.g. suburban owner-occupation, with its enormously increased demands for private means of transport and domestic consumer durables. Castells pursues the point by arguing that there is a tendency for consumption facilities to be consumed collectively rather than individually, and for management to be increasingly concentrated in the State. As a consequence, he argues, decisions in this area become increasingly politicized and lead to social movements around these issues.

The detail of this argument is not crucial for our purposes, but it does bring up the question raised earlier of how justified we are in separating

Castells's epistemological theses about urban sociology from his substantive points about the urban in advanced capitalism. In fact we need to distinguish three aspects of his work, his general orientation, the application of Marxist analysis to the urban, and urban politics in particular; secondly, his substantive hypotheses about the urban in advanced capitalism, for instance, the theory of collective consumption referred to above, and thirdly, the structuralist epistemology which binds them together. It is a major argument of this paper that this structuralism works in an ahistorical manner which serves to sever the study of the urban from its roots in the historical dynamic of class struggle. In particular, it will be argued that Castells's conceptualization of the urban is specific to one social formation, and that the elements of this conceptualization are themselves the outcome of the historical class struggle. As long as one accepts the structuralist epistemology which sees class conflict as internalized within a social formation, one fails to grasp the dynamic processes which ensure that a social formation can never have a static existence but must be in continuous transformation. Having said this, there are elements of Castells's work which are valuable for the study of nineteenth-century cities, in particular his central proposition that the relations of classes are an essential factor for the understanding of urban policy and urban politics.

The problem of structuralism does not arise with Lojkine's work, since he identifies the roots of urban policy more clearly with the contradictions of capitalist development. His starting point is Marx's concept of the general conditions of production which he interprets as the forms which mediate between the immediate process of production, and the overall process of production and circulation of capital. The urban thus refers to spatial interdependencies between parts of the capitalist economy. He argues that these interdependencies create contradictions between capitalists, which only the State is in a position to attempt to resolve.

There are three types of contradiction involved. In the first place there are limits to the capacity or willingness of individual capitalists to finance urban facilities, which are necessary but do not meet normal profitability criteria, either because of their very high organic composition (e.g. means of communication) or because possible returns are too slight (e.g. working-class education). The State, by providing devalorized capital (i.e. capital on which below-average returns are obtained) for these facilities solves the immediate problem of provision, but at the cost of creating longer-term contradictions by raising the organic composition of social capital. The second contradiction follows from the anarchic competition for space which is implicit in the capitalist system, which will not necessarily lead to a rational allocation of activities in space, even for the individual capitalist, and hence may lead to a need for state intervention to control the use of space. The third contradiction is linked to this, and follows from the private appropriation of land which Lojkine argues is increasingly tied in with monopoly capital and is an obstacle to 'socially combining the means of production, and the means of reproduction' (Lojkine 1976: 135).

It will be seen that the main difference between Castells and Lojkine is in their respective emphases on the political and economic levels of the social formation. While Castells places a very strong emphasis on political processes, Lojkine is concerned to see urban policy as 'an active reflection of the relations between the different classes and class fractions' (1972: 8) and in his empirical work he tends to assume that the relation between class interest and political practice is not problematic. Moreover he does not clarify the relation between state urban policy and the strategies of private social 'actors', and is not clear about the 'autonomy' (or otherwise) of the State from capitalist fractions. On the other hand, one of his theoretical strengths (though it is not altogether clear how it works out in practice) lies in his stress on the contradictory character of state actions, that the State is an *active* reflection of class conflict and that state attempts to resolve urban contradictions can in fact only exacerbate them. It is not entirely clear whether he means that this happens directly (i.e. that because of its class character state action necessarily exacerbates the visible economic form of the contradiction), or indirectly (i.e. that the State can only partially resolve the contradictions, again because of its class character, and so, by becoming involved, introduces the contradiction at the political level). If the former, then he provides little empirical evidence for it, and if the latter, then he is not saying anything very different from Castells (see Pickvance 1977: 244–5). Nevertheless his stress on the fact that the relation between urban policy and urban contradictions is not that of a safety valve on a smooth-running machine, but a relation which may develop in contradictory ways, is very valuable.

In evaluating the differences between Castells and Lojkine, the former seems to have a much stronger grasp of political processes, while there are serious problems in the latter's treatment of the political. On the other hand Lojkine's grasp of the economic contradictions at the heart of the urban as an object of study is much stronger, and leads him to a much more satisfactory conceptualization of the urban than Castells's. While he considers the means of reproduction of labour power, and the collective means of consumption, Lojkine criticizes Castells for focusing on these exclusively. He argues that this involves a series of confusions, stemming from the definition of production as a material process, and failing to recognize it as a social process. He argues that this

'tends in effect to reproduce the ideological division, which is dominant in classical economics, between the economic, identified with crystallised labour – and the social – identified with living labour. Within this framework, the urban appears as the domain of "non-labour", as the group of activities which take place "outside the enterprise", otherwise known as "consumption".' (1977: 146).

Having indicated the main lines of the new French urban sociology we now turn to a consideration of its usefulness for the study of nineteenth-century cities.

The problem which immediately arises is whether this school reformulates its object, the urban, as something which has only a negligible real existence in earlier periods. If the object is state intervention in the reproduction of labour power then it may indeed be of limited application, and though it would be possible to find a significant 'reality' in this area in Britain after about 1870, it would be much more difficult before that date. There is a paradox here, in that the development of facilities which Castells and Lojkine identify as urban has been coterminous with the dissolution of the real basis for the ideological distinction between urban and rural, and indeed the national development of these facilities has actively served to break down the division between town and country. The development of mass education has been as significant in rescuing a part of the population from 'the idiocy of rural life' as in reproducing urban labour power. Transport links similarly act both to aid the circulation of capital and to draw rural areas into the capitalist economy.

There is an alternative to defining the urban in these terms, which is to focus urban sociology on the political economy of spatially defined units. Pickvance, in a review of *The Urban Question* suggests:

'In brief, Castells's first conceptualization of the city as an arena for the articulation of contradictory processes and elements of the whole social structure is, in my view, not only a soundly based theoretical critique, but is also capable of explaining the specificity of the urban as a domain of experience.' (1978: 176)

This formulation in no sense diminishes the importance of the political, since Castells's second formulation (i.e. the urban as the unit of reproduction of labour-power) had merely focused on one form of the State, the interventionist State of monopoly capitalism, and the task is opened up of understanding the role of different forms of the State in the urban, particularly the laissez-faire State of the competitive phase of capitalism, and the transitional form between this and the contemporary form.

This formulation of urban sociology also makes clear, as the other does not, the relevance of the British social historians. If the class consciousness of the British working class developed only very falteringly after the middle of the nineteenth century, it is even more difficult to find evidence of political movements around 'urban' issues (i.e. issues relating to the reproduction of labour-power) which could meaningfully be a separate object of study. Thus we cannot look to the British social historians either for discussions of 'urban social movements', or, with the exception of Stedman Jones's work, for an analysis of state intervention in the 'urban', due to their tendency, indicated above, to focus on the working class as independent agents. Rather, their usefulness in an urban context comes from this reformulation of the urban as a spatial site, and the study of class conflict at the level of the local community. The importance of, for instance, Foster's study becomes apparent from this perspective.

It should be clear from the outline above of the two contrasted approaches, British social history, and French urban sociology, that they have rather complementary strengths and weaknesses. The British historians focus predominantly on the working class as independent actors, on working-class consciousness and also on the material roots of ideology. The French, on the other hand, focus on the State as the initiator of action, and on the functioning of the capitalist system as a whole. As a complement the French seem to discount independent working-class activity, while the British work, by implication, seems to see the State solely as a responsive actor. This may relate to real differences between the British and French States. Duclaud-Williams (1978) has argued that the formation of social policy (particularly housing policy) is much more strongly related to party political conflict in Britain than in France, where there is a broadly supported tradition of state intervention in the interests of efficiency. The French State may thus appear more strongly to be above the class struggle. The French urban sociologists do see the State as primarily a factor of cohesion within the social formation, and although Lojkine, in particular, pays lip-service to the idea of responses to threats from the working class, even he does not study this in a very concrete way. On the other hand, for the British writers, the State is primarily seen as a responsive aparatus, with much less autonomy from the bourgeoisie. The pitfalls for the French work are functionalism and structuralism, for the British, instrumentalism. The problem in carrying out research on nineteenth-century cities from a Marxist perspective is to know how far it is possible to be eclectic in using these two approaches. The relation between the two approaches will be explored further in the second half of this paper.

A 'production relations' approach to local state intervention

In the second part of this chapter, we shall present a case study of intervention by a local state apparatus in a nineteenth-century city. We shall consider the labour policy of the Borough of West Ham in east London during the period 1886 to 1914, and the political conflicts around this policy area, which includes policy relating to unemployment and municipal employment. This is an area of policy which falls outside Castells's definition of the urban, and would be marginal even for Lojkine. We shall argue, however, that the scope of local authority action, as well as the particular forms it took, was the subject of political conflict, and that working-class political parties, which controlled the council for some part of the period, were seeking to use the local State to effect fundamental changes in the relations of production. We shall argue further that to separate out such policy areas from intervention relating to the reproduction of labour power, or the general conditions of production is to achieve only a partial view of the nature of state intervention in the urban.

We shall begin by outlining some of the differences between the late nineteenth-century British economy and political system and the situation now, and then briefly indicate the character of West Ham at this period, before considering in more detail the case study of labour policy.

In the last third of the nineteenth century Britain was losing her position of hegemony in the world economy, in terms of manufacturing output and exports, though she still dominated the world's capital market. This was against a background of rapidly changing technologies, and an increase in the scale of production and a concentration in the ownership of capital. These tendencies, however, were in their early stages, and most manufacturing was in much smaller units, compared to the contemporary situation. London's economy was an extreme case of this, being dominated by very small-scale production, very little of it even in factories. London lacked the industrial bourgeoisie who were present in most northern cities, and its hegemonic fraction was instead a commercial and financial bourgeoisie, oriented towards the international economy.

As we have suggested earlier, labour as a political movement was relatively weak in Britain in the mid-nineteenth century, and this remained true up to 1914. Nonetheless the years from about 1880 did see a substantial advance in the political organization of the working class, including the growth of independent working-class parties, particularly the Labour Party, founded in 1900, and the development of general trade unionism, beginning from the gasworkers strike and the dock strike of 1889, as opposed to sectional craft unionism which had dominated the movement before, and had acted as a brake on political militancy.

National politics at this period were dominated by two bourgeois parties – the Liberals and the Conservatives (also known as Unionists). Although they represented different fractions of the bourgeoisie, both parties emphasized a non-class approach to politics, and sought working-class electoral support, presenting themselves as progressive parties. There was however some measure of class polarization, and the Liberals were more successful in gaining working-class support, while the Conservatives had more success with the suburban petty bourgeoisie. However other factors, notably religion, disturbed the identity between party and class.

Changes in the nature of the State are perhaps more difficult to grasp. It is clear enough that, at the level of state action, there has been an enormous expansion of intervention in social and economic policy and a corresponding increase in state expenditure, since about 1870. By 1914 the expansion was already well under way, and the State had a significant interventionist role, though well short of the situation now. In considering the character of the State and of state policy it is more difficult to know whether one is seeing real changes, or only apparent ones.

There are two points which are relevant for us here, both relating to the definition of state functions and their flexibility. In the first place there was a rhetoric of laissez-faire prevailing in the nineteenth century which implied that state intervention was in principle unjustifiable, unless it was doing something which individual citizens could not do on their own. This was breaking down gradually in the latter part of the century, and thus the question of the legitimate scope for state intervention was itself a political issue. Secondly, in contrast to the present-day ideal (or ideology) that state policy should be consistent and explicit, in the late nineteenth century for most areas of central government domestic policy (excluding fiscal policy) it would be very difficult to specify what the policy was. It may even be misleading to suggest that there was such a thing as, for instance, government housing policy at this period. There was, rather, a body of legislation, most of it permitting rather than compelling local authorities to act, without any strong support from central government behind it. Policy in most areas developed in a highly incremental manner and moreover initiative came as much from local as from central government.[1]

West Ham

West Ham was a parish on the eastern edge of London, on the north bank of the Thames. It was established as a borough (i.e. an autonomous local government unit) in 1886. It was growing very rapidly at this stage, and had already reached a substantial size. The population had grown from 18,817 in 1851 to 204,903 in 1891. After this the rate of growth slackened somewhat, but the population reached 289,030 in 1911. In spite of this large population and its proximity to the centre of London – four miles at its nearest point – it was not included in the London County Council when this was established in 1889. From an economic point of view however it was very closely integrated within the Metropolis. It was the site of the largest of the dock systems, the Royal group, and associated with this was a substantial group of factories. Indeed, in about 1907 the Borough was advertising itself as 'the factory centre of the south of England', and certainly it contained the greatest concentration of large-scale industry in London until 1918. It was thus rather atypical of London, which, as suggested, was dominated by small-scale craft production, mainly of articles for final consumption. West Ham on the other hand was dominated by such industries as chemicals, processing of imported foodstuffs (e.g. sugar refining) and capital goods industries such as shipbuilding and electrical cables, and a substantial railway locomotive-building works was located in the north of the borough. The contrast with London as a whole is thus very striking: where London was dominated by small archaic industries, West Ham was dominated by large industries, many of them in the most technologically advanced sectors of the economy.

There was a significant difference between northern and southern parts of the borough, with most of the industry concentrated into the latter, while the former contained large numbers of commuters into central London. This also led to differences in class structure, since the southern part contained a large casually employed working class associated with the docks as well as factory workers in process industries, while the north contained clerks and more skilled workers commuting into London, along with skilled workers at the railway works. As we shall see, this had consequences for the working-class political movement, creating divisions within it.

As was pointed out above, the relation between party and class at the national level was by no means simple, particularly given the slow development of the Labour Party. This was equally true of parties at the local government level, where the tradition of non-party government persisted much longer – and indeed persists to this day in some areas. Further, there was not necessarily any correspondence between national parties and parties on local councils, even where they did exist. On West Ham Council, in its first ten years after 1886, party groupings are not usually discernible in the practice of council business, but did arise during elections. Within the Council many members saw themselves as progressive, and were prepared to support trade-union issues, though in social composition the Council was overwhelmingly dominated by the petty bourgeoisie: builders, publicans and small shopkeepers.

Against this background there was a growing representation of labour[2] on the Council – this was not a single party, but a group containing members from two parties, and a number of unaffiliated councillors. Of the two parties, the Independent Labour Party (ILP) and the Social Democratic Federation (SDF), the latter were the more radical, and indeed were the only British political group at this period to be avowedly Marxist. They tended to be more successful in London than the ILP, whose strong bases were in the industrial areas of the north. In London the working-class political movement was particularly weak before the First World War, with the inner areas in particular castigated by socialists for their lumpenproletarian character. Such areas of working-class political strength as there were can be identified with areas of larger-scale industry, which had a genuine factory proletariat – such as Woolwich and Poplar to the east and Battersea to the south-west. In the period before 1914, however, the labour movement in West Ham was the most successful, and, moreover, combined this success with the dominance of the more radical SDF.

After a steady growth from 1889, labour representation on West Ham Council reached a majority in 1898 – becoming the first local authority in Britain to come under Labour control. The achievement of this labour majority, which was followed by the organization of the separate elements into a Labour Group with a fairly high level of internal discipline, did not go unchallenged. The Conservatives and many of the Liberals on the Council reorganized themselves into a body known as the Municipal Alliance, with the support of local industrialists. The Municipal Alliance was able to regain control of the Council two years later in 1900, partly as a result of a number of splits within the Labour Group, and from this date the Council was organized explicitly along party lines. The Labour Group maintained considerable political strength over the next few years, and regained control for two years in 1910, in coalition with progressive Liberals; in 1917 they achieved a control which was to prove permanent.

However, there were changes in political approaches over this period. In the first period of control many members of the Labour Group saw the Council as a base for making fundamental changes in society and they did not accept the 'rules of the game' of council practice. As the limitations of an individual local authority as an instrument of change became apparent, the Labour Group retreated to more limited aims. Although there were distinctive differences in policy between the parties in the later period, they were within the confines of the generally accepted scope for local government activity.

A 'PRODUCTION RELATIONS' APPROACH TO LOCAL POLITICS It will be argued that in many ways policy on municipal labour and unemployment was central to the concerns of the Labour Group in West Ham, and of British working-class politics in general at this time, which tended to be primarily concerned with the relations of production rather than with consumption issues or social welfare (with the possible exception of public housing). Pelling (1968) in fact argues that much of the working class viewed social reform as more of a threat than a gain to be made from political action, and that this view was born of the experience of class control that was implicit in much previous social policy. While this does not apply to all working-class organizations, it does seem that many of the socialist groups saw their task as transforming the State, rather than extracting reforms from it, and focused this transformation on the relations of production. This 'production relations' approach to politics was brought to local government, which was seen by many as the part of the state most readily captured. It follows from this that an exclusive focus on collective consumption issues, as Castells's approach would suggest, will ignore a central element of the class struggle over the local State. The 'production relations' approach also saw unemployment and municipal employment as related aspects of a single issue, the labour market, and not as separate problems, except in so far as their incidence reflected differently the state of the economy and the level of unemployment.

An example of this view comes from a socialist election manifesto in Bradford in 1897:

'Is there any reason why the municipality should not itself organize its industries so that it could supply almost every public want without the intervention of the profit-monger? Why should not a corporation have its own quarries, ironworks, printing office and workshops for various industries? The city of Bradford alone could give employment to some thousands of workers, guaranteeing good wages and reasonable conditions of work and saving the middleman's profit. . . It is to municipal employment that we must look for the beginnings of that industrial reform which will one day solve the problem of the unemployed. Here lies the hope of the worker who desires constant employment.' (Quoted in *The Times* (10 September 1902): 4)

In West Ham itself, one Labour councillor's manifesto of 1898 stated that: 'in municipal progress and development will be found the line of least resistance to the political, social and industrial emancipation of the working masses' (quoted in Hobsbawm 1974: 134). Another councillor, in a pamphlet on West Ham (Terrett 1902), made it clear that he regarded the treatment of labour questions as being the key test as to whether municipal activity was to be regarded as municipal socialism in any real sense, or communal capitalism. Proposals relating to local labour-market conditions vied with housing for the central place in Labour Group manifestos, and in most cases were predominant.

A wider range of political groups, including some radical Liberals, thought local authorities were in a position to improve the working conditions of workers in many industries. This was central to the demand for the municipalization of the transport system made in Webb's *London Programme* (1891), and the demands for the establishment of direct labour organizations. Local authorities could also provide greater regularity of work than private employers, and could thus mitigate the casual employment which was endemic in London, particularly in the docks and the building industry.

Conservatives were also concerned about the relation between the local labour market and the local authority, but from a very different perspective. They argued that municipal employment, and municipal provision of relief for the unemployed could amount to electoral bribery, if the employees and recipients were also voters. This variety of political views supports our treatment of labour policy as a single policy area, quite apart from the theoretical coherence of local authority policies affecting the local labour market, even if this does not always appear in the way the policies are undertaken.

The two elements of labour policy which will be considered here in some detail are, on the one hand policy relating to unemployment, particularly the provision of work by the local authority, and on the other hand, direct labour, i.e. the use of workers employed directly by the council on major projects, rather than putting the projects out to tender by private contractors, which was the usual practice. There are three other elements of labour policy which can only be referred to in passing: (i) the insertion of clauses in contracts for work carried out for the council specifying minimum rates of pay and conditions of employment for the contractor's workers (a practice which was aimed at exerting some influence over the local labour market outside the direct control of the council – the clauses would support trade-union struggles to raise wages and extend their bargaining power), (ii) the fixing of rates of pay and conditions of work of the council's own workers, and (iii) the general support given to local trade unions.

UNEMPLOYMENT   The important distinction we shall draw concerning policies dealing with unemployment is between a 'production relations' approach which saw the solution as restructuring the labour market, and a social welfare approach, whose objective was to provide relief for the effects of unemployment without altering the labour market. The balance of political forces was such that the latter approach had prevailed by the time of the National Insurance Act of 1911. However, socialist parties had tried to use local authority relief works as a first step towards the recognition of their policy for dealing with unemployment, which was that the State should take responsibility for providing work for the unemployed and recognize 'the right to work', which they thought would undermine the capitalist labour market. We are concerned here with conflicts at the local level over this particular issue.

In addition to the conflict over policy at this period, however, there was also a major problem in the allocation of responsibility for dealing with unemployment between various central and local government bodies. This was a result of the breakdown of the Poor Law system, established by the Poor Law Amendment Act of 1834, and, linked to this, the growing recognition of unemployment as a problem separate from poverty and pauperism.

The Poor Law system, which had been responsible for dealing with poverty, was administered at the local level by elected Boards of Guardians (separate from other parts of the local government system). The Boards could relieve poverty either by giving out-relief in the form of money payments or payments in kind, or indoor relief, confining the recipient to the workhouse. The system had been developed with a philosophy of deterrence, enshrined in the concept of 'less-eligibility', i.e. that no provision of relief should provide a greater level of comfort than could be achieved through the most poorly paid work in the open market. It was intended that the receipt of relief should carry a heavy moral stigma, and this accorded with the prevailing belief that poverty was a consequence of personal failings such as lack of thrift, idleness or drunkeness. This stigma was reinforced by the disenfranchisement of those who received poor relief in any form.

Harris (1972) argues that the middle-class consensus around the Poor Laws was breaking down from the mid-1880s and unemployment was beginning to be recognized as a major cause of poverty, and recognized as being independent of the efforts of the individual worker. This new perception, and the strains on the administrative system for dealing with unemployment may be linked to four trends.

In the first place, the persistence of the Great Depression (1873–96) led to high unemployment in many sectors, and in particular extended long-term unemployment to many groups of workers who in the past had been able to manage with accumulated savings during relatively short periods of seasonal unemployment. Such groups were highly resistant to resorting to the Poor Law system, and it was explicitly recognized in the circulars of the Local Government Board that an effort should be made to avoid identifying those groups with the normal recipients of poor relief.

Secondly, the issue was becoming highly politicized, and the fact that there were working-class parties in this period which could articulate the issue gave the unemployed much greater political force than in earlier periods. In particular there were substantial demonstrations of the unemployed in London in 1886 and 1887, which, as Stedman Jones (1971) shows, caused considerable middle-class panic.

Thirdly, the Boards of Guardians were themselves changing, and becoming more susceptible to political influences. From 1894 the franchise for the election of guardians was extended, and the property qualifications for the posts removed. In some London Boards, particularly Poplar, this allowed working-class parties to gain control, and change the directions of policy, so as to expand the level of out-relief as against indoor relief, and raise the amounts paid in relief. The West Ham Board of Guardians (which also covered East Ham, Leyton and Walthamstow) gave a relatively high level of out-relief, compared with other east London Boards, but this did not result from a high level of Labour representation, for here Labour strength was rather weak. Contemporaries (e.g. Howarth and Wilson 1907: 343) were more inclined to account for it as lax administration. As we shall see, the main conflicts over unemployment policy were in the Borough Council, rather than the Board.

A fourth cause, which may account for some of the increased concern with unemployment, though it is more difficult to find specific evidence, is a growing awareness that unemployment could impair industrial efficiency, by its effects on the health of the worker and by the socially harmful effects of the confusion between unemployment and pauperism.

The central government was thus caught between a number of contradictory impulses. On the one hand it had to deal with unemployment and find new approaches in response to the failure of the Poor Law. It was constrained both by political forces which resisted higher expenditure, and by those which resisted the removal of the subject from local competence. But on the other hand they were concerned that the subject was becoming too politicized at the local level, and that at all levels of government the issue gave too much help to working-class political parties.

The starting-point for West Ham Council's involvement in the unemployment issue was the Local Government Board's circular in the winter of 1893–4 suggesting that local authorities should programme their works so as to relieve seasonal unemployment in the winter months. A number of local trade unions sent petitions and deputations to the Council urging them to implement the circular. The Council opened a register of the unemployed and started some works, partly funded by the Council, and partly by a relief committee, whose main source of funds was Arnold Hills, the director of a large local shipbuilding firm, who had a reputation as a progressive employer.

The work done by this method proved to be very expensive, in part because it was being done during a hard frost, and when the West Ham Relief Committee renewed its offer of assistance in the following year they specified that the work should be done at piece rates, with a minimum of 4d per hour, rather than at the time rate of 6d per hour paid the previous year, which was the trade-union rate for general labour. The Council accepted this, which involved them in matching the £1,000 offered. However local trade unionists protested that this involved the cheapening of labour, and might encourage other employers to cut wages. Will Thorne, the Labour leader on the Council, organized a large deputation to the Council meeting a fortnight later, and the motion accepting Hills's offer was rescinded by a large majority, so that two relief schemes operated, one paying 4d per hour and the other, the Council's scheme, paying 6d. Hills, discussing the meeting, said that Thorne had

'brought up a contingent of his friends with bands (they brought a red flag into the Council Chamber), and a large crowd came down, consisting of about 500 men, who stood outside, and, practically, brought a good deal of terrorism to play on the Town Council, and they rescinded that resolution by exactly the same majority as that by which they had accepted the offer a fortnight before.' (Select Committee 1895: 154, Question 2535)

The whole episode raises the question of the influence of the working class on the Council in the period before 1898, when labour representation became predominant. It is perhaps surprising that the Council should have been prepared to reverse their decision. The main difference between the two schemes was that the lower rate of pay would have served to mark out-relief work more clearly from the rest of the labour market, while the payment of trade-union rates would reinforce the principle that the State was providing alternative employment. In fact, the character of the work offered, and the way in which it was managed was such as to make it quite clear that it was relief rather than employment that was being offered, whatever the rate of pay. The Labour Group was thus being optimistic in thinking that they had established a new principle. On the other hand, the rest of the Council, as well as the central government, may have failed to appreciate at this stage, that if the provision of relief works was going to be a long-term solution to the problem of unemployment, the works could no longer be organized solely in this small-scale, *ad hoc* fashion, and that larger more permanent works must have had a significant effect on the labour market. A further reason for the Council accepting the trade unions' case was that, as suggested above, party lines were much less firmly drawn than later and many Liberals saw themselves as progressives and representatives of working-class interests. The trade unions and organized labour were only just beginning to be seen as a political threat.

Alongside action at the local level, West Ham Council was also petitioning Parliament to take some of the burden of dealing with unemployment off their shoulders, claiming that the Local Government Board circulars had led to local authorities being considered the proper bodies to deal with the problem, without providing them with any powers. They argued that: 'The duty of finding employment for able-bodied persons out of work should be undertaken by the state, which alone possesses the means for dealing with a problem of such vast magnitude' (Select Committee 1895: 101, Question 1296).

Keir Hardie, the member for West Ham South, was also raising the issue in Parliament, and at a public meeting in West Ham in December 1894 at which Lord Rosebery, the Prime Minister, questioned Hardie's estimate that there were 5,000 unemployed in West Ham. As a response to this, local trade unions organized a census of the unemployed which showed that the true figure was in fact over 10,000. These and other pressures forced the government in February 1895 to establish a Select Committee to look into the subject, which concentrated on unemployment in West Ham. The results of the Committee were not very significant, partly because of internal dissension, partly because of a change of government from Liberal to Unionist in the middle of the year, and partly because the level of unemployment began to fall during 1895.

The fall in the level of unemployment from 1895 led to a reduction in interest in the subject on the part of West Ham Council, and though the unemployed register was reopened in the winter of 1895–6, it was on a reduced scale and was not reopened in the following two years. The Council, and also, it would seem, local trade unions, viewed relief works for unemployment as an exceptional measure for years of exceptional unemployment. However the state of London's labour market was such that at any time, and particularly in any winter, there was a high residual level of unemployment and a large amount of irregular work. The Labour Group recognized this, and reopened the unemployed register and provided relief works in the winters of 1898–9 and 1899–1900, which were not particularly bad years by the standards of the early 1890s. All these schemes were hampered by the small amounts of money available, and by the fact that there was no significant work being done, given that local authorities could not become involved in production for profit, and did not want to displace their permanent staffs. The total amount paid in wages was hardly significant as a solution to the problem. Thus in 1894 over a period of three months an average of just over £1 per head was paid to 2,152 applicants (£1 was less than the average weekly wage for unskilled labour). On the other hand, it was clear that large numbers of the unemployed were prepared to accept temporary employment but not poor relief, with all the stigma attached to it.

As the limitations on purely local action became apparent, the Labour movement, and particularly the SDF shifted their attention towards putting pressure on central government. From about 1902 there was pressure for grants to be made to local authorities to allow them to provide temporary work. The first, rather ineffective measure, the Unemployed Workmen's Act of 1905, provided limited finance to locally established Distress Committees to provide relief works for the unemployed. West Ham made fairly full use of this scheme, and was thus able to increase the scale of the works offered, but even so, in 1905–7 less than half the total number of applicants were offered any work. After the limited success of the Unemployed Workmen's Act, the Labour Party lost the initiative in the development of policy at the national level, and failed conspicuously to obtain any commitment by the State to provide work for the unemployed, or to establish the principle of the 'right to work'. Developments in unemployment policy focused rather on insurance and the improvement of information in the labour market through labour exchanges.

The provision of relief works by the local authority was a highly ambiguous policy. It did genuinely have implicit within it a major restructuring of the labour market, and thus seemed to be the proper objective for a 'production relations' approach to unemployment. On the other hand it would seem to have reached its limit in West Ham within the available administrative structures. It has to be questioned whether, as implemented, it was actually offering alternative employment. The policy was heavily imbued with the principle of 'less-eligibility', just as the Poor Law was, and to ensure that the applicants were not being drawn away from the regular labour market there were strict controls on the amount of work that could be offered to any one person. These limits were so low as to be insufficient to provide a subsistence without other help. Moreover, because the system was in reality no more than a disguised form of relief, it did nothing to change the character of the local labour market. On the contrary, as a contemporary study of labour in West Ham argued:

'Under the present method of administration the money spent acts largely as a subsidy to the employers of casual labour; and though it may for the moment do something to relieve the want due to irregularity of work, it actually tends to perpetuate that irregularity' (Howarth and Wilson 1907: 379).

Unemployment was thus rather a contradictory issue for the Labour movement to come to terms with. Whilst it was a highly visible issue, most policies dealing with it were palliatives, which did not deal with the problem fundamentally. Expenditure on the unemployed, which is likely to be at least partly funded out of taxation on the working class, may as suggested above simply be a subsidy to employers, releasing them from paying the full costs of reproduction of labour. Certainly, in a number of relatively skilled but irregular trades wages tended to be rather higher than in equivalently skilled but regular trades suggesting that the former contained at least an element of insurance against unemployment (e.g. Stedman Jones 1971: 39). These features did not extend to unskilled casual workers, but there would be an indirect benefit to employers, in the support for the excess of casual workers in the seasons when the market was glutted, which served to maintain the long-term surplus of labour. Stedman Jones (1971) shows that this was a condition to which the market was predisposed by other features, such as its highly local character, and the tendency for labour supply to meet the peaks in demand rather than the mean.

The corollary of this argument however tended to lead in the 1890s and 1900s to rather authoritarian approaches to the unemployed, particularly proposals to set up labour colonies away from London to remove the surplus labour. As Hay (1977) shows, British employers, like their German and American counterparts, were a significant pressure group for the development of social welfare policies in the period before 1914. The Birmingham Chamber of Commerce was an early proponent of labour exchanges, and a significant number of other chambers of commerce and individual employers supported the development of health and unemployment insurance, though they objected to employers being forced to make contributions. It seems likely that more indirect motivations such as industrial efficiency, the maintenance of social order, and the desire to preempt socialist schemes were more important than any direct subsidy they might have hoped to obtain. Support was in any case not universal, the engineering section of the London Chamber of Commerce, for instance, claimed that unemployment insurance would remove an element of industrial discipline.

As we have seen, there were proposals within a 'production relations' approach for dealing with unemployment, particularly 'the right to work,' and the extension of public employment, but as we have shown, these could only be palliatives unless pursued to a degree which was highly unlikely within a capitalist State. The manner in which local authority relief work was carried out was such as to make it indistinguishable from the social welfare approach to unemployment. Nevertheless, the fact that there was a conflict between a 'production relations' approach and a social welfare approach, whose central concern was with the reproduction of labour power, shows that to divide state policy between that relating to production and that relating to reproduction is to exclude arbitrarily some approaches to particular issues. It may indeed be to accept only the dominant definitions of these issues.

We will now consider another area of policy in which there was a conflict over a 'production relations' approach – the extension of municipal employment through direct labour – and then contrast the two policy areas.

DIRECT LABOUR   There were two senses in which direct labour was a different sort of issue from unemployment. Firstly, it was not a response to a generally perceived problem in the way that unemployment policy was, but was an individual initiative by the Labour Group. Thus, there was not a conflict between alternative policy solutions, but an attempt to implement a policy where the alternative was no action. Secondly, linked to this, the policy was specific to West Ham – though there were other authorities attempting to set up direct-labour organizations – and the central government did not intervene over the issue.

Although the conflict around direct labour was basically a question of acceptance or rejection, there was a confusion of perceptions of the issue, particularly in the early stages, as to whether it was purely a question about how work done by the Council was to be organized, or whether it implied that the Council should do significantly more work itself, rather than employing private contractors.

Initiatives for the creation of a works department came first in 1893, and related to small-scale works such as laying sewers, and to regular work such as street cleaning and refuse collection, which had previously been let out to contractors. It was suggested from a number of quarters that these jobs might be more cheaply undertaken by the Council's own employees. In addition there was a movement towards creating a Works Department for purely organizational reasons. In November 1893 the Borough Engineer drew attention to the extent of work already being carried out by the Council's own workmen on miscellaneous small projects and suggested that if the Council intended to continue this, it would be necessary to establish a works department under a qualified manager. The Borough Engineer's suggestion was accepted, but it was clear that the department was to have a limited scope, and was to be strictly subordinate to the Engineer's own department. In addition to this the progress towards establishing the department was rather slow, and debates in Council in February 1894 suggest that there was some opposition to the proposal. The Labour Group, which was quite small at this stage, supported the proposal, but viewed it in a more expansive fashion – some members saw it as a way of finding work for the unemployed.

The Works Manager was in fact appointed in May 1894, and the department was established: by November it had 299 employees. Within the first year of its operation, concern began to be expressed about the cost of the department, and more stringent financial constraints were placed upon it. In March 1895 the Council resolved that before any large works were entrusted to the Works Department, the Borough Engineer should prepare an estimate, which should be submitted to the Manager to report whether he could execute them at this cost. The situation was confused by the fact that the Works Department was also administering the Council's unemployment relief schemes and carrying out work in winter months in uneconomic conditions. Subsequently, the Labour Group disassociated themselves from this first Works Department, arguing that it was incompetently managed.

Whatever the faults of the Works Manager, it is clear that the Council placed the enterprise in a highly constrained position, particularly when it was compelled to compete on the open market for contracts. The Works Manager, in a report in June 1895, complained of the problems of this competition, pointing out that the Works Department was restricted to working only for West Ham Borough Council, and that statutory provisions hampered its purchase of materials. It thus had no permanent works to fall back on in the event of its not gaining a contract. This highlights a problem with all direct labour organizations, that to cover their establishment costs, they require a steady flow of work, which can only come from the local authority. Moreover, since one objective was to avoid the irregularity of employment normally prevalent in private building firms, there was an even greater need to maintain a regular flow of work throughout the year.

In the next few months after the Works Manager's report, the situation deteriorated, leading to an acrimonious exchange of reports between the Borough Engineer, the Works Manager and the Borough Accountant. The Engineer pointed out the cost of a number of the schemes, and the Manager asserted that he had been hampered by the Engineer. The Works Committee submitted a report to Council which criticized the Works Manager, accusing the department of 'incomplete organization and imperfect supervision of labour'. In October 1896 they recommended that the Works Department should be discontinued as a separate organization, and that small works should in future be carried out by the Borough Engineer's

Department. Over the next two years larger-scale works were put out to tender. In some areas of recurrent works however the principle of direct labour persisted, and a stabling department was established which took over refuse collection and street cleaning from private contractors.

Given the presence of a number of Council contractors within the dominant political group on the Council and the prevailing laissez-faire ideology, it should not be surprising that the first Works Department was so limited in scope and so quickly brought to an end. It is perhaps more surprising that it should have been started in the first place. As has been suggested, the explanation may lie in the growth of small-scale council work, and the need to organize this. In fact, the Works Department had the effect of expanding the Council's work, rather than simply reorganizing it. Moreover, when the Department began to become too independent of the Borough Engineer's Department, and not simply an organization for implementing its work as the Borough Engineer had envisaged, it lost its main support from among the Council's permanent officials, and its technical rationale.[3]

The second phase of direct labour began a few months before the Labour Group took control of the Council in July 1898, when it was agreed that as an experiment one of the blocks of houses being erected by the Council should be built by workmen employed directly by the Borough Engineer's Department. A more serious start was made in January 1899, after the 1898 elections brought Labour into power, when the Works Department was re-established. This time it was set up on a more ambitious scale, with a more qualified Works Manager, and was given some large projects very soon. This aroused opposition from some quarters where guarded support for the new department had initially been given. In particular, the decision that the Works Department should undertake the new Infectious Diseases Hospital, a project estimated to cost £100,000, was strenuously attacked in the local press, mainly on the grounds of the inexperience and anticipated inefficiency of the new department. (In fact, the hospital was built within the estimates, bearing in mind subsequent modifications in design.) The main work of the department was in relatively small-scale projects, particularly the laying out of streets and drains, and street repairs, but it was also involved in the Council's housing programme. The department came into conflict with some of the Council's permanent officials and the Borough Engineer was dismissed in July 1899 for allegedly hampering the activities of the Works Department, and for a lack of sympathy with some of the other aims of the Labour Group, particularly in relation to trade unions.

Once the strength of the Labour Group on the Council began to wane, the independence of the Works Department was curtailed, and in particular it was forced to tender in competition with outside contractors for the housing projects. It lost one of these to a local builder, but two other, later, housing projects were carried out by the department. After this, however, no other large works were carried out by the department, and it was used only for small-scale works. This reflects two trends: firstly, the dislike of the idea of direct labour by the Conservatives on the Council, who were in control for most of the period from 1900 to the First World War, and, secondly, the fall in the amount of large building works undertaken by the Council after about 1904. This also reflected the dominance of the Conservatives, who, for instance, halted the housing programme in 1904. In 1907 the Works Department was closed down.

Layton, discussing the current situation of direct labour in Britain in 1961, says:

'Some of the direct labour organisations are of very long standing, and date from a time when there was no party political argument concerning their use. West Ham first established its direct labour organisation in 1893 *(sic)* . . . The motive for their introduction was often a dissatisfaction with the price or quality of work carried out by private contractors, particularly painting work, and a desire to carry out such work better and more economically, without any political overtones.' (1961: 239)

As we have seen this is misleading, at least in the case of West Ham. Not only was direct labour a politically contentious issue from the start, but it provoked much more controversy at the local level than did unemployment. In particular it aroused the opposition of the bourgeois parties once the organizational question had been detached and the issue had been clarified as one of whether the Council should do more work itself. In passing, it may be noted that this high level of controversy still exists in Britain today. Direct labour is not securely established, and the building industry, supported by the Conservative Party, is mounting a strenuous campaign against further extensions, and against existing organizations.

Two reasons can be adduced for the level of controversy over direct labour. In the first place, direct labour did represent an explicit threat to one fraction of capital, the building industry, and had no particular benefits for any other fraction – unlike for example, the nationalization of the transport system, gas or electricity, where this could result in the cheapening of inputs into the production process – particularly as the political argument for direct labour was not based on using profits to reduce local authority building costs, but using them instead to improve the wages and regularity of work of the employees. Secondly, the idea that a local authority should engage in competition with the private sector in the production of a commodity was a serious threat to private capital in general, and to received laissez-faire views of the role of a local authority. These views were breaking down in cases relating to social welfare and the reproduction of labour power, but not in cases where there was a threat to the private ownership of the production process itself.

We can summarize these explanations by suggesting that the controversy over direct labour reflected the major challenge to existing interests of a 'production relations' approach to local government. If this is the case, we have to ask why direct labour should have produced more conflict as an issue on West Ham Council than unemployment, since we have argued that unemployment policy too reflected a 'production relations' approach.

The main reasons would seem to be, firstly, the limited scale of action on unemployment – as we have seen earlier, it led to few additional jobs – and, secondly, the fact that the action in effect operated as a subsidy to employers. In fact, while the Labour Group and the labour movement nationally were quite clear that the 'right to work' was an attempt to restructure the labour market, this was not apparent from the way in which local authority relief works were carried out. By contrast, the direct-labour policy not only affected capitalist interests, but was also a clear attempt to change conditions in the labour market, if only in a small section of it. Both policies were failures in terms of their long-term objectives but direct labour's failure was more clearly due to conflicts in the Council over its adoption, while the failure of the 'right to work' reflects a failure to implement a policy in a way which would effect a fundamental transformation of the labour market. The Labour Group's 'production relations' approach as a whole must also be counted a failure. The approach did not succeed in significantly changing the functions of local government or in expanding state intervention towards a modification of the relations of production. However, the fact that a 'production relations' approach was an issue shows the danger of restricting a definition of urban politics to issues concerned with the reproduction of labour power, and failing to recognize that this definition is itself the object of class struggle.

It now remains to draw some general points from this case study. We shall do this in the concluding section.

Conclusion

This paper has been concerned with the relation between state intervention and urban politics in response to urbanization and change in the city in nineteenth-century Britain on the one hand and the theoretical understanding of these processes on the other.

We have considered the relevance of two groups of writers, British Marxist social historians and French Marxist urban sociologists. These groups have different epistemological views of the relation between history and theory, and their work has different strengths. The British historians focus on class formation, and enable us to clarify the relation between classes, political parties and political practices. In general terms, the value of the French work is in its application of Marxist analysis to the study of the urban, and, following from this, their conception of urban politics and urban policy as expressions of class conflict. They also focus more clearly than do the British on the interests and actions of the capitalist class and on the contradictions between different fractions of this class. Castells in particular focuses on the importance of political analysis in understanding urban processes, though his analysis of politics tends to abstract it from the specific characteristics of class organization. Lojkine's importance is in his clearer analysis of the nature of the contradictions of capitalist urbanization and his stress on the contradictory character of state action. On the other hand, he does not analyse very clearly the relation between class interest and political practice.

There is a more general difficulty with the French work which applies most specifically to Castells. This concerns the historical specificity of the work, and can be seen most clearly in Castells's definition of his object of study, the urban. It is not simply that he is studying phenomena which only exist in one historical period, but that in his definition he abstracts the processes of reproduction of labour power from their historical development.

This point becomes clearer when we consider the case study material in the second part of this paper. Our discussion here of local government policy has had two aims. Firstly, we have documented the political struggles over the scope of local government intervention in a particular case, and shown the importance of looking at the political practices of classes that sought to transform the state apparatus, and not simply those practices relating to conflicts over specific issues. The 'production relations' approach adopted by the Labour Group in West Ham in the 1890s embodied a conception of socialist transformation which has been lost in the routinized controversies in local government today. We hope also to have shown that it is possible to draw on the separate theoretical strengths of the two groups of writers discussed in the first part: the emphasis on party and class in the case of the British, the general conception of urban politics and the contradictory character of state action in the case of the French.

Secondly, the case study allows us to draw out some general points about the analysis of state intervention at the local level. Our conclusion is that the range of local state functions cannot be taken as given. It is continuously the object of class struggle. Any approach which abstracts from the dynamic development of the capitalist mode of production retreats to the position we have called 'ahistorical theory'. More specifically, the 'production relations' approach to local state intervention we have described, even if it ultimately failed in its aims, denies the ahistorical equation of state intervention in the urban with intervention in the reproduction of labour power.

To equate urban intervention with interventions relating to 'collective consumption' may be an adequate description of the current situation, but it fails to grasp the historical contingency of this situation, and provides no reason to project the same definitions backwards into the past or forwards into the future.

Notes

[1] Cottereau (1970), one of the French Marxist urban sociologists, in the course of a discussion of the development of urban planning in Paris in the period between 1870 and 1939, makes two points which are relevant here. Firstly, he argues that there is a real change in the nature of state intervention in the urban, arguing that one can identify two types of urban planning. The earlier

of these consists solely in assuring the direct preeminence of the capitalist class in the economic system, as an example of which he cites Haussmann's rebuilding of the centre of Paris in the interests of the business class. The later type involves an intervention to resolve contradictions between fractions of capital by collectivizing the external effects of the private appropriation of space. An early example of this was the planning of the form of urban growth through the control of investment in the transport system and other public utilities (e.g. gas, water and electricity supplies). The latter involves a qualitatively new form of state action, which transcends the limits of the 'liberal' capitalist State.

Cottereau's second point is that, at least until 1914, the political structure of the local area was of considerable importance in determining the development of urban planning, and that political conflicts over planning involved the expression of local political interests detached from national political party groupings. He goes further than this, to suggest that at this stage significant developments in urban planning were only possible where there were strong local government units, expressing local interests, and were not possible where power was exercised by local representatives of the central government.

[2] Throughout, 'labour movement', 'labour representation', etc. refer to the working-class political movement etc., in West Ham or nationally, according to context. 'Labour Group' refers specifically to Labour councillors on West Ham Council, while 'Labour Party' refers to the national political party, established as the Labour Representation Committee in 1900, and renamed Labour Party in 1906.

[3] The relations between the Works Department and the Borough Engineer's Department raise a range of questions about the relevance of intra-organizational conflicts for political decision-making. In the past this has been an area of study separate from most Marxist analyses. Poulantzas (1973) for instance, argues that bureaucracies cannot be seen as having internal sources of power, but must be seen as purely a medium for the exercise of class power. Research into intra-organizational determinants of local government policy has been almost exclusively restricted to contemporary studies. There would seem to be two reasons for this. The first is pragmatic: where sources available are restricted to written ones, it is more difficult to penetrate into the real behaviour of bureaucracies. Secondly, the increase in scale of local government bureaucracies, and also their increase in professional organization, seem to have led to a shift in power away from councillors and political parties, which has made the study of local authority organizations more important. It does seem that Marxist analyses of urban policy today must take more notice of these intra-organizational determinants of policy than Poulantzas' perspective would allow.

References

Ashworth, W. (1954) *The Genesis of Modern British Town Planning.* London: Routledge.
Bendix, R. (1964) *Nation Building and Citizenship.* New York: Wiley.
Biarez, S. *et al.* (1973) *Institution Communale et Pouvoir Politique.* Paris: Mouton.
Braverman, H. (1974) *Labour and Monopoly Capital.* New York: Monthly Review Press.
Castells, M. (1977a) *The Urban Question.* London: Edward Arnold.
——(1977b) 'Towards a political urban sociology', in M. Harloe (ed.) *Captive Cities.* London: Wiley.
Castells, M. and Goddard, F. (1974) *Monopolville.* Paris: Mouton.
Cottereau, A. (1970) 'Les débuts de la planification urbaine dans l'agglomeration parisienne.' *Sociologie du Travail* 4:362–92.

Duclaud-Williams, R. H. (1978) *The Politics of Housing in Britain and France.* London: Heinemann.

Engels, F. (1973) *The Condition of the English Working Class.* Moscow: Progress Publishers (originally published 1844).

———(1975) *The Housing Question.* Moscow: Progress Publishers (originally published 1872–3).

Foster, J. (1968) 'Nineteenth-century cities: a class dimension', in H. J. Dyos (ed.) *The Study of Urban History.* London: Edward Arnold.

———(1974) *Class Struggle and the Industrial Revolution.* London: Methuen.

Freyssenet, M. (1977) *La Division Capitaliste de Travail.* Paris: Savelli.

Harris, J. (1972) *Unemployment and Politics: 1886–1914.* Oxford: Oxford University Press.

Hay, R. (1977) 'Employers and social policy in Britain: 1905–1914.' *Social History* 4:435–55.

Hays, S. P. (1964) 'The politics of reform in municipal government in the Progressive era.' *Pacific Northwest Quarterly* 55:157–69.

Hennock, E. P. (1973) *Fit and Proper Persons.* London: Edward Arnold.

Hobsbawm, E. J. (1964) *Labouring Men.* London: Weidenfield and Nicholson.

———(1970) 'Lenin and the aristocracy of labour.' *Marxism Today* 14.

———(ed.) (1974) *Labour's Turning Point: 1880–1900.* Hassocks, Nr. Brighton: Harvester.

Howarth, E. G. and Wilson, M. (1907) *West Ham.* London: Dent.

Layton, E. (1961) *Building by Local Authorities.* London: George Allen and Unwin.

Lojkine, J. (1976) 'Contribution to a Marxist theory of capitalist urbanisation', in C. G. Pickvance (ed.) *Urban Sociology: Critical Essays.* London: Tavistock Publications.

———(1977) *Le Marxisme, l'État et la Question Urbaine.* Paris: Presse Universitaire de France.

Marshall, T. H. (1950) *Citizenship and Social Class.* Cambridge: Cambridge University Press.

Marx, K. (1976) *Capital* vol. 1. Harmondsworth: Penguin Books (originally published in 1867).

Miliband, R. (1972) *Parliamentary Socialism.* London: Merlin.

Moorhouse, H. F. (1978) 'The Marxist theory of the labour aristocracy.' *Social History* 3:61–82.

Pelling, H. (1968) 'The working class and the origins of the welfare state', in *Popular Politics and Society in Late Victorian Britain.* London: Macmillan.

Pickvance, C. G. (1977) 'Marxist approaches to the study of urban politics.' *International Journal of Urban and Regional Research* 1:219–55.

———(1978) Review of M. Castells: 'The urban question.' *Sociological Review* 26:173–6.

Poulantzas, N. (1973) *Political Power and Social Classes.* London: New Left Books.

Stedman Jones, G. (1971) *Outcast London.* Oxford: Oxford University Press.

———(1975) 'Class struggle and the industrial revolution.' *New Left Reivew* 90:35–69.

———(1976) 'From historical sociology to theoretical history.' *British Journal of Sociology* 27:295–305.

Terrett, J. J. (1902) *Municipal Socialism in West Ham.* West Ham: privately printed.

Thompson, E. P. (1963) *The Making of the English Working Class.* London: Gollancz.

———(1968) 'Time, work-discipline and industrial capitalism.' *Past and Present* 38:56–97.

Wohl, A. S. (1977) *The Eternal Slum.* London: Edward Arnold.

Parliamentary Papers

Select Committee on Distress from Want of Employment. P.P. 1895, volume VIII.

20 Amnesia, integration and repression: the roots of Canadian urban political culture*

Harold Chorney

Introduction

Canada's existence as a nation is highly artificial. Because of its immense size, but geographically concentrated population, it more resembles a chain of city-states than one nation state.[1] Each link dominates its own vast hinterland and is a focus for both political culture and daily life. But though they are extended linearly across the country, these major cities are themselves organized within a hierarchy of dominance. Consequently, the rivalry among these city-states is considerable and contributes to intense regionalism.

The relative underdevelopment of class politics and class consiousness which characterizes Canadian society cannot be separated from the nature of her urban and regional structure. Despite all the prognoses that a modern industrialized, urbanized working class would result in class consciousness and class conflict, Canada and her urban culture continues to be largely devoid of class-conscious politics. What urban politics do exist have hardly moved beyond interest-group politics – sometimes militant interest-group politics to be sure, but interest-group politics, nonetheless. This has been

* I should like to dedicate this article to the memory of my teacher and friend, Rubin Simkin.

I should like to thank Kent Gerecke for giving me the opportunity to write this article. I would also like to thank Ashley Chester, Adil Cubukgil and the editors for their advice and critical contribution to making it more understandable. Any remaining confusions are entirely my own. This essay is part of a larger study in process on urbanization and class consciousness in Canada.

far more apparent in English Canada, than in Quebec where in recent years an emergent nationalist movement contains within it significant fragments of class-conscious politics.[2]

The nature of politics in Canada, then, is genuinely contradictory. Fractionalism is a direct outgrowth of the frictions between the French and English language groups, the influence of an ethnically diverse and substantial immigrant population, and the intense rivalries among the distinct regions. But the successful repression of class politics by the dominant class has integrated all political expression into established and controllable forms. This combination of fragmentation and integration has promoted widespread passivity.

Despite the current crisis of accumulation in Canada, a crisis of legitimation and cultural reproduction has yet to appear. In fact, the current economic crisis has been accompanied by a significant swing to the right in Canadian politics and the entrenchment of a 'welfare backlash' among substantial portions of the Canadian working class.[3]

In order to understand how class consciousness and the urban political culture have been moulded in Canada, and the roots of the current situation, it is necessary to delve at some length into the early history of urban settlement and economic development. We should stress that most of what we discuss is not original, but owes itself to the considerable scholarship that has been accomplished on Canadian economic, labour and urban history. But, nonetheless, we must attempt this survey, for unless one understands the historic basis of regionalism, ethnic division, economic dependence, repression and integration, it is impossible to understand the underdeveloped nature of Canadian urban political culture.

We shall begin our survey with a short recapitulation of the historical basis of the Canadian political economy. This is essential if we are to situate Canadian urban culture within the process of capital accumulation in a metropolitan-hinterland economy. The process of capital accumulation in Canada has imposed an inescapable commercial nexus of relationships upon the culture of Canadian cities. In many respects the cultural parameters of daily life in urban Canada revolve solely around earning a living. Work, achievement and material success define the limits of daily life. In this sense, urbanism in Canada carries with it little of the precapitalist urban tradition that Canadians recognize in European cities. The longing for a 'European holiday' reflects the deep-seated yearning for a daily life which transcends the instrumentality of Canadian cities.

We then turn to an examination of the role of immigration. Immigration has played a very special role in the formation and deformation of urban class consciousness in Canada. We explore this problematic at some length because we feel that it holds the key to unlocking part of the mystery of the state of urban class consciousness in Canada.

We then examine in some detail the process of the establishment of a routinized system of labour relations and the repression of radical labourism because we contend that this process of state intervention is crucial to understanding the process of integration. Indeed, one is tempted to argue that the integration of the working class through collective bargaining resembled, in certain respects, the process of urbanization itself. Urbanization can be regarded as the transcendence of the communal intimacy of the Gemeinschaft of precapitalist society. This transcendence implies the weakening of the common conscience and the capacity for collective behaviour, the substitution of instrumental reason and calculation for the emotional and natural will, and the overwhelming of the cultural traditions central to the formation of class consciousness by the philosophy of money.[4] Communal and collective institutions give way to impersonal intermediary ones. This reduction of interpersonal relationships within the emerging metropolis to instrumental means, and the mediation of the money economy and all that it entails between workers of the same class, is the paradoxical outcome of the collection and concentration of workers in space according to the principles of capital accumulation.[5]

In a similar fashion, trade-union integration has also involved the mediation of the State and the union, and the reduction of class conflict between workers and capitalists to routinized instrumental wage bargaining. The notion of interest has come to replace that of the need of a universal class for a revolutionary transformation of society. Working-class interest, however, is no substitute for working-class consciousness.[6]

Finally we explore in considerable detail, as a kind of case study, the Winnipeg General Strike. We do so, not only because of our interest in Canadian labour history, but also because we believe that the strike and its aftermath illustrates our argument.

Our goal in this somewhat diffuse paper is to explore the problematic of urban class consciousness which we see to be central to any radical analysis of the urban question. Urbanism in its modern metropolitan manifestations can be seen to be the spatial expression of modern capitalist society. As the locus for capital accumulation and the social reproduction of labour, it has usually been regarded by Marxists as the arena in which class consciousness would develop.[7] Yet, as we shall argue, despite this mysterious optimism about the city and social life, urbanization in the metropolitan sense has facilitated dialectically both the accumulation of capital and the cultural reproduction of capitalism. At the same time, it has changed the very nature of capitalism itself.

Our argument, the rudiments of which we can only sketch in this paper, is that urbanization as a mode of social control, as well as an engine of capital accumulation, has thus far successfully facilitated the synthesis of amnesia, integration and repression as a powerful bulwark against the development of class consciousness.

The origins of Canada's urban system

The principal urban centres in Canada were developed as part of a pattern of trade and settlement directed by economic forces centred in England and France, when the foundations of mercantile capitalism were being laid.[8] The earliest Canadian cities were founded either as fishing settlements, fur trading posts, military outposts or entrepôt centres designed to facilitate the trade in staple goods from the colony to the motherland.

From the very beginning of the European invasion, the rationale for urban settlement had been imposed from outside by a dominant external culture for purely commercial reasons. This pattern of settlement clearly has similarities to the historical origins of European cities. But, whereas most European cities were founded either during the classical period of the Roman empire or during the precapitalist medieval period, even the oldest of Canadian cities were founded at the birth of capitalism. This fact has certain important implications which cannot be ignored if we wish to understand the nature of urban culture in Canada. It has meant, in particular, that urban social relations have occurred in an environment wholly shaped by capitalist requirements.

By the middle of the nineteenth century a string of cities and towns had been established. As we can see from Table 20.1, the population of these towns and cities was small in comparison with comparable American cities.

By the 1850s the Canadian urban economy had begun to break the shackles of unequal colonial trade with the mother country. The economy

**Table 20.1 Population of selected Canadian and American cities,
1850 and 1851[a]**

Canada, 1850		USA, 1850	
Atlantic region			
St. John's	25,000	New York	523,000
Charlottetown[b]	5,000	Philadelphia	409,000
Halifax	21,000	Baltimore	168,000
Saint John	23,000	Boston	137,000
Fredricton	4,500	Cincinnati	117,000
		New Orleans	116,000
Lower Canada			
Quebec	42,000		
Montreal	58,000		
Trois Rivieres	5,000		
Upper Canada			
Toronto	31,000		
Hamilton	14,000		
Kingston	11,500		
London	7,000		
Ottawa	7,000		

Source: Canada: *Census* 1851, Vol. 2, pp. xvii-xix. USA: *Census* 1850, Vol. 1. G. Stelter and A. Artibise (1977). Cities in the Wilderness, in G. Stelter and A. Artibise *The Canadian City*. Toronto: McClelland and Stewart.

[a] The population of Atlantic Canada in 1851 was about 640,000, Lower Canada 890,000 and Upper Canada 952,000. The population of the United States was 23,000,000. The six largest American cities constituted 6.3% of the population. The six largest Canadian cities constituted 7.7% of the population. All numbers rounded off.

[b] 1848

passed through a brief period of 'active capitalist involution'.[9] The benefits of this growing domestic economy were allocated upon strict class lines. The local merchants and manufacturers were the principal beneficiaries. The tradesmen and unskilled labourers were the principal victims. For many impoverished tradesmen and most labourers, urban life was highly transient and harsh.[10]

Class tended to be structured along ethnic as well as socio-economic lines. In Upper Canada, for example, the Irish Catholics were largely concentrated in lower paying occupations. Indeed, H. C. Pentland has argued that the Irish Catholics were an essential ingredient in the development of a capitalist labour market in Canada.[11] Labour unrest, in particular, among labourers involved in the construction of canals and railways was quite common. Strike leaders were usually subject to arrest and employers relied upon the use of troops to break strikes. As Pentland puts it so well, 'labour relations meant troops and mounted police to overawe the labourers, government spies to learn their intentions and priests to teach them meekness'. Civic officials in Montreal, Hamilton and London, most of whom were either merchants or manufacturers, availed themselves of troops to quell labour unrest on a number of occasions during this period.[12]

By the turn of the century Canada's urban structure was firmly established. Born out of colonial development and the unequal exchange of imperial trade in staple commodities, its twentieth-century form was to be clearly rooted in commercial notions of capital accumulation. In the nineteenth century it had largely been a creature of British imperial domination. In the twentieth century it was rapidly to become an object of American imperial control.

This pattern of American penetration and control of the Canadian economy was rooted into place during the first decades following Confederation in 1867. When the initial wheat and land settlement boom in the West came to an end just prior to the start of the first World War, Canada was left with a retarded domestically controlled industrial base, a vastly overbuilt and scandalously over-subsidized railway network, a heavy financial indebtedness to British finance, a highly concentrated financial sector, and an increasingly American-controlled manufacturing sector concentrated in Southern Ontario and Montreal.[13] The basis for regional underdevelopment and metropolitan-hinterland dominance, which was to be reinforced by the process of American multinational corporation intervention into Canada, had been laid.[14]

Thus, the chronic problems of high unemployment and unstable economic development which today plague Canada more than any other leading capitalist country of comparable economic maturity, can be traced back to this period. While the marginalization of the Canadian economy has played a role in radicalizing certain strata of the working class in hinterland regions, it has also played a conservatizing role. The fragile nature of economic prosperity in Canada seen in the light of the promise of the American 'high intensity' consumption society, has made many Canadian workers highly security conscious and reluctant to experiment with radical politics at the risk of their economic livelihood. In many ways, Canada is an excellent illustration of the argument made by Michael Kalecki about the political economy of full employment policy.[15] Persistently high unemployment in the midst of a mass-consumption, achievement-principle society, has thus far been largely successful in keeping Canadians strongly attached to job security and the capitalist work ethic. The nature of Canadian urban political culture, thus, cannot be separated from the chronic underdevelopment of the Canadian economy.

The impact of immigration

The absorption and settlement of western Canada, a region over ten times the size of France, had been central to the economic strategy that Canadian capitalists had adopted in their push for Confederation. The West had become a captive hinterland of central Canada, whose merchants and manufacturers would carry on an unequal exchange in staple commodities with the waves of settlers which they would attract to the region. The trade was to be a mirror image of the motherland – colony relationship which had governed the settlement and development of central Canada itself. Only this time the imperial capital was to be Montreal or Toronto and not London or Paris.

The key to this strategy was a massive influx of immigrants, brought not only from Britain and northern Europe, but from the 'foreign' countries of central, southern and eastern Europe as well. Although this strategy had been adopted in 1867, it was not fully realized until the first decades of the twentieth century, when frontier opportunities in Canada had been thoroughly demonstrated. The world price of wheat had risen sufficiently and a boom in British capital accumulation had made possible the massive inflow of capital that was required.[16] The peak of immigration to Canada thus occurred in the years 1909 to 1914. In order to understand the impact of this wave of immigration upon the cultural life of Canadian cities, we must consider this immigration process in some detail.

Up until the last decade of the nineteenth century the proportion of the population which was born outside of Canada, France or the British Isles was quite small.

The 1851 Census of Upper and Lower Canada listed less than 1 per cent of the total population of the two Canadas as 'foreigners' falling into this category. By 1881 the 'foreign' born population was 2.9 per cent, 1891 3.2 per cent, 1901 5.0 per cent and by 1911 10.4 per cent.[17]

The enormous impact which immigration had upon Canada at the turn of the century and the two decades following is made clear by Table 20.2.

The population of Canada in 1901 was 5.4 million people. By 1921 it was 8.8 million people.[18] Since a large proportion of those immigrants who arrived from the United States were also of non-British and non-American origin, the non-Anglo-Saxon influx into Canada during this period was substantial. The impact of this wave of 'foreign' immigration was heightened by the fact that many of the new immigrants settled in

Table 20.2 Origin and number of immigrant arrivals in Canada,
1897–1920, in thousands

Period	Origin UK	US	Other	Total
1897–1902	67	76	95	238
1903–1908	420	289	272	981
1909–1914	541	569	476	1,586
1915–1920	133	320	70	523
Total	1,161	1,254	913	3,328

Source: Canada *Statistical Yearbook* 1921, p. 126. Calculated from Table 20.

the rapidly growing cities and towns of Western Canada. Tables 20.3 and 20.4 illustrate this point very well.

During the first decade of the twentieth century Canada ranked ahead of all other Western countries in terms of population increase.[19] The most significant factor in this increase was the massive wave of immigration. It is interesting to note that immigrant arrivals were nearly double the increase in the foreign-born proportion of the population. In fact, during the period 1901 to 1911, almost 900,000 immigrants left the country after having spent a period of time in it as largely unskilled labourers. This, as well as the fact that single males or males unaccompanied by families were a substantial proportion of the immigrants, suggests that the wave of immigration was an important source for a migrant workforce. This workforce was not very different from the migrant 'guest workers' that are an important phenomenon in contemporary Western European capitalist cities.[20]

As Table 20.3 indicates, the Western region attracted the bulk of the 'foreign' immigrants. The four Western Provinces in 1921 had a total of 338,000 'foreign born', as compared to 178,000 in the older eastern provinces. Alberta and Saskatchewan, and to a lesser extent British Columbia, also attracted a large number of Americans. These were mainly farmers attracted by the homesteading opportunities in the 'last best West'. But they also included miners, loggers, and industrial workers, many of whom brought with them radical syndicalist ideas, and who were to play an important role in the labour movement in the West.[21]

The influx of immigrants in this period has had a permanent impact upon the ethnic composition of Canadian cities. In 1971, the most ethnically diverse cities in Canada were all located in Western Canada.[22]

The ethnic diversity of Canadian immigration has played a complex role in the development of urban class politics. On the one hand, the immigrants constituted a large body of exploited workers. Many had had contact with socialist and radical politics in their country of origin. They were also socially ostracized and marginalized by the mainstream Anglo-Saxon ruling class. As such, they were a fertile source for class-conscious

Table 20.3 Birthplace of population by province, 1921, in percentage terms

Province	Total population	Canada, British Isles and British possessions	'Foreign' US	'Foreign' Other[a]
		Place of birth		
		%	%	%
PEI	88,615	98.5	1.4	.1
Nova Scotia	523,837	97.4	1.3	1.3
New Brunswick	387,876	97.2	2.2	.6
Quebec	2,361,199	95.8	1.8	2.4
Ontario	2,933,662	93.8	2.4	3.8
Manitoba	610,118	82.0	3.6	14.4
Saskatchewan	757,510	73.7	11.6	14.7
Alberta	588,454	70.4	17.0	12.6
British Columbia	524,582	81.0	6.6	12.4
Yukon and NWT	12,145	91.0	5.0	4.0
Canada	8,788,483	89.9	4.3	5.9[a]

Source: Canada Census 1921, Vol. 11, Table 52.
[a] As a result of rounding, the Canada figures total more than 100 per cent.

Table 20.4 Birthplace of population by urban area, 1921, in percentage terms

City	Population	Canada	British Isles and possessions	Foreign
		%	%	%
Saint John NB	47,166	89.8	6.4	3.8
Moncton NB	17,488	93.2	4.1	2.7
Halifax NS	58,372	84.5	12.1	3.4
Glace Bay NS	17,007	81.0	14.0	5.0
Sydney NS	22,545	75.5	16.5	8.0
Quebec Que.	95,193	97.0	1.3	1.7
Trois Rivieres Que.	22,367	94.8	1.4	3.8
Montreal Que.	618,506	81.3	8.9	9.8
Ottawa Ont.	107,843	83.2	11.4	5.4
Kingston Ont.	21,753	79.7	16.2	4.1
London Ont.	60,959	72.6	22.5	4.9
Kitchener Ont.	21,763	80.9	6.8	12.3
Peterborough Ont.	20,994	78.1	18.4	3.5
St. Catherine's Ont.	19,881	67.5	24.0	8.5
Sault Ste. Marie Ont.	21,092	66.6	12.4	21.0
Toronto Ont.	521,893	62.2	28.6	9.2
Fort William Ont.	20,541	58.1	21.9	20.0
Winnipeg Man.	179,087	52.4	28.3	19.3
Regina Sask.	34,432	56.3	26.3	17.4
Saskatoon Sask.	25,739	56.6	28.7	14.7
Edmonton Alta.	58,821	55.5	27.4	17.1
Calgary Alta.	63,305	52.2	33.2	14.6
Vancouver B.C.	117,217	48.9	33.0	18.1
Victoria B.C.	38,727	46.4	39.7	13.9

Source: Canada Census 1921, Vol. 11, p. 372, Table 59.

politics. On the other hand, the very diversity of the urban workforce, the multiplicity of languages and cultures, and the role of immigrants as a cheap source of surplus labour, worked in an opposite direction. The transient attachment of many of the workers to the new country, as well as the fear of some of them that they would be rejected and sent back, without the opportunity to have saved enough to make the ordeal worthwhile, also worked against the development of consciousness. The racism of some British workers, and the fairly general racism of the ruling class, often deflected attention away from class issues and retarded class-conscious politics.[23] The pressure for assimilation, depending upon the culture into which the immigrants assimilated, could reinforce either tendency. This dual nature of the role of immigration in Canadian urban politics persists to this day. There can be little doubt, however, that despite the general racism of the Canadian ruling class, there was recognition by them of the crucial role which an immigrant workforce was to play in the accumulation process in Canada.[24] On the whole, notwithstanding the distaste which the upper class expressed for the 'foreign' and sometimes 'repulsive' and 'primitive' nature of the immigrants' culture and customs,[25] the dividends which immigration paid to Canadian capitalism were substantial indeed. Yet the process of exploitation was not without its periods of crisis. The flood of immigration generated not only booms in real estate, but outbursts of class consciousness and militancy as well.

Labour radicalism and urban development

The conservative business-union practices and integrationist philosophy which dominates the majority of contemporary labour unions in Canada, is the outcome of a process of repression and state-sponsored integration which originated in the first two decades of the twentieth century. While Canadian labour radicalism was by no means a strictly urban phenomenon, there is a critical link between the urbanization process and the rise and fall of labour radicalism in Canada.

The heartland of the Canadian radical labour movement at the turn of the century was located in the West. It was centred in the coal and metal mines of British Columbia, in the growing towns which had sprung up along the lines of the Canadian Pacific Railway, and in the immigrant ghettos and working-class quarters of Winnipeg and Vancouver, and the other booming Prairie cities. It was here that radical labour, socialist, syndicalist, Marxist and anarcho-syndicalist ideas freely circulated among the miners, railway workers and newly created urban proletariat.

Business-union labourism was not absent. It had, however, to compete with radical ideas which were vigorously held by an increasing number of workers. In the older urban centres in Ontario and Quebec this kind of competition had occurred in the 1880s and the battle largely had been won by the turn of the century.[26]

In addition to the European immigrants who had contact with radical working-class and peasant movements before coming to Canada, many of the recent British working-class immigrants to Western Canada were also imbued with more militant conceptions of the workers' cause than their predecessors who had immigrated to Central Canada decades earlier.[27] Finally, the influence of radical syndicalist, anarchist and Marxist ideas which American workers brought with them as they migrated up and down the Pacific coast in search of work, were an important ingredient in Western radicalism.[28] In a very real sense, the West in Canada at the turn of the century was as much a frontier of ideas and political philosophies, as it was one of economic development and settlement.

Both of the radical industrial unions which played a critical role in Canadian industrial relations at the turn of the century were affiliated to the American Labour Union (ALU). The ALU was a socialist labour federation founded to organize industrial unions in direct opposition to the narrow craft-union business philosophy which dominated the American Federation of Labour headed by Samuel Gompers.[29] The Western Federation of Miners (WFM) and the United Brotherhood of Railway Employees (UBRE) successfully organized and mobilized large numbers of workers in British Columbia and along the route of the Canadian Pacific Railway from 1900–3. These two organizations and the workers who belonged to them were responsible for most of the class-conscious rebellion against industrial capitalism which occurred in Canada during this period.

It is a point of some significance that a large proportion of the workers who belonged to these two unions were not urban workers. While many of them did work in the new and rapidly growing cities of Western Canada, the impetus for their radicalism and rebellion originated in the isolated mining towns and resource settlements of British Columbia. This fact is not surprising. For, in such company towns the naked power of capital, symbolized by a rapacious resource company, and the exploitative nature of the wage relationship, was crystal clear.[30] It was this very clarity and the contrast which it made with the opaque and increasingly complex social and class relations in urban settlements, that was important.

Urbanization as the spatial expression of modern capital accumulation and concentration has greatly increased the heterogeneity and fragmentation of social life. This phenomenon, although sometimes falsely attributed to ecological factors, was first identified by the sociologists Durkheim, Tönnies and Simmel.[31] To the extent that class consciousness is rooted in a homogeneity of experience in the workplace, and reinforced or attenuated by the daily experience of life outside of the workplace, urbanization has played an acute role in the process of the formation of class consciousness. For, as Marx pointed out in his work *The Eighteenth Brumaire of Louis Bonaparte,* the absence of a sense of community prevented the French peasantry from developing a sense of class solidarity. They remained, as he put it, 'homologous magnitudes, much as potatoes in a sack form a sack of potatoes'. Yet Marx clearly identified urbanization in its nineteenth-century form as a force enhancing the possibilities for revolutionary class consciousness. In *The Communist Manifesto, The Civil War in France* and elsewhere, Marx celebrated the city as a revolutionary hothouse.[32]

However, perhaps in its twentieth-century form, the city, at least in North America, has changed, and the link between urbanization and the formation of class consciousness has been rendered more tenuous. To the extent that urbanization has weakened the sense of community and the capacity for collective action, and thereby the subjectivity of the working class, it can be seen to have made the formation of class consciousness more problematic.

While there were a number of important and violent urban labour struggles which occurred in eastern Canadian cities in the first two decades of the twentieth century, none could match the upheavals which occurred in the West, and in Cape Breton in 1909 and in the early 1920s, for the militancy of the struggle or the degree of class-conscious politics involved. Among the more important of these eastern urban conflicts were the street railway strikes which occurred in London in 1899, Toronto in 1902, Montreal in 1903 and Hamilton in 1906.[33] In all of these strikes, armed troops were used to defeat attempts by strikers to prevent the railways from operating. A similar strike occurred in Winnipeg in 1906, where troops armed with machine guns patrolled the streets to ensure the operation of the street railway.

On the whole, most of the labour disputes which took place in the older eastern cities revolved around disputes over wages, employment of non-union labour and working conditions. In the main, they were fought by craft unions in a fairly orthodox manner.[34] The fact that craft business-unionism came to dominate the eastern Canadian urban labour movement raises some interesting questions. To what extent was this phenomenon simply a function of the age of the cities and their more diversified workforce in which skilled labour played a greater role? To what extent was it a manifestation of the power of North American urbanism, in its more mature stages, to deflect class-conscious struggles onto the safer paths of business unionism? If the process of urbanization, indeed, effectively weakened community relationships and the capacity for collective action, while at the same time increasingly reducing daily life to the making of a living and passive participation in the spectacular world of modernity, then business unionism was certainly far better suited to this world than radical labourism.

Trade union integration

The integration of workers into 'safe' business-oriented trade unions was the outcome of a process of state integration and craft-oriented labour struggle. State integration policies in Canada were developed during the first decade of the century as a means of countering the threat of radical unionism which flourished in the West. A Royal Commission which investigated a series of general strikes among miners and railway workers in British Columbia (BC) in 1903 found the unions involved, the WFM and the UBRE to be 'secret political organizations with revolutionary objectives'. It recommended that 'legitimate trade unionism . . . be encouraged and protected but that [radical] organizations be prohibited and declared illegal'.[35]

W. L. Mackenzie King, then Canada's Deputy Minister of Labour, served as Secretary to the Commission and undoubtedly had a major hand in the writing of the final report. King and the Commission were quick to recognize the importance of channelling labour unrest onto the safer paths of business unionism, where legitimate rights could be granted in return for predictable and controlled behaviour within the limits of capitalist enterprise.

At the same time that it condemned radical unionism, the Royal Commission recognized the importance of working-class progress. It even, in the abstract, acknowledged the desirability of reducing working hours, provided it was done gradually and only if international competitive conditions permitted.[36] This liberal approach, which stressed the benefits to be won if radicalism were eschewed, was typical of Mackenzie King.

Mackenzie King served as Canada's first Deputy Minister of Labour, then Labour Minister, and then Prime Minister on three separate occasions for a period of over twenty years. His influence in Canadian political life spanned half a century. Among his many achievements was Canada's system of industrial relations, of which he was the principal architect.[37] His influence on labour relations was not restricted to Canada. He also served as a special labour-relations consultant to John D. Rockefeller, beginning with the violent Colorado coal strike in 1915, and later advising such companies as General Electric, Standard Oil and International Harvester.[38]

King helped to perfect the policy of integrating the trade union movement into business-union contract bargaining under state supervision, while at the same time weaning workers from any inclination they might have toward radicalism. It is, also, not without significance that he was a strong supporter and advocate of the town planning and urban reform movement in Canada.[39]

Drawing upon his extensive experience in engineering the defeat of radical labour organizations, and conciliating and arbitrating labour disputes, King as early as 1915 had proposed a system of labour relations which incorporated business welfare, formal grievance procedures, an employees' bill of rights, and joint councils representing workers and management to consult on subjects ranging from health and safety to recreation and education. He reserved, however, the right to hire and fire in management hands. In *Industry and Humanity* published in 1918, King urged that the system of industrial relations be routinized, so that 'guerilla warfare' could be replaced by 'rules and regulations'. He argued that 'under collective agreements a sense of equality . . . between the parties' would be established. 'Even handed justice and reasonableness' would become the basis of industrial discipline. For, 'in dealing with human nature of inferior qualities', justice needed 'to be tempered with mercy'. The right for workers to have their own trade unions recognized was of fundamental psychological importance, provided of course, such unions were to be responsible bodies.[40]

Clearly King's contribution to labour relations theory is a significant one. This is particularly evident when we consider that he was writing in 1918, considerably before such conceptions of social control were widely understood or accepted. He was an important influence in moving Canadian capitalism, albeit very gradually, beyond naked accumulation to a more sophisticated, rationalized, and therefore more powerful pattern of accumulation and legitimation, more in tune with the needs of an emerging mass consumption and increasingly urbanized society.

The Winnipeg General Strike

The culmination of the radical labour movement in Canada during the first decades of the twentieth century, if not for the whole of Canadian labour history up to the present, took place in Winnipeg during May and June of 1919. In order to advance our argument that urbanization in Canada has had a significant impact upon the formation of class consciousness, it is important that we examine this event in some detail. For the Winnipeg General Strike was in many ways the last sustained mass collective action of class-conscious labour at the urban level in Canada in the twentieth century. The strike is particularly revealing because it occurred at a key transitional point in North American economic and cultural development. As such, it embodied aspects of the old order of direct class-conscious confrontation. Yet, at the same time, it reflected the coming of the new order of integrated collective bargaining and trade-union legitimacy. The very fact that the strike was fought by workers whose emotional commitment was class conscious in nature, against employers who were vehemently hostile to unions, in order to win the right to bargain collectively is indicative of this coexistence of two orders. While the outcome did not result in the immediate legitimacy of union recognition and collective bargaining, the focus of the confrontation was irrevocably shifted from questions of social transformation to questions of reform. The threat which the strike posed to Canadian capitalism played no small role in promoting the cause of reformers like Mackenzie King.[41] Paradoxically, one of the most radical and class-conscious struggles in Canadian labour history signalled the end of an era of radicalism, and the beginning of another era, more fundamentally linked to economic and social integration.

In many ways, Winnipeg was an appropriate venue for such an event, for its history typified so much of the Canadian urban historical tradition.[42] At the time of the Winnipeg General Strike, Winnipeg was the third largest city in Canada with a population, including its suburbs, of 227,000. It was the centre of the grain trade for Canada and the metropolitan centre of Western Canada, with major transportation, wholesale, retail and manufacturing facilities. The local merchants and finance capitalists called the city the 'bull's-eye of the Dominion' and the 'Chicago of the North'.[43]

The Winnipeg business and upper class, not unlike their counterparts in English Eastern Canada, were rather uniformly Anglo-Saxon, staunch British loyalists, and often racist in their attitudes toward immigrants from the 'foreign' countries of Europe. They were also extremely laissez-faire whenever there was any question of interference in their private busi-

Table 20.5 Population growth of Winnipeg and Manitoba, 1871–1921

	Winnipeg[a]	Winnipeg City	Manitoba
1871	–	241	25,228
1881	–	7,985	62,260
1891	–	25,639	152,506
1901	47,969	42,340	255,211
1911	155,563	136,035	461,394
1921	227,200	179,087	610,118

Source: A. Artibise (1977) *Winnipeg: An Illustrated History*, Table III, p. 200. See n. 48.
[a] Winnipeg[a] includes suburbs.

Table 20.6 Population growth in principal western cities, 1901–21 (excluding suburbs)

	Winnipeg	Regina	Saskatoon	Calgary	Edmonton	Vancouver
1901	42,340	2,249	113	4,392	4,176	27,010
1911	136,035	30,213	12,004	43,704	24,900	100,401
1921	179,087	34,432	25,739	63,305	58,821	163,200

Source: A. Artibise (1977) *Winnipeg: An Illustrated History*, Table IV, p. 201. See n. 48.

ness interests.[44] They never failed to lobby for substantial state intervention on their behalf, however, whenever the need arose. Thus, for example, the monopoly of the Winnipeg Street Railway Company over the supply of electricity was broken in 1909 by the establishment of a municipally owned electric power company.[45]

Power was supplied to the city at half the cost by the municipal company and municipalization received overwhelming support by the business community. The *Manitoba Free Press* clearly established the rationale for this support. 'Public power is the force which has carried real estate values upwards in phenomenal bounds. . . .'[46]

The ethnic composition of Winnipeg by the time of the General Strike reflected the flood of immigrants which had entered Canada since the turn of the century. There was a very large Eastern European population which was residentially segregated, largely in the city's 'North End'. This pattern, in which ethnic origin overlapped with class, made for a strong sense of injustice among the city's immigrant population which has persisted to this day. Eighty-three per cent of the Slavic population of the city in 1916 lived in the North End, where they were joined by 87 per cent of the Jewish population, but less than 20 per cent of the British population of the city.[47] The North End was isolated in more than just sociological terms. The vast railway yards of the CPR separated it from the rest of the city. The district was characterized by widespread poverty, illiteracy, poor-quality and overcrowded housing, poor sanitation and relatively radical politics. The south part of the city, on the other hand, separated from the central part by the Assiniboine River and well endowed with natural beauty and good drainage, was the district of the city's wealthy ruling class. It was characterized by broad boulevards, spacious lots, restrictive zoning, impressive homes, a plenitude of parks and a largely Anglo-Saxon ethnic composition.

The wealthiest grain merchants and trading magnates reserved the attractive wooded areas bordering the Assiniboine River for their exclusive residential pleasure. One of these residential enclaves was depicted in lavish terms in a newspaper advertisement for it in 1903.

'This most desirable portion of the City is now controlled by a syndicate who have authorised us to offer a limited number of Lots for Sale, with building restrictions, ensuring the construction of handsome residences. The improvements now being made by the city and those contemplated by the syndicate with the serpentine drive . . . will make the Point not only the finest locality for artistic and stately homes, but it will become . . . the "Faubourg St. Germaine" of Winnipeg, the most fashionable drive in the city.'[48]

The cultural impact which the residential segregation of immigrants imposed on the daily lives of the working class is well captured in the following excerpt from a novel about life in Winnipeg during the 1920s by John Marlyn. The excerpt describes the experience of a young Hungarian immigrant boy crossing the river into the South End for the first time.

'Freshly washed and scrubbed, dressed in a clean blouse and his Sunday pants, Sandor sat impatiently dangling his legs in the Academy Road Streetcar. Soon, he thought, and looked imploringly across at the conductor. He had been doing so ever since they had crossed the bridge, and at every glance the conductor had winked and motioned to him to remain seated.

But this time he nodded and pressed the buzzer. Sandor sprang to his feet. The car ground to a stop. He got off, walked a short distance up the street, and suddenly he stood still.

It was though he had walked into a picture in one of his childhood books, past the painted margin to a land that lay smiling under a friendly spell where the sun always shone, and the clean-washed tint of sky and child and garden would never fade; where one could walk, but on tip-toe, and look and look but never touch and never speak to break the enchanted hush.

It grew real. There was the faint murmur of the city far in the background and overhead the whisper of the wind in the trees.

The green here was not as he had ever seen it on a leaf or weed but with the blue of the sky in it, and the air so clear that even the sky looked different here.

In a daze he moved down the street. The boulevards ran wide and spacious to the very doors of the houses. And these houses were like palaces, great and stately surrounded by their own private parks and gardens.'[49]

The hero of this novel, Sandor Hunyadi, grows up aspiring to escape from the stigma of his North End immigrant background. He later seeks assimilation through business success and the changing of his name to

Alex Hunter. That, in the end, he fails, cut down by the depression, is perhaps less important than his drive for assimilation. For it movingly illustrates the haunting problematic which confronted all immigrants to the new world from 'foreign' lands, to succeed and assimilate rather than rebel. And yet the decision was not without its tension, for the very social class which by its exploitation fostered rebellion, was no less anxious to avoid contagion by 'foreign' elements. The fact that this novel *Under the Ribs of Death* was written in the 1950s, the nadir of North American class politics, and is set in the 1920s after the defeat of the radical labour movement, is not without significance. For, just as the end of the preceding decade was one of profound energy and class struggle, the 1920s was a decade marked by passivity, defeat and integration.

The ethnic immigrant dimension played an important role in the Winnipeg General Strike. From the perspective of the Anglo-Saxon ruling class the strike was an alien bolshevik plot' which directly threatened not only their authority to accumulate capital and manage the work process at will, but the very foundations of their society. While most of the leadership of the strike was composed of British working-class immigrants, 'foreign' immigrants were represented among the leadership, and, more significantly, among the thousands of workers and their families who supported the strike.[50]

The anti-strike campaign which was orchestrated by the establishment 'Committee of 1,000' devoted considerable attention to whipping up hysteria over 'alien subversion' in their efforts to defeat the strike. Advertisements in Winnipeg's three daily newspapers sponsored by the Committee called on the federal government to deport 'the undesirable alien and land him back in the bilgewaters of European Civilization from whence he sprung and to which he properly belongs'.[51]

This anti-alien campaign struck a mixed chord of response among the strikers. During the course of the strike and later in the strikers' own history of the event the alien issue was debated in ambivalent terms.

> 'The bosses love the alien when they can use him to break strikes. In fact, in many cases he was brought here for that purpose. Who brought the aliens to Canada? It was not the workers. They opposed it by might and main. We will support all efforts on the part of the authorities to deport all the undesirable aliens.'[52] (*Western Labour News* 7 June 1919)

> 'The bosses have no quarrel with the rich alien, no quarrel with the unorganized alien. The only aliens they complain about are those who have had sense enough to join the ranks of organized labour and therefore cannot be used to scale down wages. . . . Employers, large and small, sent deputations to beg of the alien to come back to work. But the alien declined the tempting offers made them and they stuck tight as a postage stamp. For the workers of Winnipeg, the barriers of color, race and creed had been torn down and are now beyond hope of being rebuilt, which is as it should be.'[53]

Thus we can see that the issue of 'foreign' immigration to Canada played an important, but highly ambivalent, role in the formation of class consciousness.

The general strike and radicalism in the West

The First World War contributed to heightening the sense of grievance felt by Canadian workers. Widespread profiteering, rapid inflation in food prices, poor living conditions, increased working hours and reduced wages, coupled with a bloody toll on the battlefield, generated considerable militancy. This working-class militancy and unrest was paralleled in other Western countries. France, Germany, Italy, Hungary, Britain, Australia, the United States and, of course, Russia were all the scenes of strikes, widespread unrest, attempted and successful revolutions of varying degrees of importance.[54]

The radical leadership of the Western Canadian labour movement had been strongly opposed to the war effort. The socialists among them viewed the war as 'a dispute of the international capitalist class for markets (and therefore) of no real interest to the international working class'.[55] Workers in general, of course, were much more ambivalent, depending upon their degree of political consciousness. While many did enlist voluntarily (under considerable social pressure) for reasons of patriotism or simply because employers pressured them, a significant minority refused to enlist and resisted conscription when it was imposed.[56]

The disillusionment which the war had brought, and the expectations of a different kind of post-war world which it created among working people, were critical factors in setting the stage for the radical unrest which occurred.[57]

The Russian Revolution was also an event of enormous psychological importance. The hopes of many workers, in Western Canada, in particular, were raised by the overthrow of the Czar and the later succession to power of the Bolsheviks. Among a number of immigrant labour groups from Eastern Europe, sympathy and political appreciation for the Russian Revolution was widespread. Many of these immigrants were located in Winnipeg.

The optimism inspired by the Revolution, the growth of the One Big Union movement in Australia, and the growing socialist orientation of the British labour movement contributed toward an increasing interest in socialist, direct action and syndicalist solutions.[58] A series of successful strikes occurred in the West, including two short general strikes in Winnipeg and Vancouver.

During the one-day Vancouver general strike, held in August 1918 to protest the shooting of a prominent BC draft evader by the police, a mob of soldiers wrecked the Vancouver labour temple. Similar vigilante actions took place in Winnipeg where both immigrant 'aliens' and known socialists were attacked and beaten in the streets. The government and their officials did not discourage these sorts of actions. The argument was made that it was essential that 'the real sensible people of the country take into their own hands the active combating of Bolshevik propaganda'.[59] The press, on the other hand, was somewhat more legalistic. The *Toronto Telegram,* for example, editorialized about socialists counselling workers to resist conscription.

'Socialism in Winnipeg has disgraced organized labour as socialism everywhere disgraces labour. . . . The Canadian Government should fill the jails so full of these people that their feet should stick out of the windows.'[60]

An official secret investigation of radicalism in 1918 concluded that a 'bolshevik conspiracy' in which 'aliens of enemy nationality' including Russians, Ukrainians, and Finns affiliated to the IWW were conspiring to foment revolution in western Canada.[61] On the basis of this report, a Department of Public Safety was created and a series of repressive measures undertaken. These included severe curtailment of civil liberties, a restriction on the freedom of the press and the outlawing of a number of radical organizations. The use of 'foreign' languages at public meetings was also prohibited. All strikes and lockouts which would interfere with the war effort were banned.[62]

The Western labour movement responded with a campaign of meetings and publicity against the measures. Labour councils in Victoria, Vancouver, Calgary, Edmonton, Regina and Winnipeg called for general strikes, if necessary, to force the government to rescind the measure. The government was also denounced for its participation in the armed intervention in the Russian civil war on the side of the Whites.[63]

Political radicalism was gaining momentum. At a series of large meetings in Victoria, Vancouver, Calgary and Winnipeg, the federal government's actions, as well as war-time corruption and profiteering, were attacked. Resolutions were passed calling for the repeal of the repressive legislation, the release of political prisoners and the withdrawal of Canadian forces from Russia. At the Winnipeg and Calgary meetings fraternal messages of greeting were sent to the Bolsheviks and the German Spartacists.[64]

In March 1919, 250 delegates met at the Western Canadian Labour Conference in Calgary. They endorsed the creation of a new labour body, the One Big Union, which was to reorganize workers along industrial lines and to advance the workers' cause through the use of the general strike. The Conference agreed to a set of proposals by which the new organization was to be created through a referendum in the four western provinces. The referendum asked workers if they wished to leave their respective American-affiliated craft unions and affiliate to the new body. The by now familiar calls for the repeal of the repressive orders-in-council, the release of political prisoners, withdrawal of troops from Russia, and expressions of support for the Bolsheviks and Spartacists were all adopted, as well as a resolution endorsing the 'principle of proletarian dictatorship'. The Conference also called for a six-hour day and the five-day week.[65]

Not all of the delegates, and certainly not the majority of workers in Western Canada, endorsed such revolutionary postures. It must be remembered that organized labour was a very small proportion of the workforce. The radicals were an even smaller fraction. Nevertheless, for the time being, the radicals had considerable influence. In the heady atmosphere which prevailed at the end of the war, with revolutionary currents swirling about, and the climate of generally rising expectations for the promise of a new society, workers in Western Canada were prepared to allow the radicals to take the lead.

This militancy and radicalism was definitely not shared by the eastern labour movement. The Trades and Labour Congress of Canada viewed western radicalism as a direct threat to their control over the Canadian labour movement, at a time when their influence, and moreover, respectability was growing. The leadership, particularly those with close links to the American Federation of Labour, therefore conspired with the federal government to crush the radical labour movement and the Winnipeg General Strike, using every means at their disposal.

The general strike in Winnipeg began with a strike over union recognition and wage increases by the building and metal trade unions. This occurred in a general climate of labour unrest and employer intransigence. Because of the refusal of the employers to bargain with the unions, the Winnipeg Trades and Labour Council called for and received overwhelming support for a general strike. The general strike began on 15 May 1919 and within hours received the support of ninety-four out of a total of ninety-six unions, including police and firemen, some 12,000 organized workers and at least as many unorganized workers. During the height of the strike the economic life of the city was paralysed with some 35,000 workers on strike out of a total work force of about 60,000. If we include the families of the strikers, close to one half the city's population was on strike. Only essential services, such as milk and other food deliveries were operational, and these, only because of the co-operation of the strike committee.

The strike was supported by many, though not all of the returned soldiers. There were marches by returned soldiers for and against the strike. One month after the strike had begun, the Royal Northwest Mounted Police, acting on federal orders, arrested strike leaders at their homes in the middle of the night, raided other homes and labour temples and seized documents and radical literature. A silent parade by veterans to protest the arrest of the strike leaders, the federal government's legislation to deport 'undesirable aliens', including Canadian citizens, and the city's resumption of streetcar service was scheduled for the downtown area in the vicinity of the city hall. The city banned the parade. When the crowds who had gathered to watch it, blocked the passage of a streetcar, they were attacked by mounted police armed with clubs and small arms, and by special armed police recruited by the Citizens' Committee to replace the regular police who had gone on strike. Some of the equipment which the 'specials' used, including their clubs, were supplied courtesy of local department stores. At least one demonstrator was killed and more than thirty injured. The use of armed troops and armoured cars (machine guns had been secretly sent to Winnipeg by the government), as well as the occupation of the city by heavily armed militia, brought the strike to its knees. On 26 June 1919 the strike was officially ended.[66]

The defeat of the Winnipeg General Strike marked an important watershed in Canadian labour and urban history. The strike was the last mass battle fought by urban workers affiliated to or influenced by the radical labour and socialist movement in Canada. In many respects, it was the high point of working-class political consciousness at the urban level. Even during the bitter class conflict engendered by the depression of the 1930s, the degree of class consciousness displayed during the strike was never equalled.

Although only three members of the official strike leadership belonged to the committee organizing the One Big Union, many of the workers on strike in Winnipeg supported the OBU idea. The One Big Union, despite the defeat of the strike, managed to recruit close to 50,000 workers from across Western Canada. These included coal and metal miners, carpenters, railwaymen, textile workers, teamsters and lumber workers. The Vancouver, Victoria and Winnipeg Trades and Labour Councils affiliated with it. Yet by the end of 1921 the OBU had been largely defeated. Because of its radicalism, the State and the conservative national Trades and Labour Congress had marked it for destruction. The federal government intervened in labour disputes wherever the OBU was involved, declaring it to be an illegitimate organization and pressuring workers to desert it for government-sanctioned TLC affiliates. In the face of this organized hostility, with a rising unemployment rate, and strategically weakened by its association with the now discredited notion of the general strike, the OBU was doomed to fail.[67]

Many interpreters of the Winnipeg General Strike argue that it was essentially a confrontation with employers over the question of union recognition, collective bargaining rights and wages. However, in the light of the times in which it took place, and the scope and intensity of the strike itself, clearly it involved, at the very minimum, a widespread aspiration for a radically different kind of social order. To be sure, this aspiration and the class consciousness of which it was an expression, were ambivalent, already influenced by integrative tendencies. Nevertheless, the contrast which the militancy and consciousness of these urban workers makes with their contemporary counterparts and the nature of urban labour struggles today, is a dramatic one. For, we have moved from class-based struggles, with definite if somewhat ambivalent class consciousness, to narrowly-based interest group struggles. These current struggles, on the whole, lack significant political content, let alone a sense of social transformation. They occur in an environment in which the majority of urbanized workers view strikes as competitive with their own interests. Solidarity, as such, is largely absent. Not only has the object of labour struggles been reduced to purely job-related demands, but these struggles are thoroughly integrated into an achievement-oriented competitive treadmill. The transcendence of universal need by interest, of conscious solidarity by instrumental calculation, and of working-class culture and public space by capitalist hegemony and privatization has been thorough.

The defeat of the radical labour movement in 1919 coincided with the rise of an increasingly urban-based, mass-consumption society. The newly created mass advertising industry, much like the mass-production economy it was designed to support, created a new product – the mass consumer. This era also coincided with the growing depoliticization of the working class. The substitution of shopping and soap opera, the cinema and radio, for politics and social interaction, reinforced the passivity that resulted from repression and defeat.[68] The public space in which radical politics had been possible increasingly fell victim to a conception of social life in which private achievement and personal consumption became primary goals.

> 'By the 1920s, the development of American consciousness industries, particularly advertising had begun to materialize as a result of the same social decisions by which mass industrialism was being developed. The priorities by which a population was being mobilized and privatized as wage earners, gave rise of necessity to a new capitalist cosmos.
>
> The pleasures of the marketplace, like the labour process from which they emanated [were] individuated, couched in a vision of a social world made up of anonymous and or hostile strangers.'[69]

It is not surprising that the reduction of work to abstract labour, which was the essence of the deskilling process involved in the decline of the crafts, paralleled the depoliticization of the urban worker and his conversion to a member of the mass society of emerging metropolitan life. For these two tendencies were central to the emergence of the urban way of life as a mode of social control, in which mass consumption of goods and cultural services ensured the decline of politics and therefore of radicalism.[70] This estrangement and atomization which is bound up in the social reproduction of labour as members of a mass-consumer society has come to be the most powerful influence in modern urban life in Canada. The kind of militancy and political activity which characterized the Winnipeg General Strike seems difficult to imagine in contemporary urban Canada.

In fact, even in Winnipeg, where the events took place, they seem only a fading memory, largely unknown among the population at large, and of interest only to urban and labour historians. Except for occasional rhetorical allusions by labour politicians, the mythological memory of the events among some of the descendants of the strikers, and the persistence of strains of radical politics in the North End of the city, 'social amnesia', the forcible repression of the collective memory, has been remarkably successful. It is this power of social amnesia and the cultural homogenization of capitalist society at the urban level in North America which must be understood, if we are to make sense of the nature of urban political culture in Canada and the social reproduction of capitalism which it entails.[71]

The decline of the radical labour movement was followed by the decline of the labour movement, in general. In 1919, the high point of radicalism, there were 387,000 workers affiliated to trade unions in Canada. After the destruction of the OBU, membership in unions declined below the 300,000 level and except for 1928 and 1929 did not surpass 300,000 until after 1935. The 1919 high-point was not surpassed until 1937 when the unions affiliated to the CIO were organized.[72]

During the 1920s a number of factors contributed to the stagnation of the labour movement. The continuing importance of immigration and out-migration of labour, as well as the movement from rural to urban areas, increased the transient character of labour which made union organizing difficult. The rise of a managerial bureaucracy and the larger scale of firms which accompanied the growing Americanization of Canadian business brought corporatist managerial industrial relations into vogue. On the whole, during this period, economic growth was centred around new high-technology industries making organizing difficult. With urban-based radicalism firmly defeated, and the mass-production/mass-consumption phase of North American capitalism under way, attention was turned toward integrating workers without the mediation of unions.[73]

The 1920s were thus the heyday of scientific management ideology which emphasized management technique and employee manipulation as the solution to labour problems. Trade unions were comfortably regarded as anachronisms. For Canadian workers the 1920s were a decade of retrenchment and defeat.

Conclusion

This survey of some aspects of the economic, labour and social history of Canada's cities was undertaken to increase our understanding of the nature of contemporary urban politics and culture. For the roots of these contemporary phenomena, like all social phenomena, lie embedded in the past. Many of the characteristic features of Canadian urbanization were established by the 1920s. These features include: uneven regional economic development in which the Canadian urban hierarchy is situated, with both regional underdevelopment and hierarchy mutually reinforcing each other; the orientation of the city as a locus for capital accumulation and commercial expansion with dominant external linkages; the critical role of immigration and migration in the process of urban development and capital accumulation; the impact of immigration and migration upon political culture; the marginalization of certain ethnic groups and the privileged positions of Anglo-Saxons and northern Europeans; the repression of radical politics and the elevation of reform; the depoliticization of urban culture; and finally the relative absence of working-class political content in urban reform movements with the possible exception of Montreal.

The 'social amnesia' which, on the whole, prevails among the urban working class in Canada has its roots in the first two decades of the twentieth century.

People make their own history, it is true. But they do so subject to the struggles, victories, defeats and repressions of the past. If, as a result of these events, they come to lack consciousness then history makes them. And it does so under the dominant influence of that class which does possess consciousness. That is not to say that the present is rigidly fixed or the past forever impressed upon the present or the future.

Manuel Castells has written critically of the notion of urban culture, which he argues is used as an ideological smokescreen, a 'myth' behind which the true nature of the city in capitalist society is obscured.[74] While his approach has the merit of redirecting attention away from the classless ecological models of American urban sociology towards the concrete articulations of capital formation and collective consumption within the spatial relations of the city, from our perspective he has narrowed the focus too much. If we are to make sense of the formation and deformation of class consciousness in urban North America, we must seek the answer to our questions at the level of the culture of everyday life. The fact that a police station and a parking garage stand on the site of the Old Market Square in Winnipeg, where mass meetings of thousands of politicized urban workers once took place, is symbolic of what has transpired in urban political life in Canada.

The process of transformation cannot be solely rooted in concrete economic factors. Canadian capitalism, like North American capitalism of which it is an integral part, has been largely able at the spatial environmental level to shape the city to meet its needs. It has been able to do so without being forced to incorporate or to destroy precapitalist spatial forms. As such, the numbing impact of the placelessness of much of contemporary urban design and its functioning as a mode of social control has operated without dilution or resistance.[75]

Despite the presence of an earlier urban working class, with significant strains of radicalism and class consciousness, the contemporary Canadian city has largely been cleansed of such influences. Once the actual repression of radicalism was accomplished, the stage was set for the construction of the city and the establishment of the rhythm of city life as a celebration devoted to accumulation and consumption.

The city and its streets were converted from a place where politics and public life were central to life, to a place where consumption and spectacle became central.[76] Politics was pushed aside, indoors, where it has come ultimately to be reconstituted as spectacle itself, and consumed passively as television. Public space is less and less available to facilitate the social interaction, beyond the confines of the workplace itself, which is necessary for the formation of class consciousness.

The repression of radicalism was followed by amnesia – an amnesia which was made more powerful by an urbanism dedicated to forgetting its past. The street has become in Lefebvre's words 'a sequence of showcases, an exposition of objects for sale. . . . Thus we can speak of the colonization of urban space brought about in the street through the image, the publicity, the spectacle of objects. . . .'[77]

It is the contrast which this image of contemporary city life makes with the pictures and discourses of the past to which we wish to draw attention. For the contrast raises quite fundamental questions about how the city and city life in Canada have changed. A sea of faces in a crowd of thousands on a street, in a park or on a square, men and women upon platforms speaking to the crowd about politics, about good and evil, about the future of humanity. The romantic and utopian unreality of these scenes seem clear when we try to imagine them in the contemporary

Canadian city, not simply as special events, but as an ongoing part of political life.

The roar of automobile traffic, the crowds of shoppers, the streams of neon and glass and plastic, the McDonalds and the gas stations – can this be the same urban world?

In his final address to the jury at his trial following the Winnipeg General Strike, William Pritchard, one of the leaders of the western radical movement, spoke of a new order which he saw on the horizon.

> 'Reason, wisdom, intelligence, forces of the mind and heart, whom I have always devoutly invoked, come to me, aid me, sustain my feeble voice, carry it, if that may be, to all the peoples of the world and diffuse it everywhere where there are men of goodwill to hear the beneficent truth. A new order of things is born, the powers of evil die poisoned by their crime. The greedy and the cruel, the devourers of people are bursting with an indigestion of blood. However, sorely stricken by the sins of their blind or corrupt masters, mutilated, decimated, the proletarians remain erect; they will unite to form one universal proletariat and we shall see fulfilled the great Socialist prophecy, "The union of the workers will be the peace of the world".'[78]

Spoken then, the words even if romantic and utopian, had the ring of hopeful authenticity for many urban workers. Spoken now, they have a naïve and hopelessly idealistic quality. Their present inauthenticity is a measure of how much the Canadian urban political culture has declined.

Notes

[1] Canada is a continental land mass, second in size to the Soviet Union. Its population of twenty-three million is 76 per cent urban. Its population is less than one-half that of France which is one-eighteenth the area of Canada. In 1971, 47 per cent of the Canadian population resided in twelve metropolitan areas of over 220,000 each.

Source: D. M. Ray (ed.) (1976) *Canadian Urban Trends: National Perspective,* Vol. 1. Toronto: Copp Clark, Table 1.7: 19.

[2] For a discussion of this radical urban movement and its relationship to the Quebec nationalist movement, see S. Shecter (1978) *The Politics of Urban Liberation.* Montreal: Black Rose Press. S. Shecter (1975) 'Urban politics in capitalist society: a revolutionary strategy.' *Our Generation* (Fall) 11, 1: 28–41. H. Milner (1975) 'City politics – some possibilities.' *Our Generation* (Winter) 10, 4: 47–60. A. Limonchik (1977) 'The colonization of the urban economy.' *Our Generation* (Fall) 12, 2: 5–24. The editors (1974) 'The Montreal Citizens movement.' *Our Generation* (Fall) 10, 3: 3–22. M. Raboy (1978) 'The future of Montreal and the MCM.' *Our Generation* (Fall) 12, 4: 5–18. M. Castells (1972) *The Urban Question.* London: Edward Arnold, pp. 348–60. 'Reflections and a retrospective: an interview with Yvon Charbonneau.' *This Magazine* (July-August 1979), 13, 3.

[3] See G. Janosik and R. Voline (1978) 'Bloodbath in the Red River Valley.' *This Magazine* (July-August) 12, 3. H. Chorney and P. Hansen (1980) 'The falling rate of legitimation; problems of legitimation of the capitalist State in Canada.' *Studies in Political Economy* 4 (Fall).

4 For the critical role of cultural institutions in the formation of class consciousness see E. P. Thompson (1968) *The Making of the English Working Class*. Harmondsworth: Penguin Books, p. 781 ff.

5 The process of urbanization and its impact upon class consciousness is the subject of a somewhat forgotten debate in urban social theory. On the one hand the works of F. Tonnies, *Gemeinschaft und Gesellschaft*, G. Simmel, 'The Metropolis and Mental Life' and *The Philosophy of Money*, and to a lesser extent E. Durkheim, *The Division of Labour in Society*, all focus on alienated social relations in the metropolis which are linked, to a greater or lesser extent, to the rise of capitalism itself. In social theory these insights, and particularly the work of Simmel, led directly to the work of G. Lukacs – (1971) *History and Class Consciousness*. London: Merlin Press – in which the notion of reification is introduced as a means of explaining the failure of class consciousness to emerge. Lukacs however cast his argument in terms quite different from Simmel's concern with metropolitan life, and in Marxists subsequent to Lukacs the critical connection of reification and metropolitan life was largely forgotten. Instead, these ideas have surfaced in a badly deformed way stripped of their original radicalism and connection to class and consciousness in the urban sociology of the Chicago school. I am presently engaged in an effort to rescue these long-submerged themes and resynthesize them in an analysis of the problem of class consciousness and urban culture.

6 See A. Heller (1974) *The Theory of Need in Marx*. London: Allison and Busby.

7 See for example K. Marx and F. Engels (1973) 'The Communist Manifesto', in *The Revolutions of 1848*. Harmondsworth: Penguin Books, pp. 71–2.

8 Mandel refers to this period as one of primitive accumulation of commercial capital. E. Mandel (1962) *Marxist Economic Theory*. London: Merlin Press, pp. 106–9.

9 For a discussion of this concept see A. G. Frank (1969) *Capitalism and Underdevelopment in Latin America*. Harmondsworth: Penguin Books.

10 See, for example, M. Katz (1976) *The People of Hamilton, Canada West: Family and Class in a Mid-Nineteenth Century City*. Cambridge, Mass.: Harvard University Press.

11 H. C. Pentland (1960) 'Labour and the development of industrial capitalism in Canada.' Ph.D. thesis, University of Toronto.

12 H. C. Pentland 1960: 400–22, 408.

13 See R. T. Naylor (1975) *The History of Canadian Business: 1867–1914*. Toronto: James Lorimer. Vols. 1 and 2. See also I. Lumsden (ed.) (1970) *Close the 49th Parallel Etc. The Americanization of Canada*. Toronto: University of Toronto Press; G. Teeple (ed.) (1972) *Capitalism and the National Question in Canada*. Toronto: University of Toronto Press; R. Laxer (ed.) (1973) *Canada Ltd. The Political Economy of Dependency*. Toronto: McClelland and Stewart Ltd; C. Heron (ed.) (1977) *Imperialism, Nationalism and Canada*. Toronto and Kitchner: New Hogtown Press-Between the Lines; K. Levitt (1967) *Silent Surrender*. Toronto: Macmillan.

14 For a discussion of the impact of American investment upon regional underdevelopment in Canada see H. Chorney (1977) 'Regional underdevelopment and cultural decay', in C. Heron (ed.) (1977).

15 See M. Kalecki (1972) 'Political aspects of full employment', in M. Kalecki

The Last Phase in the Transformation of Capitalism. New York: Monthly Review Press. The phrase 'high intensity' is Bill Leiss's; see W. Leiss (1976) *The Limits to Satisfaction.* Toronto: University of Toronto Press.

[16] K. Buckley (1974) *Capital Formation in Canada 1896–1930.* Toronto: McClelland and Stewart Ltd, pp. 4–6.

[17] Canada, *Census* 1851, Vol. 1, pp. 36–7. Canada, *Census* 1921, Vol. 1, p. xv.

[18] Canada, *Census* 1901, 1921, Vol. 1.

[19] J. Weaver (1977) *Shaping the Canadian City: Essays on Urban Politics and Policy 1890–1920.* Toronto: The Institute of Public Administration of Canada, p. 6.

[20] J. Weaver 1977: 7–8. See also D. Avery (1972) 'Continental European immigration in Canada, 1890–1919: from stalwart peasant to radical proletarian'. Canadian Historical Association, *Historical Papers*; R. Harney and H. Troper (1975) *Immigrants: A Portrait of the Urban Experience, 1890–1930.* Toronto: Van Nostrand Rheinhold; D. Avery (1979) *Dangerous Foreigners.* Toronto: McClelland and Stewart.

[21] P. Phillips (1967) *No Power Greater: A Century of Labour in British Columbia.* Vancouver: BC Federation of Labour. A. R. McCormack (1977) *Reformers, Rebels and Revolutionaries: The Western Canadian Radical Movement 1899–1919.* Toronto: University of Toronto Press.

[22] The only exception is Thunder Bay located in North Western Ontario and as closely connected to Western Canada as it is to Central Canada. See D. M. Ray (ed.) (1976) *Canadian Urban Trends: National Perspective,* Vol. 2. Toronto: Copp Clark, Table 5.13, p. 115.

[23] See for example, T. Peterson (1978) 'Ethnic and class politics in Manitoba', in M. Robin (ed.) *Provincial Politics.* Toronto: Prentice-Hall.

[24] For an analysis of the role of immigration in shaping American class consciousness, see S. Aronowitz (1973) *False Promises, The Shaping of American Working Class Consciousness.* New York: McGraw-Hill.
   In 1971 there were close to four million out of a total of twelve million people who resided in Canada's metropolitan areas whose ethnic origin was neither English nor French.

[25] See A. Artibise (1977) 'Divided city: the immigrant in Winnipeg society, 1874–1921', in A. Artibise and G. Stelter (eds) *The Canadian City.* Toronto: McClelland and Stewart. R. Harney and H. Troper, *Immigrants* (1975).

[26] In the main, this competition had occurred between 'all-in' industrial unions like the Knights of Labour and the craft unions. See R. Hann (1976) 'Brainworkers and the Knights of Labour: E. E. Sheppard, Phillips Thompson and the Toronto News, 1883–1887', in G. Kealey and P. Warrian (eds) *Essays in Canadian Working Class History.* Toronto: McClelland and Stewart Ltd. C. Lipton (1968) *The Trade Union Movement of Canada 1827–1959.* Montreal: Canadian Social Publications Ltd. See also M. Piva (1979) *The Condition of the Working Class in Toronto 1900–1921.* Ottawa: University of Ottawa Press. Piva argues that the decline in the labour movement in terms of effectiveness, relative size, if not militancy, began at the turn of the century and continued through the first two decades with the brief exception of 1919–20.

27 A. R. McCormack 1977: 13–14; H. Pelling (1963) *A History of British Trade Unionism.* Harmondsworth: Penguin Books, Chapter 6, pp. 93–122.

28 A. R. McCormack 1977: 98–117; P. Phillips (1968) 'No Power Greater', in M. Robin *Radical Politics and Canadian Labour 1880–1930.*Kingston: Queen's University Press, Chapter IV.

29 A. R. McCormack 1977: 41–4.

30 This relative clarity and the radical unionism which it engenders persists. Most of the major one-company towns in Canada are noted for their more militant unions.

31 G. Simmel (1958) 'Metropolis and Mental Life', in *The Sociology of George Simmel.* Glencoe, Ill.: Free Press; F. Tönnies (1957) *Community and Society.* East Lansing: Michigan State University Press; E. Durkheim (1955) *The Division of Labour in Society.* Glencoe: Free Press.

32 K. Marx (1972) *The 18th Brumaire of Louis Bonaparte.* Moscow: Progress Publishers, p. 106; Marx and Engels 1973: 71–2; K. Marx (1968) *The Class Struggles in France: 1848–1850.* Moscow: Progress Publishers, pp. 37–8; K. Marx (1974) 'The Civil War in France', in *The First International and After.* New York: Random House, pp. 206–21.

33 S. Jaimeson (1965) *Times of Trouble: Labour Unrest and Industrial Conflict in Canada 1900–1966.* Task Force on Labour Relations, Study Number 22, Ottawa: Information Canada. See also *Canada Labour Gazette* 1900–20.

34 *Canada Labour Gazette* 1900–20.

35 *Canada Labour Gazette* 1903, Vol. 3, pp. 133–6.

36 *Canada Labour Gazette* 1903, Vol. 3, pp. 33–6.

37 B. Rudin (n.d.) 'Mackenzie King and the writing of Canada's Labour Relations Acts.' *Canadian Dimension* mimeo. V. Levant (1977) *Capital and Labour: Partners? Two Classes Two Views.* Toronto: Steel Rail Publishing.

38 W. L. M. King (1973) *Industry and Humanity.* Toronto: University of Toronto Press, introduction by D. J. Bercuson, pp. x–xi.

39 W. L. Mackenzie King 1973: 227–32.

40 W. L. Mackenzie King 1973: 133–6.

41 N. Penner (ed.) (1973) *Winnipeg 1919: The Strikers' Own History of the Winnipeg General Strike.* Toronto: James, Lewis and Samuel, p. xxi.

42 Winnipeg had originated as a fur-trading post founded by La Vérendrye in 1738. The first settlers, as opposed to fur traders, arrived in 1812 as part of a colony founded by the Scottish Earl of Selkirk, who acquired a land grant by buying into the Hudson's Bay Company who, in turn had received the lands by a grant of Charles II of England in 1670. An uneasy cohabitation of the area by European settlers and native and Metis fur traders persisted until Manitoba was annexed by Canada in 1870. An unsuccessful rebellion led by the Metis leader Louis Riel in 1870 briefly established a Metis provisional government. At the time of the rebellion the population of the settlements in the Red River valley, including Winnipeg, was about 12,000, including about 5,700 Metis of French-native descent, 4,000 of Anglo-Scottish-native descent and 1,600 Europeans, Canadians and Americans. See J. Howard (1973) *Strange Empire: Louis Riel and the Metis People.* Toronto: James, Lewis and Samuel; A. Artibise (1975) *Winnipeg: A Social History.* Montreal: McGill-Queens University Press.

[43] A. Artibise (1975) *Winnipeg: A Social History.* Montreal: McGill-Queens University Press.

[44] Artibise 1975. See also H. C. Pentland (1968) 'The socio-economic background to Canadian industrial relations', background study for *Canadian Industrial Relations: Task Force on Labour Relations.* Ottawa: Privy Council.

[45] Artibise 1975: 101.

[46] Artibise 1975. See also H. V. Nelles (1974) *The Politics of Development.* Toronto: Macmillan of Canada.

[47] Artibise 1975: 163.

[48] A. Artibise (1977) *Winnipeg: An Illustrated History.* Toronto: James Lorimer and Company, p. 66.

[49] J. Marlyn (1957) *Under the Ribs of Death.* Toronto: McClelland and Stewart, p. 64.

[50] See N. Penner (ed.) (1973) *Winnipeg 1919: The Strikers' own History of the General Strike.* Toronto: James, Lewis and Samuel.

[51] From an advertisement in the *Winnipeg Telegram,* 6 June 1919, quoted in N. Penner (1973), 'Introduction', p. xviii.

[52] Quoted in T. Peterson (1978) 'Ethnic and class politics in Manitoba,' in M. Robin (ed.) (1978), p. 74.

[53] N. Penner (ed.) 1973: 78.

[54] A. Wolfe (1977) *The Limits of Legitimacy.* New York: Free Press, Collier Macmillan, pp. 117–19. See also G. Williams (1975) *Proletarian Order, Antonio Gramsci, Factory Councils and the Origins of Communism in Italy 1911–1921.* London: Pluto Press.

[55] *British Columbia Federationist,* 7 August 1914, quoted in A. R. McCormack 1977: 119.

[56] Robin 1968: 131–9.

[57] For a literary statement of this disillusionment see J. D. Passos (1961) *1919.* New York: Washington Square. See also D. Reed (ed.) (1978) *The Great War and Canadian Society: An Oral History.* Toronto: New Hogtown Press.

[58] The importance of the Russian Revolution, its perversion, and the anti-communist hysteria it aroused, cannot be underestimated in its contribution to the defeat of radicalism in North America. For the importance of British radicalism see N. Penner (1977) *The Canadian Left: A Critical Analysis.* Scarborough: Prentice-Hall. See also R. K. Murray (1955) *Red Scare: A Study in National Hysteria, 1919–1920.* New York: McGraw-Hill.

[59] Letter from the chief Canadian censor to a publisher. Quoted in McCormack 1977: 162.

[60] Quoted in Penner 1977: 70.

[61] Robin 1968: 16–19; McCormack 1977: 150–1. The security services had attributed labour unrest to the activities of the Industrial Workers of the World (IWW). The IWW had been active in Western Canada since its founding in 1905 among transient workers who were employed as railway labourers, loggers, miners and longshoremen. By 1912 they had a membership of about 5,000. They openly and vigorously espoused revolutionary syndicalist ideas and they

were an important force among the marginal unskilled and largely unurbanized workers. Their propaganda played an important role in spreading the notion of direct action and the general strike. But their actual direct influence upon Western radicalism and working-class discontent was undoubtedly exaggerated by the security service. See also J. Scott (1975) *Plunderbund and Proletariat: A History of the IWW, in BC.* Vancouver: New Star Books.

[62] McCormack 1977.

[63] McCormack 1977.

[64] McCormack 1977: 158; see also D. C. Masters (1973) *The Winnipeg General Strike.* Toronto: University of Toronto Press, pp. 29–39.

[65] Masters 1973.

[66] For extensive treatment of the Winnipeg General Strike on which this account is based, see N. Penner (1973); Masters (1973); Bercuson (1974) *Confrontation at Winnipeg.* Montreal: McGill-Queens University Press; H. C. Pentland (1969) 'The Winnipeg General Strike fifty years after.' *Canadian Dimension* (July).

[67] D. J. Bercuson (1978) *Fools and Wise Men: The Rise and Fall of the One Big Union.* Toronto: McGraw-Hill Ryerson. McCormack 1977: 167–8.

[68] See H. Innis (1951) *The Bias of Communication.* Toronto: University of Toronto Press, pp. 81–2, on the cultural and political importance of radio and other mass media.

[69] S. and E. Ewen (1978) 'Americanization and Consumption.' *Telos* 37 (Fall): 49.

[70] For the social significance of the mediation of abstract categories in the construction of metropolitan life see Simmel (1958).

[71] For the notion of social amnesia in the psychoanalytic sense, see R. Jacoby (1975) *Social Amnesia.* Boston: Deacon Press.

[72] *Canada Statistical Yearbook,* 1921, 1931, 1941.

[73] R. Bendix (1956) *Work and Authority in Industry.* Berkeley: University of California Press, Chapter 5. See also Pentland (1968).

[74] M. Castells (1977) *The Urban Question.* London: Edward Arnold.

[75] For a discussion of the phenomenological importance of placelessness see E. Relph (1977) *Place and Placelessness* London: Pion.

[76] G. Debord (1970) *Society of the Spectacle.* Detroit: Black and Red.

[77] H. Lefebvre (1970) *La Revolution Urbaine.* Paris: Editions Gallimard, p. 32.

[78] W. Pritchard 'Address to the jury' (24 March 1920) in N. Penner (ed.) (1973), p. 283.

21 The relative autonomy of the State or state mode of production?*

Ivan Szelenyi

Introduction

In this paper I shall propose that it is increasingly insufficient to explain the role state interventionism plays in urban and regional development in terms of the 'relative autonomy of the State'. Under the conditions of the present stag-flationary crisis, it is becoming more and more obvious that the State is not merely an instrument of capital or the dominant class and that *structural conflicts* emerge between the State and capital. On the other hand, one cannot explain these structural conflicts as conflicts emerging out of the problems of distribution (distribution of power or scarce resources), but they reflect the increasing contradictions of the reproduction process under state monopoly capitalism and they foreshadow the emergence of a new socio-economic formation under which state interventionism will not be limited to the sphere of distribution. I will suggest that *one* possible solution of the present stag-flationary crisis of Western capitalism is a new compromise between capital and State, a new 'New Deal', which would allow a significant control of the production process by the State. This new compromise might lead to the emergence of a

* This paper was presented at the 3rd session of the Research Committee of Regional and Urban Development, ISA at the 9th World Congress of Sociology, Uppsala, 1978.

I should like to express my thanks to my students at Flinders University with whom I was able to discuss earlier versions of this paper. I also found papers written by students in my courses most helpful. I am especially indebted to Stan Aungles, Wayne Pash, Ushi Sombietzki, Frank Regan, John Cook, Rolf Neeman, Ray Jureidini and Bill Lumsden for their comments, stimulation and information.

'state mode of production' and the resulting new socio-economic formation could be more accurately described as the co-existence of the capitalist mode of production with a state mode of production; thus a socio-economic formation of late capitalism which is beyond the present state monopoly capitalism and its 'Welfare State'. But if such a state mode of production is just about to emerge, one can retrospectively re-interpret the sociological significance of the institutions of state interventionism of the 'relatively autonomous State' in socio-economic formations dominated by the capitalist mode of production, classical capitalism or state monopoly capitalism. One can define the 'relatively autonomous' institutions and actions of state interventionism as pre-formations of the state mode of production, and I shall try to demonstrate that this new approach might be helpful in explaining the diversities of urban and regional conflicts among different capitalist societies. Finally, I shall also propose that the emerging new socio-economic formation is pregnant of conflicts from the minute of its conception. It is likely to produce a new and complex system of class conflicts, and while solving some of the contradictions of the present urban crisis in the Western world, it is likely to produce urban and regional conflicts of a new nature, which will reflect the new class structure of the new emerging formation.

Capitalism, socialism and the State

The main contribution of Marxist structuralism to the new urban sociology has been the discovery that the present crisis of Western cities are reflections of the crisis of state monopoly capitalism. Thus, if we want to understand the process of urbanization, we have to start our investigation with the analysis of internal contradictions of the capitalist mode of production. Castells and Godard, Lojkine, Lamarche and others demonstrated convincingly in French, Canadian and American examples that beyond urban and regional processes one can detect class conflicts and, at least 'in the last instance', the intervening State will serve the interests of capital, since this State is not neutral; it is also not purely a 'bureaucratic agency', but it is the 'State of Monopoly Capital'. An orthodox Marxist theory of the State, especially as it was reformulated by Poulantzas, was skilfully and powerfully applied to the analysis of the Paris agglomeration, Dunkerque, Montreal, and more recently by Castells to the explanation of urban decline in the US.

Ray Pahl launched attacks against this position on a number of occasions during the past five years. First of all he questioned whether the 'mode of production' was a useful concept at all to explain the diversities of urban structures in the contemporary world. He pointed out that urban planners in the Soviet Union and in the West were confronted from time to time with strikingly similar tasks and suggested that the distinction between socialist and capitalist modes of production might not be that helpful after all in understanding the differences and similarities (he usually emphasized the similarities) between East European and Western patterns of urban and regional development. Pahl proposed that the 'mode of production' (capitalist vs. socialist) might be too abstract and too general to explain urban processes, and he thought that the 'level of production' might be more directly relevant to understanding the peculiarities of urban problems. But if this is true, then the Marxist theory of the State is also not sufficient to explain the role of state intervention. We need a new theory of the State. In his search for this new theory, he relied on the neo-Weberian theory of bureaucracy and power. In his most penetrating paper on the 'corporate state' he argued that 'corporatism' emerges in the UK as the State gains 'control' while ownership remains private. Thus the nature of the State is being understood in terms of the power the State possesses rather than by the underlying productive system. More recently he adopted Runciman's typology of 'industrial societies' based on the analysis of 'the distribution of power'. Thus, he reversed the whole logic of Marxist sociological analysis which treats the productive system as an 'independent variable' and tries to explain the phenomena of distribution of material goods and power as reflections of contradictions in the sphere of production, and, more specifically, in the relations of production. Hill summed up the differences between Pahl's approach and the Marxist analysis accurately:

'. . . for Pahl the key problem to be investigated is the relationship between the expansion of state activity and patterns of social inequality. He, therefore, points to the influence of the state on economic differentials among class segments in a social formation as evidence for his thesis that the state is increasingly determining class relations. The marxian paradigm, on the other hand, is a theory of the social relations of production and their impact on human development. Patterns of distribution and relative consumption emerge from the class relations of production. The impact of state policy on economic differentials among classes and class fractions reflects of intra-class competition and inter-class struggle.' (Hill 1977: 42).

One almost could declare the whole debate about the nature of the State as closed. The 'neo-Weberians'[1] and Marxists are interested in different questions and the counter-arguments of the neo-Weberians can easily be incorporated into modern Marxist theory, which regards the State anyway as the instrument of the 'hegemony' of the dominant class. The concept of hegemony does allow a significant degree of autonomy in the sphere of distribution to the State. Thus when Pahl states: 'Les questions pratiques en ce qui concerne qui recoit quoi, qui détermine qui recoit quoi, qu'est-ce qui détermine qui recoit quoi, ne sont pas en mesure d'être répondues d'une facon monocausale', he will hardly surprise any Marxist who is familiar with the theory of hegemony of Gramsci; and it is unlikely that he can convince them that they need a new, or reformulated, theory of the State (Pahl 1977: 6).

On the other hand, some of the questions Pahl raised (and further problems which logically would follow from his questions) remained largely unanswered by the Marxist urban sociology. I would like to reformulate these questions, not in order to reject a Marxist approach to the analysis of the State, but rather to show what tasks a Marxist theory of State is confronted with in the analysis of urban conflicts in contemporary state socialist societies, and especially under the conditions of the stag-flationary crisis of state monopoly capitalism.

Pahl's Marxist critics always carefully avoid commenting on the nature of urbanization under 'capitalist and socialist modes of production'. I agree with Pahl that urban problems in contemporary Eastern Europe and in the West cannot be explained with the socialist/capitalist mode of production dichotomy. On the other hand I disagree with his view that the differences in urban processes are the results of different 'levels of production'. I also do not accept that urban and regional structures, thus, will become increasingly similar as the East European economies approach the Western 'levels of production'. It is clear from all available research evidence that urban and regional social conflicts in Eastern Europe are significantly different from the urban conflicts the Western world experienced two or three decades ago, i.e. while they were at a similar stage of economic growth. Any serious East-West comparison would show that the differences are significant and the similarities superficial. True, the urban and regional systems are unequal both in the East in the West, in terms of housing allocation, urban segregation, regional differences, etc., but what is important is that inequalities in the two different parts of the world are of a different nature. They are created by different mechanisms. They emerge in different class contexts and they are divergent even in ecological-geographical terms. (There are slums, for example, in most of the East European cities, but if compared with the West one can establish that they are created by different mechanisms. They occur in different geographical locations, etc.). How can one explain, then, the *newly emerging system* of urban-regional conflicts and inequalities in contemporary East European societies, in societies which do not know the private ownership of means of production and urban land, which are centrally planned and in which the State keeps under control not only the sphere of distribution, but also the whole reproduction process, including the allocation of surplus among branches of the economy and in space? Whose State is this? Can the question of 'relative autonomy' be asked in any meaningful way? I suggest that under those circumstances, when the State brings the reproduction process under direct control, the notion of 'relative autonomy' does not apply any more, since 'relative autonomy' assumes the existence of a 'civil society' and the relative separation of the economic and political system. Thus for socio-economic formations like the contemporary East European societies, where the State dominates the production process, we need a new *Marxist* theory of the State. I emphasize the word Marxist, since in principle it is possible to develop such a theory from the analysis of the social reproduction process, to base it on conflicts which are emerging from the relations of production and we do not have to escape into the description of the sphere of distribution. On the contrary, if we limit our analysis to the question of distribution (how to allocate schools, medical services, etc., in space) then we shall not be able to understand the qualitative difference between East European and Western societies. (Indeed, in this respect planners are confronted with technically similar problems, such as norms of adequate services for 1,000 or 10,000 inhabitants, etc.) To foreshadow my future arguments I would suggest that the East European societies can best be described as socialist socio-economic formations which are based on centrally planned, state redistributive economies, thus dominated by a state mode of production, which co-exists with other modes of production (with certain embryonic forms of 'self management' in co-operatives and with a declining 'capitalist mode of production', as represented by the still-surviving small entrepreneurs, etc.). Such a definition of the East European socialist socio-economic formations allows us to look at them as complex conflict systems (thus the existence and even emergence of urban-regional conflicts and inequalities should not surprise us any more and they should not be regarded as indications of the 'hegemony of a state mode of production'). It is not the State but rather 'society' that might be 'relatively autonomous'.

The purpose of this paper is not to describe urban-regional conflicts in East European socialist socio-economic formations (I attempted to do this in a number of previous publications). I used the above analysis on state socialism more as an heuristic device in order to map the dimensions of our task if we attempt to interpret possible qualitative changes in the role of the State in contemporary capitalism. And here again I should like to refer back to Pahl's theory about the 'corporate State', which did not attract sufficient attention from his Marxist critics. While discussing the 'coming age of corporatism' in Britain, he described with good insight how labourite governments are almost ready to take over tasks that are beyond the usual role of the State. Here again, unfortunately, his analysis lacks depth in political economy. Let me quote his more recent definition of the problems of 'corporatism':

'The corporatist strategy is to manipulate tax concessions and price controls in return for power to control and to channel investment. Private ownership remains, although we should remember that perhaps up to three-quarters of the equity in British industry is controlled by institutional investors. Overwhelmingly, it is the pension funds, investment trusts and insurance companies who are the 'owners' in capitalist societies. It is crucially important to recognize that the goals of such investors are different from those of the traditional capitalist. The latter would be interested in long-term growth and capital appreciation. The former, however, are more interested in high dividends in order to pay out the pensions and insurance benefits to their clients. This means that more capital must come out of industry to pay the pensions of the majority of the population who are, of course, the workers. The inevitable emphasis on return on capital rather than capital appreciation is the paradoxical consequence of the decline in private individual ownership of equity capital. The State is bound to grow in importance as a source of capital for long-term growth. This will inevitably add to the power of the State.' (Pahl 1977: 14–15)

I have difficulties in establishing from this quotation what is the qualitatively new element in the 'corporate State'. Are there structural conflicts
under these new circumstances between capital and the State? And if
yes, along what lines? The main deficiency of the above analysis is that
it does not analytically distinguish between the 'welfare' and 'productive'
functions of the State, and it does not draw a clear distinction between
'capital' and the 'State'. But if 'corporatism' does not refer (at least mainly)
to the replacement of individual private ownership with corporate private
ownership, but basically to an increased responsibility of the 'corporate
State' in *investment decisions* ('The corporate strategy is . . . to channel
investment . . . The State is bound to grow in importance as a source
of capital for long-term growth'), then following Pahl's analysis we can
indeed identify a qualitatively new role for the State, and the State which
plays or intends to play this role cannot be in such harmony with capital
as one might suggest from the above quotations, since it will strive to
intrude the capitalist logic of the production process and, in an obviously
antagonistic way, to overrule the logic of profit-oriented capitalist reproduction. But is it likely that this will happen? Are there signs of a structural
conflict (conflicts bearing signs of an antagonism) between State and capital
in the present stag-flationary crisis?

Structural conflicts between State and capital

I suggest that such a structural conflict between the State and capital
does emerge, in fact this conflict is one of the most important features
of the present stag-flationary crisis and it is quite possible that at least
in certain countries its 'solution' will be provided by a further increase
of state interventionism.

One can look upon the present stag-flationary crisis as the crisis of
the Welfare State of state monopoly capitalism. Following the Great Depression an increasing sphere of economic activities (more characteristically
those which are directly related to the reproduction of the labour power,
the ones which require long-term capital investments like housing, provision of urban infrastructure, education and health) were redefined as 'non-
profitable' for private business. Governments had to take responsibility
for these tasks and they did so, if in exchange capital accepted the system
of progressive taxation. Over the last three decades this 'non-profitable'
sphere has continuously increased, which was partially the cause, partially
the reason, for the permanent increase of state budgets. I would suggest
that the contradiction between the ever-increasing 'socialized', government-
financed, non-profitable 'welfare' sector and the continuously shrinking
privately profitable 'productive' sector, is the major conflict behind the
present stag-flationary crisis and it is the link between the present urban
crisis and the general crisis of state monopoly capitalism. The post-Great
Depression era is a classical illustration of the relative autonomy of the
State: the State intervenes in order to guarantee the capitalist nature of
the reproduction process; without state intervention the capitalist reproduc-

tion would not be possible at all. During recent decades, on the other hand, state intervention, quantitatively, has gradually increased and has probably reached the point when the accumulated contradictions might explode and produce a qualitatively different structure. I should like to illustrate this process and the emerging structural conflict between capital and the State in a case study, with the example of Whyalla, where I am conducting a community study at present. (Whyalla is in South Australia, 380 km from Adelaide).

Whyalla[2] is probably the largest one-company town in the world, a city with 34,000 inhabitants which relies almost exclusively for employment on the biggest Australian monopoly company, the BHP. Whyalla, on the other hand, is also the town that has attracted more public capital investment per capita than any other city in Australia. The state of South Australia accepted an extraordinary role in financing the development of the city.

Prior to the Second World War Whyalla was a small outpost without any economic or other significance. As the war broke out, the state government and BHP agreed that the company would establish a shipyard in the town with government assistance. The state government, in fact, attempted to attract BHP earlier to the town, but at that stage these plans were defeated, partially due to the resistance of British shipbuilding interests, who did not want Australian competition on the market, partially because BHP was also not particularly keen to invest in shipbuilding. But the war gave new stimuli, Britain changed her mind and the state government offered such assistance to BHP that the company could hardly afford to turn it down. From the very beginning the state took the lion's share in covering the expenses of the whole operation. Such a state involvement was not only part of the war-effort, but an integral part of the official policy of the liberal government under the premiership of Playford. Playford was personally committed to the project of the industrialization of South Australia, and although a conservative in politics at large, he took a very social-democratic stand in terms of economic policy. He was prepared to commit large government expenditures in order to attract private business to the state. The South Australian Housing Trust (SAHT) was also created as one of the major instruments of the government for achieving the goal of rapid industrialization. When, not long after the establishment of SAHT, its general manager A. Ramsey summed up the tasks of the organization, he explicitly stated that the purpose of public housing in this state was to attract private business by guaranteeing cheap labour for industrial development. Whyalla was from the very beginning, and remained until the end of the 1960s, a classical example of such a 'harmonious' co-operation between State and monopoly capital. The state government deepened the harbour, provided sweet water, built most of the housing, and the government financed the development of the collective infrastructure (such as building schools, a large hospital and other fairly decent community facilities). The government also assisted the passage of migrants who were recruited by BHP abroad, especially in Britain.

The whole venture was so successful that during the 1960s BHP established a steelworks in the town which grew rapidly and quickly surpassed the shipyard in importance. To indicate the importance of the state's involvement in the whole operation, it is probably sufficient to mention that SAHT built about 70 per cent of the housing (about 7,000 houses), and about half of the workforce employed in the shipyard came from the United Kingdom, with paid passages.

It would be very difficult to estimate the investment the state and federal governments and BHP made in the town, but probably we do not exaggerate when we suggest that the main investor was the state government (the capital invested only in housing should be near $200 million in current prices and this is only a portion of the total commitment of the state government in the city). BHP on the other hand was fortunate enough to need to invest only in productive activities and did not have to bear the costs of the development of the 'nonproductive' infrastructure, or in other words the costs of the reproduction of the labour-power employed by the BHP factories. This is true to the extent that BHP did not have to pay rates for the industrial sites upon which the shipyard and the steelworks were built, and the company provided only about 1–2 per cent of the local government budget (which in the Australian urban management system is insignificant anyway, since local governments are not responsible for any significant infrastructural development). It is another question of how significant BHP's contribution is in the forms of different taxes to the revenues of the federal and state government, but we can solidly establish that BHP is not bound economically in any way to Whyalla.

This obviously becomes a problem with the world-wide decline in shipbuilding industry which followed the oil crisis. The decline in demand for new ships especially hit those countries where production costs were higher, and Australia, due to relatively high real-wage levels, was among these. It is not sufficient to explain the decline of the Whyalla shipbuilding industry exclusively in terms of high wage levels – and even less to do so in terms of 'industrial unrest'. (In fact, industrial disputes did not play a very significant role in the Whyalla shipyard, contrary to quite common belief spread by the mass media and shared by many, even locally). Two other factors have to be taken into consideration: the extremely cautious policy of BHP towards investment in shipbuilding in Whyalla for at least ten years and the reluctance of the federal government to subsidize the shipbuilding industry in Australia.

It is again almost impossible to get reliable figures on BHP's investment policy, but there are at least some grounds for believing that the Whyalla shipyard is quite under-capitalized at present. In a small pilot-study, carried out by the Sociology Discipline at Flinders University in 1977 among shipyard workers in Whyalla, we found that shipyard workers thought that the shipyard, if kept in operation, would urgently require the updating of facilities, and there is no indication of any major investment on behalf of BHP in shipbuilding during recent years. It is also important to remember that about half of the ships built in Whyalla were never sold. BHP used this yard to build up its own fleet, basically to build ships that carried iron ore from Whyalla to Newcastle and coal from Newcastle to Whyalla. It would certainly be interesting to have access to BHP archives and to establish if BHP ever had a serious interest in shipbuilding or whether from an early stage it regarded the whole operation as a minor and basically 'domestic' affair of the company and in fact was planning the coming closure of this production profile.

My present task is not to find the scapegoat for the present crisis of Whyalla; thus with the above comments I do not intend to 'blame' BHP in any way. BHP is a private company and its main purpose is to maximize profit. Looking at BHP's policies concerning Whyalla and shipbuilding from the stockholder's point of view, one might even praise them and suggest that the general management of the company planned this operation wisely, having the interests of its shareholders in its mind all the time. From BHP and its shareholders' standpoint, the story of the Whyalla shipyard can be interpreted as an extremely successful example of long-term planning. BHP is not a philanthropic institution, but a business venture. It has to invest capital where the expected profit is the highest, and if management could foresee ten years ago that shipbuilding in the long run would not be profitable, management should not be blamed, but on the contrary it should be praised for its caution in investing in the Whyalla yard. If the gradual, cautious withdrawal was planned by management, then certainly events of the last two or three years have proved their business competence. Profits in shipbuilding in Australia have continued to fall, and at one point the federal government decided to cut further subsidies back – even if it led to the closure of shipbuilding in Australia.

Trade unions and the state government have fought a losing battle since mid-1976 to save the Whyalla shipyard. In mid-1976 the IAC[3] prepared a report to the federal government and recommended the abandonment of shipbuilding, since the committee found that there was no reasonable chance to make Australian shipbuilding competitive on the world market, and continuing activity in this area would require further increases in federal subsidies. Trade-union and state government experts tried to challenge the IAC report, basically arguing that shipbuilding was not sufficiently subsidized in Australia. Shipbuilding elsewhere in the world, even in Japan, receives more state subsidy, and this is the main reason why, for example, Japanese shipbuilding remains competitive. A complex conflict system developed in Whyalla, the BHP and the federal government on one side, the trade unions and the state government on the other side. To put it crudely, BHP and the federal government had no interest whatsoever in shipbuilding in Whyalla unless it could become profitable again in the foreseeable future. Trade unions and the state government basically wanted to keep the shipyard open at any price, but unfortunately it was only the federal government which could 'pay this price' and, as indicated above, it did not have any major interest in doing so. Thus, if one takes into account the sociological factors, one can establish the fact that by

1976 the Whyalla shipyard did not have the slightest chance of survival. I would just like to add that a comparison with the Newcastle shipyard would be interesting, since it is state-owned and the sociological forces at work are quite different, and shipbuilding might have more of a chance of survival in Newcastle than in Whyalla.

The closure of the Whyalla shipyard was a major threat both to the state government and the trade unions. The state government, as indicated above, had a major investment interest in the city. It was keen to avoid a significant drop in the population which would have meant a large number of Housing Trust homes being left vacant, under-utilization of community facilities (schools, etc.) and increased welfare problems arising from large-scale, long-term unemployment. Trade unions or 'labour' also had a strong vested interest in avoiding the closure, for two reasons: unemployment arising from the closure would affect people who were highly qualified, but whose skills were not easily transferable (such as shipwrights, etc.); so even if they could find jobs in the steelworks or in other industries outside Whyalla, they would have to take significantly lower wages and to accept less-qualified positions. Furthermore, as mentioned above, about half of the shipbuilders were migrants from Britain, most of them not only dedicated to shipbuilding (their main motive in migrating to Australia was to be able to continue in their profession), but they also did not have much experience of Australia beyond Whyalla and for them to leave the town would have had much more traumatic effects than for the Australian born. Not surprisingly, the submission prepared by the State to the federal government concerning the expected consequences of the shipyard closure painted a gloomy picture, suggesting that about 3,000 jobs would be lost and a dramatic decline of the population had to be expected. But even this dramatized – and as it turned out later, exaggerated – report could not change the course of events and the closure of the shipyard went gradually and quite smoothly ahead. During 1977 employment in the shipyard continuously declined, but until mid-1977 BHP were able to decrease employment by 'natural wastage' only, simply not replacing those who left voluntarily. They only had to make workers redundant after mid-1977. At present it looks as though the shipyard will be completely closed down by July 1978. It is still early to predict what kind of impact the closure will have on the town and on its working class. (I am carrying out a survey to estimate this.) It is, on the other hand, quite likely that the closure will have serious effects: the unemployment rate has already reached 20 per cent and might stabilize around 15 per cent in the long run, and some drop in the population, around 10 per cent, can also be expected.

The Whyalla story began as a fairy-tale about the 'relative autonomous' State which subsidized private business, but finally ended up with a structural conflict between the State and monopoly capital. The State and BHP clearly had different kinds of commitments to the town. Capital will move out when profitability falls as measured against investments in the narrowly defined productive sector. Capital is moving out of Whyalla shipbuilding with costs recovered and with substantial gains made over the last decades: BHP did a good job from the point of view of its stockholders. The State, on the other hand, is stuck with investments it can no longer utilize. The central core of this structural conflict is between state responsibility for infrastructural development and, generally speaking, for the reproduction of labour power and the interest of capital to make profit only in the relationship of investments in the productive sphere. In other words, 'benefits' from the whole operation are not measured against the total costs, but only against a fraction of the costs.

To put it very pragmatically one could suggest that the state government 'made a mistake' when it did not work out a 'contract' with BHP and did not oblige the company to take into consideration the state investments as well in its policy towards shipbuilding and the town. But the only way that this could have worked effectively would have been if the State had had control over investment decisions; the State would have had a say as to what a private company did with its profit; the State would have had the right to overrule the immediate profit considerations of private capital. It might not have seemed unreasonable that, for example, in the case of Whyalla, the State could have obliged BHP to keep investing in the shipyard many years ago (when investment elsewhere in other branches of industry, for example, in mining in Papua New Guinea or in Western Australia, seemed to be a safer and better proposition for the company) simply on the grounds that the employment level had to be maintained in the long run in the city in order to guarantee the return of capital investment by the State in the infrastructural development. But such powers certainly would be beyond the rights of the Welfare State. They would assume that the State is active not only in the sphere of distribution but intervenes *directly* into the production process and more specifically into the process of shaping the speed and proportions of the extended reproduction. Such an economic system would be beyond the Welfare State of state monopoly capitalism and probably could be described as the co-existence of a capitalist and a state mode of production.

Welfare State and state mode of production

I suggest that from the case of Whyalla, one might attempt to work out a model that explains some of the mechanisms behind urban decline, and beyond that those structural contradictions of state monopoly capitalism that brought about and maintain with such persistency the present stagflationary crisis. On the following pages I shall attempt to sketch out such a model and to interpret sociologically the changing nature of state interventionism under different stages of the development of the capitalist mode of production.

'State interventionism' is a rather vague and analytically not very useful concept. Since the emergence of 'civil society' (the existence of civil society is the analytical precondition for conceptualizing the 'State' as a separate identity from 'society' or 'economy') the State always 'intervened' in one way or another. Therefore the theoretically relevant question is to attempt to develop a typology of the qualitatively different involvement of the State in the economy. For the purposes of present analysis I suggest that we can differentiate between the 'liberal period of state interventionism', the 'labourite period of state interventionism' and a 'eurocommunist period of state interventionism'; the last of these will basically represent a socio-economic formation that is still dominated by the capitalist mode of production (already in decline) at the same time as a new, state mode of production is emerging. The main purpose of the following analysis is to distinguish analytically between the 'classical social-democratic' and euro-communist strategies of state interventionism; thus the 'liberal period' will be described only very superficially. The analysis will also remain fairly abstract and theoretical and in this paper I shall not be able to do much more than merely indicate summarily why the understanding of the nature of the state mode of production is relevant to contemporary urban research.

With a gross over-simplification one can call the era preceding the Great Depression the 'liberal period of state interventionism'. In this period the State played a strictly regulatory role, ideally only an insignificant proportion of the GNP was concentrated in state budgets and budgetary intervention of the State was limited. The State played the role of the 'referee' in the free game of laissez-faire capitalism, thus setting the rules of the game with legislation, export and import duties, but basically leaving both production and distribution on their own, in self-regulating mechanisms. (This descripion – as suggested – is extremely simplified. There are enormous differences between different countries, let us say England and West Germany, the United States and Australia, and there are important changes occurring over time in any given country.)

The Great Depression indicated a qualitative change in the nature of state interventionism and after the Depression, or even more clearly following the Second World War, the States in most if not all Western capitalist societies began to play a role that the revisionist social-democratic parties of the Second International formulated for the first time around the turn of the century – this is why I would call this era the social-democratic or labourite period of state interventionism. The first historical act of this new era was the New Deal. One can interpret the New Deal as a compromise between national capital and the nation state in which the capital granted the rights of progressive taxation to the nation state. During the decade which followed the New Deal, the nation state redistributed an ever-increasing proportion of real incomes through the federal budget and this redistribution of real income was legitimated with values of social policy, with the values of 'social justice' and 'equality'. These ideologies were directly borrowed from the turn-of-the-century, revisionist social-democratic parties. Ideologues, like Bernstein, believed that a more just distribution of incomes was in fact possible without the abolishment of the institution of private ownership, if social-democratic parties gained political power via parliamentary democracy. At the turn of the century this proposition was regarded as a reactionary utopia by those who were on the left of revisionist social democracy, and certainly representatives of the interests of capital were less than ready to believe that capital could ever accept a significant redistribution of income without undermining the very foundations of the capitalist 'free enterprise'. People who would predict at the turn of the century that tax rates on high incomes might reach 60, 70 or 80 per cent would have been referred to the care of psychiatrists. Curiously enough these 'utopian dreams' became reality practically everywhere in the Western capitalist world following the Second World War.

It should be noted that this basically social-democratic programme in many countries was not introduced by social-democratic parties: in the United States, due to the weakness of even the revisionist socialist movement, the Democratic Party made this New Deal with capital; in West Germany, the Christian Democrats laid the foundations of the 'Welfare State' after the Second world War, keeping the Social Democratic Party in opposition; and the story is very similar in Australia where the Liberal Party – despite its ideological commitment to laissez-faire capitalism – gradually built up the new institutions of state interventionism. On the other hand it is still true that the classical model of the Welfare State of state monopoly capitalism is the Swedish model, based explicitly on the turn-of-the-century revisionist ideologies. It might also be worth noting that in countries where conservatives – against their own ideologies, but forced by the requirements of the 'times' – moved towards the 'big States with large budgets' as a consequence of their ideological commitments, they usually lagged behind the times; thus if they were politically defeated by their social-democratic opposition the incoming labourite governments always had a lot to catch up with in order to move towards the 'Swedish model' (an explicit goal of both the incoming Labour governments in the early 1970s in New Zealand and Australia). Furthermore, as the state institutions of redistribution of real income strengthened, it became evident that increased state budgets were necessary to guarantee the capitalist reproduction process, but the ideological justification of labourite state interventionism became more and more dubious. It turned out that the social-democratic Welfare State was not all that efficient in its 'wars against poverty' and welfare interventionism also did not seem to lead to a more equal society. During the late 1960s poverty was rediscovered even in Sweden, and Swedish sociologists proved that the range of social inequalities had remained practically the same since the Second World War. The same applied to Australia. Henderson demonstrated that after decades of the Welfare State, about 10 per cent of the Australian population still lived below the 'poverty line'. It was documented by Encel and by others that the share of labour and capital from the national income was not altered in the interest of labour in any way. This had significant political implications: social democratic parties in power could not be regarded any more as 'workers' parties' properly speaking; they were rather parties of 'state interventionism', and as their political base they had to rely more and more on those who had an immediate vested interest in social-democratic state interventionism. The new 'constituency' of these parties increasingly became the rapidly expanding state bureaucracy and all those who make their living out of state budgets (including academics, etc.). This 'new constituency' was increasingly the 'new petty bourgeoisie', to use Poulantzas's terminology.

I suggest that under labourite state interventionism it would be highly inaccurate to speak of a 'state mode of production'. State intervention in *its logic* is limited to the sphere of distribution. The State redistributes *real income which is generated on a price regulating market* (this market is certainly not 'free' in the traditional sense, but still it sets the prices in principle, thus if the price of labour-power is systematically pushed above its 'value' the economy will respond with a decreasing demand for labour and with increasing unemployment). The Welfare State does perform in a certain sense 'productive functions' as well, but a more detailed analysis can show that in its productive activity the State does not overrule in any sense the logic of the capitalist reproduction process. The productive functions the Welfare State is allowed to perform are activities that do not produce profit any more under the economic growth strategies of state monopoly capitalism. The classical example is public housing. The State takes responsibility for the provision of a certain kind of housing because it cannot be provided profitably by private business. Public housing is more a way to reallocate means rather than a 'productive activity' as it is defined by a capitalist mode of production. It is also symptomatic that quite often the actual construction tasks are 'reprivatized', i.e. commissioned to private constructors. These private developers can make a profit out of the construction of cheap housing because it is subsidized by the State. The same applies to those instances when the State takes over certain branches of production, properly speaking, like coalmining in Britain. As a rule nationalization will occur only when the nationalized industry is not profitable for private business. It is also important to note that in nationalized factories the State will follow the logic of capitalist reproduction in shaping wage scales, in working out investment policies. It will fit into the capitalist environment, thus even nationalized companies can hardly be interpreted as the institutions of a new mode of production. The notion of the mode of production refers to the logic of the reproduction process, and the Welfare State in its productive operations will obey the 'iron laws of profit'. This also means that nationalized companies, when they become profitable again, are quite often re-privatized, and as a rule the State usually does not enter areas where it will compete with private business. I find it quite paradigmatic that, for example, when the South Australian State Government considered creating jobs with state investments in Whyalla to ease unemployment there, the most serious proposition government advisers could come up with was to establish a state clothing factory which would produce police uniforms and similar products. Thus it would not present competition to private business.

The main proposition of this paper is that the present stag-flationary crisis of state monopoly capitalism is the crisis of the above-outlined

Labourite state interventionism. Labourite state interventionism was based on the continuous increase of state expenditures; the increase of budgetary intervention created the need for further intervention with a further increase in expenditure. (Public housing in Britain is a good example: as the public-housing sector increased it undermined the rate revenues of local governments and from then on the State had increasingly to subsidize local government budgets as well.) The main mechanism that has guaranteed the steady increase of state expenditure has been the Keynesian inflationary policy (here I rely heavily on O'Connor's analysis on the fiscal crisis of the State) which at the early stages of the inflationary economy served both the interests of capital, oriented towards mass consumption economy and interested in 'hot' money and a high level of consumer expenditure, and certainly the interest of the State, which under the circumstances of progressive taxation could only gain in absolute terms from inflation. For a while this inflationary economy could accommodate the interests of organized labour as well, which through collective bargaining could maintain real-wage levels above inflation rates, and worked only against the politically least-organized sections of the society, against small business, pensioners, etc. (see the problems of de-valorization of small capital). As on the other hand, it is known by now that the vicious circle of ever-increasing state expenditure and inflation at a later stage works against the interest of labour and even against the interests of capital. With the 1970s the counter-offensive of capital began against the 'big State', and this counter-offensive found its ideologues in economists like Friedman. This counter-offensive proved to be highly successful, the interventionist social democrats could not rely politically on the forces of labour either, and the 'new petty bourgeoisie' was not powerful enough to keep them in power. A significant proportion of the working class was ready to cast their votes for conservatives who promised to cut state expenditure and to bring inflation down, therefore social democrats had to change platforms and behave like conservatives (this has happened in the United Kingdom and in West Germany at the top of the social democratic political movement), or they were disastrously defeated as in Australia and in New Zealand. (The ALP 'learned its lesson' and 'opened up to the right' at its Perth convention in 1977, but this was not only too late – why should the Australian vote into power a conservative social democratic party, when they have in power a genuinely conservative force? – but if the following analysis is correct it was also a strategic mistake.) The Friedmanite ideologies about the 'small State' and propositions to cut state expenditure drastically, were, on the other hand, probably only romantic illusions. The huge apparatus of state interventionism had been already created. These institutions are also necessary for the whole reproduction process and more specifically to the reproduction of labour-power in modern capitalism. If the fiscal bases of these institutions are undermined, the whole reproduction process is threatened. Conservatives moving into government find themselves in an impossible situation and they are unable to keep the promises they made during the election campaign. Relatively small cut-backs in government expenditures have spectacular and dramatic effects, and this can be most clearly demonstrated in the current processes of urban decline. Certainly urban redevelopment plans and regional decentralization programmes are the most obvious targets of budgetary cut-backs. But they lead at once to the most dramatic urban crises of the modern age, as in the case of New York City. It seems reasonable to assume that there is no way back from labourite interventionism into liberal interventionism.

This might be the reason why in countries that are the most seriously affected by the stag-flationary crisis such as Italy, eurocommunist parties came so close to power. It is not very easy to reconstruct what is precisely the economic strategy of the eurocommunist parties (one even might argue that they probably have none, they have only worked out a political strategy of how to get into power). It is still justified to argue that the eurocommunists attempted to go beyond the usual social-democratic strategy – politically and, even more so, economically. The eurocommunists claim new economic functions for the State by maintaining the co-existence with private capital. They are not prepared any more to limit the power of the State to the redistribution of real income, which is generated and defined on a price-regulating market, but they want to intervene directly into the production process. (Remember: the social-democratic Welfare State had powers to redistribute real income, but in exchange *it left the production process alone. It* did not alter the capitalist nature of reproduction.) The eurocommunist – usually without large-scale nationalization – would like to bring under state control the *allocation of surplus.* It is more than likely that a eurocommunist government in power would establish a central state-planning bureaucracy which will claim powers to decide about productive investments, about the movements of capital among different branches of the economy, among different regions and among countries, systematically overruling the logic of the maximization of profit in the name of 'long-term' and 'societal' interests. To put it very theoretically: the eurocommunist would like to extend state power beyond the redistribution of real income towards the redistribution of surplus which can be used for the purposes of extended reproduction. But if such a programme becomes operative then the Western world will move towards a new socio-economic formation which cannot be described any more as state monopoly capitalism which allows only a 'relative autonomy' to the State, but it will become a social formation where the State has real autonomy, represents a new logic in issues related to extended reproduction which will compete with the 'iron law of profit maximization' of the capitalist mode of production.

I found Carrillo's proposition in his recent book, *Eurocommunism and the State,* quite accurate and stimulating. Carrillo suggests that the difference between traditional social-democratic and eurocommunist power will be in the nature of the State. According to him, the social democrats only 'administered the state of the capital', but the eurocommunists intend to change the nature of the State. As he puts it: under eurocommunism the State will represent the 'hegemony of the forces of labour and culture'. This is certainly the most apologetical formulation – but what else can one expect from the founding father of eurocommunism? But if, beyond the rather poetic concept of the 'forces of culture', with a bit more rigid sociological analysis, we identify the 'technocratically well trained', 'politically aware', 'teleological or evangelistical' planning bureaucracy, then one could suggest that Carrillo's definition of the essence of eurocommunism is not that far from my own analysis of the new state mode of production. From the point of view of political sociology, Carrillo's position again makes a good deal of sense: the eurocommunists can move into the positions of governmental power only if they are able to maintain their working-class votes. They have to build their political strategy on the coalition of interests of the 'forces of labour and culture' (and on a somewhat benevolent neutrality of the forces of capital, which they intend to guarantee by not pushing large-scale programmes of nationalization).

It is certainly highly questionable whether the 'forces of capital' are ready to accept this benevolently 'neutral' role. Is capital ready to make a new New Deal with the State? And, if yes, which will be the political force capital will find more reliable and trustworthy? It looks as if in Italy even multinational capital is considering reaching such a new compromise with the Italian Communist Party (ICP). In other countries with strong social-democratic movements and weak communist parties (especially where these CPs were too much compromised in the past by their unconditional obedience to Moscow) one could expect the social democrats to move to the left and introduce basically eurocommunist programmes. In fact in countries like Britain, West Germany or Australia, it is already quite difficult to distinguish between a eurocommunist platform and the platform of the left-wing social democrats (who by the way prefer to call themselves 'democratic socialists'). I do not want to suggest with these qualifications that there is no difference between traditional Second International social democracy and eurocommunism. The main thrust of my previous argument was that eurocommunism, by foreshadowing the emergence of a state mode of production, indicates a qualitatively new stage in the development of Western capitalism. Therefore it is qualitatively different from traditional social democracy, but under given social and historical circumstances it is quite possible that this qualitatively new political and economic strategy will be carried out by social democratic parties, but they have to undergo a significant change in order to qualify for this 'new job'. Therefore, I do not think the critics of the eurocommunists from the Left are correct when they accuse them of being the same as the 'revisionist social democrats' were. On the other hand I think right-wing critics of eurocommunism – or for that matter of left-wing labourites – are equally wrong when they pretend that eurocommunism or left-wing social democracy is simply the Trojan horse that will smuggle the Moscow line, East-European style, totalitarian state socialism, into the West. If a state mode of production does emerge in the West co-existing with the capitalist mode of production maintaining the political and cultural institutions of the 'civil society', then this new social formation will be as different from East European state socialism as it will be from state monopoly capitalism. I also would be quite hesitant to make any ideological statement about the nature of this newly emerging social formation (if we are talking about the same thing, Pahl calls it ironically 'fascism with a human face'); it is certainly a newer, but probably slightly more complex class society than classical capitalism. It certainly does not promise us the radical emancipation of the working class and it is pregnant from the time of its conception with antagonisms. Or in other words: I am far from suggesting that the state mode of production *should* come; the main thrust of my analysis is that from the understanding of the nature of the present stag-flationary crisis one can predict that it is more likely that the extent of state interventionism will increase, rather than decrease. Therefore, the State *might* move beyond a simple 'relatively autonomous' role and around the already existing institutions of state redistribution, a new state mode of production might emerge which will represent a new challenge *both* to the forces of capital and labour.

Ecological problems (the need for more decentralized energy and production systems), syndicalist movements (a move towards co-operatives, workers' control, etc.) or the overwhelming power of multinational capital certainly can set completely different patterns for the future, but at least in the light of the present urban crisis a state mode of production seems to be a possible (and quite near) future. From Whyalla to New York City (if I may use this quite incredible extrapolation) it seems to be clear that states simply cannot afford to keep financing urban infrastructure unless they gain some control over the process of how jobs are created and abolished, how productive investments flow in and out of cities and regions – unless they can move towards a 'new accounting system' in which 'economic efficiency' or 'profitability' is calculated not only in the relationship of private investments in the narrowly defined 'productive sphere', but in a relationship to the total investment, private and public, which is necessary to the total reproduction process.

A retrospective view: the role of the State in urban and regional development in capitalist societies

This paper was not intended to be an exercise in futurology. The main purpose of extrapolating present trends into the future is to gain a better understanding of present and past processes. If we have a vision of the possible future we might be moving towards, we can probably re-interpret the sociological significance of some phenomena of the present or even those of the past.

If we have, for example, a good understanding of the nature of the capitalist mode of production, we can quite meaningfully reinterpret some of the processes of European feudalism as pre-formations of capitalist institutions, as forces leading towards capitalism. (If I am not wrong, this is what Weber meant when he proposed that we have to search for the 'cultural meaning' of phenomena under investigation, and that the same phenomena can have different 'cultural meanings' depending on the point of view we adopt in approaching its analysis.) This means that we can explain aspects of society under feudalism that we could not understand from the analysis of feudalism alone.

One main analytical weakness of the 'relative autonomy' thesis is that it is not very helpful in explaining the areas where the State is 'relatively autonomous'. I find it a distressing intellectual perspective if I have to keep repeating that since the State under capitalism is a capitalist State, it will always serve the interests of capital, and, with an elegant gesture, I just throw all that I cannot explain this way into the garbage can of 'relative autonomy'. (Proposing the empirical content of 'relative autonomy' to be that the ruling class is in itself divided, and that, therefore, the State will always represent the interest of sections of the ruling class, is intellectually certainly not more stimulating, and is empirically unworkable.) If the notion of 'relative autonomy' has any meaning at all, it has to suggest that certain actions of the State cannot be explained in terms of class antagonisms, that the State represents its own interests; but then, certainly, we have to ask the unpleasant question as to what kinds of interests these are. If my proposition about the 'coming age' of the state mode of production is correct, then we have suddenly gained analytical tools to explain what is relatively autonomous, namely as preformations of a future state mode of production, elements, forces, leading towards a state mode of production.

Let me illustrate this point first with a general example, before I try to indicate the strength of this kind of argumentation for a comparative analysis of urbanization. I find Poulantzas's suggestion that is is more meaningful to define white-collar workers as 'a new petty bourgeoisie' (or in other words, a new middle class) rather than a 'new working class', quite imaginative and intriguing. It also makes a lot of sense to propose that the existence of a middle class always indicates that in the social formations we do study, more than one mode of production exists; and the 'middle class', or 'in-between class', or 'middle strata' are products of the 'other', namely the non-dominating, mode of production. This is certainly a handy way to explain the nature of the 'old petty bourgeoisie', the self-employed or especially the peasantry which can be looked upon as the remains of a declining, pre-capitalist mode of production which will be gradually destroyed as capitalism gains ground. But what about the 'new petty bourgeoisie'? And here certainly Poulantzas does not provide an answer. We just do not know what is the 'new mode of production' that gives birth to this new middle class. Again, if my forecast about the emergence of a state mode of production is correct, one can define this 'new middle class' – and its hard core is without doubt the state bureaucracy – as the representative of the new mode of production, and quite possibly as a class that aspires to become a new dominant class, or at least in the short run to share the dominant class position with the 'old dominant class'. In the light of this analysis, it is not surprising at all that Poulantzas conducts his debate with the ideologues of the French CP which was just turning eurocommunist at the time Poulantzas emerged as a theorist. If the French CP as a eurocommunist party represent basically the interests of the new state mode of production, certainly their ideologues will have to emphasize the coalition of interests of the new upcoming class with the working class. They have to suggest – to use again Carrillo's formulation – that they strive for the hegemony of 'the forces of labour and culture'. By extrapolating the present trends retrospectively we can reinterpret the nature of contemporary social structure and we can gain a better understanding of the underlying ideological assumptions beyond different sociological theories of class structure.

Such a methodology might be helpful to explain the diversities of urban and regional structures among different capitalist countries. When Pahl rejects the concept of 'capitalist mode of production' as a useful tool to explain urban structures, he quite convincingly points out that there is a wide variety of urban and regional structures and patterns of urbanization among countries that are capitalist. Both Hill and Lojkine give the answer one can expect from a Marxist structuralist, well armed with Althusserian theories; Pahl confuses the concept of the mode of production with the concept of social formations. When we analyse empirical patterns of urbanization we analyse social formations, and they are composed by different modes of production (e.g. petty commodity production and capitalism). No wonder their urban system will be different. While methodologically I agree with Hill and Lojkine, I am not sure that some of the features of urban processes can be explained from the 'still surviving' pre-capitalist modes of production. I propose that one can look at different government structures as the pre-formations of a coming state mode of production. Thus, from the nature of local governments, mechanisms of local government finance, relationships of local governments to state or federal governments, etc., one will be able to explain *some* of the characteristic features of an empirically observable urban and regional system. I should like to illustrate this point with a short and necessarily superficial comparison between patterns of urbanization and nature of urban conflicts between the United States and Australia. The reason for such a comparison is quite evident: we are comparing societies that do not have a 'feudal past'; their urban systems developed quite recently and the geographical and climatic conditions are not that dissimilar (especially if we focus our attention on the West Coast of the US). On the other hand, the role that the State played in the development of capitalism is very distinctly different in the two countries and their urban and regional structures are also strikingly different. Since I am comparing two capitalist societies, my main hypothesis is that in their urban and regional development the State played only a 'relatively autonomous' role. In both cases urban and regional development is subordinated to the interest of the development of the capitalist mode of production. But I will suggest that the Australian urban system differs from the American to a large extent because specific features of government structures, because the pre-formations of a state mode of production appeared much earlier in Australia than they did in the United States.

The State played different roles in the colonialization of North America and Australia. North America is more an example of a 'private colonialization'. Australia is one of the examples of 'modernization from above', its settlement was from the very beginning state controlled, managed and financed. There were convict settlements in early North America as well, but the role convicts performed in Australia, how the State used the convict system and later alternative systems (like the method of the so-called 'systematic colonialization') in order to promote the development of a capitalist mode of production are quite distinct and quite suggestive concerning the role of the State in the development of Australian capitalism.

I do not want to enter the debate here as to whether or not the convict system in early Australia can be interpreted as a 'slave mode of production', but I assume that both those who accept this proposition and those who reject it would accept that convict labour very early became a form of state subsidy to private enterprise. Private enterprise emerged very quickly in New South Wales as land grants were given to retiring officers or to convicts who earned a 'ticket of leave' and within a decade private enterprise played a comparable role with the state-managed prison system. On the other hand, the emerging private enterprise was very much controlled and subsidized by the State. The State assigned convicts to work for pastoralist-entrepreneurs and covered a significant proportion of the costs of their maintenance. For a very long time the State had a total monopoly of the market, bought up the products of the private enterprises and sold them consumer goods and the means of production at administratively fixed prices. The State played a crucial role in providing labour-power at a subsidized price to private capital and also in accelerating the accumulation of capital. This went much beyond the assignment of convict labour to pastoralists, since with its land grant and price policy the State also guaranteed that former convicts on 'tickets of leave' could not become self-employed small farmers or peasants (land grants to the 'ticket-of-leavers' were just too small to allow the development of a subsistence economy and prices, administratively set, systematically served the interests of the larger landholders). Elsewhere in Australia where convict labour did not exist or did not play such a significant role as in New South Wales or in Van Diemen's Land, the State performed a similar function but through different mechanisms. The Wakefield scheme in South Australia is quite suggestive again: here, following the philosphy of 'systematic colonialization', the State set a high enough price for the land in order to pay for the passages of the less well-to-do, who after their arrival were then unable to afford to buy land for themselves and so had to become wage labourers. The role that the State played in urban and regional development served very much the same purposes: the fastest possible accumulation of capital in the private sector of the economy. To reach this goal the State took over the tasks of infrastructural development and communal infrastructure was financed from state resources (or in the case of South Australia to some extent from the revenues from land sales). Since the provision of infrastructure is a state task the settlement of the continent was in a way 'planned', state-controlled. The state apparatus moved together with the pioneers and permissions had to be sought from the authorities for the establishment of a new settlement (even if on some important occasions, as in the case of Melbourne, this permission was granted after de facto settlement; these were exceptions to the general rule).

This kind of urban and regional development is very different from that in the United States. In the United States, at least up to the mid-nineteenth century, the pioneers settled the new areas first; the government followed them, sometimes years later. The pioneers built up their communities from their own resources. They built their own schools and churches. They had to learn how to govern themselves. They elected their own sheriffs and judges. The three or four-dozen Mormon families that laid the foundations of Salt Lake City were in fact escaping state authority. They successfully undertook the seemingly impossible task of creating a city under climatic conditions that were unfavourable even by Australian standards. In the United States an urban management system developed that was based on self-financing, politically-autonomous local governments, and this guaranteed that part of the locally produced surplus was directly channelled into infrastructural development. The American settlers created *communities* with communal infrastructures of a decent standard and this gave a certain stability to the regional system and produced a relatively even distribution of population in space. The American 'small town' became part of the Jeffersonian vision of America, an integral part of the American way of life in the nineteenth century.

In contrast, those who settled the Australian continent relied on state provision and did not care much about building cities or creating communities. The Australian pioneers came to this continent to become rich as quickly as possible and to accumulate capital as fast as possible. They did not mind that the State took the responsibility for infrastructural development, since this relieved them of having to tie up part of their profits locally in non-productive investments, when they often were also not that keen to settle in the 'outback' where they made their profit. The Australian system of 'squatting' was an integral part of this regional management system. Squatting also existed on the West Coast of the United States, but there mostly the poor squatted – those who could not afford to buy land. In New South Wales or in Queensland, on the other hand, the squatters were the aristocrats – the 'squattocracy'. Squatting as a specifically Australian system of land tenure was again a way to assure that as much capital as possible was used for 'productive investments'. It is suggestive that land was 'leased' to squatters according to the amount of capital they possessed; thus the more capital you had, the more cattle or sheep you could buy (or owned already), the more land you could squat on. This way the ground-rent mechanisms were bypassed and again the rapid accumulation of productive capital assured. But from the point of view of urban and regional development squatting meant that pastoralists were not tied down to any locality, and they in fact moved across the continent following their immediate profit interests.

Consequently, the self-regulating, self-financing urban management system that was the system of laissez-faire capitalism in the United States (and for that matter in England as well) never developed in Australia. The Australian urban management system from the very beginning was very much a state grant economy, with very limited financial or political autonomy of the local governments. Curiously enough the British-trained civil servants, running the Australian bureaucracy, found this system so unusual that they made an attempt to change it according to the English pattern, and, paradoxically, in the 1840s there were serious attempts in New South Wales, at least, initiated again 'from above', to grant more autonomy (and also certainly more financial responsibility) to the local governments. But these moves were defeated by the resistance of the local 'communities'. The local governments just did not want to govern themselves, if that meant that they had to pay the costs of their own community development. Quite a paradoxical story: elsewhere in the world it is the aggressive central authority that gradually breaks up local autonomy; here in Australia it is the local community that resists the initiatives coming from above for more local autonomy. But the lack of development of autonomous local government structure and the domination of the regional management system by a state grant economy meant that a direct link between production of surplus and infrastructural development never existed in Australia, and I would suggest that this is responsible for two important features of the Australian regional and urban system: the over-concentration of the regional structure, and the relative underdevelopment of the 'collective means of consumption', of the collectively used urban non-productive infrastructure.

The over-concentration of the Australian regional system (about two-thirds of the population lives in six metropolitan areas, in the case of South Australia almost four-fifths of the population lives in the capital city) is partially the product of climatic conditions, but at least partially it is the reflection of the over-concentration of power. Since all decisions concerning community development were and are made by state governments, it is hardly surprising that capital cities were always in privileged positions, and they just became more and more privileged as their power monopoly grew. It is also interesting to note that almost all the successful cases of regional decentralization, the creation of new urban centres, are cases where the driving force was the decentralization of administration, or the creation of centres for administrative purposes, for example, Canberra, Darwin, Alice Springs. Industrial development in itself rarely leads to lasting urban development outside the main metropolitan areas. It might be quite illuminating to compare, for example, from this point of view, Port Augusta with Whyalla. Port Augusta got a better share-out of administrative functions and certainly seems to have a safer future than Whyalla, which was mostly just the product of industrial development. During the Whitlam era, DURD[4] experts committed to the idea of regional decentralization were also probably aware, at least to some extent, that the over-concentration of the Australian regional system was the product of over-concentration of political power in the regional management system. Although this was another 'revolution from above', at least DURD attempted to bypass state governments and to encourage the development of regional centres of counter-power. Clearly, successful decentralization in Australia is not purely an economic, even less a climatic problem. It depends on the decentralization of power in the regional management system; regional counter-poles can only be based on centres of counter-power.

The other important consequence of the Australian urban-grant economy is the relative underdevelopment of the collectively used urban non-productive infrastructure. I proposed earlier that under the Australian centralized urban-grant economy no direct link exists between the production of surplus and the growth of urban non-productive infrastructure. Investments in the development of non-productive infrastructures are functions of budgetary decisions and governments are inclined to 'economize' with these kinds of investments. Local groups usually do not have a great deal of political bargaining power. Consequently, if infrastructural development is exclusively the subject of budgetary decisions this usually leads to a delayed infrastructural growth. If we were able to develop an index that would measure capital investments in 'private infrastructure' and to relate it to investments in the 'publicly used infrastructure' I am sure we would find that Australia is probably the most underdeveloped country in the Western capitalist world in terms of 'collectively used non-productive urban infrastructure'. This society spends more on private housing, private swimming pools and tennis courts, fences, sheds, etc. than it spends on public parks, schools, libraries, etc. These relatively scarce 'means of collective consumption' are also quite unevenly distributed in space, and due to the domination of metropolitan centres by the state governments they are usually concentrated in the city centres. The relative scarcity of the 'means of collective consumption' and their distribution in urban space explains some of the ecological processes in Australian cities, and some of the differences in terms of urban segregation and change compared with American cities. The scarce urban non-productive infrastructure, concentrated in urban centres, becomes the main object of 'urban class struggle' in Australian cities, and this might explain why the upper-middle class has been moving back during the last decade towards the central areas and pushing the working class out into infrastructurally extremely deprived far-outer suburbs. This is certainly pretty much the opposite pattern to urban segregation and ecological change in American cities, where the middle class still continues to leave the city centre, in order to segregate itself into its middle-class or upper-middle-class ghettos. It 'pays off' for the well-to-do in America to move into the outer suburbs, since with higher property values from higher rates it can provide better community services for itself under the still to some extent self-financing urban management system, and can leave the financing of the declining urban centres to the urban poor. Urban deprivation in the Australian cities works the other way around. Here, since there is no self-financing urban management system, the well-to-do are not motivated to move out, but, on the contrary, they try to monopolize the centrally financed, better urban facilities in the urban centres and to push the poor as far away from it as possible. Certainly one thing is common to both urban systems: both add to the privilege of the already privileged, but they do it in a significantly different way. Here again, if we want to understand the actual mechanisms of urban deprivation we have to look into the actual structure of the urban management system, and as we can see, the differences in government structures are quite crucial in explaining the empirical differences.

Finally, I should like to add that despite the general underdevelopment in terms of the collectively used infrastructure, the Australian urban system is significantly more egalitarian than the American one, and the urban 'crisis' of the Australian cities is consequently less dramatic. If my proposition is correct, and in the Australian urban and regional management system one can identify a somewhat more developed pre-formation of a state mode of production, then one has to acknowledge that this system seems to be at least somewhat more efficient in handling inequalities.

If it is true that a new state mode of production is emerging in Western capitalism, then it will be increasingly important in analysing urban processes to look at governmental structures and the ways in which different state institutions participate in the financing of urban and regional development, since the differences in urban and regional structures among different capitalist countries can be probably explained more and more according to these criteria. This kind of analysis is also important in order to map the emerging new contradictions in urban and regional systems, since, as mentioned above, the emerging state mode of production is likely to produce new dimensions of contradictions while solving some of the old conflicts. It is beyond the task of this paper to map these new contradictions, but let me raise just one crucial issue. The state mode of production is only in *statu nascendi* but it is already an anachronism: capital is increasingly multinational, and it is highly questionable whether the nation states can make any 'deal' with multinational capital and whether they can exercise any effective control over the movements of multinational companies and capital. This is certainly an issue relevant to the present urban crisis: the large cities are smaller than the nation states, but in a way they are above nations and nation states. New York City is a world metropolis, and the United States is only a nation state. Increasing intervention by the nation state might not only meet the successful resistance of multinational capital, but it might also have problems in dealing with the largest cities, and these world metropolises might attempt to challenge the authority of the nation states above themselves. The relationship of New York City to the federal government is already quite a hostile one: the contribution of New York City in the different taxes paid by the city to the federal budget is about ten times more than the 'subsidies' it receives from the federal government. In a way New York's fiscal crisis is an imaginary one, and more control by the nation state does not seem to be the answer. I think Hill defined this problem quite clearly when he stated: '. . . what we need is a theory of how the dynamics of capital accumulated on a world scale embodied most visibly in the multinational corporation intersect with the expanding role of nation states such as to generate contradictions leading to transformations in social formations as well as in the capitalist system as a whole' (Hill 1977: 43). I agree with Hill, but I have had to limit this paper to the questions of the 'expanding role of nation states', and the vital questions concerning the complex conflict system between nation states, multinational capital and world cities are beyond the scope of my present analysis.

Notes

[1] I call this approach 'neo-Weberian' since I believe it is based on an extremely one-sided interpretation – to say the least – of Weber's theory of capitalism. It is highly questionable that Runciman, for example – and Pahl relies heavily on his analysis – does enough justice to the complexity of Weber's argument. With few exceptions (certainly Giddens is such an exception) contemporary Anglo-Saxon sociology still operates with the functionalist, parsonian understanding of Weber and does not acknowledge Weber as a historical sociologist.

[2] The following analysis on Whyalla is only a summary of a larger paper 'Structural conflicts between the State, local government and monopoly capital – the case of Whyalla in South Australia' presented at the meetings of the Research Committee on Community Research, ISA at the 9th World Congress of Sociology, Uppsala, 1978.

[3] Industries Assistance Commission Report (1976) *Shipbuilding.* Canberra.

[4] The Department of Urban and Regional Development was created during the early 1970s as the Labour Party formed the Federal Government. It was abolished shortly after the fall of the Federal Labour Government.

References

Castells, M. and Godard, F. (1974) *Monopolville.* Paris: Mouton.
Hill, R. S. (1977) 'Two divergent theories of the State.' *International Journal of Urban and Regional Research* 1: 37–43.
Lamarche, F. (1976) 'Property development and the economic foundations of the urban question', in C. G. Pickvance (ed.) *Urban Sociology: Critical Essays.* London: Tavistock Publications, pp. 85–118.
Lefebvre, Henri (1977) *De l'état. Tome III, Le mode de production étatique.* Paris: Inedit.
Lojkine, J. (1973) *La politique urbaine dans la region parisienne.* Paris: Mouton.
—— (1977) 'L'analyse marxiste de l'état.' *International Journal of Urban and Regional Research* 1: 19–23.
—— (1977) *Le marxism, l'état et la question urbaine.* Paris: PUF.
Mingione, E. 'Pahl and Lojkine on the State – a comment.' *International Journal of Urban and Regional Research* 1: 24–36.
O'Connor, J. (1973) *The Fiscal Crisis of the State.* New York: St. Martin's Press.

Offe, C. (1975) 'The capitalist State and policy formation', in L. Linkberg (ed.) *Stress and Contradiction in Modern Capitalism.* Lexington, Mass.: Lexington Books.

Offe, C. and Ronge, V. (1975) 'Theses on the theory of the State.' *New German Critique* 6.

Pahl, R. (1977) 'Collective consumption and the State in capitalist and state socialist societies', in R. Scase (ed.) *Cleavage and Constraints – Studies in Industrial Society.* London: George Allen and Unwin.

_____ (1977) 'Stratification, the relation between States and urban and regional development.' *International Journal of Urban and Regional Research* 1: 6–18.

Pahl, R. and Winkler, J. (1974) 'The coming corporatism.' *New Society,* 10 October 1974.

Poulantzas, N. (1976) *Classes in Contemporary Capitalism.* London: New Left Books.

Szelenyi, I. (1976) 'Gestion regionale et classes sociales – le cas de l'Europe de l'est.' *Revue Francaise de Sociologie* 1.

_____ (1978) 'Social inequalities in state socialist redistributive economies.' *International Journal of Comparative Sociology* 1–2.

Winkler, J. T. (1977) 'The corporate economy – theory and administration', in R. Scase (ed.) *Cleavage and Constraints – Studies in Industrial Society.* London: George Allen and Unwin.

22 The apparatus of the State, the reproduction of capital and urban conflicts*

Joachim Hirsch

I

From the materialist point of view, state interventions (or the process of the political apparatus of bourgeois society) are essentially determined by the crisis-laden character of capitalist society and by related class confrontations.

In principle, capital itself can temporarily avoid its own inherent tendency to collapse by speeding up improvements in production technology, and by opening up new spheres of production, thus continually conquering new markets and outlets. Contradictions within the capitalist mode of production thus manifest themselves in an accelerating tendency to violent restructuration (on a world scale) in every sphere of society; for example, in technological change in production processes, in the constant shifting around and setting free of living labour-power, in labour intensification, in the destruction of traditional forms of life and social relations, in the subjugation of broad areas of society to the laws of commodity production and exchange, in the massive concentration of people and capital in metropolitan centres (while remote and backward regions are exploited and laid waste), and in the progressive exploitation and destruction of the natural foundations of production and reproduction.

Now, what is decisive for the changing role of the State is that these social restructuring processes, accompanied by progressive human socialization, require to an increasing extent the *intercession of the State*. More

---

* Translated from the German by Linda J. Dietrick.

and more, capital requires the organized social authority and power of the State. With the growing speed of the technical and social processes of upheaval, and the progressively thorough capitalization of society, the State is pressed increasingly into the function of (i) initiating and executing a comprehensive socio-economic restructuring process; (ii) simultaneously intercepting and compensating for the social consequences of these developments; and (iii) regulating the resulting conflicts. The activities of the State and the motion of its apparatus thus become an integral component of the capitalist mechanism of crisis. The national State does not merely *react* to socio-economic crises, but it also *executes* them under conditions which it cannot fundamentally influence, and through which capital creates surplus value on the world-market level. The concomitant changes in the function of the State lead to certain shifts in its internal structure, and also modify the shape and outcome of social conflicts and class confrontations.

It is essential that the 'crises of the cities', and connected forms of social conflict, be seen in this total context.

II

From what has been said above, it follows that the notion of the State as an authority which stands *outside* the economic context of reproduction is inexact, if not misleading. Even the related Marxist basis-superstructure concept, which contrasts the State as a component of the political 'superstructure' with the economic 'basis', tends to hide the fact that in capitalist society the State is always 'present in the economy'.[1] To be sure, it is a basic characteristic of the bourgeois State that it represents an institutionalization of political power which is formally distinct from social class and the immediate process of production.[2] Nevertheless, the capitalist production process, in the total social context of reproduction, has always been politically mediated by the State; in the strict sense, there never has been a 'government-free' sphere of the economy or of society. The State as an institutionalization of political power is directly and constitutively involved in the relations of capitalist production. These have no existence without the State; on the contrary, the State must be viewed as a component, subject to specific formal determinations, within the social context of production and reproduction. Its inner structure is essentially determined by the existing social relationships of class and the division of labour; it is *their* reproduction which the State mediates. This means, however, that the inner structure of the State, and the way it functions, can undergo specific changes in the course of capitalist development, even if its basic social determinations ('relative autonomy' and institutional 'apartness' *vis-à-vis* classes, including the ruling classes) are not fundamentally affected.

Changes which occur in the course of capitalism's historical development have the effect of moving the State increasingly into the centre of the social context of production and reproduction. The State's mediation of the processes of social production and reproduction becomes more comprehensive and intensified, although this does not immediately imply that the State's socio-economic controlling power increases. We must start from the fact that (under the structural conditions of bourgeois society) it is always the reproduction laws of capital which fundamentally determine the social dynamics, and thus also the motion of the apparatus of the State. Nevertheless, state regulation becomes increasingly a factor in the operation of these laws, such that their appearance and the way in which they are realized are at the same time modified.

This development is based on specific historical changes, produced by the crisis-mediated process of capital accumulation in the social conditions of production and reproduction. It is primarily a question of the *thorough capitalization of society,* i.e. the tendency for capital to assimilate all social spheres into the production of surplus value through the universalizing of commodity relationships. This capitalization is ensured by capital in its struggle over the rate of profit and in the accelerated development of productive forces. The latter manifests itself as a progressive revolutionizing of production technology and labour processes, and of the social structures of the division of labour and class.[3] This has various consequences. First, this process has a tendency (which Marx anticipated) toward the development of ever more complex production machinery, and toward the evolution of ever more differentiated production systems (rationalization, automation). This complex production apparatus, in which a considerable intensification of the social division of labour is expressed, becomes (simply for technical and organizational reasons), more susceptible to unforeseen disturbances and interruptions. Hence, in the highly developed capitalist societies, *the frictionless continuity of the production process becomes a decisive condition for capital's creation of surplus value.*[4] At the same time, the progressively thorough capitalization of society (i.e. the disappearance of non-capitalist sectors) has the effect of making the vast majority of the population directly and materially dependent on the capitalist reproduction process. This holds not only for the labouring class, but also in a comparable manner for the wage-dependent 'new middle classes' that are becoming more and more significant in the capitalist metropolises. The labouring class (or the wage-dependent class in general) gains objectively in strength both with the progressive development of productive forces and as a result of its economic and political organization. To be sure, this class is still involved in the accumulation and the crisis of capital, and is subject to capital. But at the same time, it constitutes itself in a contradictory manner as a subject acting *against* capital.[5] This means that the continuity of the production process becomes a problematic, in a dual sense: the process of production and surplus-value creation by capital becomes more susceptible to disturbance by the labouring classes' incalculable 'autonomous' actions, which break through the boundaries of the surplus-value creation process; and at the same time, the effect of economic reproduction crises becomes more far-reaching, involving the whole social system in its ramifications.

The progressively thorough capitalization of society and the capitalist form of productive development have the ultimate effect of fundamentally modifying the production/reproduction complex of society as a whole. The destruction of traditional 'pre-capitalist' social structures and forms of production (family, community, various forms of subsistence economies), and the increasing commodity-like character of social relations (i.e. the total dependence of the reproduction of productive power on the flow of commodity circulation and thus on the process of surplus-value production by capital), make developments within the 'reproduction sphere' closely dependent on conditions in the production sphere. In a certain sense, however, the same development also involves a critical rupturing of this bond, wherever individuals and social groups are excluded from the economic context of the reproduction of capital. Such conditions may arise, for example, through unemployment, illness, age, various forms of 'marginalization', etc. The original, indigenous process of reproduction and of social cohesion becomes more and more undermined by the anarchy of commodity relationships and thus is itself increasingly characterized by crisis. At the same time, the capitalist form of productive development leads to the familiar phenomena of destroyed cities and landscapes, to pollution and poisoning of the 'environment', and to destruction of the natural bases of production and reproduction in general. All this taken together indicates a *tendency for the social manifestations of crisis in the 'reproduction sphere' to become magnified.* Moreover, these manifestations, called forth by the dynamics of the capitalist production process, can, in turn, variously affect this process, i.e. the continuity of the surplus-value creation process of capital. This contradictory linkage between the production and reproduction spheres, resulting from thorough capitalization and capitalist productive employment, leads to a multiplication of the forms and levels in which an economic crisis manifests itself, so that the latter can no longer be described simply as the cyclical crisis of production. Hence there is a magnified possibility that crises in certain limited sectors may spread into other areas of society, and that initially 'disparate' crisis phenomena may accumulate, insofar as they cannot be suppressed, intercepted, or compensated for.

All in all, the developmental tendencies presented here have the consequence that the production and reproduction processes of capitalist society condense increasingly into a (crisis-defined) system of interrelations, whose continued existence is, to a decisive degree, dependent on the continuity of production. Because of the anarchy of capitalist socialization (i.e. the inability of competing individual capitalists to guarantee their own conditions of production and surplus-value creation) there is an increase in state administrative regulations – a tendency which appears as increasing 'state interference' *(Durchstaatlichung)* throughout society, bringing with it fundamental changes in the political administrative system. State regulation must concern itself more and more with preventatively intercepting (whenever possible) 'disturbances' in the socio-economic reproduction process. From an institutional standpoint, this manifests itself in the expansion of planning control, and in the development of an array of economic and social planning techniques. The structural interconnections and developments of society on the global scale become an explicit and continuing focus of politics, such that (under capitalism, of course) state actions must essentially always be viewed as *reactions* to conditions set by the reproduction process of capital on the world level. Thus the State does not become a 'planner of society' (as argued, e.g., by Tronti, Negri and Agnoli), but a contradictory apparatus of central importance for the continued existence and the reproduction of the capitalist system. With this as background, the traditional distinction between 'political' and 'economic' crisis loses credibility. More and more directly, economic crises appear as political crises and vice versa; political and economic crises overlap; and crises in limited sectors threaten to involve all society.

In order for the State to guarantee the capital reproduction process, there has long been a need for more than a *Konjunkturpolitik,* a policy to regulate economic cycles. Under pressure from the tendency of the rate of profit to fall, and from the resultant sharpening of competition on a world scale, the reorganization of social structures is assimilated into the sphere of state activities. The whole complex of state action becomes ever more strongly subordinated to this imperative. Individual capitalists (even monopolistic ones) become unable to cope organizationally and financially with the changes in production technology (along with all their social consequences) which have been forced on them by the pressure of world competition. This calls for increasing state involvement in the field of scientific and technological development. The resulting changes in labour processes, and in the international division of labour, have consequences which themselves demand further state regulations and interventions: for instance, the protection of foreign investments and markets by military as well as 'peaceful' means; the accommodation of regional structures and infrastructures to altered production conditions; the qualifying/disqualifying, repair, and mobilization of labour-power, etc.

A peculiarity of this development is that the State (while it increasingly appears as the agent of social structural change, without removing the crisis mechanisms rooted in capitalism) also appears as the originator and executor of processes of crisis (e.g. unemployment, structural crises). This leads to changes in the socio-political background of conflict, and makes the process of securing mass loyalty much more problematic. The success of state-mediated *mass-integration* is increasingly determined by the complex interconnections of 'production continuity and restructuring'. To assure that the course of capitalist reproduction will be disturbance-free, the system of modern 'mass-integrative' apparatuses (large bureaucratic parties, mass alliances, integrative labour unions, and especially those organizations which relate to the labouring class) gains central importance. It is their function (i) to guarantee institutionalization and 'constitutionalization', i.e. a control of the social conflicts and labour struggles; and (ii) at the same time, to allow for material demands, while also filtering and reducing them to a size which conforms to capital's requirements. The institutionalization of corporate interests, and the constitutionalization of social conflicts, becomes indispensible to the continuity of a production system which is subject to a continual technological restructuring. This presupposes wide-reaching bureaucratization and centralization in the party and labour union apparatuses. Moreover, it leads to specific forms of institutional involvement (particularly by the unions) in the state administrative system, the development of new forms of a 'corporatist' structure, and tendencies toward a 'nationalization' of parties and unions.[6]

Out of this development there arises a significant change in political conflict: there is a tendency for (i) conflicts to intensify within the individual apparatuses, and for a problematic alienation to develop between the apparatuses and their clienteles; while (ii) the divergences between organizations diminish as the latter unite to form a bureaucratic complex.[7] This structure of conflict reinforces the tendency of the State to intervene in order to safeguard the continued existence of the central mass-integrative apparatuses, and their ability to function. This occurs through legislation controlling unions and other associations, the sanctioning of exclusions and the prohibition of factions, the political surveillance of members, loyalty checks, the repression of extra-institutional forms of direct-interest representation, etc. Nevertheless, it must be noted that the complexity of production processes in advanced capitalism suggests that mass-integration based strictly on violent repression will hardly be of long-term duration. Mass-integration requires a certain degree of passive-voluntary compliance which (though occasionally repressively enforced) is still tied to the guarantee of certain material concessions. Without this compliance, the function of the late capitalist mass-integration system, operating as it does through large-scale bureaucratic organizations with voluntary membership and formally democratic legitimation procedures, would be fundamentally threatened.

Out of this arises a basic political dilemma: the amount of latitude for material concessions by capital and the State is essentially dependent on a relatively 'disturbance-free' operation of the accumulation and growth process. This however, presupposes increasingly profound changes in the structure of production and society. Such changes (e.g. structural unemployment, the disqualification of workers, social disintegration, environmental destruction) tend to undermine the basis of integration and consensus. A further important aspect of the modern mass-integration problem arises from the fact that, with increasing capitalist socialization and structural changes of production technology, more and *more spheres of social reproduction must be placed immediately under the direction of the State*. This affects social groups which fall outside the context of capital reproduction and commodity circulation (the unemployed, the aged, the ill, and other marginal groups), as well as the whole sphere of nature and the environment, whose progressive destruction is not only carried out and sanctioned, but also partially compensated for, by the State. It is precisely the successful establishment of capitalist production conditions on a broad scale that requires an increase in state regulation and control of the reproduction complex, and an administrative assimilation of whole sectors of society. The disintegration of indigenous social reproduction relations as a result of the anarchical form of capitalist socialization leads to progressive 'state interference' *(Durchstaatlichung)* in society. Even here there is a basic contradiction, in that it is just those state-aided changes in the production structure which confront the State with the job of crisis regulation. Furthermore, the availability of the necessary material means depends on continued capitalist growth – and thus on the corresponding, conflict-breeding process of destruction. The problem of mass-integration in the state-regulated sectors of the reproduction sphere is that the mediating function of mass-integrative apparatuses recedes into the background, and the confrontation between the apparatus of the State and those affected by its intervention becomes more direct. At the same time, however, the conflict salients are divided and fragmented, which facilitates the division, isolation and suppression of conflicts, as well as hindering its spread to other social spheres.

Because of changes in the reproduction conditions in advanced capitalist societies, there arise several notable modifications in the structure of the state apparatus.[8] There is a tendency toward the unification and homogenization of the collective political apparatus (tendencies to centralize, 'nationalization' of parties and labour unions); the 'overdetermination' of the complex of total state activity by the imperatives of socio-economic restructuring; progressive 'state interference' *(Durchstaatlichung)* throughout society; and bureaucratic institutionalization/constitutionalization of social conflicts and class struggles. This homogenization process rests less on the fact that traditional institutions of the political system are being eliminated than on the fact that their inner structure, the relations between them, and the distribution of competences are changing. This is not conducive to the formation of an organic, contradiction-free state machinery. Now as ever, the various political apparatuses remain specifically based on classes and fractions of bourgeois society, and from this basis they derive their dynamic.[9] This means that it is primarily social conflicts and class oppositions which determine the processes of the state apparatus, and which lead to inner frictions, conflicts and ruptures in this apparatus. These appear, however, in altered form and, as it were, tangentially to the mechanisms of conflict regulation in the traditional parliamentary system (perhaps in the form of 'vertical' conflicts within the mass-integrative apparatuses, or in bureaucratic divergences between functionally or regionally differentiated administrative agencies). As a consequence, the State appears more encompassing, more omnipresent, 'stronger', and at the same time more unstable – suffused with inner ruptures and crises which can barely find expression in the traditional bourgeois parliamentary forms. On this basis, there arises the modern form of 'authoritarian state control' *(étatisme)* (Poulantzas). Its character is such that the *unity* of the state administrative system in the broadest sense (including the mass-integrative and 'ideological' apparatuses, the educational system and mass media) can only be guaranteed through the accelerated expansion of diverse supervisory and control networks. The same holds for the State's mass-integrative *stability* in the presence of intensifying social conflict, and a loss of function in the traditional organs of decision-forming and representation. This is mirrored in the growing significance of the various 'State protection' agencies, e.g. intelligence gathering, the police, and social control, as well as in an increasing 'overdetermination' of traditional bureaucracies by this type of function. The ability of the political apparatus to function depends upon these formal and informal control networks (which rely on an ever more perfected information and data-processing technology). Their stability and cohesion, and the repressive ideological integration of the masses, are also predicated on the effectiveness of these networks.

The more state politics are restructured by 'objective laws' to conform to capital's requirements, the less functional become the traditional forms of conflict regulation, such as group lobbying, wage/price-setting autonomy *(Tarifautonomie)*, elections, parliamentarianism and party rivalry. The system of political apparatuses, which includes mass-integrative and ideological functions, tend to condense into a homogeneous bureaucratic structure – still suffused with ruptures and conflicts – against the mass of the people. In the mass-integrative apparatus, the function of mediating interests is (to a significant degree) moved *into* centres of state decision-making by the mediation of political decisions *to* those affected. The consequence is that opposition movements tend to direct their activities against the political institutions of the bourgeois State as a whole, leaving themselves increasingly open to loss of legitimacy and to accusations of criminal behaviour. Thus, political opposition to the logic of a state politics which has been restructured by logic of world capitalism is, as it were, forced to take the form of pathological or deviant behaviour, and becomes the object of repressive surveillance and control.

To summarize, the political structure of crisis in developed capitalist societies may be described as follows.

The production of surplus value by capital, and the stability of the political administrative system, depend closely on the *continuity of production,* which itself presupposes that far-reaching socio-economic restructuring processes (tending to affect all social spheres) will be executed. The economic and political reproduction process is becoming more susceptible to 'disturbances' which originate from autonomous, non-institutionalized forms of interest representation, unplanned labour struggles, and a 'deviation' of individual institutions from the capital-oriented functional mode of the political-administrative system. These developments are provoked again and again by the deterioration and destruction which accompany the restructuring process. They are intensified as the established system of mass-integrative apparatuses is absorbed into the functional logic of capitalist restructuring, and as its capacity to resolve conflicts thus becomes restricted. Because of the socially pervasive character of the restructuring processes, conflicts in the 'production sphere' or 'reproduction sphere' are objectively less and less isolatable from one another. Reproduction conflicts spread, and have a tendency to penetrate into the centre of surplus-value production, or at least directly to affect the production of surplus value (as in the case of environmental protection actions or of social resistance to industrial location and transport development projects). The need to reinforce socio-economic restructuring through state mediation brings with it the need to neutralize its effects through compensatory or repressive intervention, to intercept conflicts through institutionalization, or to isolate them repressively. This leads to an expansion and a rigidification of the whole political apparatus, however, which (because of this development) also becomes more unstable and susceptible to disturbance. The result is a considerable modification of the shape and course of socio-economic confrontations, which no longer coincide with the traditional notions of class struggle.

III

The development of urban and regional crises and conflicts in advanced capitalism must be seen in connection with these general conditions.[10] It has been in the cities, and in the problems of cities, that political and social conflicts have particularly escalated in recent years. This can be traced back to specific tendencies inherent in the capitalist development and restructuring process. As competition for world markets intensifies, pressure is exerted by the capitalist crisis-mechanism to expand labour productivity at an accelerated pace, and to affect rapid technical revolution in labour processes. This brings with it the necessity of a *concentration of capital*, not just in terms of value, but also *spatially*, and thus also a concentration of labour-forces.[11] This leads to the familiar phenomenon of the polarization of space into regions of extremely high and extremely low density, in the face of which state attempts to achieve a more uniform regional development have thus far remained largely unsuccessful. However, it is the State itself which supports and implements these polarization processes, while attempting to compensate, in a makeshift manner, for their most serious consequences. Under these social conditions, the term 'city' refers to the extreme spatial concentration of production facilities, labour and political-economic centres of management. The development of a whole region is determined by the political-economic processes concentrated there. The relationship between 'city' and 'country' can no longer be seen as a simple opposition; it is the expression of a process of continual reproduction of socio-economically dissimilar developments. At the same time, the regional distribution pattern of agglomeration remains quite unstable over time. It changes according to the technological conditions of capital production and reproduction (change in the natural and raw-material bases, in transport technologies, in the international division of labour, and in labour requirements). The result is devastation in the older urbanized areas (in West Germany, for example, the Saar and Ruhr regions): the forced industrial development of new areas (e.g. the Upper Rhine and Lower Elbe regions); and the formation of management and 'service-oriented' centres specifically concerned with the financial aspect of capitalism.

Under the anarchy of capitalist socialization, agglomeration also means the increased destruction of nature (air and water pollution, overdevelopment *(Zersiedlung)*, arterial highway construction) and is regularly accompanied by serious infrastructural deficiencies (housing, social services *(Versorgung)*, recreation). In a peculiar way, however, the same process operates in the depopulated and marginal regions produced by this development: in the shutting down of 'unprofitable' transport connections, the curtailment of spending on infrastructure, the relation of particularly dirty or dangerous production units outside high-density regions (e.g. atomic waste storage and processing). This phenomenon of *spatial segregation*, however, is not only concerned with the broad-scale relationship between high-density and depopulated regions, but is duplicated *within* the agglomeration centres themselves, in the form of 'functional differentiation' via the mechanism of ground rent (industrial suburbs, central business districts, bedroom communities, slums). The result is additional deficiencies in maintenance and recreation facilities, and an increase in traffic volumes which borders on collapse. The expansion of mass transport, which at the same time represents a significant sphere of investment for capital, is based not only on this 'functional differentiation' within the cities, but also on a broad-scale segregation of high-density and depopulated regions. This creates for many city-dwellers the need to flee periodically to the 'country' for purposes of recreation. Hence the countryside is also exposed to specific destruction processes (highway construction, tourist colonies, and the 'opening up' of areas to tourism). There is a contradiction in the fact that natural space is at once 'free productive power' for capital (through the presence of raw materials, convenient transport facilities such as waterways, and the possibilities for extremely intensive industrial and commercial utilization in high density regions) *and* the necessary basis for the reproduction of labour-power.[12] This contradiction grows all the more acute as the restructuring process of capital accelerates. With it, the processes of agglomeration, of continual changes in utilization, and of uneven development accelerate as well. Natural space, its utilization and its development, thus becomes an increasingly explosive problem for social reproduction.

The 'crisis of the cities' essentially derives from the late capitalist tendency to shift social crises into the reproduction sphere, and the effects naturally accumulate where the density of manufacturing establishments, transport facilities and people is greatest. It is, in this sense, imprecise to talk of a specific 'crisis of the city' as a limited social field of inquiry. Rather, the crisis of the city is an especially acute manifestation of the general crisis of development of advanced capitalist societies. It is where the effects of social 'marginalization' (unemployment, inability to work, criminalization, etc.) culminate. It is where all the implications of the large-scale exclusion of individuals and social groups from the direct reproduction process of capital come together. It is the locus of the destruction of natural and social conditions essential to life, where the disintegration of traditional residential milieus and massive relocation processes, and the separation of 'work' and 'reproduction' take on particularly sharp forms. In the meantime, the production of compensatory infrastructural facilities for social services *(Versorgung)*, recreation and 'leisure' limps hopelessly behind the development of still new, more high-density centres and the dislocation of masses of population. It should be borne in mind that the resulting deterioration in the quality of life can only be compensated in a limited way by material 'transfer-payments'.

The destruction of the city as 'living space' (which brings with it a corresponding damaging restructuring of the 'country') will have more serious effects as progressive rationalization and labour-intensification in the production and service sectors place higher demands on the reproduction of labour power. At the same time, required 'use values' (everything from the objects of material consumption, to housing, to cultural services) are progressively reduced to a commodity form, and thus become useless for the satisfying of needs. The modern capitalist city is, in a way, *the* expression of an increasingly commodity-like character of use values and social relations. It emphasizes that an immense concentration of social wealth goes hand in hand with the progressive reduction of the possibilities for human self-fulfilment, and that under capitalism, poverty and the abundance of commodities are ironically linked.[13] It is probably the particular clarity and prevalence of oppositions and contradictions which make social conflicts in the city more pressing and explosive.

The *reactions of the State* to the crisis of the cities are first determined by the contradiction that the State, as the necessary guarantor of surplus-value creation by capital, supports and itself implements the crisis-producing restructuring process. At the same time, however, it is also compelled (in order to prevent disruptive social conflict) to intervene through regulation, repression and material compensation. This contradiction in state activity expresses itself in a specific way on the level of state finance. With decreasing economic growth rates (accompanying the tendential falling rate of profit), the financial flexibility of the State is reduced, while resource demands for restructuring (transportation and technical development, etc.) and for capital subsidy increase. The result is a tendency toward a restriction on the finance available to the State for 'social policy' *(Sozialpolitik)* in the broadest sense, including 'social overhead capital', especially since this is labour-intensive and allows only limited rationalization. Increasing demands for compensatory action by the State are thus faced with diminished material means because of the same restructuring process which caused them to arise. This is an important basis of the modern financial crisis of the State, which finds specific expression in the fiscal crisis of the cities.[14] The fiscal crisis of the cities is, at least partly the result of a conflict-diversification strategy, which shifts the effects of the structural crisis in the state financial system (in the form of taxation and spending policies) to the community and regional levels. That is, it deprives local authorities of the means to attack their particular infrastructural and social tasks, in order to benefit the central State's strategies for protecting capital's pursuit of profit. In this manner, local administrations function as a filter and buffer against the demands of reproduction. The financial crisis of the cities is thus primarily an expression of an administrative and financial-political strategy which regionalizes political conflict affecting *all* of society, and which (temporarily) takes the burden off the central political apparatus.

This is the background for the tendencies in many developed industrial cities toward administrative centralization – a centralization that is formally and effectively undermining the autonomy of local and regional administrations. These tendencies are evident in a constant reduction of the number of local political units, in financial restrictions, and in the creation of intermediate planning units adjacent to the vertical hierarchy of local and central administrations. Thus, the number of local municipalities *(Gemeinden)* in West Germany, for example, fell from 24,357 to 8,518 between 1967 and 1978. At least one purpose of this centralization process consists of creating political, planning and decision-making bodies which are isolated from disturbing influences, especially from local and regional authorities. This effect is probably much more significant than the frequently cited reasons of administrative rationalization. Administrative centralization is a specific instance of the 'homogenization' of the political administrative apparatus. It is the prerequisite for planned uneven regional development, since it makes it possible to neutralize any resistance from the local institutional level to the effects of the socio-economic restructuring process, because the local authorities are effectively 'disempowered'. Nevertheless, this does not imply that the local and regional political administrative units can be completely eliminated. Formally, they continue to exist in reduced form, along with their representation and decision-making mechanisms, but their field of influence is decisively narrowed by standards which are set centrally and by centrally imposed financial restrictions. In this way, local politics tends to take on the character of a symbolic conflict-processing mechanism. In essence, it serves to diversify conflict and to filter out demands systematically.

This process of 'rigidifying' and 'homogenizing' the system of state institutions produces a tendency toward the centralization and bureaucratization of the mass-integrative apparatuses (parties) on the local level. The latter thus become less able to concern themselves with concrete interests, needs and problems. This produces specific conflicts between the local and central levels within these organizations. In West Germany, for instance, certain regional party organizations have come out against the atomic energy policies of the national party, but they have failed to gain any credibility as representatives of various interests, because they have no influence on the central authorities.

Political conflicts at the local and regional level are substantially determined by the inability of the mass-integrative apparatuses (implicated as they are in the centralized decisions regarding restructuring) to react to concrete and goal-directed interests which are not related to the centralized processes of bargaining and compromise. Moreover, the State, as both the supporter of restructuring processes and the repressive/compensating guarantor of social reproduction, increasingly becomes the direct guarantor and opponent of reproduction interests. This may be as the originator of environmental destruction, or as the repressive authority in the social problems of unemployment or social marginalization. On the other hand, the State, in order to guarantee the interests of capital, may be forced *not* to make use of its ability to regulate (e.g. in the case of environmental protection, the construction industry, or the provision of housing).

The structure of these phenomena may be summarized as follows. The established system of political apparatuses, at a given level of socialization, is increasingly forced to guarantee the reproduction of labour-power at the established level of productivity, while (at the same time) it is actually less and less in a position to do so. It is obliged, to a large extent, to bow to the imperatives of capitalist restructuring, and thus it becomes increasingly less able to respond to concrete needs and interests and to permit them to influence political decision-making. This growing rigidity results in a tendency towards a loss of legitimation and a structural crisis of mass-integration. This is expressed in an increasing number of persons who make extra-institutional, autonomous forms of interest representation independent of, or even directed against, the system of established apparatuses. The protection of their own interests by those affected persons implies, in a certain sense, a rejection of the structure of class division, delegation and abstraction which are peculiar to the bourgeois political process. It involves a decentralization and reestablishment of the unity between everyday life experience and politics. This implies a new understanding of politics and society in general, and is in fundamental contradiction to bourgeois social and political forms.

This, then, is the background for the growing occurrence of 'grass-roots' movements *(Basisbewegungen)* outside traditional political organizations. Such a background also explains a number of characteristics of these movements, especially their relation to the 'reproduction sphere', and their peculiar diffuseness and multiplicity. These characteristics are due to several factors. (i) The movements are decentralized, based on differing interests and problems and directed toward diverse conflict situations, yet are usually without a firm inner structure.[15] (ii) The movements possess an heterogeneous class basis, because the effects of developments in the 'reproduction sphere' reach across all classes. Frequently, they are strongly middle-class oriented, which reflects the particular sensitivity, articulateness and organization of this social group. In any case, these social movements and conflicts cannot be clearly traced to the basic class contradictions of capitalism, but are only indirectly connected with it. (iii) Ultimately, there is a certain ambiguity about the content of these social movements, for their concrete, openly articulated demands often convey much more than a single purpose. Their manifest goals serve more as starting-points for a continuing disquiet about, and alternative conceptions of, society. This probably contributes more than anything else – apart from the very limited latitude for concession by the political apparatus – to the fact that these movements cannot simply be appeased by partial concessions.

The structure of the grass-roots movements has both strengths and weaknesses. Because of their decentralized character and their lack of stable organizational structures, they are easy to split, to play off against each other, to starve out and to suppress violently. On the other hand, it is precisely their class heterogeneity, and their challenging of the apparatus of the State which allows for a certain stabilizing unity on a 'populist' level.[16] What is more important is that the success of this new form of social movement and conflict, which may well become characteristic for the developed capitalist countries, is not so much connected to the fulfilment or non-fulfilment of individual concrete demands and goals. These must remain limited no matter what, and they are repeatedly called into question by the socio-economic restructuring processes.

Nevertheless, in the face of these developments in the 'reproduction sphere' in the highly developed capitalist countries we find relatively stable mass-integrative mechanisms in the production sector, mediated by integrative labour-union organizations. The basis of this stability is, indeed, the mechanism for isolating socio-economic crises and shifting them into the 'reproduction sphere', and for successfully splitting production and reproduction interests within individuals themselves (which is encouraged by the existing system of mass-integrative apparatuses). What will be decisive for future developments is whether, and in what way, this split can be overcome; and whether, and to what extent, it is possible to consolidate and broaden such experiences with self-organization, with autonomous interest representation, and with the collective reappropriation of the practical concerns of life. If these processes of learning and experience, which must of necessity originate in the reproduction sphere, should penetrate into the production sphere, then the integration and repression strategies of the state apparatus will reverberate for a long time. It is probably this social and political explosiveness which explains the frequent severity of state strategies of repression and criminalization with regard to these movements.

Notes

[1] Poulantzas, N. (1978) *L'Etat, le Pouvoir et le Socialisme*. Paris: Presses Universitaires de France, pp. 18 ff.

[2] On the recent discussion about a Marxist theory of the bourgeois State ('debate on the derivation of the State'), cf. Holloway, J. and Picciotto, S. (eds.) (1978) *State and Capital: A Marxist Debate*. London: Edward Arnold. And J. Hirsch (1976) 'Bemerkungen zum theoretischen Ansatz einer Analyse des Bürgerlichen Staates', in *Gesellschaft: Beiträge zur Marx'schen Theorie*, Frankfurt/M.: Suhrkamp Verlag, Vol. 8/9. pp. 99 ff.

[3] Cf. Hirsch, J. (1974) *Staatsapparat und Reproduktion des Kapitals*. Frankfurt/M.: Suhrkamp Verlag.

[4] Agnoli, J. (1975) *Uberlegungen zum burgerlichen Staat*. Berlin: Wugenbach Verlag.

[5] Cf. Tronti, M. (1971) *Operai e Capitale*. Torino: Giulio Einauli Editore; Negri, T. (1973) *Krise des Plan-Staats*. Berlin: Merve Verlag; Agnoli (1975). For a

criticism of these variants of the 'planner-State', see Altvater, E. (1977) 'Staat und Gesellschaftliche Reproduction', in V. Brandes *et al.* (eds.) *Staat: Handbuch 5.* Frankfurt/M. and Cologne: Europäische Verlagsaustalt, pp. 74 ff.

6 On the 'neo-corporativism debate' cf. Jessop, B. (1978) 'Corporativism, Fascism and Social Democracy', manuscript; Panich, L. (1978) 'Recent Theoretizations of Corporativism: Reflections on a Growth Industry', manuscript.

7 Cf. Agnoli, J. and Brücker, P. (1967) *Die Transormation der Demokratie.* Berlin: Voltaire Verlag; Poulantzas 1978.

8 Cf. Poulantzas and the neo-corporativism literature.

9 Hirsch, 'Bemerkungen'.

10 In this section I refer mainly to the situation in the Federal Republic of Germany, particularly the development since the mid-1960s of various extra-institutional grass-roots movements *(Basisbewegungen)* or 'citizens' initiatives. Despite some 'national' particularities, the tendencies visible here should have general significance for the developed capitalist industrial states.

11 Lapple, D. (1978) 'Gesellschaftlicher Reproduktionsprozess und Stadtstrukturen', in M. Mayer, R. Roth and V. Brandes (eds.) *Stadtkrise und soziale Bewegungen.* Frankfurt/M. and Cologne: Europäische Verlagsaustalt, pp. 23 ff.

12 Lapple (1978: 43).

13 Gorz, A. (1977) *Okologie und Politik.* Reinbek: Rowohlt Verlag, p. 25 ff.

14 O'Connor, J. (1973) *The Fiscal Crisis of the State.* New York: St. Martin's Press. And Hirsch, J. (1978) 'Was heisst eigentlich "Krise der Staatsfinanzen"? Zur politischen Funktion der Staatsfinanzkrise', in Grauhaun, R. R. and Hickel, R. (eds) *Krise des Steuerstaates? Leviathan,* Sonderheft 1. Opladen: Westdeutricher Verlag, pp. 34 ff.

15 On the extra-institutional grass-roots movements in W. Germany, cf. Mayer-Tasch, P. C. (1977) *Die Burgerinitiativbewegung.* Reinbek: Rowohlt Verlag; Grossmann, H. (ed.) (1972) *Burgerinitiativen – Schritte zur Veranderung?* Frankfurt/M.: Fischer Verlag; Bundesminister für Forschung and Technologie (ed.) (1977) *Burgerinitiativen im Bereich von Kernkraftwerken.* Bonn; and 'Bürgerinitiativen-Bürgerprotest – eine neue vierte Gewalt?' *Kurbuch* 50, Berlin: Rotbuch Verlag, 1977.

16 On the theory of 'populist' movements cf. Laclau, E. (1977) *Politics and Ideology in Marxist Theory.* London: New Left Books.

Author index

Subject index

Accumulation, xv, xvi, xxi, 29, 30, 38, 39, 40, 103, 104, 319–24, 331, 332, 335, 357, 358, 384, 385, 409, 536; and Conflict, 434–6, 438, 442, 443; Contradictions of, 32, 33; Crises of, 536, 537; Cycles of, xxii, 33, 104, 106–8, 405–9; Laws of, 93–6, 119
Action, 73
Advanced Corporate Services, xx, 288, 290, 294, 297–300
Africa, 288
Akron, 302, 303
America, xiii, xix, 29, 124, 148, 490, 566, 584, 585
Amsterdam-Rotterdam, 307, 308
Artisans and Labourers Dwelling Improvement Act, 342, 367
Australia, 461, 571–88

Bahamas, 290, 308
Bahrain, 290, 293
Baltimore, 403
Barlow Report, 234, 235
Birmingham, 209, 224, 304, 345, 346, 502, 504
Boston, 301, 302, 386
Bristol, 268
Buffalo, 435

Built Environment, xv, xvii, 384, 393, 439; Contradictions of, 112–14; for Consumption, 96, 105; for Production, 96, 105; Investment in, 104, 114, 118, 384, 402, 403, 408

Canada, xxv, 126, 201, 306, 490, 535–8, 540–4, 551–4, 556; Urban System, 537–40
Capital, 22, 23, 205, 222, 388, 593, 601; Accumulation, xxii; Banking, 241, 242, 258; British, 352, 353, 359; Circuits of, 93–9, 269–73, 384, 389, 393, 408; Circulation of, xxi, 93–100, 118, 280, 281, 384, 385, 389; Commercial, xx, 269–71, 273, 276, 279–82; Finance, 132; Fixed, 96, 101, 106, 111, 112, 113, 206, 212, 384, 408; Industrial, xx, 241, 242, 245, 246, 247, 257, 267–72, 279, 281, 284; Interest Bearing, xx, 267, 270–3, 279; Monopoly, 181, 184, 232, 247; Property, 279, 280, 282, 434–42; Requirements of, 38; Restructuring of, 200, 201, 208, 210–26, 242, 245, 267, 268, 272–9, 288–92, 294–300, 602, 603; Variable, xx, 199, 205